FATIGUE DESIGN HANDBOOK

Second Edition

Prepared under the auspices of the
Design Handbook Division of the
SAE Fatigue Design and Evaluation Technical Committee

Richard C. Rice, Editor, Battelle Columbus Division
Brian N. Leis, Associate Editor, Battelle Columbus Division
Drew V. Nelson, Associate Editor, Stanford University
Henry D. Berns, Handbook Division Chairman, Deere & Co.
Dan Lingenfelser, Manuscript Reviewer, Caterpillar Tractor Co.
M. R. Mitchell, Manuscript Reviewer, Rockwell International

Published by: Society of Automotive Engineers, Inc.
400 Commonwealth Drive
Warrendale, PA 15096-0001

(c) 1988 Society of Automotive Engineers
Library of Congress Catalog Card Number: 88-63724
ISBN 0-89883-011-7
Printed in U.S.A.

TABLE OF CONTENTS

Chapter 1 Overview 1
1.1 Introduction 1
    1.1.1 Scope 1
    1.1.2 Purpose 1
    1.1.3 Structure 1

Chapter 2 General Fatigue Design Considerations 5
2.1 Introduction 5
2.2 Design Goals and Product Planning 5
2.3 Analysis of Product Usage 6
2.4 Consideration of Failure Modes 6
2.5 Characterization of Materials 8
2.6 Analysis Procedures 8
2.7 An Example 8
    2.7.1 Service Loads, Noise and Vibration 8
    2.7.2 Stress Analysis 8
    2.7.3 Material Properties 8
    2.7.4 Cumulative Damage Analysis 9
    2.7.5 Component Test 11
2.8 References 11

Chapter 3 Material Properties 15
3.1 Introduction 15
    3.1.1 Scope 15
3.2 Material Properties and Tests Relevant to Fatigue Design 15
    3.2.1 Monotonic Tension Properties 18
    3.2.2 Stress-Cycling Properties 18
    3.2.3 Strain-Cycling Properties 19
    3.2.4 Fatigue-Crack Growth/Threshold Concepts . 21
    3.2.5 Fracture Toughness Tests 23
    3.2.6 Service Simulation Tests 24
3.3 Fatigue Behavior of Representative Alloys 24
    3.3.1 Cyclic and Crack Initiation Properties 24
        3.3.1.1 Wrought Carbon and Alloy Steels . 24
        3.3.1.2 High Strength, Low Alloy Steels .. 26
        3.3.1.3 Cast Steels 27
        3.3.1.4 Cast Irons 27
        3.3.1.5 Wrought Aluminum 29
    3.3.2 Crack Growth Properties 29
        3.3.2.1 Wrought Carbon and Alloy Steels . 29
        3.3.2.2 High Strength, Low Alloy Steels .. 32
        3.3.2.3 Cast Steels 34
        3.3.2.4 Cast Irons 35
        3.3.2.5 Wrought Aluminum Alloys 36
3.4 Additional Considerations in Material Fatigue Behavior 36
    3.4.1 Effects on Crack Initiation 36
        3.4.1.1 Notches 36
        3.4.1.2 Stress Relaxation 40
        3.4.1.3 Cyclic Creep 40
        3.4.1.4 Periodic Overstrain 42
        3.4.1.5 Material Variability 43
        3.4.1.6 Environment 44
    3.4.2 Effects on Crack Propagation 45
        3.4.2.1 Crack Closure/Retardation 45
        3.4.2.2 Environment 46
3.5 References 47

Chapter 4 Effects of Processing on Fatigue Performance 63
4.1 Introduction 63
    4.1.1 Scope 63
    4.1.2 Relationship to Other Chapters 64
    4.1.3 Chapter Plan 64
4.2 Considerations in Evaluating the Effects of Fabrication and Processing 64
    4.2.1 Limitations of Standard Specimen Fatigue Data 64
    4.2.2 Fabrication and Processing Effects 67
4.3 Fatigue Performance Data on the Effects of Fabrication and Processing 69
4.4 Estimating Fabrication and Processing Effects on Fatigue Performance 70
    4.4.1 Example No. 1 72
    4.4.2 Example No. 2 72
4.5 Summary 73
4.6 References 73

Appendix 4   Supporting Data 74

Appendix 4A   Mechanical Prestressing 75
    4A.1 General 75
    4A.2 Shot Peening 76
    4A.3 Surface Rolling 76
    4A.4 Overloading 76
    4A.5 Prestressing and Preloading of Holes in Sheets 77
    4A.6 Conclusion 78
    4A.7 References 78

Appendix 4B   Heat Treatment 80
    4B.1 General 80
    4B.2 Furnace Hardening Through Heat and Quench 80
    4B.3 Induction and Flame Hardening: Surface Heat and Quench 81
    4B.4 Carburizing 82
    4B.5 Nitriding 82
    4B.6 Stress Relief 82
    4B.7 References 82

Appendix 4C	Weldments 85
4C.1	Linear Weldments and Surface (Notch) Effects 85
4C.2	Spot Welds 86
4C.3	Residual Stresses 87
4C.4	Other Methods for Improvement in Weld Fatigue Performance 87
4C.5	References 88

Appendix 4D   The Effect of Machining on Residual Stresses and Fatigue Resistance 89
- 4D.1 Introduction 89
- 4D.2 Metal Removal Parameters 90
- 4D.3 Residual Stresses 90
- 4D.4 Observed Correlations 90
- 4D.5 Summary 91
- 4D.6 References 92

Appendix 4E   The Effects of Forming on the Fatigue Strength of Metals 93
- 4E.1 Introduction 93
- 4E.2 Observation and Analysis 93
- 4E.3 References 95

Appendix 4F   Fatigue Design Considerations for Forgings 97
- 4F.1 Bulk Effects 97
- 4F.2 Surface Effects 97

Appendix 4G   Plating and Coating 98
- 4G.1 Effects on Fatigue Resistance ... 98
- 4G.2 References 98

Chapter 5   Service History Determination 99
- 5.1 Introduction 99
  - 5.1.1 Scope 99
  - 5.1.2 Relationship to Other Chapters 99
  - 5.1.3 Chapter Plan 100
- 5.2 Measurement Objectives 100
  - 5.2.1 Setting Measurement Objectives 100
  - 5.2.2 Customer Environment Sampling 100
  - 5.2.3 Test Site Sampling 100
- 5.3 Service History Measurement 100
  - 5.3.1 Types of Data 100
  - 5.3.2 Measurement Transducers 101
  - 5.3.3 Special Problems 101
- 5.4 Data Acquisition Techniques 103
  - 5.4.1 Signal Conditioning 103
    - 5.4.1.1 Shielding Practices 103
    - 5.4.1.2 Suppression of Automotive Transients 104
    - 5.4.1.3 Preventing Ground Noise 104
  - 5.4.2 Signal Transmission 107
    - 5.4.2.1 Wires and Cables 107
      - 5.4.2.1.1 Magnetic Noise 107
      - 5.4.2.1.2 Static Noise 108
      - 5.4.2.1.3 Crosstalk Noise 108
      - 5.4.2.1.4 Ground-Loop Noise 108
      - 5.4.2.1.5 Common Mode Noise ... 108
    - 5.4.2.2 Telemetry 108
  - 5.4.3 Digital Techniques 109
    - 5.4.3.1 General Principles 109
    - 5.4.3.2 Computer Facilities 109
- 5.5 Signal Recording, Storage and Display 110
  - 5.5.1 Oscilloscopes 111
  - 5.5.2 Strip Chart Recorders 111
  - 5.5.3 X-Y Recorders 111
  - 5.5.4 Tape Recorders 111
  - 5.5.5 On-Board Data Collectors 112
- 5.6 Data Reduction Techniques 112
  - 5.6.1 Filtering 112
    - 5.6.1.1 Sample Rate Definition 113
    - 5.6.1.2 Aliasing Problems 115
    - 5.6.1.3 Prevention of Aliasing 115
  - 5.6.2 Spectrum Analysis 116
    - 5.6.2.1 Time Compression Analyzers 117
    - 5.6.2.2 FFT Analyzers 117
- 5.7 Data Evaluation Techniques 117
  - 5.7.1 Statistical Summarizations (Spectrum Analysis) 118
  - 5.7.2 Amplitude Based Summarization 119
    - 5.7.2.1 Peak-Valley Sequencing 120
    - 5.7.2.2 Cycle Counting Procedures 120
      - 5.7.2.2.1 Level Crossing Counting 121
      - 5.7.2.2.2 Peak Counting 121
      - 5.7.2.2.3 Simple Range Counting 121
      - 5.7.2.2.4 Rainflow Counting and Related Methods 122
- 5.8 References 122

Appendix 5A   Sample Programs for Cycle Counting .. 124

Appendix 5B   Examples of Cycle Counting 133

Chapter 6   Vehicle Simulation 137
- 6.1 Introduction 137
  - 6.1.1 Scope 137
  - 6.1.2 Relationship to Other Chapters 137
  - 6.1.3 Chapter Plan 137
- 6.2 Suspension Design Cycle 138
  - 6.2.1 Selection of Front or Rear Suspension ... 138
  - 6.2.2 Suspension Geometry Optimization 138
  - 6.2.3 Kinematic Modeling of Front or Rear Suspensions 138
  - 6.2.4 Dynamic Modeling of Front or Rear Suspensions 138
  - 6.2.5 The Full Vehicle Model 139
- 6.3 Simulation Types 139

6.4	Full Vehicle Model Simulations 140
6.4.1	Suspension Abuse Test—Lab Simulations 140
6.4.2	Full Vehicle Model Road-Load Simulation 140

6.5 Summary/Remarks 140

6.6 References 142

Chapter 7 Strain Measurement and Flaw Detection 143

7.1 Introduction 143
    7.1.1 Scope 143
    7.1.2 Relationship to Other Chapters 143
    7.1.3 Chapter Plan 143

7.2 Techniques to Define Fatigue Critical Location .. 144
    7.2.1 Brittle Coatings 144
    7.2.2 Strain Gages 144
    7.2.3 Photoelasticity 145
    7.2.4 Holographic and Speckle Interferometry . 146
    7.2.5 Thermoelasticity 148

7.3 Flaw Detection Techniques 148
    7.3.1 Ultrasonic 148
    7.3.2 Dye Penetrant 149
    7.3.3 Magnetic Particle 149
    7.3.4 Radiography 149
    7.3.5 Eddy Current 150
    7.3.6 Holography 150
    7.3.7 Acoustic Emission 151
    7.3.8 Potential Drop 152
    7.3.9 Gel Electrode 152

7.4 Residual "Self-Stress" Determination 153
    7.4.1 X-ray Diffraction Method 153
    7.4.2 Sectioning 156
    7.4.3 Blind Hole Drilling 158
    7.4.4 Ultrasonic Method 159
    7.4.5 Barkhausen Noise 160
    7.4.6 Other Methods 160

7.5 Advantages and Limitations of Various Techniques 160

7.6 References 166

Chapter 8 Numerical Analysis Methods 171

8.1 Introduction 171
    8.1.1 Scope 171
    8.1.2 Relationship to Other Chapters 172
    8.1.3 Chapter Plan 172

8.2 Closed Form Methods 172
    8.2.1 Background 172
    8.2.2 Nominal Stress Estimation 172
        8.2.2.1 Strength of Materials 172
        8.2.2.2 Theory of Elasticity 173
        8.2.2.3 Energy Methods 173
    8.2.3 Local Stress Estimation 173
        8.2.3.1 The Handbook Approach 173
    8.2.4 Linear Elastic Fracture Mechanics 174
    8.2.5 Compensation for Plasticity Effects 174
    8.2.6 Topical Bibliography 175

8.3 The Finite Element Method 175
    8.3.1 Background 175
    8.3.2 Theoretical Basis 175
        8.3.2.1 The Element Stiffness Matrix 175
        8.3.2.2 Structural Stiffness Matrix 178
        8.3.2.3 Isoparametric Elements 178
        8.3.2.4 Modelling Considerations 180
    8.3.3 Elastic-Plastic Analysis 181
        8.3.3.1 Background 181
        8.3.3.2 Incremental or Flow Theory of Plasticity 182
        8.3.3.3 Deformation Theory of Plasticity . 184
        8.3.3.4 Numerical Analysis Procedures .. 185
    8.3.4 General Applications and Examples 186
        8.3.4.1 Engine Block Analysis 186
        8.3.4.2 SAE Keyhole Specimen 187
        8.3.4.3 Uniformly Loaded Sheet with a Central Crack 187
    8.3.5 Application to Linear Elastic Fracture Mechanics 188
        8.3.5.1 The Stress Intensity Factor 188
        8.3.5.2 Calculation of SIF Using FE Methods 190
    8.3.6 Available Software 192
    8.3.7 Topical Bibliography 192

8.4 The Finite Difference Method 192
    8.4.1 Background 192
        8.4.1.1 Formulations and Types of Boundary Value and Initial Value Problems 192
        8.4.1.2 The Basic Idea of the FDM 193
        8.4.1.3 Advantages and Limitations 193
    8.4.2 Theoretical Basis 194
        8.4.2.1 The Classical FDM 194
        8.4.2.2 The Generalized FDM 195
    8.4.3 The GFDM Approach to Nonlinear Problems 198
    8.4.4 General Applications and Examples 199
        8.4.4.1 Deflection of a Beam 200
        8.4.4.2 Shear Stress in a Prismatic Bar Subjected to Torsion 200
    8.4.5 Application to Linear Elastic Fracture 202
    8.4.6 Available Software 203
    8.4.7 Summary 203
    8.4.8 Topical Bibliography 203

8.5 The Boundary Integral Equation and Boundary Element Methods 203
    8.5.1 Background 203

		8.5.2	Theoretical Basis	204
			8.5.2.1 Governing Integral Equation for Elastostatics	204
			8.5.2.2 Boundary Point	205
			8.5.2.3 Boundary Elements	205
			8.5.2.4 Internal Results	206
			8.5.2.5 Higher Order Elements	206
			8.5.2.6 Non-Smooth Boundaries	206
		8.5.3	Other Considerations	206
		8.5.4	Available Software	207
		8.5.5	Summary	207
		8.5.6	Topical Bibliography	207
	8.6	Comparison of the Methods		207
		8.6.1	Historical Background	207
		8.6.2	The Finite Element Method	207
		8.6.3	The Finite Difference Method	207
			8.6.3.1 Derivative Approximation	207
			8.6.3.2 Variational Formulation	208
		8.6.4	The Boundary Integral Equation Method	208
	8.8	References		208

Chapter 9 Structural Life Evaluation 217

- 9.1 Introduction 217
  - 9.1.1 Scope 217
  - 9.1.2 Relationship to Other Chapters 217
  - 9.1.3 Chapter Plan 217
- 9.2 The Role of Testing in Design 217
  - 9.2.1 Support of Analysis 218
    - 9.2.1.1 Material Properties 218
    - 9.2.1.2 Model Parameters 219
    - 9.2.1.3 Environmental Parameters 219
  - 9.2.2 Prototype Evaluation 219
    - 9.2.2.1 Functional Evaluation 219
    - 9.2.2.2 Parameter Evaluation 219
    - 9.2.2.3 Environmental Response Evaluation 220
    - 9.2.2.4 Durability Evaluation 220
  - 9.2.3 Validation of the Production Process 220
  - 9.2.4 In-Service Monitoring 220
    - 9.2.4.1 Service Use Monitoring 220
    - 9.2.4.2 Service Failure Monitoring 220
    - 9.2.4.3 Customer Satisfaction 221
    - 9.2.4.4 QC Testing 221
- 9.3 Analytical and Experimental Tools for Structural Life Evaluation 221
  - 9.3.1 Level 1: System Control 222
  - 9.3.2 Level 2: Test Site Control 223
  - 9.3.3 Level 3: Data Analysis 223
  - 9.3.4 Level 4: Laboratory Management 223
- 9.4 Vehicle Durability Testing 224
  - 9.4.1 Test Objective 224
  - 9.4.2 Test Specimen 224
  - 9.4.3 Environmental Input Exciters 225
  - 9.4.4 Data Recording Locations 225
  - 9.4.5 Data Collection Techniques 226
  - 9.4.6 Test System Programming and Control 227
  - 9.4.7 Durability Test Monitoring and Analysis 227
- 9.5 Modal Testing and Analysis 228
- 9.6 Summary 229
- 9.7 References 229

Chapter 10 Fatigue Life Prediction 231

- 10.1 Introduction 231
  - 10.1.1 Scope 232
  - 10.1.2 Relationship to Other Chapters 232
  - 10.1.3 Chapter Plan 233
- 10.2 Background Considerations in Life Prediction 234
  - 10.2.1 Background 234
  - 10.2.2 Crack Initiation and Crack Propagation Approaches to Life Prediction 235
- 10.3 Crack Initiation Approach 236
  - 10.3.1 Overview 236
  - 10.3.2 Notch Stress Analysis 236
  - 10.3.3 Damage Analysis 238
  - 10.3.4 Methods of Crack Initiation Life Prediction 240
    - 10.3.4.1 Load-Life Method, Constant Amplitude Loading 240
    - 10.3.4.2 Load-Life Method, Variable Amplitude Loading 242
    - 10.3.4.3 Stress-Life Method, Variable Amplitude Cycling 243
    - 10.3.4.4 Strain-Life Method, Variable Amplitude Cycling 245
  - 10.3.5 Selection of a Crack Initiation Life Prediction Method 249
- 10.4 Crack Propagation Approach 249
  - 10.4.1 Stress Intensity Factors 250
  - 10.4.2 Fracture Toughness 252
  - 10.4.3 Critical Crack Size 253
  - 10.4.4 Fatigue Crack Growth 253
  - 10.4.5 Examples of Crack Propagation Life Prediction for Constant Amplitude Loading 255
  - 10.4.6 Crack Propagation Life Prediction for Variable Amplitude Loading 256
  - 10.4.7 Summary 258
- 10.5 Practical Aspects of Life Prediction 258
  - 10.5.1 Crack Initiation vs. Crack Propagation Analysis/Small Cracks 258
  - 10.5.2 Multiaxial Effects 260
    - 10.5.2.1 Multiaxial Loading - Initiation Effects 260
    - 10.5.2.2 Multiaxial Conditions - Initiation Effects 260
    - 10.5.2.3 Multiaxial Conditions - Propagation Effects 261

| | | 10.5.3 Environmental Effects 261
| | | 10.5.4 Fabrication Effects 261
| | | 10.5.5 Processing Effects 262
| | | 10.5.6 Load Sequence and Crack Closure Effects 264
| | | 10.5.7 Residual Stress Effects 266
| | | 10.5.8 Cracks Growing from Notches 268
| | | 10.5.9 Cracks at Welds 270

10.6 Applications of Life Prediction in Design Analysis 272
    10.6.1 Fatigue Assessment of Cast Axle Housings 272
    10.6.2 Fatigue Assessment of a Piston Rod in a Forging Hammer 273
    10.6.3 Analysis of Crack Growth in a 1025 HR Welded Frame Structure 273
        10.6.3.1 Background 273
        10.6.3.2 Test Description 274
        10.6.3.3 Test Results 274
        10.6.3.4 Crack Growth Prediction Methodology 274
        10.6.3.5 Prediction 274
        10.6.3.6 Discussion 276

10.7 Conclusion 277

10.8 References 278
    Appendix 10A SAE Cumulative Fatigue Damage Test Program 285
        10A.1 Load Histories 285
        10A.2 Specimens and Materials 285
        10A.3 References 289
    Appendix 10B Stress-Strain Simulation at a Notch 292
        10B.1 Introduction 292
        10B.2 Notch Analysis 293
        10B.3 References 295

Chapter 11 Failure Analysis 297

11.1 Introduction 297
    11.1.1 Scope 297
    11.1.2 Relationship to Other Chapters 297
    11.1.3 Chapter Plan 298

11.2 Why Things Break 298
    11.2.1 Cracking and Fracture 298
    11.2.2 Strength in the Case of Cracks 299
    11.2.3 "Brittle" vs. "Ductile" Fractures 300
    11.2.4 The Reasons for Failures 302
    11.2.5 The Elements of Failure Analysis 304

11.3 Fractography for the Engineer 305
    11.3.1 Scope 305
    11.3.2 Fractographic Tools 306
    11.3.3 Cracking Mechanisms and Microscopic Features 306
    11.3.4 Fracture Mechanisms and Microscopic Features 308
    11.3.5 Brittle and Ductile Fracture Revisited .. 309
    11.3.6 Cleavage vs. Rupture 311
    11.3.7 Other Fracture Surface Analysis Tools . 312
    11.3.8 Quantitative Fractography 312

11.4 The Search for the Origin 317
    11.4.1 Scope 317
    11.4.2 Distinguishing the Crack and the Fracture 317
    11.4.3 Crack Features: Beach Marks, Mussel Shells and Sunrise Markings 318
    11.4.4 Fracture Features: Chevrons, Branches, Shattering 319
    11.4.5 Crack Origins 320

11.5 Failure Analysis Procedure 322
    11.5.1 Preliminaries and General Considerations 322
    11.5.2 Preliminary-Failure-Analysis Plan and Failure Hypothesis 324
    11.5.3 Secondary-Failure-Analysis Plan: Verification 324

11.6 Possible Actions Based on Failure Analysis 328
    11.6.1 Scope 328
    11.6.2 Failures During Design Development or Design Verification Testing 328
    11.6.3 Service Failures 329

11.7 Summary 329

11.8 References 329

11.9 Topical Bibliography 330
    11.9.1 General Failure Analysis 330
    11.9.2 Fracture Surface Analysis 330

Chapter 12 Case Histories 331

12.1 Introduction 331
    12.1.1 Scope 331
    12.1.2 Relationship to Other Chapters 331
    12.1.3 Chapter Plan 331

12.2 Case No. 1 - Automotive Wheel Assembly Design Using High Strength Sheet Steel 331
    12.2.1 Introduction 331
    12.2.2 Case History 331

12.3 Case No. 2 - Design and Development of Components for a Suspension System 334
    12.3.1 Introduction 334
    12.3.2 Case History 335

12.4 Case No. 3 - Heavy Duty Cast Axle Housing ... 338
    12.4.1 Introduction 338
    12.4.2 Case History 338

12.5 Case No. 4 - Forged Connecting Rod 342
    12.5.1 Introduction 342
    12.5.2 Case History 342

12.6 Case No. 5 - Cast Steel Axle Box for a Railway
     Vehicle 343
     12.6.1 Introduction 343
     12.6.2 Case History 344

12.7 Case No. 6 - Axle Shaft Problem on a Scraper
     Vehicle 345
     12.7.1 Introduction 345
     12.7.2 Case History 345

12.8 References 346

Appendix A   Definitions 349

FOREWORD

Our second edition of this handbook is as unique as the original edition published in 1968. It is different than any textbook on the subject of fatigue. The majority of contributions have been made by people working in the ground vehicle industry, which gives the material particular relevance to those who will be the principal users of the book. The many contributors are noted in the Preface and Acknowledgment.

Because most fatigue design textbooks do not cover the latest developments from the perspective of the ground vehicle designer, the Handbook Division of the Fatigue Design and Evaluation Committee was organized in the early 1980's to "capture" the spirit of the latest fatigue design technology. The major difficulty with users writing a book lies in completion and the editing of the final draft so that it can be printed. A subcommittee of the Handbook Division screened several good proposals and selected Richard Rice and the Battelle Columbus organization to do the final editing of the assembled material. Many improvements were made in the book while working with the many contributors. Readers should bear in mind, however, that this handbook remains an assembly of contributions from many sources that were written in different styles. As a result, some chapters are not completely consistent in their organization. This is a small inconvenience for input from the best sources of the latest techniques being used in fatigue design technology.

The work of the FD&E Committee is actually never complete, since new developments are occurring continuously. In that regard, some topics in the Handbook could already be updated. We must, however, stop and publish this second edition just as was done with the first in 1968, when low cycle fatigue prediction techniques were just emerging. While this topic was mentioned in the first edition, it was not addressed at length. This did not stop the first edition from being a useful and popular handbook — more than 4000 copies have been sold to date. Because of the introduction of much new material we expect that the second edition of the SAE Fatigue Design Handbook will be as popular as the first.

Henry D. Berns, Chairman for the Handbook Division

PREFACE AND ACKNOWLEDGMENT

In 1968 the Society of Automotive Engineers published the AE-4, Fatigue Design Handbook: A Guide for Product Design and Development Engineers. Since that time thousands of copies of AE-4 have been distributed throughout the world. It has proven itself to be a valuable resource. However, as time has passed, the technology of fatigue design has changed significantly. In the early 1980's it became apparent that a revision to AE-4 was necessary.

The SAE Fatigue Design and Evaluation (FD&E) Committee took responsibility for revision of this document. Volunteer efforts by a number of people got the project started. However, after several years, it was determined by the FD&E Committee that it would be necessary to identify a contractor who would complete the document, and coordinate the editing and production of the Handbook with the SAE Publications Group. Proposals were solicited and the Battelle Columbus Division was selected for this task. Funding to support this effort was solicited by the FD&E Committee and financial support was obtained from the following organizations:

- J. I. Case Co., Components Division, Racine, Wisconsin
- Caterpillar Inc., Peoria, Illinois
- Deere & Co., Technical Center, Moline, Illinois
- Eaton Corp., Engineering and Research Center, Southfield, Michigan
- Ford Motor Co., Dearborn, Michigan
- General Motors Corp., Fisher Body Division, Warren, Michigan
- Ingersoll Rand Co., Woodcliff Lake, New Jersey
- Metal Improvement Co., Paramus, New Jersey
- MTS Systems Corp., Minneapolis, Minnesota
- Radian Corp., Milwaukee, Wisconsin
- Rockwell International Automotive Operations, Troy, Michigan
- SAE FD&E/The University of Iowa—Fatigue Concepts and Design Short Course
- A. O. Smith Co., Milwaukee, Wisconsin

The draft material for the Handbook was screened by Battelle and suggested revisions to the content of the Handbook were reviewed with and approved by the FD&E Committee. Contributors, reviewers and editors were identified for all of the chapters. After several years of effort on the part of many people the task was completed. The following people, in particular, played important roles in completion of this document.

Reviewers

- M. R. Mitchell, Rockwell International, Thousand Oaks, California
- Dan Lingenfelser, Caterpillar Inc., Peoria, Illinois
- Henry Berns, Deere & Co., Moline, Illinois
- Drew Nelson, Stanford University, Stanford, California

Chapter 1—Overview

Contributors

Henry Berns

Reviewers

Raj Thakker, A. O. Smith Co., Milwaukee, Wisconsin
Henry Jaeckel, Nevada City, California

Chapter 2—General Fatigue Design Considerations

Contributors

Brian Dabell, GKN Technology Ltd., Wolverhampton, England
Henry Berns

Reviewers

Gail Leese, MTS Systems Corp., Minneapolis, Minnesota
Dan Lingenfelser

Chapter 3—Material Properties

Contributors

Ravi Rungta, Battelle Columbus Division, Columbus, Ohio
Ron Landgraf, Ford Motor Co., Dearborn, Michigan

Reviewers

Henry Berns
Gary Mauritzson, Deere & Co., Dubuque, Iowa
Mike Mitchell
Steve Tipton, University of Tulsa, Tulsa, Oklahoma

Chapter 4—Processing Effects on Fatigue Performance

Contributors

Brian Leis, Battelle Columbus Division, Columbus, Ohio
John Cammet, Metcut Research, Cincinnati, Ohio
Henry Fuchs, Stanford University, Stanford, California
Harold Reemsnyder, Bethlehem Steel Corp., Bethlehem, Pennsylvania
R. Ricklefs, Caterpillar Tractor Co., Peoria, Illinois
Bob Testin, General Motors Corp., Indianapolis, Indiana

Reviewers

Charlie Barrett, Metal Improvement Co., Carlstadt, New Jersey
Henry Fuchs
Bob Gillespie, Metal Improvement Co., Paramus, New Jersey

Chapter 5—Service History Determination

Contributors

Al Conle, Ford Motor Co., Dearborn, Michigan
Mike Morton, J. I. Case Co., Hinsdale, Illinois

Reviewers

Jay Fash, MTS Systems Corp., Minneapolis, Minnesota
Tom M. Johnson, General Motors Corp., Milford, Michigan

Chapter 6—Vehicle Simulation

Contributors

Dan Thomas, Rockwell International Corp., Troy, Michigan

Reviewers

Ron Landgraf
Jim McConville, Mechanical Dynamics, Ann Arbor, Michigan
Chuck Stanton, Farmington Hills, Michigan

Chapter 7—Strain Measurement and Flaw Detection

Contributors

Drew Nelson
Kenneth Anderson, Stress Measurements, Inc., Minneapolis, Minnesota
Gordon Hayward, Hayward Hydraulic Grip System, Palo Alto, California
John McCrickerd, Newport Corp., Fountain Valley, California

Reviewers

Mike James, Rockwell International Corp., Thousand Oaks, California
Dave Lineback, Measurements Group, Inc., Raleigh, North Carolina
Mike Morton
Michael Resch, Stanford University, Stanford, California
Bruce Boardman, Deere & Co., Moline, Illinois

Chapter 8—Numerical Analysis Methods

Contributors

Ken Mormon, Ford Motor Co., Dearborn, Michigan
Janusz Orkisz, Technical University of Cracow, Warsaw, Poland
Mike Rosenfeld, Battelle Columbus Division, Columbus, Ohio
C. A. Brebbia, Computational Mechanics Institute, Southhampton, England

Reviewers

Ken Mormon
John Hakala, General Motors Corp., Warren, Michigan
Sing Tang, Ford Motor Company, Dearborn, Michigan
Donald Dewhirst, Ford Motor Company, Dearborn, Michigan
Daniel Anderson, Ford Motor Company, Dearborn, Michigan

Chapter 9—Structural Life Evaluation

Contributors

Kelly Donaldson, MTS Systems Corp., Minneapolis, Minnesota
Raj Thakker

Reviewers

Russell Cherenkoff, Ford Motor Co., Dearborn, Michigan
Tom Cordes, John Deere Product Engineering Center, Cedar Falls, Iowa
Raj Thakker

Chapter 10—Fatigue Life Prediction

Contributors

Brian Leis
Drew Nelson
Darrell Socie, University of Illinois, Urbana, Illinois

Reviewers

Bruce Boardman
Ron Landgraf
Mike Mitchell
Ralph Stephens, University of Iowa, Iowa City, Iowa

Chapter 11—Failure Analysis

Contributors

David Broek, FractuResearch, Inc., Galena, Ohio
Brian Leis

Reviewers

Steve Hopkins, Failure Analysis Associates, Palo Alto, California

Chapter 12—Case Histories

Contributors

Brian Dabell
Drew Nelson
Henry Fuchs

Reviewers

Henry Fuchs
Mike Mitchell

Appendix A—Definitions

Reviewers

Graham Markes II, Eaton Corp., Southfield, Michigan
Phil Dindinger, Radian Corp., Milwaukee, Wisconsin

The inputs and comments of these individuals were invaluable in the compilation of this broad-based document. To the extent that the Second Edition of the Fatigue Design Handbook meets the needs of a diverse group of individuals concerned with the design of fatigue resistant vehicles and structures, it is directly attributable to the efforts of these individuals. Considerable effort has gone into the correct transcription and editing of this material. Undoubtedly, errors and omissions have occurred. Hopefully they are few and do not distort the message being presented.

In addition to these individuals, the efforts of Leslie-Anne Boss of the SAE Publications Group were very significant in the successful completion of this Handbook. The editors of this edition of the SAE Fatigue Design Handbook are grateful to all of these individuals and companies for their support.

SYSTEM OF UNITS

The Society of Automotive Engineers has adopted the practice of publishing data in both metric and customary U.S. units of measure, where possible. In preparing this Handbook, the editors have attempted to present data primarily in metric units, with secondary mention of the corresponding values in customary U.S. units. The decision to use SI as the primary system of units was based on the widespread use of metric units throughout the world, and the expectation that the use of metric units in the United States will increase substantially during the anticipated lifetime of this Handbook.

For the most part, numerical engineering data in the text and in tables are presented in SI-based units with the customary U.S. equivalents in parentheses (in the text) or adjoining columns (in tables).

In most graphs and charts, grids correspond to SI-based units, which appear along the left and bottom edges; where appropriate, corresponding customary U.S. units appear along the top and right edges. (In some cases, where graphs were utilized from other sources, only customary U.S. units are shown.)

Data pertaining to a specification published by a specification-writing group may be given in only the units used in that specification or in dual units, depending on the nature of the data. For example, the typical yield strength of aluminum sheet made to a particular specification written in customary U.S. units would be presented in dual units, but the thickness specified in that specification might be presented only in inches.

Dual units of measure also are not provided in some articles that cite original data from other sources. For example, stress levels utilized in a fatigue data analysis are given in either SI or customary U.S. units, depending on the measurement system used in the original test program. Conversion factors for stress and other units of measure can be found in Tables 1 through 3.

The policy on units of measure in this Handbook contains several exceptions to strict conformance to ASTM E 380; in each instance, the exception has been made to improve the clarity of the Handbook. The most notable exception is the use of $MPa\sqrt{m}$ rather than $MPa \cdot m^{-3/2}$ as the SI unit of measure for fracture toughness.

SI practice requires that only one virgule (diagonal) appear in units formed by the combination of several basic units. Therefore, all of the units preceding the virgule may be considered to be in the numerator and all units following the virgule may be considered to be in the denominator of the expression; no parentheses are required to prevent ambiguity.

These tables are intended as a guide for expressing weights and measures in the Systeme International d'Unites (SI) for use in mechanical testing. The purpose of SI units, developed and maintained by the General Conference of Weights and Measures, is to provide a basis for world-wide standardization of units and measure. For more information on metric conversions, the reader should consult the following references:

- "Standard for Metric Practice," E 380, 1984, ASTM, 1916 Race Street, Philadelphia, PA 19103

- "Metric Practice," ANSI/IEEE 268-1982, American National Standards Institute, 1430 Broadway, New York, NY 10018

- Metric Practice Guide—Units and Conversion Factors for the Steel Industry, 1978, American Iron and Steel Institute, 1000 16th Street NW, Washington, DC 20036

- The International System of Units, SP 330, 1981, National Bureau of Standards. Order from Superintendent of Documents, U.S. Government Printing Office, Washington, DC 20402

- Metric Editorial Guide, 4th ed., 1984, American National Metric Council, 1625 Massachusetts Ave. NW, Washington, DC 20036

- ASME Orientation and Guide for Use of SI (Metric) Units, ASME Guide SI 1, 9th ed., 1982, The American Society of Mechanical Engineers, 345 East 47th Street, New York, NY 10017

NOMENCLATURE

a	1. Crack size, crack length 2. Material constant	m	Coefficient of mean stress influence
A	1. Area 2. Stress Ratio	M	Bending moment
A_f	Area after fracture	n	1. Number of cycles endured 2. Sample size 3. Number of standard deviations
A_o	Initial area	$n_1, n_2 \ldots$	Number of load, or strain cycles observed at each level of a history
b	Fatigue strength exponent	N	1. Fatigue life in cycles 2. Revolutions per minute
BHN, (HB)	Brinnell hardness number		
c	1. Fatigue ductility exponent 2. Distance from centroid to outermost fiber	N_f	Cycles to failure
		$2N_f$	Reversals to failure
C	Cycle ratio	$2N_t$	Transition fatigue life in reversals
D or d	Diameter	p	Probability
DPH	Diamond pyramid hardness	P	Load
da/dN	Fatigue crack growth rate	q	Notch sensitivity factor
e	Engineering strain	r	Notch root radius
E	Modulus of elasticity	R	1. Load (stress) ratio 2. Fillet radius
F	Force		
G	Modulus of elasticity in shear	R	Strain ratio
HRC	Hardness, Rockwell "C" scale	s	Sample standard deviation
I	Moment of inertia	s^2	Sample variance
J	Polar moment of inertia	S	Engineering or nominal stress
K	Stress intensity factor	S_a	Nominal alternating stress
K_f	Fatigue notch factor	S_f	Fatigue limit
K_t	Theoretical stress concentration factor	S_m	Mean stress
K_1, K_2, K_3	Stress intensity factor in Mode 1, 2 or 3	S_{max}	Maximum stress
K_{max}	Maximum stress intensity factor	S_{min}	Minimum stress
K_{min}	Minimum stress intensity factor	S_N	Fatigue strength at N cycles
ΔK	Stress intensity factor range	S_r	Range of stress
K_ϵ	Strain concentration factor	S_u	Ultimate strength
K_σ	Stress concentration factor	S_y	Yield strength
L or l	Length	S_1, S_2, S_3	Principal stresses

NOMENCLATURE (Cont)

S_x, S_y, S_z	Normal stresses in x, y and z directions	$\epsilon p/2$	Plastic strain amplitude
SF	Safety factor	$\epsilon f'$	Fatigue ductility coefficient
T	1. Torque 2. Temperature	τ	Shear strain
		ν	Poisson's ratio
t	Time	σ	1. True or local stress 2. Standard deviation
x,y,z	Cartesian coordinates		
δ	Deflection	σ_x	Standard deviation of x
ϵ	Total local strain	σ_x^2	Variance of x
$\Delta\epsilon$	Total local strain range		
$\epsilon/2$	Total local strain amplitude	σf	True fracture strength
ϵe	Local Elastic strain	σa	Alternating stress amplitude
ϵe	Elastic strain range	$\sigma f'$	Fatigue strength coefficient
$\epsilon e/2$	Elastic strain amplitude	γ	Engineering shear stress
ϵp	Local plastic strain	γoct	Octahedral shear stress
ϵp	Plastic strain range	$\gamma xy, xz, yz$	Shear stresses on specified cartesian planes
		% RA	Percent reduction in area

TABLE 1

Base, supplementary, and derived SI units

Measure	Unit	Symbol	Measure	Unit	Symbol
Base units			Electric resistance	ohm	Ω
Electric current	ampere	A	Energy, work, quantity of heat	joule	J
Length	meter	m	Energy density joule per cubic	meter	J/m³
Mass	kilogram	kg	Force	newton	N
Thermodynamic temperature	kelvin	K	Frequency	hertz	Hz
Time	second	s	Moment of force	newton meter	Nm
			Power, radiant	fluxwatt	W
Supplementary units			Pressure, stress	pascal	Pa
Plane angle	radian	rad	Quantity of electricity, electric charge	coulomb	C
Derived units			Specific volume	cubic meter per kilogram	m³/kg
Acceleration	meter per second squared	m/s²			
Angular acceleration	radian per second squared	rad/s²	Surface tension	newton per meter	N/m
Angular velocity	radian per second	rad/s	Thermal conductivity	watt per meter kelvin	W/m K
Area	square meter	m²	Velocity	meter per second	m/s
Current density	ampere per square meter	A/m²	Volume	cubic meter	m³
Density, mass	kilogram per cubic meter	kg/m³			
Electric potential, potential difference, electromotive force	volt	V			

TABLE 2

SI prefixes—names and symbols

Exponential expression	Multiplication factor	Prefix	Symbol
10^{18}	1 000 000 000 000 000 000	exa	E
10^{15}	1 000 000 000 000 000	peta	P
10^{12}	1 000 000 000 000	tera	T
10^{9}	1 000 000 000	giga	G
10^{6}	1 000 000	mega	M
10^{3}	1 000	kilo	K
10^{2}	100	hecto[a]	h
10^{1}	10	deka[a]	da
10^{0}	1	BASE UNIT	
10^{-1}	0.1	deci[a]	d
10^{-2}	0.01	centi[a]	c
10^{-3}	0.001	milli	m
10^{-6}	0.000 001	micro	μ
10^{-9}	0.000 000 001	nano	n
10^{-12}	0.000 000 000 001	pico	p
10^{-15}	0.000 000 000 000 001	femto	f
10^{-18}	0.000 000 000 000 000 001	atto	a

[a] Nonpreferred. Prefixes should be selected in steps of 10^3 so that the resultant number before the prefix is between 0.1 and 1000. These prefixes should not be used for units of linear measurement, but may be used for higher order units. For example, the linear measurement, decimeter, is nonpreferred, but square decimeter is acceptable.

Chapter 1 Overview

1.1 Introduction
    1.1.1 Scope
    1.1.2 Purpose
    1.1.3 Structure

1.1 Introduction

Fatigue is an important consideration in the design of most ground vehicle products. The literature on fatigue design and analysis is voluminous and can be overwhelming to a design or product engineer who must deal with many issues besides fatigue. Even someone well trained in specific disciplines, such as load-data acquisition or numerical analysis, does not necessarily have the time to learn the intricacies of other parts of the total fatigue-design process. This handbook is designed to provide easily usable, yet important information on each of the major elements of fatigue design.

The first two chapters provide introductory material useful to any user of the handbook. The first chapter addresses the scope, purpose, and structure of the handbook. The second chapter discusses general fatigue-design considerations; specifically, it describes why the designer should be concerned about fatigue and reviews some of the basic tradeoffs and potential problem areas.

The remaining ten chapters cover specific aspects of the fatigue design process. Appendix A provides definitions for all important fatigue design terminology.

1.1.1 *Scope*

This handbook describes the major elements of the fatigue-design process and how those elements must be tied together in a comprehensive product evaluation. It is *not a textbook* on fatigue; *it is a handbook* for industrial use. The fatigue-design process is broken down into a series of interrelated topics—each topic is discussed in a separate chapter. Each chapter begins with a statement of its scope, a description of its relationship to other chapters, and an overview of its contents. The technical material which is presented is substantial but not exhaustive. The presentations are amplified with illustrations and photographs where appropriate. A topical bibliography (or simply a list of references) is provided at the end of each chapter to provide the interested reader with additional, detailed information.

1.1.2 *Purpose*

The purpose of this *Fatigue Design Handbook* is to offer a single source that highlights the various current technologies and procedures of importance in a fatigue-design evaluation. The broad coverage is included within this single document with an aim to save the design engineer time, while concurrently ensuring that she or he understands the important elements of the fatigue design process and the potential impact of tradeoffs that may have been made or are being contemplated.

1.1.3 *Structure*

The structure of the handbook is illustrated in Fig. 1-1. Each of the chapters are written to address a specific element of the fatigue design process as shown in this figure.

Chapters 1 and 2 can be used to help define the scope of a specific problem and identify the logical steps to a solution. These introductory chapters suggest a different emphasis on particular parts of the process depending on whether or not the product of current interest is in the initial design stage or in a later stage of development, production, or service.

Chapter 3 reviews the factors which are important in defining the basic material properties of a particular product. Data on unnotched laboratory specimens are emphasized. It describes the properties of interest, how they are generated, and how they should be interpreted. Representative data tabulations and fatigue curves are included. Other sources are cited which contain detailed data compilations and descriptions of pertinent test procedures and analysis methods.

Chapter 4 is an extension of the preceding chapter. It describes the various processing methods that are applied to ground vehicle components and the influence these processes typically have on fatigue performance relative to baseline material fatigue properties. In a fatigue design situation, this chapter can be reviewed to identify what properties can be expected for a given part (compared to baseline data) if that part is processed in a certain way. This chapter also offers some guidance on what alternative processes or procedures can be considered to reduce the likelihood of, or eliminate a fatigue problem in a component.

Even after the material properties and processing effects are identified, typically there are a series of steps that must be taken to properly evaluate the adequacy of a particular design for service operations. In a very simplistic sense, it is a matter of (1) defining the loads which a part will see, (2) translating the loads into stresses and strains, and then (3) using known material fatigue properties, estimate the component's fatigue life. For each of these steps there are accepted analytical and experimental approaches which can be taken. In some critical applications, both approaches are pursued in parallel.

The choice between an analytical approach, an experimental approach, or a combined approach will depend on a number of factors including the complexity of the part and the service environment, the time and funds available for the evaluation, the criticality of the application, and the amount of prior experience with similar components in analagous applications.

— Define the Problem and Logical Steps to a Solution

— Evaluate Basic Material Properties

— Choose Analytical or Experimental Approach (Or a Combination)

— Consider How the Fatigue Properties of the Real Part Might Differ

— Define Forces Acting on the Structure

— Translate Loads into Stresses and/or Strains and Likely Sites for Crack Initiation

— Evaluate Fatigue Life, Failure Location

— Have Failures Occurred or Are They Predicted? If So, Consider Alternatives

— An Examination of Documented Case Histories May Suggest a Course of Action

— A Failure Analysis May Clarify the Source(s) of the Problem

— Evaluate the Need to Make Changes in the Design and/or Analysis

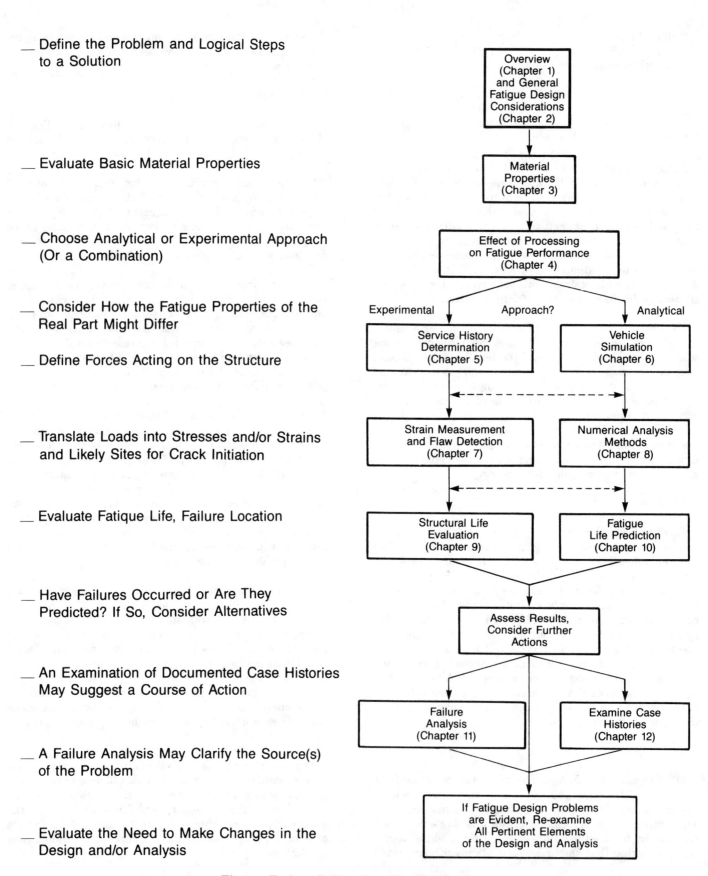

Fig. 1-1 Fatigue Design Handbook outline.

Chapters 5 and 6 address the problem of defining the forces that will act on a part in service. Chapter 5 focuses on the experimental approach for service history determinations. Chapter 6 addresses the analytical procedures which can be used to estimate vehicle loads through simulation. Both chapters address the pitfalls that should be avoided in defining a duty cycle or operating envelope for a system or component. The tradeoffs that must be made in what is measured or modeled, how much is measured or modeled, and how the data are recorded and subsequently digested to define a part's service history are discussed.

Chapters 7 and 8 deal with the translation of loads into stresses and/or strains and the identification of likely sites for crack initiation. Chapter 7 offers guidance for the individual interested in the experimental approach. The techniques commonly used to experimentally determine high-strain locations in a structure are reviewed, along with commonly used crack detection and monitoring techniques. Finally, techniques for determination of residual stresses are described. The focus of this chapter is on physically measured and/or observed quantities, e.g., strains and crack sizes.

In contrast, Chapter 8 focuses on numerically derived quantities. In particular, this chapter describes how numerical analysis procedures can be used as part of a total product-design evaluation. Linear elastic and small deformation elastic-plastic stress analysis procedures are described and some practical examples are included. The chapter deals with stress and strain estimation in uncracked components, as well as stress-intensity approximation in components assumed to contain cracks or defects.

Chapter 8 provides only limited details of the analytical development of the numerical procedures. Instead, it focuses on the background necessary to allow the interested reader to make logical choices concerning which methodology should be pursued in a given design situation. Numerous other sources that provide more detailed information on specific topics are cited.

Chapters 9 and 10 deal with the evaluation of the fatigue performance of a structure or component. Chapter 9 deals with the experimental evaluation of a structure; Chapter 10 addresses analytical fatigue-life prediction. These two chapters are the focal point for most of the other chapters in the handbook. The introductory paragraphs describe how the other chapters of the handbook contribute to the process of performing either an analytical or experimental fatigue-life evaluation. The introduction also addresses the circumstances under which an analytical approach would be appropriate as compared to a purely experimental evaluation or a combined analytical and experimental investigation.

Chapter 9 addresses the laboratory or field evaluation of components and structures under simulated service conditions. The pros and cons of laboratory evaluations compared to field tests are discussed and the practical pitfalls to avoid in each approach are reviewed. Various sources are cited which provide more detailed discussions of specific issues.

The tradeoffs between a durability and damage tolerance approach to life prediction are discussed in Chapter 10. Linear damage accumulation analysis procedures are emphasized for crack initiation life prediction. Crack-growth procedures are described which range from the simple linear elastic—no history accountability methods, to the more complex procedures which incorporate influences of retardation and other history-dependent factors.

Chapter 11 deals with failure analysis. Failures commonly occur in field tests and laboratory evaluations of new products. In fact, the conditions are sometimes extrapolated beyond the expected service environment so that potential real-life failure modes might be assessed. Of course, component failures do occur in service as well, and such occurrences normally require immediate investigation.

The chapter on failure analysis is written to serve three functions: (1) to define what steps can be taken to better understand the features of the failure itself, (2) to offer suggestions regarding other related things which should be examined that may give additional insight into the cause of the failure, and (3) to suggest logical steps to take to ensure that additional recurring failures do not take place.

Chapter 12 describes pertinent case histories that can be reviewed to see "how it was done before" or to gain some additional insights into how the techniques and procedures described in previous chapters can be applied. Case histories are included which cover product design evaluations involving aluminum alloys, cast materials, sheet steel, and forged steel. Applications which are considered include wheel assemblies, suspension system components, axle housings, connecting rods, and axle boxes and shafts.

Chapter 2 General Fatigue Design Considerations

2.1 Introduction

2.2 Design Goals and Product Planning

2.3 Analysis of Product Usage

2.4 Consideration of Failure Modes

2.5 Characterization of Materials

2.6 Analysis Procedures

2.7 An Example
    2.7.1 Service Loads, Noise and Vibration
    2.7.2 Stress Analysis
    2.7.3 Material Properties
    2.7.4 Cumulative Damage Analysis
    2.7.5 Component Test

2.8 References

2.1 Introduction

The engineering procedures used by the ground vehicle industry to promote structural integrity in vehicles is continuously evolving. Early engineers used many cut-and-try practices in their machine designs. This led to the development of experimentally oriented tests and development procedures. As the engineering profession developed further, analytical techniques were also introduced to make improvements in design procedures. Significant changes are still taking place in instrumentation, analysis procedures and methods of handling and analyzing data. As long as these techniques continue to evolve, engineers will have the opportunity to utilize these new resources to make improvements in their fatigue design procedures.

Present practice in this area, and future developments can best be understood by reviewing the total product development process that is used by many companies in the ground vehicle industry. Fig. 2-1 shows a typical series of steps that are used in the development of a new product. The central element of Fig. 2-1 indicates that both material properties and field service histories are necessary in order to complete analyses that can be used to reliably predict structural integrity. The upper left element in outer ring suggests that the design process should begin with a clear perception of the needs of the customer. The next step of product planning is necessary to ensure that the family of products which is produced is cost-competitive but profitable and to ensure that the needs of the customer in all parts of the world where the products will be marketed will be met. For example, a company may offer a full range of tractor sizes to cover small farms, groundskeeping, orchard operations, etc., as well as the large operations for row crop or small grain farms, where many of the tractors are large four-wheel drive models. In a similar manner, a range of automobile models may be manufactured to meet a variety of customer needs.

The next step in Fig. 2-1 is the one with which this Handbook deals extensively, the process of engineering design and development. The next two blocks in the figure address the planning of manufacturing facilities needed to make the components and difficulties which may be encountered in the production and assembly of the machine. The final step in the cycle is marketing of the product to the customer to fulfill the needs which were defined at the start of the cycle. Each of these steps are shown in series in Fig. 2-1. This is not meant to imply that some of these steps cannot be accomplished in parallel. In fact, many progressive companies are moving toward an integration of design, analysis and testing efforts in the product development process. An effective integration of these steps can reduce development cycle time and costs, while assuring optimized structures from the design and manufacturing viewpoints without compromise concerning durability requirements.

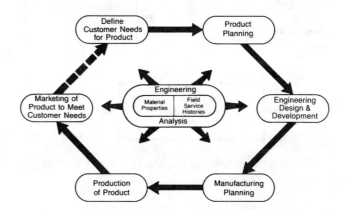

Fig. 2-1 Product development cycle.

Fig. 2-2 is a block diagram which shows how field service histories and material properties are the two basic inputs necessary to do either analytical or experimental analyses on a proposed design during the engineering development phase. (Experimental methods of load and strain measurement are discussed in detail in Chapters 5 and 7 of this handbook.)

Fig. 2-3 provides another illustration of the fatigue design process. It includes the same basic building blocks, but the terminology is different and perhaps more familiar in some engineering circles.

2.2 Design Goals and Product Planning

To properly plan a product, the engineering department may need to have a survey completed to determine the needs of the customer. This is frequently done through dealerships, technical society or trade related organizations, university

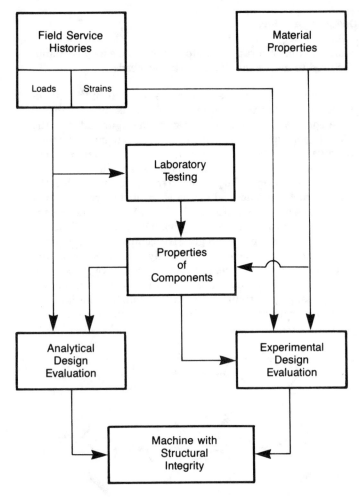

Fig. 2-2. Engineering design and development process.

reports or other customer related organizations. In such a survey it is important to consider the different functions a product may be expected to perform in its sales region (which may be worldwide). A survey of customers in each area should be made to determine the range of expected usages. For example, if a designer is concerned with the possible range of functions of an agricultural machine, he or she would want to determine customer needs representing harvesting techniques for various crops in different countries or localities. The same basic machine may be used to harvest wheat or other small grains in the central states, row crop grains such as corn or soybeans in the corn belt and rice in southern states. In a similar manner, automobile usage may vary in different parts of the country and with different types of operator, i.e., personal use, fleet use, police car, etc.

Frequently, usage data on the amount of time a product performs different functions can be adequately collected through dealerships and past sales records. The results of such a survey can be used by an engineering department to specify a family of machines to cover the various needs of different customers. This estimate should also include the desired service life or engineering design life for the various types of customers using the product. After an engineering department has established the design life goals it is then possible to begin the process of developing a machine with the needed structural integrity.

2.3 Analysis of Product Usage

The usage pattern of many products primarily results from the different applications the product is designed to perform. It is essential that the designer understand how physical loads are produced on the product as it is being used by the customer. The manner in which a product interacts with the environment during its use produces forces on it which, in turn, load its internal structure. Before the designer can analyze the significance of these loads, it is necessary to characterize their magnitude, frequency, rate of change, etc.

To evaluate the durability of a product for any specific configuration, usage area and type of customer, the designer must obtain and evaluate load information for the proper series of operations. (Analytical and experimental techniques for estimating loads in a structure are discussed in detail in Chapters 5 and 6.)

Several sets of operations may be completed with the same product design depending on the multiplicity of functions, usage area and types of customers. If data are available for each type of operation, the problem can be reduced to considering the appropriate set of operations and the corresponding percentages of total usage for each.

2.4 Consideration of Failure Modes

Knowledge of how a product is loaded will assist in understanding the potential failure modes that must be considered in the design analysis phase. Overloads may cause yielding and/or buckling. Cyclic loads can eventually lead to the initiation of fatigue cracks at highly stressed areas. Additional cyclic loading may cause these fatigue cracks to propagate and eventually cause failure of the part.

Depending on the road, soil, crop, weather conditions, etc., wear and corrosion may also be important failure modes to consider. These damage mechanisms are also cumulative, and continued use may lead to eventual failure.

The principal mode of failure that designers of ground vehicles typically need to be concerned with is *fatigue*. For example, a study by a large construction equipment company showed that about 60 percent of their field failures were caused by fatigue problems [1]. Other failures were attributed to other damage mechanisms, e.g. creep, buckling, wear and corrosion.

In total, the economic effects of fracture in the United States are staggering, with an estimated cost of *99 billion dollars annually* [2]. In this same study it was also estimated that these costs could be reduced by 29.1 billion dollars by the

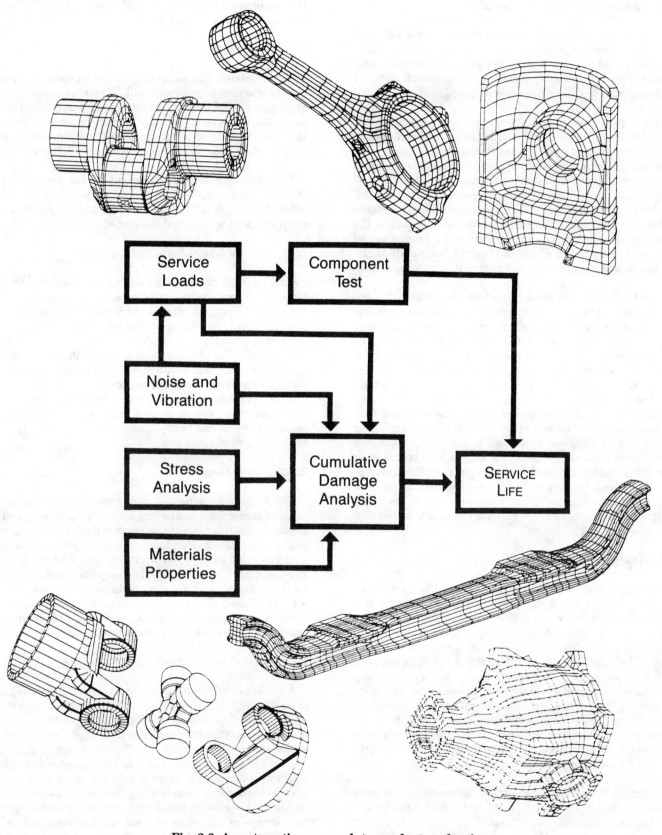

Fig. 2-3 A systematic approach to product evaluation.

application of presently known technology, such as that described in this handbook.

2.5 Characterization of Materials

Historically, engineers have used the monotonic yield and ultimate strength of materials during initial machine design. This was done to prevent static failures that may otherwise have resulted when an experimental product was field tested and subjected to the maximum anticipated loads.

Use of such static design procedures does not address the often important influence of smaller amplitude cyclic loads. The fatigue damage resulting from these cyclic loads is generally cumulative and a "lifetime" of testing may be required to determine if crack initiation will occur. As a result, it is essential to characterize the fatigue properties of the materials that may be used, as well as the intended usage patterns of the component or assembly. An estimation of fatigue life can then be made based on appropriate fatigue properties and the load history that is expected. (Various fatigue-life analysis procedures are discussed in Chapter 10 of this handbook.)

Different fatigue analysis procedures can be used depending upon the anticipated failure modes and loading, as well as the characteristics of the material or component. Depending on the anticipated mode of failure, an engineer can make material tradeoffs to improve the design. For example, components which are subjected to a large number of low stress cycles, but no high stress cycles, can be designed of a high hardness material which has good long life fatigue resistance. If the component is part of a frame structure that experiences a considerable range of loadings, the controlling factor may be sufficient toughness to avoid fracture from a single large overload. A component may be subjected to wear due to interaction with soil, crop materials, or other environmental factors. Various material selections can be made to resist wear. An engineer must accurately assess the likely failure mechanisms and make a material selection that will optimize the design.

2.6 Analysis Procedures

The engineering design and development block presented earlier in Fig. 2-1 is a key point in the development of a machine. An engineer can utilize a number of different analysis tools to facilitate this process. (Chapters 6, 8, 9 and 10 discuss many different analysis procedures which can be used.) The engineer may use all, or only a few of these procedures depending on how complicated the design may be, and how far along he or she is in the design process. As previously mentioned, the accuracy of these procedures are highly dependant upon knowledge of the field loads and the material or component properties. In some cases an engineer may have to complete the same step more than once (experimental field testing for example) in order to optimize a design.

A proper balance of field service information, material properties and design development techniques must be utilized by an engineering department. If an engineer is lacking either service data or material and component property data, the analysis will be incomplete and the results questionable. An engineer must use a proper balance of service data, material properties and analysis in order to develop a machine that will have adequate structural integrity and still be competitive in the marketplace.

2.7 An Example

Before delving into more of the details on specific aspects of the fatigue design process it will be helpful to some design engineers to read through the following, largely pictorial and schematic example of the fatigue design process as applied to the steering axle of a "heavy goods" vehicle. Other, more detailed case histories are also included in Chapter 12.

The basic elements of the fatigue design process are reviewed in Fig. 2-4. What was done with respect to each of these blocks to develop a fatigue life prediction for the example component will be reviewed briefly.

2.7.1 *Service Loads, Noise and Vibration*

First, a description of the service environment was obtained. The goal was to develop an accurate representation of the loads, deflections, strains, etc. that would likely be experienced during the total operating life of the component. This activity is illustrated and further described in Figs. 2-5 and 2-6. (Chapter 5 in this handbook describes the process of field load measurement in detail. Chapter 6 describes analytical methods which can be used to predict loads in a structure that does not yet exist or cannot readily be field tested.) In this particular case, the noise and vibration environment for specific parts of the heavy goods vehicle were analytically modeled as illustrated in Fig. 2-7.

2.7.2 *Stress Analysis*

The shape of a component or structure dictates how it will respond to service loads in terms of stresses, strains and deflections. Analytical and experimental methods are available to quantify this behavior. Examples of each approach, applied to the steering axle, are shown in Fig. 2-8. (Chapter 7 in this handbook provides more details on experimental techniques for strain measurement and crack detection. Chapter 8 covers a range of analytical procedures that can be applied for the purpose of translating external loads and deflections on a component into estimates of local stress and strain.)

2.7.3 *Material Properties*

A fundamental requirement for any durability assessment is a knowledge of the relationship between stress and strain and fatigue life for the material under consideration. Fatigue is a highly localized phenomenon that depends very heavily on the stresses and strains experienced in critical regions of a component or structure. The relationship between uniaxial stress and strain for a given material is unique, consistent and, in most cases, largely independent of location. Therefore, a small specimen tested under simple axial conditions in the laboratory can often be used to adequately reflect the behavior of an element of the same material at a critical area

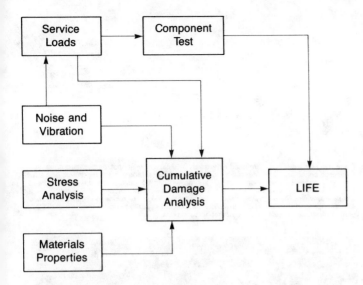

Fig. 2-4 Basic elements of the fatigue design process.

Service loads determination.
Typical Time/Load History from the Steering Axle of a Heavy Goods Vehicle.
Loading sequences experienced by components are rarely simple. These sequences, based on an understanding of overall usage, are developed from loads measured and recorded during specific operations. Consequently, all potential uses should be characterized by load measurement; including both normal and abnormal use. The most useful service load data are recordings of the outputs of strain gauges which are strategically positioned to directly reflect the input loads experienced by the component or structure. These load inputs are not generally influenced by either geometry or material changes. Consequently, this data provides a sound basis on which to assess the effects on component durability of design modifications and evaluate potential new designes at the drawing board stage.

Fig. 2-5 Service loads determination.

in a component or structure. It is important to remember, however, that most critical locations are at notches where stresses are multiaxial, even when loading is uniaxial. This multiaxiality, combined with potential out-of-phase variable amplitude loadings, can greatly affect fatigue resistance. In any case, a test specimen that is subjected to a strain history which is equivalent to that experienced by a component at a critical region, can normally be expected to display a similar stress response and have very nearly the same life to fatigue crack initiation as the component. (Differences in processing, residual stress and microstructure can complicate such comparisons, see Chapter 4 for further details.)

The critical relationship between stress and strain that must be modeled is the cyclic material response rather than static. The cyclic stress-strain curve takes into account the softening or hardening behavior of materials when subjected to cyclic loading. Several different methods exist for characterizing the cyclic stress-strain behavior of a material. In general, these approaches define the change in stress amplitude that can be expected to result from a known change in strain amplitude, as shown schematically in Fig. 2-9.

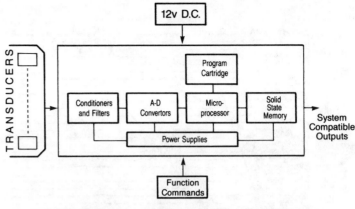

Simplified Schematic Diagram of an On Vehicle Data Analysis System

The load data is usually recorded on magnetic tape in either analog or digital form. Signal outputs from transducers are conditioned before recording on multi-channel instrumentation recorders to establish a permanent record suitable for subsequent computer analysis. The recorded load data is stored as multi-channel files in a computer data bank and is then readily accessible for analysis.

The requirements for analog or digital tape recordings necessarily restrict duration, but for longer term operations, the scope of the data acquisition phase can be extended significantly by the use of microprocessor based units.

Fig. 2-6 Analysis of service loads data.

Fatigue data (stress or strain range versus cycles or reversals to failure) should normally be developed over the total life range, from the apparent fatigue limit at long life to the low cycle fatigue regime. Fatigue data covering a range of lives are important to properly analyze most modern structures, since most design concepts based solely on fatigue limit stresses are overly conservative. A series of tests carried out at different strain amplitudes can be completed to define a strain life curve as shown in Fig. 2-10. (Further information on the fatigue properties of various engineering materials and guidelines on how to develop such data are provided in Chapter 3 of this handbook. The effects of processing on fatigue performance are discussed in Chapter 4. This last chapter will help the designer understand the ramifications of different choices in product form, heat treatment, surface condition and other processing factors on component fatigue resistance.)

2.7.4 Cumulative Damage Analysis

The fatigue life prediction process or cumulative damage analysis for a critical region in a component or structure consists of several closely interrelated steps. A combination of the load history (Service Loads), stress concentration factors (Stress Analysis) and the materials cyclic stress-strain properties (Material Properties) can be used to simulate the local uniaxial stress-strain response in critical areas. Through this process it is possible to develop good estimates of local stress amplitudes, mean stresses and elastic and plastic strain components for each excursion in the load history. Rainflow counting, as illustrated in Fig. 2-11 can be used to identify local cyclic events in a manner consistent with the basic material behavior. The damage contribution of these events is calculated by comparison with material fatigue data generated in laboratory tests on small specimens. The dam-

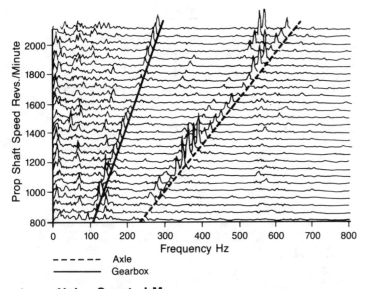

a) Noise Spectral Map

Noise and vibration studies are used to provide an insight in the modes and mechanics of component and structural behavior. An objective description of vibrating systems by frequency and amplitude is available by a signature analysis which forms a complete description of NVH behavior. Spectral mapping, for example, shows noise and vibration sources and their relative importance and can indicate the effect of structural resonance. A common technique for noise reduction is to alter the system response by detuning, using modal analysis.

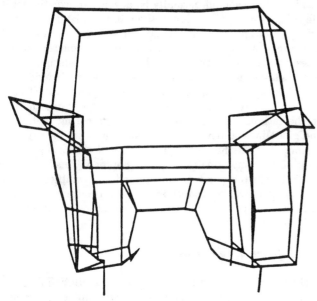

b) Tractor Cab Model

By building dynamic computer models, animated displays of component systems and structures can be obtained by simulating excitation inputs. These displays clearly show structural behavior and permit ready understanding of noise and vibration problems to facilitate engineering solutions. Structural modification software can predict dynamic changes brought about by adding or removing stiffness or weight and can predict modifications necessary to solve a problem.

Fig. 2-7 Noise and vibration analysis.

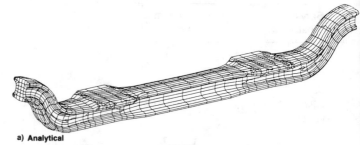

a) Analytical

Finite element techniques are used extensively to establish relationships between nominal measured conditions and localized stress/strain responses throughout components and structures (e.g. stress concentration factors). This method can be employed at the drawing board stage to identify areas of both high stress, where there may be potential fatigue problems, and low stress, where there may be potential for reducing weight. Combination of this information with service load data is used to establish the total stress/strain behavior throughout the component or structure under service operating conditions.

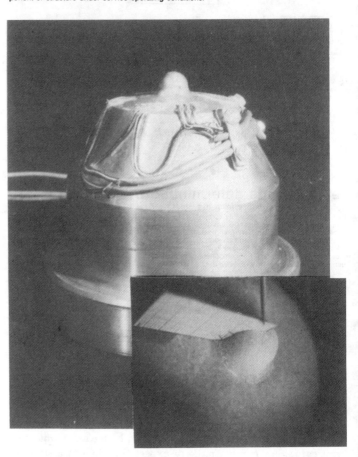

b) Experimental

Experimental methods such as brittle lacquer and photoelastic coating techniques can be used in situations where components or structures actually exist. Observation of coating behavior when the component or structure is loaded can illustrate areas of high and low strain on a subjective basis. Strain gauges strategically located can be used to quantify strains at critical areas, complex strain behavior (rosette gauges) and input loads.

Illustration shows a strain gaged and coated diesel engine injector cup; Inset show cracks in brittle lacquer at high stress sites.

Fig. 2-8 Stress analysis options.

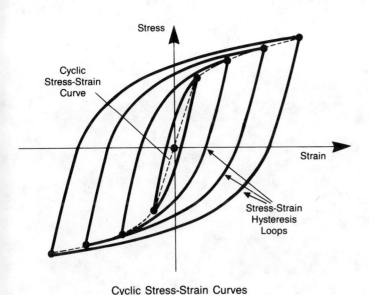

Fig. 2-9 Characterization of material properties.

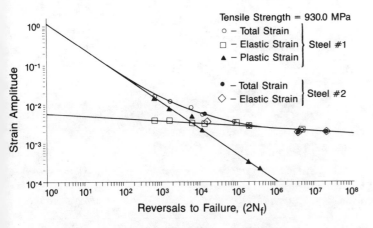

Fig. 2-10 Strain-life curves for two steels.

age fractions are summed linearly to give an estimate of the total damage for a particular load history. A comparison of the equivalent field time of the load history with the calculated damage fraction allowed the estimation of component life prior to initiation of a fatigue crack as shown in Fig. 2-12 for a steering axle component. (Chapter 10 in this handbook describes the tradeoffs in fatigue life prediction in considerable detail.)

2.7.5 *Component Test*

Full scale durability testing in the laboratory is an essential aspect of any fatigue life evaluation of a component or structure. It must be carried out at some stage in the development of a product to gain confidence in its ultimate service performance. Component testing is particularly important in today's highly competitive industries where the desire to reduce weight and production costs must be balanced with the necessity to avoid expensive service failures. A schematic of the multi-axis test system used to evaluate the fatigue performance of the steering axle is shown in Fig. 2-13.

The stage at which testing is done and the degree of sophistication that is employed should be based on available knowledge concerning the components function and criticality. Simple, rapid tests towards the end of the development phase are often possible in situations where the operating environments and products are comprehensively understood and proven analytical assessment methods have been used for preliminary evaluation work. In situations where the analytical methods are still under development or the failure mechanisms, materials, stress distributions or load histories are complex, it may be appropriate to perform laboratory experiments earlier in the development phase. It may also be necessary to increase the complexity of the test setup to improve the simulation of field service conditions. (Further details and tradeoffs in component testing are presented in Chapter 9 of this handbook.)

2.8 References

[1] Ohchuda, H., "Analysis of Service Failures of Hitachi Products (1970-1975)," July 1979.

[2] Duga, J. J., et. al., "The Economic Effects of Fracture in the United States," Part 2—A Report to NBS by Battelle Columbus Laboratories, March 1983.

Addresses of Publishers

U.S. Dept. of Commerce, National Bureau of Standards, Gaithersburg, MD 20899

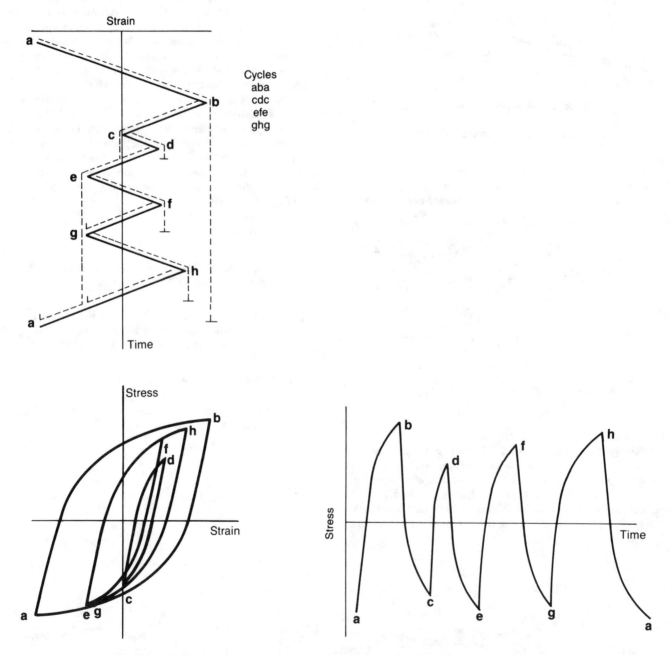

Fig. 2-11 Rainflow cycle counting to estimate local stress-strain response.

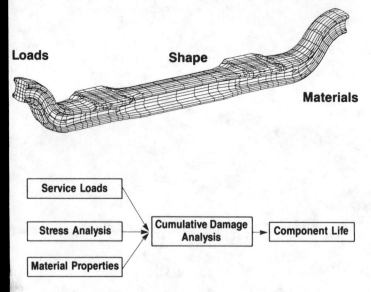

Fig. 2-12 The cumulative damage analysis process.

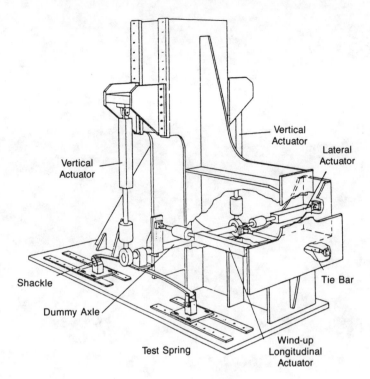

Multi Axis Test System Used of Evaluate Performance of a Steering Axle

Fig. 2-13 Multi-axis test system used to evaluate performance of a steering axle.

Chapter 3 Material Properties

3.1 Introduction
    3.1.1 Scope

3.2 Material Properties and Tests Relevant to Fatigue Design
    3.2.1 Monotonic Tension Properties
    3.2.2 Stress-Cycling Properties
    3.2.3 Strain-Cycling Properties
    3.2.4 Fatigue-Crack Growth/Threshold Concepts
    3.2.5 Fracture Toughness Tests
    3.2.6 Service Simulation Tests

3.3 Fatigue Behavior of Representative Alloys
    3.3.1 Cyclic and Crack Initiation Properties
        3.3.1.1 Wrought Carbon and Alloy Steels
        3.3.1.2 High Strength, Low Alloy Steels
        3.3.1.3 Cast Steels
        3.3.1.4 Cast Irons
        3.3.1.5 Wrought Aluminum
    3.3.2 Crack Growth Properties
        3.3.2.1 Wrought Carbon and Alloy Steels
        3.3.2.2 High Strength, Low Alloy Steels
        3.3.2.3 Cast Steels
        3.3.2.4 Cast Irons
        3.3.2.5 Wrought Aluminum Alloys

3.4 Additional Considerations in Material Fatigue Behavior
    3.4.1 Effects on Crack Initiation
        3.4.1.1 Notches
        3.4.1.2 Stress Relaxation
        3.4.1.3 Cyclic Creep
        3.4.1.4 Periodic Overstrain
        3.4.1.5 Material Variability
        3.4.1.6 Environment
    3.4.2 Effects on Crack Propagation
        3.4.2.1 Crack Closure/Retardation
        3.4.2.2 Environment

3.5 References

3.1 Introduction

Material properties data appropriate for fatigue and fracture analysis have evolved over the past century from simple life diagrams, defining safe ranges of operating stresses, to detailed portrayals of a material's resistance to fatigue crack initiation and propagation, as well as complete cyclic constitutive relations for use in structural analysis. Such data find a variety of applications in, for example, alloy development, material/process selection, design analysis, and failure analysis. Strain-life analysis has become a very effective tool for the ground vehicle industry for life prediction on experimental machines subjected to typical field conditions.

Establishing a sound basis for designing a component for a specified life or to investigate the cause of unexpected failure of a component requires a greater reliance on analytical procedures in modern day engineering practice. Through the use of computer aided design, an engineer can consider a greater variety of design possibilities prior to building prototype hardware. This of course, is a significant departure from the "build it and bust it" methodology of another era. Such analytical procedures rely very heavily on reliable materials properties data.

3.1.1 *Scope*

The intent of this chapter of the Fatigue Design Handbook is to present an overview of materials properties and tests that can be performed to derive the properties that characterize the room temperature fatigue and fracture behavior of engineering materials. Topics included are monotonic stress-strain behavior, cyclic stress-strain behavior, mean stress/strain effects, stress-life and strain-life behavior, stress relaxation behavior, fatigue crack growth and threshold stress intensity concepts as well as fracture toughness criteria for design analyses. Environmental influences on fatigue crack initiation and propagation are also addressed because of the deleterious effects of environment on these properties.

3.2 Material Properties and Tests Relevant to Fatigue Design

Wohler's [1] classic fatigue studies of over a century ago were the first attempt to quantify material fatigue performance. Using the rotating bending test machine shown in Fig. 3-1 [2], he performed tests on smooth and notched specimens plotting the results in the form of stress amplitude versus cycles to failure diagrams. He found that as the stress amplitude was lowered his specimens no longer failed in a specified number of cycles. A fatigue (or endurance) limit of safe operating stress was defined based on this no failure condition (Fig. 3-1).

Extensive efforts have been undertaken over the years in characterizing the fatigue behavior of metals along the lines of Wohler. Many, more recent studies have also explored the influence of other variables, such as mean stresses, notches, environment, and surface finish. Attempts have also been made [3,4] to correlate fatigue performance with more easily determined monotonic tension properties. One such approach for estimating stress-cycling fatigue resistance is shown in Fig. 3-2 [5]. An approximate stress-life curve for completely reversed cycling is estimated from the material's ultimate tensile strength. The long life portion of the fatigue curve can then be reduced by factors related to loading mode, geometry and environment [5-7]. Mean stresses are taken into account through constant life diagrams that provide combinations of mean and alternating stresses that are supposed to result in equal fatigue lives [5,7]. The empirical nature of this stress-based approach could not be generalized, however, and a higher demand on reliability and durability of complex structures required that a more fundamental understanding of the process be developed.

It is now understood that *fatigue occurs because of cyclic plastic deformation* [8-26] that causes irreversible changes in

Stress-Based Analysis

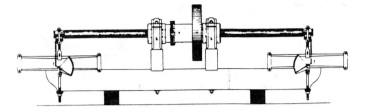

a) Rotating Bending Test Machine

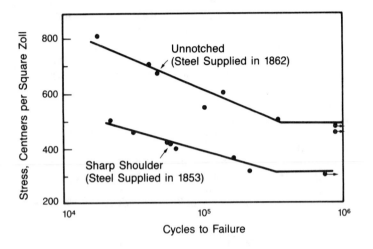

b) Fatigue Test Results

Fig. 3-1 Wohler's classic fatigue studies showed the influence of notches on fatigue resistance.

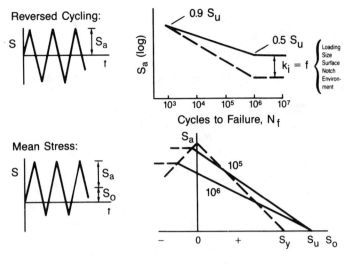

Estimation of Fatigue performance from Monotonic Tension Properties

Fig. 3-2 Estimation of fatigue performance from monotonic tension properties.

the materials dislocation substructure. The type of irreversible change in the material's substructure changes as the fatigue process progresses. Based on these changes, it is possible for most materials to divide the process into partially overlapping stages as follows:

(a) cyclic hardening and/or softening, depending on the initial condition of the material and the magnitude of stress or strain amplitude. The entire volume of the material may be affected by the change in the substructure.

(b) microcrack nucleation in the subsurface layer due to stress concentration effects at extrusions and intrusions.

(c) propagation of small cracks (crack sizes of the order of the grain size of the material).

(d) crack propagation (crack sizes significantly larger than the grain size of the material), ending in final fracture. The magnitude of cyclic plastic deformation concentrated in the plastic zone at the crack tip controls the rate of growth of such cracks.

A variety of specimen configurations and test techniques have been devised to generate fatigue data. As suggested by the fracture surface diagram presented in Fig. 3-3 [6], the results obtained will depend upon:

- the type of specimen (smooth, notched)

- the loading mode (axial, bending, torsion)

- the range of cyclic stresses or strains.

- the level of mean stress

- grain direction within the specimen (rolled sheet, plate and bar stock can exhibit drastically different cyclic and fatigue properties in different planes relative to the primary direction of cold work)

An example of different bending and axial fatigue data trends for an alloy steel is shown in Fig. 3-4 [22]. The apparent higher fatigue resistance in bending is due partially to the fact that only the surface of the bending specimen is exposed to the maximum stresses. More importantly, however, the apparent difference in fatigue properties results because the elastic bending relation routinely used to calculate stress is prone to increasing errors at high stresses (due to plasticity effects). Such concerns were less important in the past when similar relations (equally incorrect) were used to calculate component stresses. Modern design tools allow a more accurate determination of stresses in structures, thus it is important that the material fatigue properties are likewise presented in terms of actual stresses.

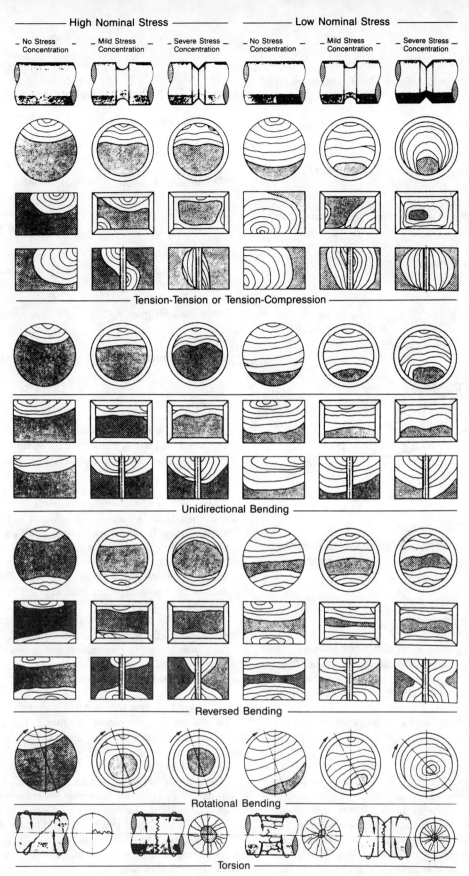

Typical Fracture Surfaces for Laboratory Test Specimens
Subjected to a Range of Different Loading Conditions.

Fig. 3-3 Typical fracture surfaces for laboratory test specimens subjected to a range of different loading conditions.

In addition to the traditional fatigue-life curves, modern fatigue analyses make use of:

- cyclic stress-strain relations for structural analysis and material modeling

- cyclic stress relaxation curves for assessing stability of mean/residual stresses

- fatigue crack growth curves for estimating propagation rates in structures.

Some of these common tests and their applications in fatigue analysis are listed in Table 3-1.

**Table 3-1
Common Tests and Applications in Fatigue Analysis**

Test	Use
Smooth specimen:	
Monotonic tension	Material comparison/screening
Stress cycling	Stress-life curves Mean stress effects
Strain cycling	Cyclic stress-strain curve Strain-life curve Mean stress relaxation
Notched fatigue	Fatigue notch sensitivity ($K_f = f(K_t)$)
Fatigue crack growth	Cyclic growth rate vs stress intensity range

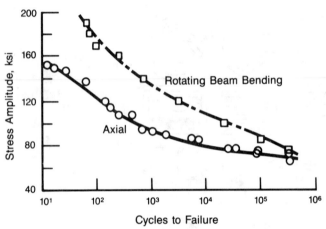

Fig. 3-4 An example of differing fatigue data trends for axial and rotating beam loading conditions.

3.2.1 Monotonic Tension Properties

Tension tests are widely used to provide basic design data such as strength, ductility and toughness of materials. They are also used as quality control tests to check if a material meets the specification for a given application. In fatigue analyses, it provides a base line stress-strain curve for evaluating the nature and extent of any subsequent cyclically-induced changes in deformation resistance (such as cyclic hardening or softening).

In a tension test, a smooth specimen with a cylindrical gage section (or a flat sheet specimen) is subjected to a continually increasing uniaxial tensile force while simultaneous observation is made of the elongation of the specimen [25-25]. Fig. 3-5 and Table 3-2 summarize the engineering and true stress-strain properties obtained from a tension test [26]. Also listed in Table 3-2 are the ASTM standards pertaining to tensile testing of metallic materials [27-30]. In addition to ASTM, SAE standard J416 specifies specimen sizes and geometry that should normally be employed in tensile tests [31]. Typical ranges in tensile properties of metals are presented in Table 3-3 [24] while properties for some selected materials of interest to the automotive industry are presented in Table 3-4 [24].

Ductility of engineering materials as measured by smooth specimen tests is sometimes reduced in the presence of discontinuities. Notched specimens have been tested to determine the notch sensitivity of materials [23]. Typically a circumferential notch is introduced in a round specimen (or a double edge notch in a sheet specimen). The notch strength is defined as the maximum load divided by the original cross sectional area at the root of the notch. The notch sensitivity is defined by the ratio of the notched strength to the unnotched tensile strength of the material. If this ratio is less than one, then the material is notch sensitive [23].

3.2.2 Stress-Cycling Properties

As noted earlier, this method of defining the fatigue behavior of a metal is the oldest technique. The stress versus cycles to failure curve may be generated using push-pull (axial) loading, rotating bending, or alternate bending technique. Different results are obtained depending on the test technique employed as shown in Fig. 3-6 [32]. The difference in behavior is believed to occur because of difference in the stress gradient, the equations by which the imposed stresses are calculated, and the cross sectional area of the material experiencing the imposed stresses [33]. The axial specimen, for example, experience a uniform stress gradient across the whole section of the specimen, while a bending specimen experiences a steep stress gradient from the outer surface of the specimen to the inner surface of the specimen.

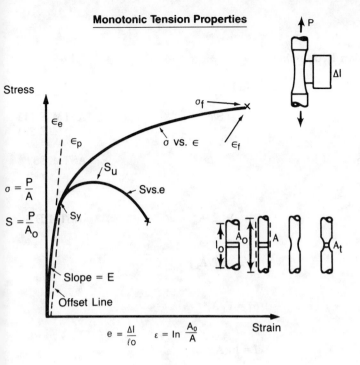

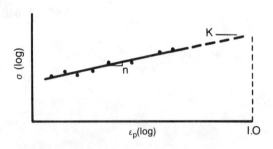

Typical Measured and Derived Monotonic Tension Properties

Fig. 3-5 Typical measured and derived monotonic tension properties.

If the finite life portion of the fatigue curve (life below the fatigue limit) is plotted as true stress versus reversals to failure and plotted on a log-log scale, the curve is linearized as shown in Fig. 3-7 [34] for two structural steels. This behavior is characterized by an equation [35]:

$$\sigma_a = \sigma_f' (2N_f)^b \qquad (1)$$

where b, the slope of the curve, and σ_f, the intercept, are considered fatigue properties of the material. Morrow [36,37] modified Eq. 3-1 to account for a mean stress, σ_o, effect as follows:

$$\sigma_a = (\sigma_f' - \sigma_o)(2N_f)^b \qquad (2)$$

The prediction based on Eq. 3-2 is shown in Fig. 3-8 [38] for a 1045 steel. It is now well established that tensile mean stress in the case of stress controlled cyclic tests shortens fatigue life, while compressive mean stress increases fatigue life. A tensile mean stress can thus be considered as an effective reduction in the fatigue strength coefficient, while a compressive mean stress an increase in the coefficient.

Data like these can be used in conjunction with stress-based structural analyses to estimate component fatigue life. Such an approach is useful for only long life problems where elastic strains are dominant. Low cycle fatigue regimes where large plastic strains determine the rate of damage accumulation, stress based analyses become highly inaccurate. Strain controlled fatigue behavior is the most suitable method under such conditions.

3.2.3 Strain-Cycling Properties

One of the more recent innovations which is used extensively in the ground vehicle industry in fatigue testing and which has gained reasonable acceptance involves using strain, rather than stress, as the controlling mode [24-26,35-37,39-42]. As indicated in Fig. 3-9 [26,39], a strain transducer attached to the specimen is used to sense and control the appropriate limits. Of particular significance is that both stress and strain can be monitored simultaneously throughout a test thus completely documenting the deformation response of a material from initial cyclic shake down through mid-life steady-state response to the formation and growth of a critical fatigue crack.

For cyclic property determination, smooth samples are subjected to completely reversed strain control, as illustrated in Fig. 3-10 [37].

After the initial hardening or softening (depending on the initial material condition), essentially the same size and shape of the hysterisis loop will be produced cycle after cycle. The tip of these stabilized loops for several specimens tested over a broad range of strain generates the cyclic stress-strain curve, as illustrated in Fig. 3-11 [37].

The same specimens tested to failure produce the strain versus reversals to failure. An example of this process is shown in Fig. 3-12 [39] for a SAE 950 steel. The resulting portrayal yields:

(a) a cyclic stress-strain curve that, when compared with the original monotonic curve, indicates the nature and extent of cyclically-induced changes in deformation resistance

(b) a strain-life curve comprised of a plastic strain component (related to material ductility) that dominates resistance at short lives and an elastic strain component (related to material strength) that dominates high cycle resistance.

The cyclic stress-strain behavior is generally characterized by an equation of the type:

$$\Delta\epsilon/2 = \Delta\epsilon_e/2 + \Delta\epsilon_p/2$$

$$= \Delta\sigma/2E + (\Delta\sigma/2K')^{1/n'} \qquad (3)$$

Table 3-2
Monotonic Tension Properties

Property	Determination	Relation	Refs.*
E, Elastic modulus	Slope of stress-strain curve at low stress.	$E = S/e$	ASTM E 111 E 231
S_y, Yield strength (offset)	Stress at which an offset line with slope E intersects the stress-strain curve.		ASTM E 8
S_u, Ultimate tensile strength	Stress at maximum load.	$S_u = P_{max}/A_o$	ASTM E 8
%RA, Percent reduction in area		$\%RA = [(A_o - A_f)/A_o] \times 100$	
σ_f, True fracture strength	Load at fracture divided minimum fracture area.	$\sigma_f = P_f/A_f$**	
ϵ_f, True fracture ductility	True plastic strain at fracture.	$\epsilon_f = \ln(A_o/A_f)$	
n, Strain hardening exponent	Slope of log σ-log E_p plot.		ASTM E 646
K, Strength coefficient	Stress intercept at $E_p = 1$ on log σ-log Ep plot.	$\sigma = K(E_p)^n$	

*Annual book of ASTM standards, Vol. 10 [27-30].
**Usually corrected for stress triaxiality, Ref. [23], pp. 252.

The delta in the above equation indicates completely reversed stresses and strains and subscripts e and p denote elastic and plastic strain, respectively. The two material properties K' and n' are the cyclic strength coefficient and cyclic strain hardening exponent respectively.

A technique for approximating cyclic stress-strain behavior using one specimen is illustrated in Fig. 3-13 [37].

In this technique, one specimen is subjected to several large, completely reversed, strain cycles, and then the strain is gradually reduced in about forty steps to the point of zero stress and strain. The strain is then incrementally increased to the same large value, then decreased, and so forth until the specimen fails. After the initial hardening or softening, the plot of stress versus strain provides an approximation to the locus of tips of stable loops [37]. Although Eq. 3-3 still describes the stress-strain behavior, the constants K' and n' obtained by this method may have a slightly different value [37]. An example of such difference is presented in Fig. 3-14 for a RQC-100 steel [37]. In addition to differences due to test method, it has been suggested that [43-45] the cyclic stress-strain behavior must be represented separately for long-life and short-life fatigue regimes with differing K' and n'. Mughrabi [43] has shown that in many cases the cyclic stress-strain curve can be divided into three regimes representing differing deformation behavior, similar to the monotonic curve of a single crystal. The three regimes can be differentiated by plotting the stress-strain data on a semi-log plot.

The strain versus life behavior is characterized by the strain-life [46,47] relationship:

$$\Delta\epsilon/2 = (\sigma_f'/E)(2N_f)^b + \epsilon_f'(2N_f)^c$$

The properties and the various terms are summarized in Table 3-5. The procedure for conducting such experiments is detailed in ASTM standard E606 [48]. Table 3-6 shows a material characterization sheet format adopted by the SAE Fatigue Design and Evaluation Committee for documenting the monotonic and cyclic properties of engineering alloys [39].

The difference in fatigue behavior as indicated by a stress controlled test as compared to a strain controlled test becomes apparent in Fig. 3-15 [34]. Stress-cycling curves for the two structural steels shown previously in Fig. 3-7 are seen to differ from their strain-cycling counterparts. In particular,

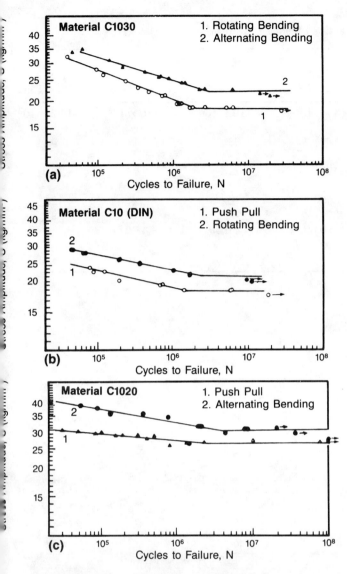

Experimental Results from Different Types of Fatigue Testing:
(a) C1030 (0.25%C) Steel — Alternating and Rotating Bending Tests
(b) C10 (0.9095%C) Steel — Rotating Bending and Push-Pull Tests
(c) C1020 (0.12%C) Steel — Alternating Bending and Push-Pull Tests

Fig. 3-6 Experimental results for three fatigue test modes.

material ductility comes into play in the low-cycle region resulting in a cross over of the curves in the mid life range. Another major difference lies in the stability of tests run under a positive mean stress. The stress controlled tests tend to show instability as cyclic plastic creep occurs. An important aspect not revealed by Fig. 3-15 is that the concept of "endurance limit" is assumed in the stress-life technique of defining the fatigue behavior of a material. The strain-life technique on the other hand does not assume the existance of an "endurance limit." It is now established that under service spectrum loading conditions the "endurance limit" *does not exist* for many engineering materials. It has been argued that

**Table 3-3
Ranges in Monotonic Stress-Strain Properties of Metals**

Monotonic property	Typical range for engineering metals
Modulus of elasticity, E, ksi	10 to 80 × 10^3
Tensile yield strength, $S_{0.2\%y}$, ksi	1 to 3 × 10^2
Ultimate tensile strength, S_u, ksi	10 to 400
Percent reduction in area, % RA	Nil to 90%
True fracture strength, σ_f, ksi	0.5 to 5 × 10^2
True fracture ductility, ϵ_f	Nil to 2
Strain-hardening exponent, n	Nil to 0.5

strain-cycling more closely simulates the control condition at a local region in a structure. This not-with-standing, a far more complete description of all aspects of a material's cyclic response is obtained with this approach.

The local strain approach to fatigue analysis using the strain-life relation essentially assumes that the amount of cracking which occurs during uniaxial, baseline strain-controlled tests is roughly equivalent to the amount that occurs in a larger engineering component during the initiation of a crack. While failure in a laboratory specimen may be defined as complete fracture of the specimen or appearance of a visible crack or some other indication such as instability in control parameters, the failure of an engineering component will be defined in terms of ability to fulfill its intended function. For a spring, for example, cyclic softening and mean or residual stress relaxation producing excessive deformation or loss of stiffness could constitute failure even though no fatigue cracks may be present. The important idea is that the estimated fatigue life of the engineering component must relate to the ability of a component to fulfill its function irrespective of the failure criteria chosen for the laboratory specimen. In general, automotive design relies primarily on crack initiation criteria.

3.2.4 *Fatigue-Crack Growth/Threshold Concepts*

Smooth specimen life curves provide indications of a material's resistance to crack initiation and early growth. Crack growth resistance is determined by continuously monitoring the growth of a crack from a notch as the specimen is cyclically loaded. Linear elastic and elastic-plastic fracture mechanics methodology is employed in studying fatigue crack growth behavior [49-54].

In linear elastic fracture mechanics (LEFM), the idea is to relate the elastic stress field developed in the vicinity of a crack tip to the applied nominal stress on the structure, taking into consideration the geometry and size of the defect. As shown in Fig. 3-16 for Mode I loading, the crack tip stress intensity factor, K_I, uniquely relates to the far field stresses for a given geometry under linear elastic conditions. Under

Table 3-4
Monotonic Stress-Strain Properties of Selected Metals

Typical monotonic property	Steels A			Aluminum Alloys		
	USS T-1 (258 HB)	MAN-TEN (150 HB)	1045 (390 HB)	7075-T73	6061-T6	2024-T351
Modulus of elasticity, E, ksi	30.2×10^3	30×10^3	30×10^3	10×10^3	10×10^3	10.2×10^3
Tensile yield strength, $S_{0.2\%y}$, ksi	105	46	185	63	43	44
Ultimate tensile strength, S_u, ksi	117	82	195	74	46	79
Percent reduction in area, % RA	66	69	59	32	43	35
True fracture strength, σ_f, ksi	176	145	270	89	64	92
True fracture ductility, ϵ_f	1.08	1.19	0.89	0.39	0.56	0.43
Strain-hardening exponent, n	0.088	0.21	0.044	0.066	0.09	0.90

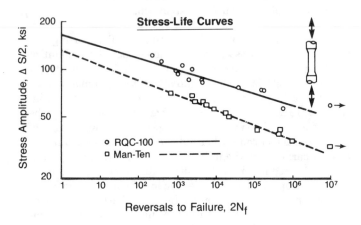

Stress-Life Fatigue Curves for Two Structural Steels

Fig. 3-7 Stress-life fatigue curves for two structural steels.

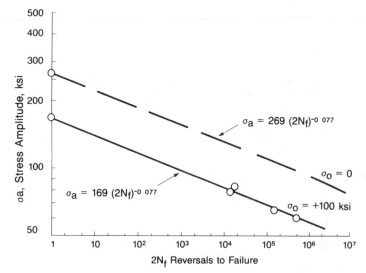

Fig. 3-8 Fatigue strength vs. life relations for 1045 steel (386 BHN) as influenced by mean stress.

cyclic conditions, the applied stress intensity factor range, delta K_I, is then considered the driving force for the growth of a crack in a geometry.

Fig. 3-17 illustrates the use of LEFM technology in the design against fatigue failure [55]. The detailed techniques for generating fatigue crack growth rate data and their analyses are discussed in detail in Refs. 56 and 57. Briefly it may be noted that the curve fitting routine used to convert the experimentally measured crack length versus number of cycles data to fatigue crack growth rate versus applied stress intensity factor range has a strong influence on the trend exhibited by the crack growth rate data. Although no clear guidelines are available for choosing a curve fitting routine, care must be excercised in such analyses so as not to mask any real trends in data.

The fatigue crack growth rate curve illustrated in Fig. 3-17 shows only one portion of the fatigue behavior. A typical crack growth curve has three distinct regions as schematically shown in Fig. 3-18 [58]. The region labeled II is characterized by the Paris relationship [59]:

$$da/dN = C\Delta K^m \quad (5)$$

where C and m are constants. The value of m typically lies in the range of 2 to 5 for structural materials. Fatigue cracking occurs in this region by the striation mechanism [60-62]. The region labeled I represents what is known as the threshold for fatigue crack growth, ΔK_{th}. Below this value of ΔK, fatigue crack growth rates become extremely low [63-65]. The region

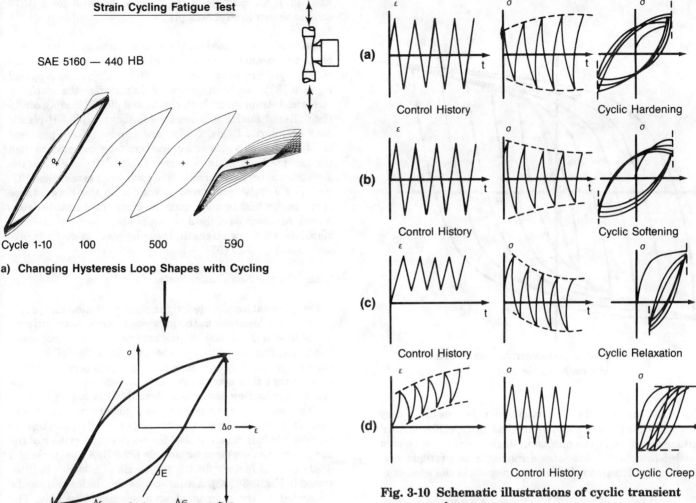

Fig. 3-9 Hysteresis loops measured in a strain cycling fatigue test.

Fig. 3-10 Schematic illustrations of cyclic transient phenomena.

labeled III shows very rapid fatigue crack growth rates because the applied ΔK approaches the fracture toughness, K_{IC}, of the material.

Recommended procedures for the measurement of fatigue crack growth thresholds are available [65]. The recommended practice defines threshold as that ΔK for which the corresponding fatigue crack growth rate is 1×10^{-10} m/cycle [66]. In practice the definition of this threshold depends on the accuracy of the crack monitoring technique and the number of cycles over which such small increments in crack length are measured. There is no consensus on the precise mechanism of near-threshold fatigue crack growth [63-65], but it is now well understood that crack growth in this region is extremely sensitive to microstructure, mean stress, and environment. These variables, in general, have less of an influence on crack growth rates in the Paris law region. A typical example of fatigue crack growth covering all three regions is presented in Fig. 3-19 [67] for an SAE 4340 low alloy, high strength steel. Fatigue crack thresholds for a number of engineering alloys are listed in Table 3-7.

3.2.5 Fracture Toughness Tests

The criteria for failure in the presence of a crack-like defect is that instability will occur whenever the magnitude of the crack tip stresses exceed some critical conditions [68,69]. Since the magnitude of the crack tip stresses can be described in terms of the stress intensity factor, K, a critical value of the stress intensity factor can be used to define the critical crack tip stress conditions for failure under linear elastic conditions. For Mode I loading under plain strain conditions, the critical stress intensity factor for fracture instability is designated as K_{IC}. Referred to as fracture toughness, K_{IC} is considered a material property that represents the material's inherent resistance to failure in the presence of a crack like defect [70-73].

The current standard method of fracture toughness evaluation using the K parameter (ASTM E399) is based on initiation of crack extension for plane strain conditions [74].

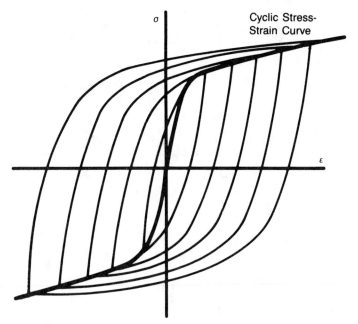

Fig. 3-11 Cyclic stress-strain curve drawn through stable loop tips.

Fig. 3-20 illustrates the various specimen designs covered by this specification. The standard method also specifies critical parameters such as the minimum thickness of the specimen needed to achieve plane-strain condition and fatigue precracking conditions to obtain a sharp crack in the specimen.

Typical values of fracture toughness for a number of structural materials are presented in Table 3-8 [75]. Example of fracture toughness behavior for an aluminum alloy as a function of specimen thickness is shown in Fig. 3-21 [50,76]. A much higher value of toughness (called plane stress toughness) is obtained at lower thickness of the specimen. It should be noted that plane strain fracture toughness is *not* strongly related to impact toughness as measured in a Charpy test. While K_{IC} type toughness is based on LEFM concepts, Charpy toughness is based on a specimen design that has a blunt notch, small size, and does not differentiate between initiation and propagation energies. Charpy toughness is widely used because it is very fast to conduct, inexpensive, simple to use, and has many years of correlation with service performance.

In specimens with thin sections (plane stress conditions), slow stable cracking is often large enough that a single parameter is not adequate to characterize the fracture toughness behavior of the material. Under such conditions, the R-curve method becomes appropriate [77-79]. R-Curve characterizes the resistance to fracture of a material during slow-stable crack extension. It consists of a plot of K_R versus Δa, where K_R represents the driving force for stable crack extension prior to unstable crack growth at K_c. The procedure for determining the R-curve of a material is outlined in ASTM E561 [80]. An example of the R-curve behavior for a A572 steel is shown in Fig. 3-22 [81].

The techniques based on the stress intensity factor as the crack tip parameter are valid only under conditions of constrained yielding at the crack tip. The J-integral, as proposed by Rice [82], has been shown to characterize the crack tip stress and strain under both elastic and plastic loading conditions. Based on Hutchinson-Rice-Rosengren [83,84] plastic zone singularity fields, Begley and Landes [85,86] proposed the J-integral as a measure of fracture toughness. Since then the test methods for the determination of J_{IC} as well as the J-resistance curve determination have been established [87-90]. The J-resistance curve is similar to the K-resistance curve except that now J versus the crack increment is determined. An example of the J-resistance behavior and the associated J_{IC} for a low strength, low alloy steel at 400°F is presented in Fig. 3-23 [91].

3.2.6 Service Simulation Tests

The constant amplitude fatigue crack initiation and propagation tests discussed in the previous sections bear little resemblance to the actual loading experienced by a component in service. But such data provide input for design of the component against fatigue damage. Once design is completed, it is important that accelerated tests be conducted that simulate service loading patterns (as described in Chapter 9). Such tests not only serve to qualify the design but can also serve to assess the fatigue analysis of the component. For such tests to be valuable it is required that the service load history of the component be developed accurately [92-97]. An example of a digitized load history for the front axle of a loader is illustrated in Fig. 3-24 [98]. Actual service simulation test can be conducted on the component or a complete assembly. The basic principle of complete vehicle test is illustrated in Fig. 3-25 [99].

Sometimes it is simpler to conduct programmed fatigue loading instead of the actual service history waveform. The programmed loading is generated by the cycle counting methods [100-104]. Some of the types of laboratory simulation of service loading are depicted in Fig. 3-26 [99]. The foremost advantage of programmed fatigue testing is that the tests can be accelerated easily. This can be achieved by (a) eliminating the small amplitude stress cycles that are believed to be innocuous in terms of fatigue damage, (b) the stress amplitudes may be increased to reduce test duration, and (c) the test frequency may be increased.

3.3 Fatigue Behavior of Representative Alloys

3.3.1 *Cyclic and Crack Initiation Properties*

3.3.1.1 *Wrought Carbon and Alloy Steels*

Cyclic response of a variety of low carbon steels are shown in Fig. 3-27 [105] comparing SAE 1008, 1020, and Ferrovac E, cyclic strength and strain hardening rate increase with increasing carbon content. Annealed Ferrovac E exhibits sig-

CONSTANT AMPLITUDE FATIGUE TEST RESULTS

MATERIAL	SAE 950X	TEST OBJECTIVE	Characterization
CONDITION	Hot rolled	SPECIMEN DESCRIPTION	Sheet
HARDNESS	150 HB	EQUIPMENT DESCRIPTION	906.15
OTHER SPECS.	—	TEST TEMP.	Ambient
MATERIAL SOURCE	Vendor	CONTROL MODE	Strain
		FREQUENCY & WAVEFORM	Triang.
		FAILURE DEFINITION	Separation

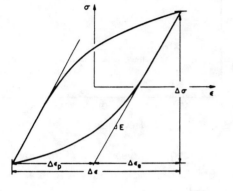

Specimen No.	Strain Amplitude, $\Delta\epsilon/2$	Fatigue Life, Reversals	STABLE or HALF-LIFE VALUES			
			Stress Amplitude, $\Delta\sigma/2$, MPa (ksi)	Plastic Strain Amplitude, $\Delta\epsilon_p/2$	Elastic Strain Amplitude, $\Delta\epsilon_e/2$	Modulus of Elasticity, E, GPa (ksi)×10³
1	0.0200	350	502 (73)	0.0175	0.0025	201 (29.2)
2	0.0100	810	448 (65)	0.0078	0.0022	203 (29.5)
3	0.0050	4,040	386 (56)	0.0031	0.0019	203 (29.5)
4	0.0025	32,000	324 (47)	0.0009	0.0016	203 (29.5)
5	0.0020	70,200	303 (44)	0.0005	0.0015	202 (29.3)
6	0.0015	289,000	265 (38.5)	0.0002	0.0013	204 (29.6)
7	0.0012	1,130,000	241 (35)	—	0.0012	201 (29.2)

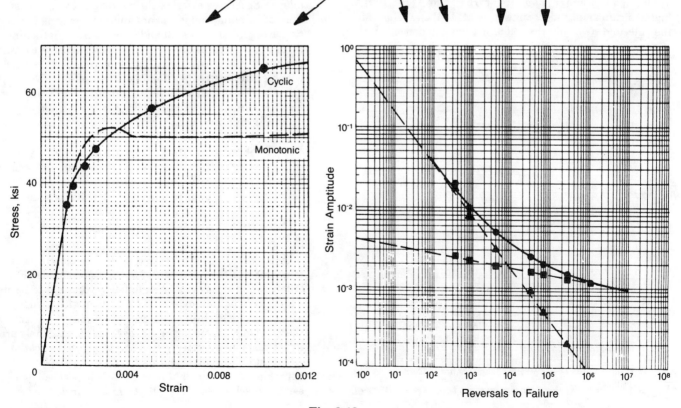

Fig. 3-12

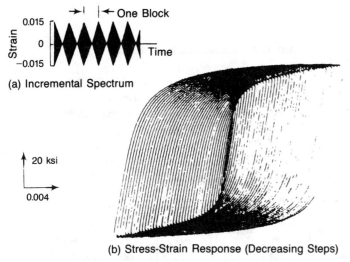

Fig. 3-13 Incremental step straining of sheet steel.

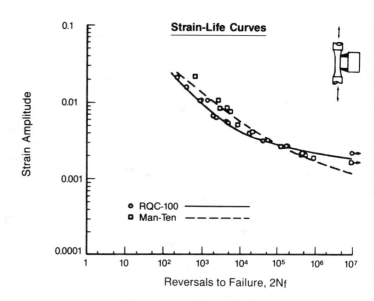

Strain-Life Fatigue Curves for Two Structural Steels

Fig. 3-15 Strain-life fatigue curves for two structural steels.

nificant cyclic hardening while the two carbon steels soften at low strains and harden at high strains [105]. Microstructure has a strong influence on cyclic response of carbon steels as shown in Fig. 3-28 for a pearlitic eutectoid steel [106]. The fine, medium, and coarse interlamellar spacing determines the cyclic stress-strain response of this material. The fine spacing material showed only cyclic softening. The medium spacing material displayed cyclic softening at low strain amplitudes and hardening at higher strain amplitudes [106]. The coarse pearlite exhibited strain hardening over the entire range of strain amplitudes. Dislocation cell structures are formed in iron [107] and iron-carbon alloys [108,109] at higher strain amplitudes. Sunwoo, et al [106] have shown for the eutectoid steel that the cell formation will depend on the interlamellar spacing as well. Cell formation becomes more likely as the interlamellar spacing or the strain amplitude increases [106].

In the higher strength regime, two high strength low alloy steels (SAE 950X and 980X) are compared with a quenched and tempered boron steel (Fig. 3-27). The high strength low alloy (HSLA) steels derive their strength from modest additions of niobium, vanadium or titanium resulting in a combination of grain refinement and dispersion strengthening. The resulting structures are cyclically stable [110] in contrast to tempered martensitic structures of comparable hardness which exhibit extreme softening. In recent studies [111,112] it has been shown that cyclic softening occurs in HSLA steels for strain amplitudes in the yield point region and hardening at higher amplitudes.

Fig. 3-29 presents fatigue data for some selected alloys in this category [113,114]. It is apparent that high hardness material is superior in the long life regime while softer material is more fatigue resistant in the low life regime.

3.3.1.2 *High Strength, Low Alloy Steels*

Cyclic data for two medium carbon steels quenched and tempered to different hardness levels is presented in Fig. 3-30 [105]. Cyclic strength is seen to be proportional to hardness while strain hardening rates change little with hardness. The range of cyclic stability in these steels coincides closely with the first stage of tempering suggesting that the finely precipitated carbides effectively stabilize the high dislocation density formed during transformation [105]. In Fe-Ni-C systems, approximately 0.1 weight percent carbon has been found necessary to achieve cyclic stability in as-transformed marten-

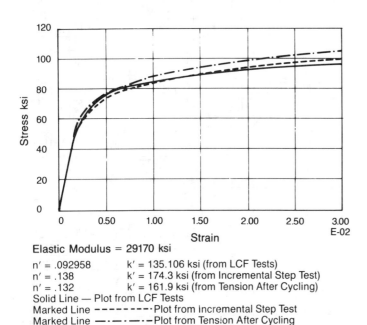

Fig. 3-14 Cyclic stress-strain curves for RQC-100 steel.

Table 3-5
Smooth Specimen Cyclic Properties

Property	Determination	Relation	Refs.*
S'_y, Cyclic yield strength (0.2%)	Stress at which an 0.2% offset line with slope E intersects the cyclic stress-strain curve.		
n', Cyclic strain hardening exponent	Slope of log $\Delta\sigma/2$ – log $\Delta\epsilon_p/2$ plot.		ASTM E 606*
K', Cyclic strength coefficient	Stress intercept at $\Delta\epsilon_p/2 = 1$ on log $\Delta\sigma/2$ – log $\Delta\epsilon_p/2$ plot.	$\dfrac{\Delta\sigma}{2} = K' \left(\dfrac{\Delta\epsilon_p}{2}\right)^{n'}$	"
σ'_f, Fatigue strength coefficient	Stress intercept at $2N_f = 1$ on log /2 – log $2N_f$ plot.		
b, Fatigue strength exponent	Slope of log $\Delta\epsilon_e/2$ – log $2N_f$ plot.	$\dfrac{\Delta\epsilon_e}{2} = \sigma'_f (2N_f)^b$	"
ϵ'_f, Fatigue ductility coefficient	Plastic strain intercept at $2N_f = 1$ on log $\Delta\epsilon p/2$ – log $2N_f$ plot.	$\dfrac{\Delta\epsilon_e}{2} = \sigma'_f (2N_f)^b$	
c, Fatigue ductility exponent	Slope of log $\Delta\epsilon p/2$ – log $2N_f$ plot.	$\Delta\epsilon p/2 = \sigma'_f (2N_f)^c$	"

*Annual Book of ASTM Standards [48].

site [115]. At later stages of tempering, with the development of more massive carbides, such stabilization can no longer occur and softening ensues [105].

Fig. 3-31 presents the behavior of several other high strength steels [105]. The higher carbon quenched and tempered grades, SAE 10B62 and 52100, behave similar to medium carbon steels showing significant cyclic softening and comparable strain hardening rates. Ausformed H-11 tool steel has exceedingly high deformation resistance and a low strain hardening rate. The precipitation hardening, 18 percent nickel maraging steel shows comparable strain hardening rates in both the annealed and the aged conditions [105]. Although aging increases cyclic strength significantly, both conditions exhibit appreciable softening. Cyclic softening of the annealed structure is associated with the formation of a dislocation cell structure; softening of the aged structure is attributed to the rearrangement of dislocation networks around precipitates [116].

Fig. 3-32 presents monotonic and cyclic stress-strain curves and strain life curves for a cross section of steels [39,114].

3.3.1.3 Cast Steels

Mitchell [117] investigated the monotonic and cyclic stress-strain response of SAE 9262 steel in the pearlitic and two martensitic conditions. The results presented in Figs. 3-33 through 3-35 indicate that the pearlitic matrix cyclically hardens while the higher hardness martensite cyclically softens. The lower hardness martensite shows mixed behavior. The strain-life curves for these materials are presented in Figs. 3-36 through 3-38 [117]. Pearlite having the lower fracture ductility, has an inferior short life fatigue resistance compared to the martensites. The harder martensite exhibits superior long life fatigue resistance as compared to the softer martensite.

The strain-life curves for cast ASTM A27-65-35 steel at various hardness levels (Figs. 3-39 through 3-41) similarly show that the higher hardness material has a superior long life fatigue resistance but a inferior short life fatigue resistance [118].

3.3.1.4 Cast Irons

Mitchell [117] investigated the fatigue behavior of pearlitic and martensitic cast irons with three graphite morphologies, as well as the low hardness martensite with the mixed graphite structure. The results are presented in Figs. 3-42 through 3-44. It seems that coarser graphite gray iron is inferior to the finer graphite gray irons. The strain-life behavior of the high hardness martensitic iron with coarse graphite is essentially the same as its softer pearlitic complement even though the martensite has approximately a 20 percent greater ultimate strength. The most drastic reduction in fatigue resistance is with the mixed graphite, high hardness martensite, that is approximately 40 percent less resistant in strain than its pearlitic complement [117]. Conversely, the mixed graphite, low hardness martensite and its

Table 3-6
Material Characterization Data

MONOTONIC PROPERTIES:		
Mod of Elast E	203 GPa ($29.5 \text{ a } 10^3$ ksi)	
Yield Strength 0.2% S_y	345 MPa (50 ksi)	
Ultimate Strength S_u	441 MPa (64 ksi)	
Strength Coeff K	675 MPa (98 ksi)	
Strain Hard Exp n	0.16	
Red in Area % RA	65	
True Frac Strength σ_f	758 MPa (110 ksi)	
True Frac Ductility ϵ_f	1.05	

Material	SAE 950X	
Condition	Hot rolled	
Hardness	150	HB
Material Source	Vendor	
Specimen Orientation	Longitudinal	

CYCLIC PROPERTIES:

		Corrosion Coeff c
Yield Strength 0.2% S_y	365 MPa (53 ksi)	
Strength Coeff K'	896 MPa (130 ksi)	0.998
Strain Hard Exp n'	0.143	
Fatigue Strength Coeff σ'_f	827 MPa (120 ksi)	−0.998
Fatigue Strength Exp b	−0.09	
Fatigue Ductility Coeff ϵ'_f	0.68	−0.997
Fatigue Ductility Exp c	−0.65	

Other Specs _____

Composition
C - 0.12
Mn - 0.65
P - 0.006
S - 0.023
Si - 0.046

MICROSTRUCTURE:

Grain Size   ASTM 10.5

---

pearlitic complement have approximately the same fatigue resistance. The best strain fatigue resistance is obtained from fine graphite types and the softer pearlite matrices [117].

Still better fatigue behavior can be obtained in gray iron when the graphite morphology is of the compacted type [119]. Molinaro, et. al. [119] investigated two variations of compacted graphite with pearlitic matrix for their fatigue behavior (Fig. 3-45). The iron designated CG(A) contained significantly less spheroidal shaped graphite than that contained in iron CG(B). It is apparent in Fig. 3-45 that compacted graphite iron exhibits a superior fatigue behavior as compared to a pearlitic gray iron. Between the two grades of the compacted iron, the one with higher spheroidal shaped graphite shows the best resistance.

Radon, et. al. [120] compared the fatigue behavior of nodular and gray cast irons with cast steel. The material examined are catalogued in Table 3-9 [120] and the results are presented in Fig. 3-46. The results suggest that the cast steels

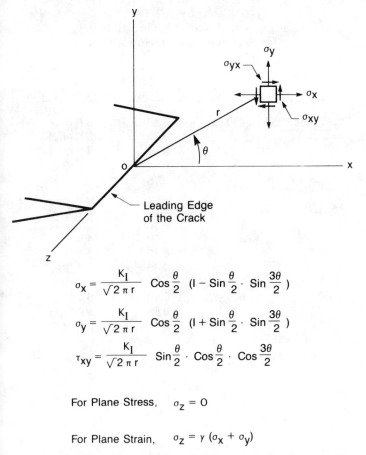

Fig. 3-16 Stress components in the crack tip stress field for Mode I loading.

$$\sigma_x = \frac{K_I}{\sqrt{2\pi r}} \cos\frac{\theta}{2}\left(1 - \sin\frac{\theta}{2}\cdot\sin\frac{3\theta}{2}\right)$$

$$\sigma_y = \frac{K_I}{\sqrt{2\pi r}} \cos\frac{\theta}{2}\left(1 + \sin\frac{\theta}{2}\cdot\sin\frac{3\theta}{2}\right)$$

$$\tau_{xy} = \frac{K_I}{\sqrt{2\pi r}} \sin\frac{\theta}{2}\cdot\cos\frac{\theta}{2}\cdot\cos\frac{3\theta}{2}$$

For Plane Stress, $\sigma_z = 0$

For Plane Strain, $\sigma_z = \gamma(\sigma_x + \sigma_y)$

C and S have the best strain cycling properties, followed by the nodular cast irons A and F, and that flake graphite cast irons E and B have the lowest cyclic strain resistance [120]. The wrought steel results included in Fig. 3-46, of course, showed the best fatigue resistance as compared to all the other materials examined.

3.3.1.5 Wrought Aluminum

Cyclic stress-strain plots for a series of aluminum alloys are shown in Fig. 3-47 [114]. Using the 1100 alloys as the base line, it can be seen that sizable increases in cyclic strength, nearly an order of magnitude, are attainable through alloying. All of the strengthened alloys exhibit some degree of hardening and comparable strain hardening rates. Cyclic hardening is particularly pronounced in the 2024 and 5456 alloys, each showing a 50 percent increase in yield strength. The unheat treated 5456 alloy, hardens to a level above the aged 6061 alloy. Cyclic responses in these systems are greatly affected by precipitate type and distribution [121-123]. Naturally aged alloys (T4 condition) containing GP zones exhibit a pronounced hardening response. Artificially aged alloys (T6 condition), that contain additional transition phases, show a stable or modest hardening behavior. The latter alloys also exhibit a large Bauschinger strain,

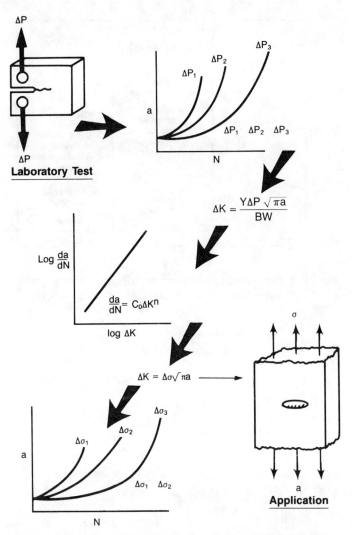

Fig. 3-17 Schematic illustration of the use of fracture-mechanics technology in the design against fatigue failure.

indicative of high internal stress [124]. Treatments involving cold work, while increasing monotonic yield strengths, have little effect on cyclic properties [114,125,126].

Fig. 3-48 presents the cyclic stress-strain as well as the strain-life curves for several aluminum alloys [39].

3.3.2 Crack Growth Properties

3.3.2.1 Wrought Carbon and Alloy Steels

Fatigue crack growth behavior of a high nitrogen mild steel is presented in Fig. 3-49 [127] for a range of stress ratios.

The fine grained mild steel showed no significant influence of stress ratio on the propagation rate and all data fell within a scatter band of slope 3.6. Cracking occurred in this material by the striation mechanism. In contrast, the coarse grained material exhibited cleavage fracture mode depending on the

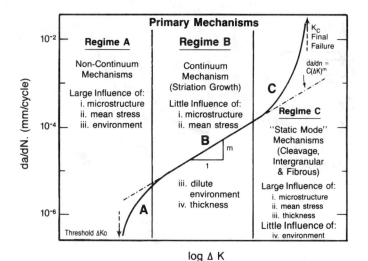

Fig. 3-18 Schematic variation of fatigue-crack growth rates (da/dN) with alternating stress intensity (ΔK).

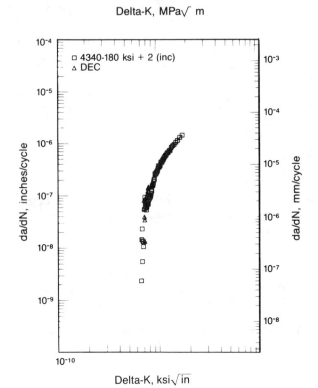

Fig. 3-19 Near-threshold fatigue crack growth behavior of 180 ksi yield strength 4340 steel at R = 0.1.

Table 3-7
Fatigue Crack Growth Threshold ΔK_{th} for Selected Engineering Alloys

Material	S_u MPa (ksi)	$R = \dfrac{K_{min}}{K_{max}}$	K_{th} MPa$\sqrt{m}$ (ksi$\sqrt{in}$)
Mild Steel[a]	430 (62)	0.13	6.6 (6.0)
		0.35	5.2 (4.7)
		0.49	4.3 (3.9)
		0.64	3.2 (2.9)
		0.75	3.8 (3.5)
A533B[b]	—	0.10	8.0 (7.3)
		0.30	5.7 (5.2)
		0.50	4.8 (4.4)
		0.70	3.1 (2.8)
		0.80	3.1 (2.8)
A508[b]	606 (88)	0.10	6.7 (6.1)
		0.50	5.6 (5.1)
		0.70	3.1 (2.8)
18/8 Stainless[a]	665 (97)	0	6.0 (5.5)
		0.33	5.9 (5.4)
		0.62	4.6 (4.2)
		0.74	4.1 (3.7)
2219-T8[c]	—	0.10	2.7 (2.5)
		0.50	1.4 (1.3)
		0.80	1.3 (1.2)

[a] N. E. Frost, K. J. Marsh and L. P. Pool, *Metal Fatigue*, Oxford Univ. Press, London, 1974.

[b] P. C. Paris, R. J. Bucci, E. T. Wessel, W. G. Clark and T. R. Mager, in *Stress Analysis and Growth of Cracks*, Part I, ASTM STP 513, 1972.

[c] S. J. Hudak, A. Saxena, R. J. Bucci and R. C. Malcom, "Development of Standard Methods of Testing and Analysing Fatigue Crack Growth Data," Westinghouse Research Labs, March 1977.

stress intensity factor, and the growth rate in the material was dependent on the stress ratio [127]. The slope of the fatigue crack growth rate curve increased with increasing stress ratio for the coarse grained material. A similar behavior was noted for En 30A steel that was heat treated to obtain

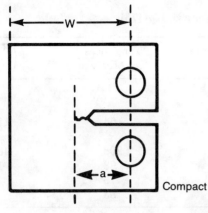

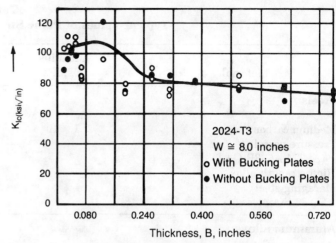

Fig. 3-21 Effect of thickness on plane stress fracture toughness of Al-Cu-Mg alloy.

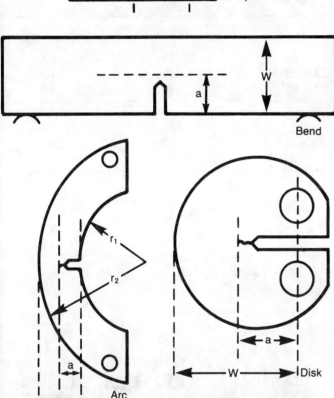

Fig. 3-20 Plate-type specimen shapes used for K_{IC} measurement.

a temper embrittled and unembrittled condition [127,128]. The results presented in Fig. 3-50 indicate that the unembrittled steel showed no dependence of crack growth rates on stress ratio while the embrittled steel exhibited a strong dependence on the stress ratio. The unembrittled steel cracked by the striation mechanism while the embrittled steel showed intergranular cracking mixed with striation growth [127,128].

Richards and Lindley [129] examined a broad range of ferrous materials and concluded that crack propagation rates associated with striation formation were insensitive to stress ratio (except at very low stress intensities). Departure from striation formation to include microcleavage, void coalescence or intergranular cracking were found to result in accelerated crack growth rates. It is generally accepted that the crack growth rates in the Paris law region (region II) is independent of stress ratio for medium strength steels. Exceptions arise when microstructural condition of the material leads to low toughness.

In contrast to region II, crack growth rates in region I are strongly dependent on the stress ratio. Results presented in Fig. 3-51 for a hot rolled carbon steel shows that the fatigue crack propagation threshold decreases with increasing stress ratio [130].

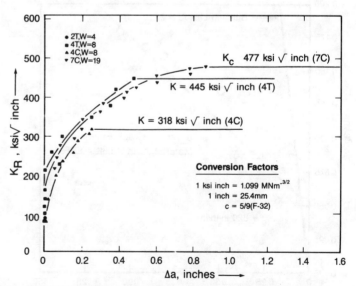

Fig. 3-22 R-curve and K_c results for full-thickness (B = 1.5 in.) specimens of A572 Grade 50 steel tested at + 72 F.

Table 3-8
Typical Values of Plane Strain Fracture Toughness, K_{IC} [75]

Material	Modulus E, MPa	Yield Stress σ_y, MPa	Toughness K_{Ic}, MPa$\sqrt{m}$	$2.5\left(\dfrac{K_{Ic}}{\sigma_y}\right)^2$ mm
Steels				
Medium carbon	2.1×10^5	2.6×10^2	54	110
Pressure vessel				
(ASTM A533B Q+T)		4.7×10^2	208	487
High strength alloy		14.6×10^2	98	11
Maraging steel		18×10^2	76	4.4
AFC 77 Stainless		15.3×10^2	83	7.9
Aluminum alloys				
2024 TB	72×10^4	4.2×10^2	27	10.4
7075 T6		5.4×10^2	30	7.9
7178 T6		5.6×10^2	23	4.2
Titanium alloys				
Ti-6Al-4V	1.08×10^5	10.6×10^2	73	12.6
(High Yield)		11×10^2	38	3.1
For comparison				
Concrete	4×10^4	80	0.2 – 1.4	4.7
WC-Co composites	1×10^5	3×10^2	13	4.7
PMMA	3×10^3	30	1	2.8

Note: Representative values only—not to be employed as design data.

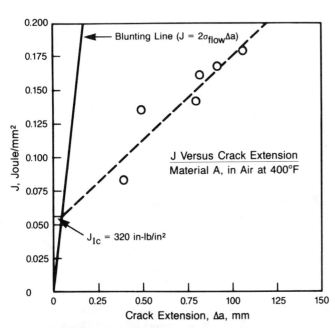

Fig. 3-23 Single specimen J_{IC} data for a low alloy steel at 205 C (400 F).

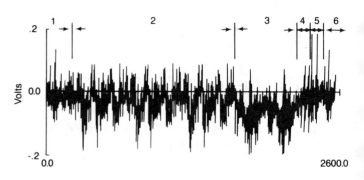

1...Empty Running Across Quarry
2...3 x Load and Dump Sequences
3...2 x Load Carrying Across Quarry Sequences
4...2 x Carrying Load Up Grade Sequences
5...2 x Carrying Load Down Grade with Severe Brake
6...Empty Running Across Quarry

Fig. 3-24 Sample of digitized load history.

3.3.2.2 *High Strength, Low Alloy Steels*

The effect of stress ratio on region II fatigue crack growth rates for HSLA steels is similar to that of the medium

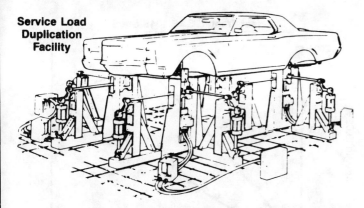

Fig. 3-25 Basic principles of service duplication test with magnetic tape control (10 servo-hydraulic cylinders applying loads).

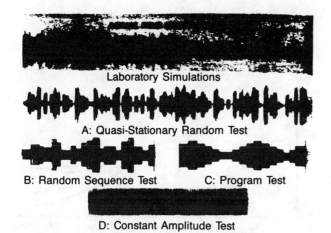

Fig. 3-26 Types of laboratory simulation of service loading.

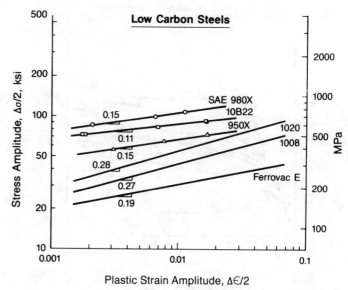

Fig. 3-27 Cyclic stress-strain responses in low carbon steels.

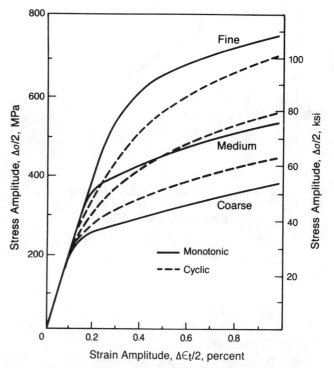

Fig. 3-28 Monotonic and cyclic stress-strain curves for fine, medium, and coarse interlamellar spacing pearlite specimens.

strength steels. This is best illustrated by the overlapping bands in crack growth behavior presented in Fig. 3-52 [114,131,132]. Actual data for several martensitic steels are presented in Fig. 3-53 [133].

The fatigue crack growth behavior of ultra high strength steels are sensitive to the heat treatment and thus the microstructure of the material. Fig. 3-54 compares the crack growth behavior of an isothermally transformed martensitic 300M steel containing 12 percent austenite with quenched and tempered structure containing no austenite [65]. It is apparent that the isothermally transformed martensitic structure offers poor near threshold fatigue crack growth resistance as compared to the quenched and tempered structure at low stress ratio. The influence of the structure is diminished at higher crack growth rates. At high stress ratio, the structure has no influence over the entire range of crack growth rates. As such the stress ratio has a rather large effect on the fatigue crack growth behavior of the materials.

Prior austenite grain size influences the near threshold fatigue crack growth behavior, but has no influence on the threshold for fatigue crack growth, as shown in Fig. 3-55 [65].

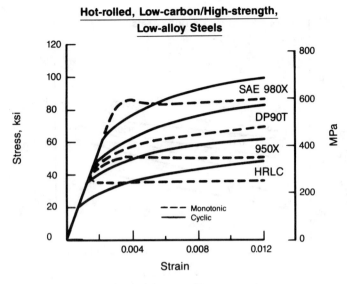

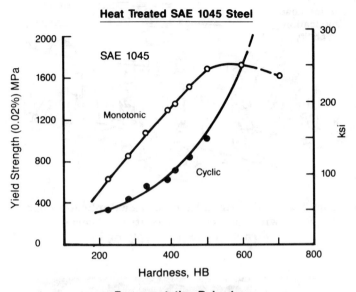

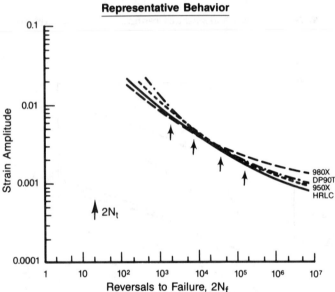

Fig. 3-29 Representative fatigue behavior of carbon and low alloy steels. (A-B)

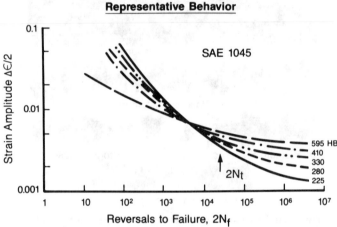

Fig. 3-29 Representative fatigue behavior of carbon and low alloy steels. (C-D)

3.3.2.3 *Cast Steels*

Fig. 3-56 presents the crack growth behavior of a low carbon cast steel in the normalized and tempered condition and equiaxed structure [134]. The band shown in the figure represents the crack growth behavior of wrought ferrite-pearlite or fully pearlitic structural steels [134,135,136]. The crack growth rate data for the cast steel fall within or slightly below the band for wrought steels.

Fig. 3-57 presents the results for the same material tested at a higher stress ratio and for two different grain structures. The crack growth rates are slightly higher as compared to that at a stress ratio of 0.1, but still fall within the wrought steel band. There is no difference in crack growth behavior for the equiaxed and columnar grain structure [134].

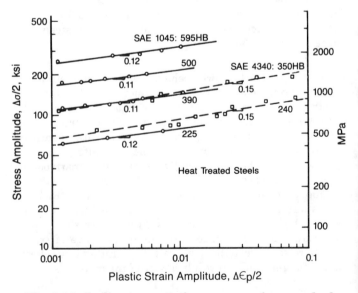

Fig. 3-30 Cyclic stress-strain responses in quenched and tempered steels.

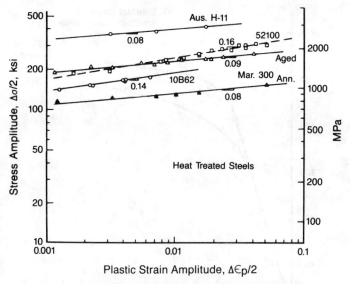

Fig. 3-31 Cyclic stress-strain responses in high strength steels.

Spectrum of Carbon and Alloy Steels

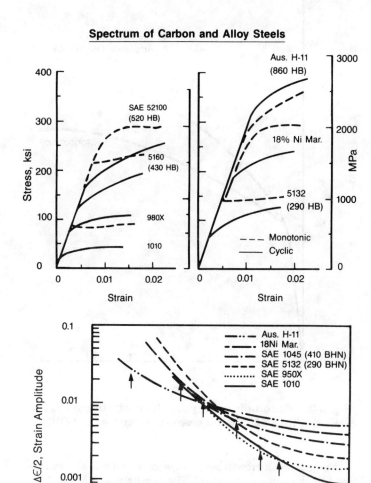

Fig. 3-32 Fatigue behavior of several carbon and alloy steels.

The crack growth behavior of the normalized and tempered medium-carbon low alloy cast steel was very similar to that of the low carbon cast steel [134]. The tempered bainite structure of the medium-carbon steel apparently had no influence on crack growth rates as compared to the ferrite-pearlite microstructure of the low carbon steel. The similarity in crack growth rates existed for significantly different tempering temperatures of 900 and 1200 F (480° and 650°C), as well as different stress ratios (0.1, 0.5, and 0.7), and grain structures (equiaxed and columnar) [134].

The crack growth behavior of the high carbon cast steel in the spheroidized annealed condition at a stress ratio of 0.1 was also similar to the low and medium carbon steels. As shown in Fig. 3-58, the data fell within the scatter band for wrought ferrite-pearlite steels [134]. The crack growth rates, however, are slightly higher at a stress ratio of 0.5 as compared to the other two cast steels. The crack growth behavior changes significantly when the high carbon cast steel is heat treated to a pearlitic-annealed condition. The results presented in Fig. 3-59 show a change in the slope of the crack growth curve (a slope of 5.0 for the cast steel compared to 3.3 for the wrought steel) [134].

Stephens, et. al. [137] investigated the fatigue crack growth behavior of cast SAE 0030, hot rolled SAE 1020, hot rolled ASTM A572-76, quenched and tempered SAE 4140 at three different tempering temperatures, and two case hardened SAE 8620H steels. The results presented in Figs. 3-60 and 3-61 for stress ratios of 0 and -1 respectively suggest that the case hardened steels exhibit the best fatigue crack growth resistance [137]. The cast 0030, wrought 1020, and A572 steels show a distinctly higher slope and are the worst performers at high stress intensities. At lower stress intensity, the high strength 4140 steel (200 ksi) exhibits the worst crack growth resistance. The comparative results of the eight steels at the two stress ratios are essentially the same, but the crack growth rates are higher at R = −1 [137].

3.3.2.4 Cast Irons

Fatigue crack growth rate data at R = 0.1 for normalized and tempered ductile iron (pearlitic) are shown in Fig. 3-62 [134]. The slope of the scatter band is 5.3 as compared to the slope of 3.3 for the wrought steels. Increasing the stress ratio to 0.5 significantly increased the crack growth rates and also its dependance on ΔK [134]. As shown in Fig. 3-63, the slope of the crack growth rate curve increased to 8.3 compared with 5.3 at R = 0.1. Changing the microstructure to ferritic annealed condition had no influence on crack growth rates at R = 0.1. The results presented in Fig. 3-64 show that the data fall along the lower bound of crack growth behavior of the pearlitic material [134].

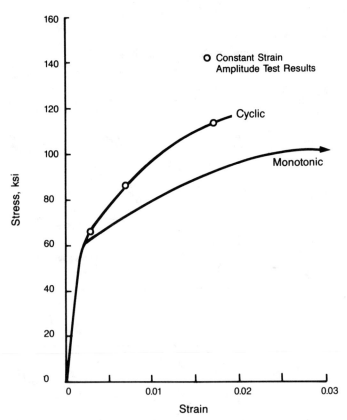

Fig. 3-33 Monotonic and cyclic stress-strain curves for SAE 9262 steel, pearlitic matrix at 272 DPH.

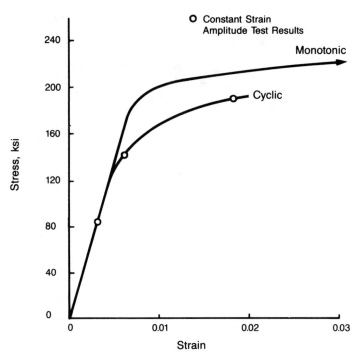

Fig. 3-34 Monotonic and cyclic stress-strain curves for SAE 9262 steel, martensitic matrix at 434 DPH.

Fatigue crack growth behavior of gray iron was also examined in the same study [134]. The results presented in Fig. 3-65 suggest that the crack growth rates in gray cast iron are higher than in ductile iron or cast steel.

3.3.2.5 *Wrought Aluminum Alloys*

Fig. 3-66 presents the fatigue crack growth behavior of a 2XXX series Al-Cu-Mg alloy (BS2L93), and a 7075, Al-Zn-Mg alloy in peak hardness (T651 temper) and over-aged (T7351 temper) condition [138]. It is apparent that stress ratio has a strong influence on crack growth rates for all three alloys, especially in the threshold region. Fatigue crack growth rates increase and the threshold decreases with increasing stress ratio for all three alloys. For the 7075 alloy, peak hardness condition seems to provide better crack growth resistance as compared to the over aged condition, especially at low crack growth rates. In the threshold region at all stress ratios, the 7075-T651 alloy gave the highest value of threshold, the 2L93 alloy an intermediate value, and the 7075-T7351 the lowest value. But at higher crack growth rates, the 2L93 alloy exhibited the best fatigue crack growth resistance [138].

For a 7150 alloy, Zaiken and Ritchie [139] also noted that the under aged alloy exhibited the highest fatigue crack growth threshold and the overaged alloy the lowest threshold with the peak aged alloy falling in between. Their results presented in Fig. 3-67 show that the crack growth rates above about 1×10^{-6} mm/cycle were very similar for the three aging conditions and that stress ratio had a strong influence on crack growth rates. Based on micro machining in the wake of the crack tip, the authors concluded that the thresholds in the three aging conditions were related to crack closure effects. The highest level of crack closure was measured in the underaged microstructure that resulted in the highest threshold value [139].

A detailed study of microstructural influences on fatigue crack growth behavior of 2XXX and 7XXX series of high strength aluminum alloys was undertaken by Truckner et. al. [140] and Bretz, et. al. [141] respectively. Their results summarized in Tables 3-10 and 3-11 suggest that variables such as precipitate type, dislocation density, and the amount of dispersoids are important in determining the fatigue crack growth behavior of aluminum alloys.

Finally, Fig. 3-68 presents a summary plot of fatigue crack growth behavior of several series of aluminum alloys [142]. The crack growth bands suggest that growth rates in 7075-T6 alloy are generally higher than in the 2024-T3 alloy. Data for alloys such as 5456-H321 and 6061T651 fall in between.

3.4 Additional Considerations in Material Fatigue Behavior

3.4.1 *Effects on Crack Initiation*

3.4.1.1 *Notches*

In dealing with structural components containing geometrical discontinuities, it becomes necessary to relate the nominal stresses and strains to maximum stresses and strains at critical locations in the component [37,143-145].

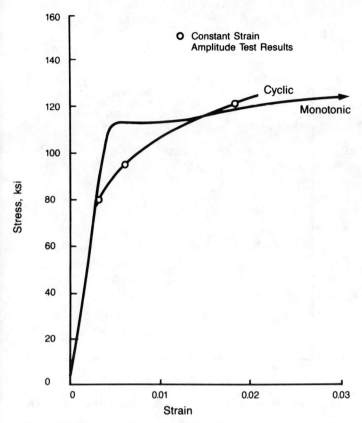

Fig. 3-35 Monotonic and cyclic stress-strain curves for SAE 9262 steel, martensitic matrix at 286 DPH.

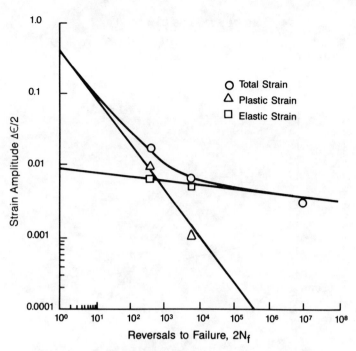

Fig. 3-37 Strain amplitude vs. reversals to failure curve for SAE 9262 steel, martensitic matrix at 434 DPH.

Neuber [146] first introduced the concept of relating the theoretical stress concentration factor to the geometric mean of stress concentration factor to the geometric mean of the stress

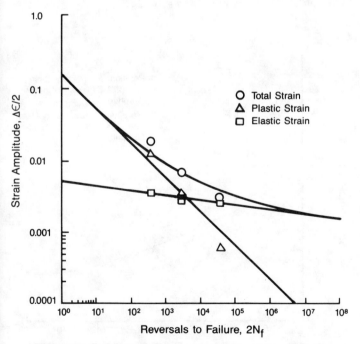

Fig. 3-36 Strain amplitude vs. reversals to failure curve for SAE 9262 steel, pearlitic matrix at 272 DPH.

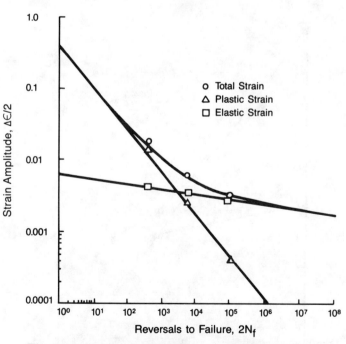

Fig. 3-38 Strain amplitude vs. reversals to failure curve for SAE 9262 steel, martensitic matrix at 286 DPH.

ASTM A27-65-35 Cast Steel

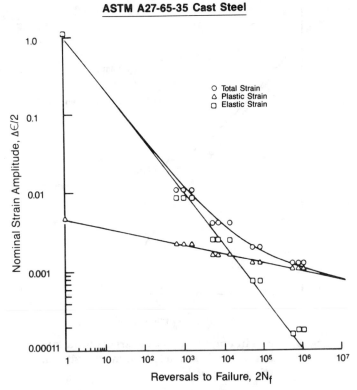

Fig. 3-39 Strain-life curve for cast steel 135 ± 5 HB.

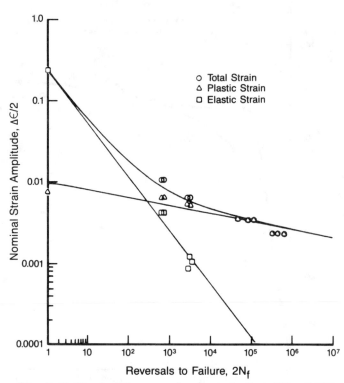

Fig. 3-41 Strain-life curve for cast steel at 410 ± 7 HB.

and strain concentration factors. Topper [147] extended this concept to the conditions of cyclic loading by introducing the fatigue notch factor, K_f, in place of the theoretical stress concentration factor. The form of the equation is:

$$K_f^2 \, \Delta S \, \Delta e = \Delta\sigma \, \Delta\epsilon$$

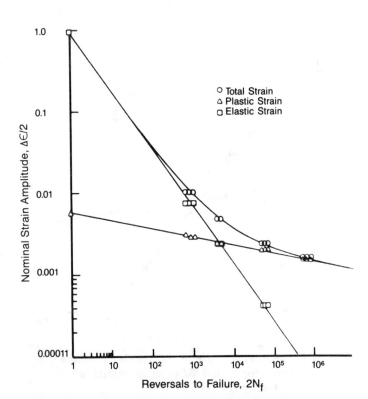

Fig. 3-40 Strain-life curve for cast steel 228 ± 7 HB.

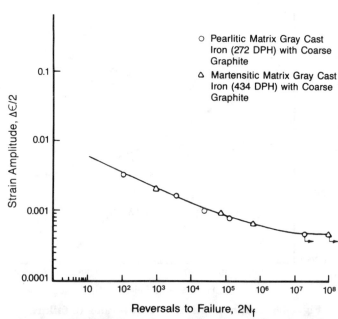

Fig. 3-42 Strain-life curve for pearlitic (272 DPH) and martensitic (434 DPH) gray cast iron with coarse graphite.

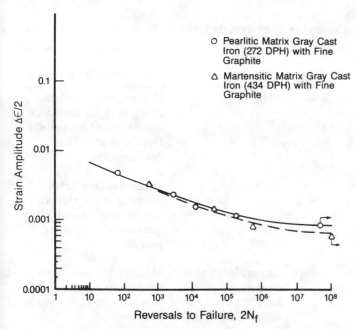

Fig. 3-43 Strain-life curves for pearlitic (272 DPH) and martensitic (434 DPH) graph cast iron with fine graphite.

where ΔS and Δe are nominal stress and strain range respectively, and $\Delta \sigma$ and $\Delta \epsilon$ are the local stress and strain range at the root of a notch, respectively. The fatigue notch factor is always less than or equal to the theoretical stress concentration factor [37]. The appropriate value of K_f depends on the geometry, material, surface finish, and stress gradient in the

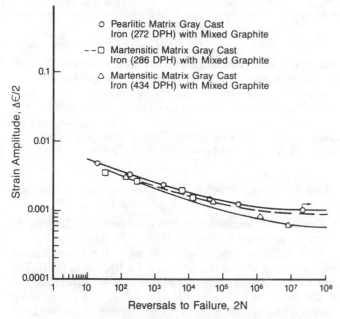

Fig. 3-44 Strain-life curves for pearlitic (272 DPH) and martensitic (286 and 434 DPH) gray cast iron

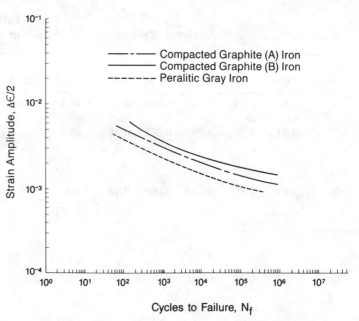

Fig. 3-45 Strain-life curves for three irons.

vicinity of the notch [37]. For notched geometries, one estimate of K_f is available from Peterson [148]:

$$K_f = 1 + (K_t - 1)/(1 + a/r)$$

where r is notch root radius and a is a material constant depending on strength and ductility. For notched members subjected to completely reversed, constant amplitude loading, life predictions can be made from smooth specimen data by determining the parameter on the right hand side of the Neuber's equation as a function of life from smooth specimen stress-life and strain-life data. The resulting master life curve can be entered with the appropriate value of $K_f (\Delta S \Delta e)^{0.5}$ to determine the life at which a detectable crack

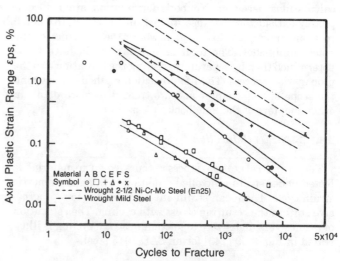

Fig. 3-46 Strain-cycling of plain specimens at room

Table 3-9
Analyses, Treatment and Metallographic Condition of Materials Used [120]

Code letter	C, %	Si, %	P, %	S, %	Mn, %	Ni, %	Cr, %	Mo, %	Al, %	Heat treatment	Metallographical report
A	2.6 to 2.21	1.73 to 1.92	0.026 to 0.040	0.005 to 0.011	0.25 to 0.48	34.4 to 35.4	—	—	—	None	Predominently nodular graphite in an austenitic matrix; small amounts of carbide at the grain boundaries
B	3.62	1.55	0.018	0.066	0.49	—	—	0.90	—	None	Medium/coarse flake-graphite in a matrix containing about equal amounts of ferrite and pearlite
C	0.265	0.38	0.026	0.012	0.64	0.31	1.19	0.29	0.080	Anneal 925°C, furnace cool, normalize 880°C, AC; temper 700°C, AC	Moderately coarse feathery bainite with segregation banding at the grain boundaries
E	3.29	1.10	0.12	0.088	0.69	1.18	—	0.39	—	6 h at 500°C	Medium flake-graphite in a completely pearlitic matrix containing very small amounts of phosphide eutectic
F	3.5	2.64	—	0.019	0.36	1.26	—	—	—	4 h at 950°C, AC; 2h at 625°C, AC	Well formed nodular graphite in a pearlitic matrix with small amounts of ferrite associated with the graphite
S	0.215	0.37	0.024	0.013	0.70	0.16	0.14	0.06	0.052	Anneal 925°C; furnace cool	Fine lamellar pearlite in a ferritic matrix

should be observed in the notched member [143]. Fatigue life calculations based on Neuber's formulation are not completely satisfactory because a crack nucleates at the notch root but propagates in a region where the influence of the notch diminishes. The total fatigue life would have to be determined then by separately treating the initiation and the propagation process. The life prediction methods are beyond the scope of this chapter and are treated in more detail in other chapters of the handbook.

3.4.1.2 Stress Relaxation

The stress relaxation process occurs when a material is tested under strain controlled condition with an asymmetric strain cycle. The mean strain effect occurs due to a progressive shift in the resulting stress with cycling. The relaxation of mean stress under zero to maximum strain cycling is illustrated in Fig. 3-69 [143] for an SAE 1045 steel.

The tensile mean stress present in the first cycle relaxes to zero upon continued cycling. Tensile mean strain in strain controlled tests reduces fatigue life [16]. The effect of mean strain is weak when mean stress relaxes quickly, but the influence may be strong when relaxation is slow. The rate of relaxation depends on the material and strain amplitude; the higher the strain amplitude the higher the relaxation rate and, therefore, the smaller the effect of mean strain [16]. This is shown in Fig. 3-70 for several steels [39].

3.4.1.3 Cyclic Creep

Cyclic creep occurs under stress controlled cycling conditions for asymmetric loading. If a tensile mean stress is superimposed upon an alternating stress, cycle dependent creep may result as shown in Fig. 3-71 for a 1045 steel [143]. The mean stress causes a large tensile mean strain to accumulate, eventually leading to a ductile tensile failure. Whether creep ceases after a certain number of cycles or continues until final failure depends on the material, stress amplitude, mean stress, and temperature [16]. This is one reason why strain controlled testing has become standard.

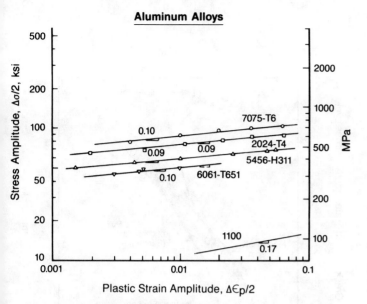

Fig. 3-47 Cyclic stress-strain responses in aluminum-base alloys.

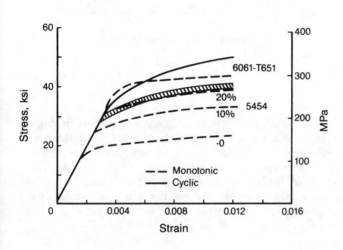

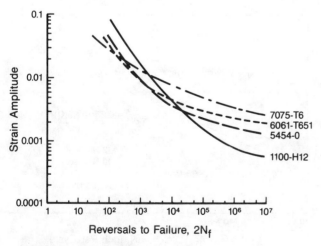

Fig. 3-48 Fatigue behavior of several aluminum alloys.

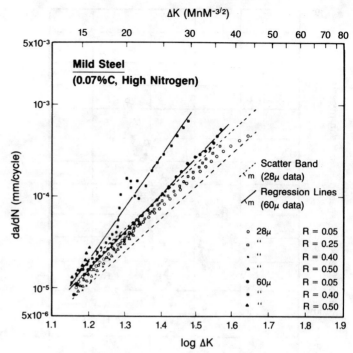

Fig. 3-49 Variation of crack growth rate (da/dN) with alternating stress intensity (ΔK) for high-nitrogen mild steel of 28 and 60 μ grain sizes. Range of R studied 0.05 - 0.60. (Numbers indicate slopes "m" of regression lines

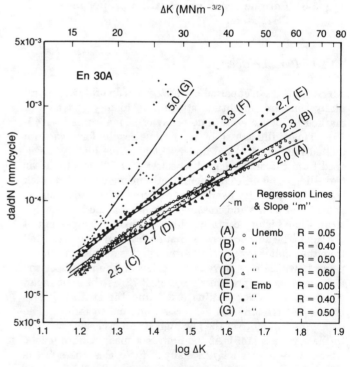

Variation in Crack Growth Rate (da/dN) with Load Ratio for Unembrittled and Embrittled E_m 30A [127]

Fig. 3-50 Variation in crack growth rate (da/dN) with load ratio for unembrittled and embrittled EM30A [127].

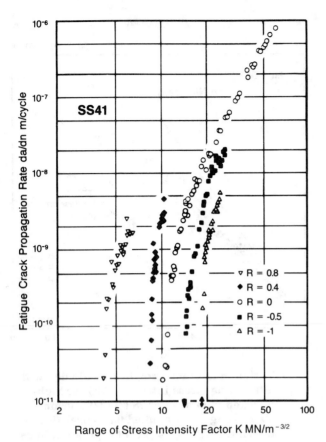

Fig. 3-51 Fatigue crack propagation properties of JIS SS41 steel.

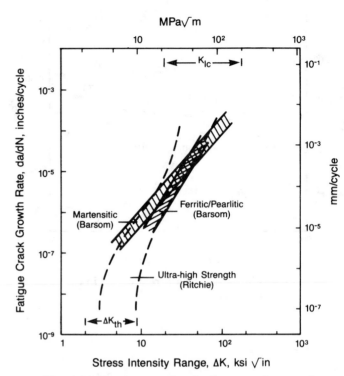

Fig. 3-52 Representative fatigue crack growth behavior for various classes of steel.

3.4.1.4 *Periodic Overstrain*

The application of periodic overstrain in an otherwise nominally constant amplitude strain-time history can have a marked influence on fatigue behavior of materials [37,149-151]. This is illustrated by the fatigue results for a vacuum melted mild steel tested under a strain-time history developed from the strain gaged wheelset of a 100 ton tank car travelling at 60 miles per hour over continuous welded track [149]. The periodic overstrain occurred when the vehicle passed over points and crossings in the track and the selected strain-time history contained high strains for 0.6 percent of the time. The authors specify that no residual stress resulted from the applied overstrains, although it is not clear from the paper how this may have been achieved. The results of the fatigue tests are presented in Fig. 3-72 as the root mean square of the track strains (excluding the strains due to points and crossing) as a function of time to failure. The damaging influence of the few high strain cycles is apparent in these results [149]. At the high root mean square levels, where there is plasticity in many of the cycles, the effect is small. However, as the root mean square levels decrease there is a significant reduction in life for the tests that included the high points-and-crossing strains [149]. Based on the examination of the specimen surface and the fracture surface, the authors concluded that the periodic over strains over low overall strains influence primarily the early damage

and microcrack propagation stages. A lesser effect on stage II crack growth rates was noted. In general the influence of periodic overstrains on the fatigue behavior of the material will depend upon the residual stresses such excursions would impart. A loading history that produces a tensile residual stress would lower the fatigue life while one imparting a compressive residual stress would increase the fatigue life.

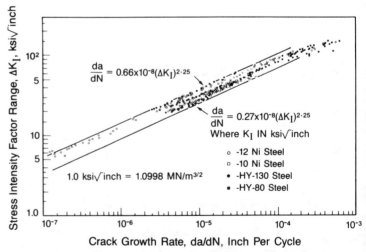

Fig. 3-53 Summary of fatigue crack propagation data for martensitic steels.

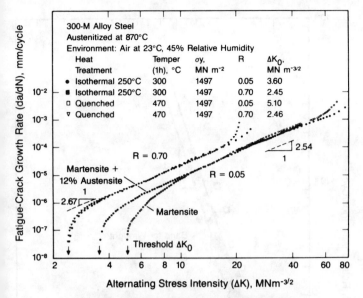

Fig. 3-54 Comparison in martensitic 300-M high-strength steel at constant monotonic strength of isothermal transformed structure, containing 12% retained austenite, with quenched and tempered structure containing no austenite, at R = 0.05 and 0.70 in moist laboratory air; quenched and tempered structure undergoes cyclic softening ($\sigma_y' = 1200$ MN m^{-2}) whereas isothermally transformed structure run cyclically stable ($\sigma_y' = 1500$ MN m^{-2}).

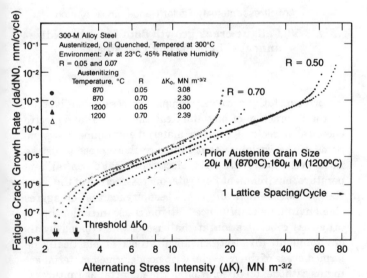

Fig. 3-55 Effect of prior austenite grain size (20-160 m) on fatigue crack growth in cyclic softening 300-M ultrahigh-strength steel ($\sigma_y = 1700$ MN m^{-2}) at R = 0.05 and 0.70 in moist laboratory air.

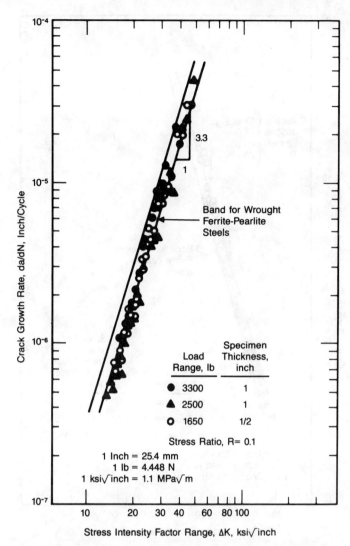

Fig. 3-56 Fatigue crack growth data for normalized and tempered (1200 F) low-carbon cast steel.

3.4.1.5 *Material Variability*

Since design decisions can rarely be based on average material properties, it is important to determine the degree of variability to be expected within a material class. When sufficient data are available, standard statistical procedures can be used to fit the data to the assumed life relation [152-154]. Some considerations for the analysis procedures are construction of confidence limits on the mean curve and on the equation parameters, construction of tolerance limits on the fatigue response of a material, probability-stress-life curve construction, treatment of runouts, characterization of the fatigue strength or fatigue limit of a material, comparison of the fatigue behavior and fatigue strength of two or more materials, and consolidation of fatigue data generated at different conditions (the different conditions may include sources or heats, mean stresses or strains, notch concentrations, and temperature) [152]. An example of the mean fatigue crack growth behavior and the 95 percent confidence

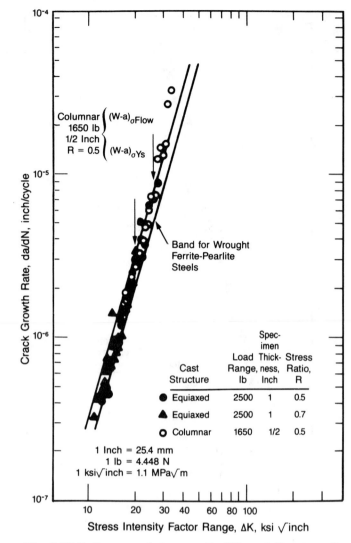

Fig. 3-57 Influence of stress ratio (R) on fatigue crack growth rate in normalized and tempered (1200 F) low-carbon cast steel.

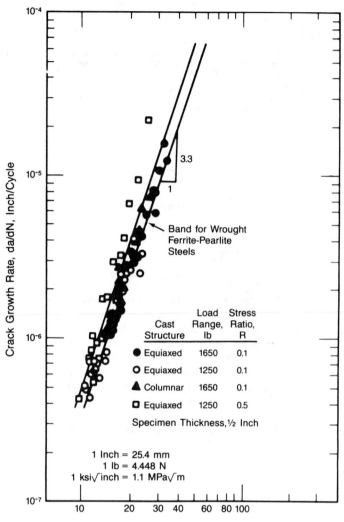

Fig. 3-58 Fatigue crack growth data for spheroidize-annealed high-carbon cast steel.

interval on the data for an SAE 980X steel is shown in Fig. 3-73 [155].

3.4.1.6 Environment

Extensive research has been conducted on the influence of aggressive environments on fatigue behavior of materials [156-161]. Most of the research effort with regard to fatigue crack initiation behavior of materials is based on stress-life results. An aggressive environment, such as salt water, has a very deleterious effect on the fatigue behavior of a structural steel. An example of such an influence is presented in Fig. 3-74 for a 7075-T6 aluminum alloy [162]. Since corrosion is a time dependent process, frequency of loading has a significant effect on crack initiation life as shown in Fig. 3-75 for a rolled steel [163]. The number of cycles to initiate a crack decreases with decreasing frequency. Similar data under strain control conditions are shown in Fig. 3-76 for a high strength steel [164].

Corrosion fatigue is a very complex process in that it occurs by the interaction of metallurgical, mechanical and electrochemical aspects of a given material/environment system. Some of the variables that influence the process are stress ratio, cyclic load wave form, electrochemical potential, temperature, environment flow rate, microstructure of the material, etc. Mechanistically it has been argued that an aggressive environment influences fatigue crack initiation life by (a) stress concentration at the base of pits created by the environment; (b) electrochemical attack at plastically deformed areas of the material with nondeformed material acting as cathode; (c) dissolution of the material at ruptures in an otherwise protective surface film; and (d) lowering of the surface energy of the material due to environmental adsorption, thereby facilitating the crack initiation process. A detailed review of the mechanistic aspects of corrosion fatigue crack initiation process can be found in a paper by Laird and Duquette [165].

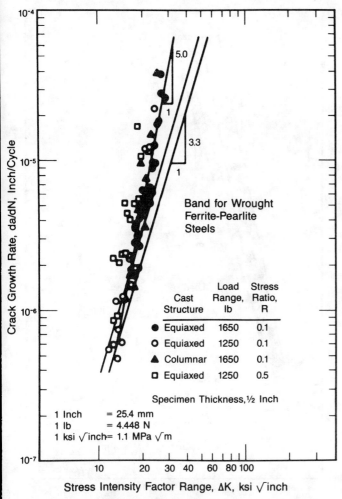

Fig. 3-59 Fatigue crack growth data for pearlitic-annealed high-carbon cast steel.

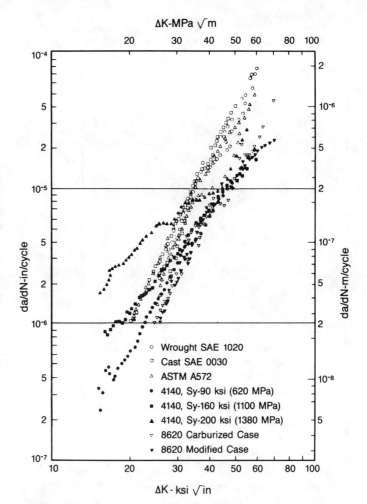

Fig. 3-60 Constant amplitude fatigue crack growth summary with R = 0.

3.4.2 Effects on Crack Propagation

3.4.2.1 Crack Closure/Retardation

Under variable amplitude loading, there is a large interaction effect due to cycles of different amplitudes. The retardation effect due to tensile overloads in an otherwise constant amplitude test is shown in Fig. 3-77 [166]. The crack growth rates remain low for a certain distance and then assume the same rate of growth as before the overload was applied. Von Euw, et. al. [167] demonstrated that the distance over which the retardation effect occurred was related to the plastic zone dimension of the overload. Therefore, once the crack grew through the overload plastic zone, resumption of normal crack propagation is expected. It has been noted that the extent of crack growth rate retardation increases with decreasing yield strength for a given set of overload conditions [168]. Several models have been proposed to explain load interaction phenomena [168]. Schjive, et. al. [166], for example, have argued that crack growth retardation is caused by residual compressive stresses generated by the overload cycle at the crack tip. Rice [169] has attributed the retardation

has suggested crack closure as the cause of crack growth retardation. It is argued that permanent tensile displacements resulting from the overload cycle generate crack surface interference in the wake of the advancing crack front.

Crack closure phenomena has been noted in constant amplitude fatigue crack growth experiments as well and is believed to account for the stress ratio effect on crack propagation rates [168]. The effect becomes more important in the determination of fatigue crack propagation threshold [171]. Crack closure can occur due to several mechanisms even under constant amplitude loading. These mechanisms include plasticity induced closure, oxide induced closure, roughness induced closure, viscous fluid induced closure, and phase transformation induced closure. The primary influence of crack closure is that the effective stress intensity factor range, ΔK, experienced by the crack tip may be far less than the nominal applied ΔK. In such a case, the threshold for fatigue crack growth indicated based on the nominal ΔK analysis will be non-conservative. The strong influence of factors such as strength, grain size, and environment on fatigue crack growth rates at low stress ratios and long crack

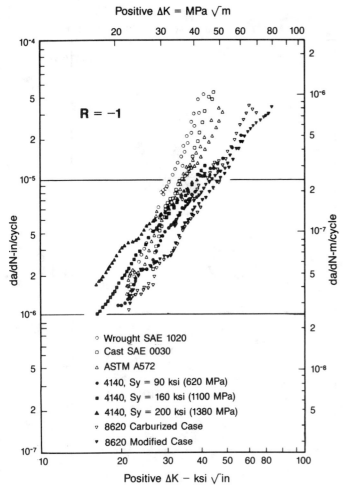

Fig. 3-61 Constant amplitude fatigue crack growth summary with R = −1.

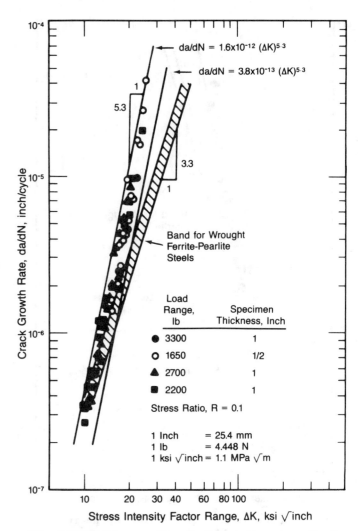

Fig. 3-62 Fatigue crack growth data for normalized and tempered ductile iron (pearlitic).

lengths is also believed to be related to the crack closure effect [171].

3.4.2.2 Environment

Aggressive environments have a strong influence on fatigue crack growth rates as well as crack initiation [156-161,172,173]. An example of the influence of environments on crack growth behavior of a 7079-T651 aluminum alloy is presented in Fig. 3-78 [174].

It is apparent that the crack growth rates are lowest in argon. Water considerably accelerates the crack growth rates and the crack growth rates are enhanced further by the addition of chloride, bromide and iodide at open circuit potential. Some of the factors that influence corrosion fatigue crack initiation behavior also influence the crack growth behavior. The influence of frequency on crack growth rates, for example, is shown in Fig. 3-79 for a 4340 steel [175]. The crack growth rates are accelerated with decreasing frequency. The influence of electrochemical potential on the crack growth rates of an HY-100 steel is shown in Fig. 3-80 [176]. The crack growth rates are increased by an order of magnitude at cathodic potentials as compared to the crack growth rate in air. Generally cathodic potentials are used to reduce the corrosion rates of the material, but hydrogen is generated at such potentials. Materials that are susceptible to hydrogen embrittlement show increased cracking rates under these conditions.

Aggressive environments influence the threshold for fatigue crack growth as well. The threshold decreases in the presence of an environment as shown in Fig. 3-81 for an E-460 steel [177]. As noted in the previous section, Suresh, et. al. [171] have argued that the threshold measurement can be influenced by the crack closure effect and that nominal ΔK should not be employed in defining the threshold. They suggest using an effective ΔK reflecting the mechanical conditions at the crack tip.

In addition to the mechanical conditions, electrochemical conditions can also be different at the crack tip as compared to the bulk conditions [178-180]. Such electrochemical conditions include solution chemistry, pH, and electrochemical potential. It has been observed that the pH at the tip of a crack

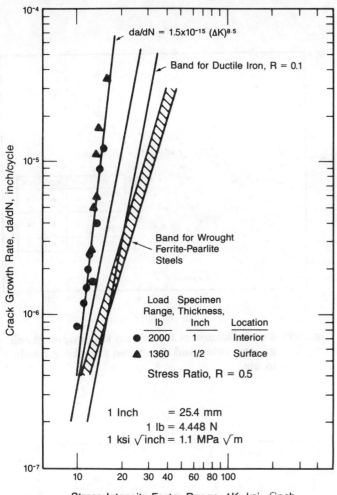

Fig. 3-63 Influence of stress ratio (R) on fatigue crack growth rate in normalized and tempered ductile iron (pearlitic).

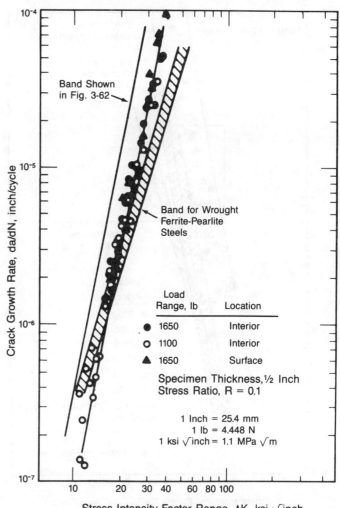

Fig. 3-64 Fatigue crack growth data for ferritic-annealed ductile iron.

is always acidic irrespective of the bulk solution chemistry under freely corroding conditions [181]. Such a variation in chemistry will not only influence the crack growth rates but also the mechanism of cracking and, therefore, has to be taken into consideration. The matter is further complicated for cyclic loading conditions where pumping of the fluid occurs due to alternate opening and closing of the crack tip. Such pumping promotes mixing of the solution the degree of that depends on the geometry of the specimen including the size of the crack, frequency of loading, applied stress intensity range, viscosity of the electrolyte, and bulk solution flow rate.

3.5 References

[1] A. Wohler, "Uber die Festigkeitversuche mit Eisen und Stahl," Zeitschrift fur Bauwesen, Vol. VIII, X, XIII, XVI, and XX, 1860/70. English account of this work is in Engineering, Vol. XI, 1871.

[2] R. E. Peterson, "Fatigue of Metals—Engineering and Design Aspects," *Materials Research & Standards*, Vol. 3, No. 2, 1963, pp. 122-139.

[3] J. Goodman, "Mechanics Applied to Engineering," Longmans, Green and Co., 1899.

[4] C. R. Soderberg, ASME Transactions, Vol. 52, APM-52-2, pp. 13-28, 1930.

[5] C. Lipson and R. C. Juvinall, "Handbook of Stress and Strength," Macmillan, 1963.

[6] Metals Handbook, "Failure Analysis and Prevention," Vol. 10, Eighth Edition, ASM, 1975, pp. 95-125.

[7] D. H. Breen and E. M. Wene, "Fatigue in Machines and Structures-Ground Vehicles, in Fatigue and Microstructure," ASM, 1978, pp. 57-99.

[8] J. Lankford, D. L. Davidson, W. L. Morris, and R. P. Wei, "Fatigue Mechanisms: Advances in Quantitative Measurement of Physical Damage," ASTM STP 811, ASTM, 1983.

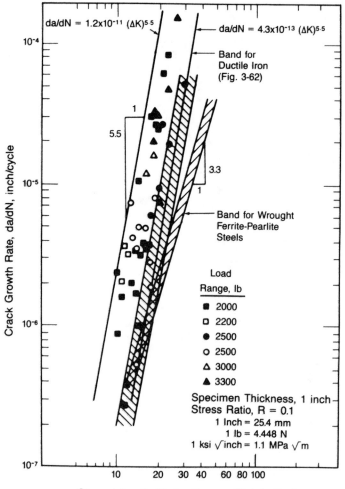

Fig. 3-65 Fatigue crack growth data for gray cast iron investigated.

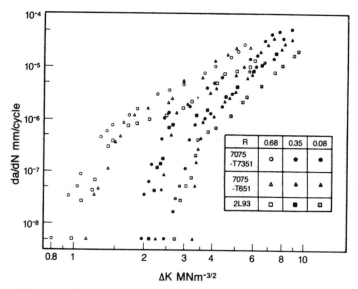

Fig. 3-66 The influence of R ratio on the fatigue crack growth rates and measured threshold levels in air.

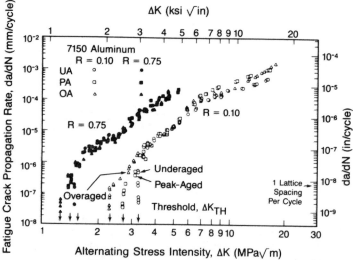

Fig. 3-67 Variation in fatigue crack growth rate (da/dN) with stress intensity range (K) for I/M 7150 aluminum alloy tested at R = 0.10 and 0.75 in controlled moist air. Data are shown for long cracks (a 25mm) in underaged (UA), peak aged (PA), and overaged (OA) microstructure.

[9] "Cyclic Stress-Strain and Plastic Deformation Aspects of Fatigue Crack Growth," ASTM STP 637, ASTM, 1977.

[10] J. T. Fong, "Fatigue Mechanisms," ASTM STP 675, ASTM, 1979.

[11] "Proceedings of the SAE Fatigue Conference," P-109, SAE, 1982.

[12] "Fatigue Resistance Testing and Forecasting," SP-448, SAE, 1979.

[13] P. R. Abelkis and C. M. Hudson, "Design of Fatigue and Fracture Resistant Structures," ASTM STP 761, ASTM, 1982.

[14] P. J. E. Forsyth, "The Physical Basis of Metal Fatigue," Elsevier Science Publishing Co., 1969.

[15] S. Kocanda, "Fatigue Failure of Metals," Sijthoff & Noordhoff International Publishers, Leyden, The Netherlands, 1978.

[16] M. Klesnil and P. Lukas, "Fatigue of Metallic Materials," Elsevier Science Publishing Co., 1980.

[17] W. J. Ostergren and J.R. Whitehead, "Methods for Predicting Material Life in Fatigue," ASME, 1979.

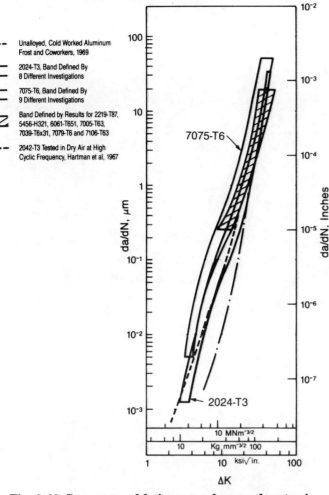

Fig. 3-68 Summary of fatigue crack growth rates in aluminum alloys.

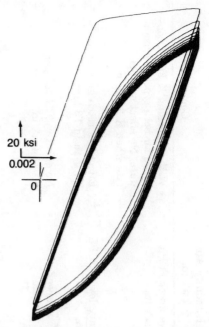

Fig. 3-69 Cycle-dependent stress relaxation of SAE 1045 steel (20 HRC) under 0-max strain cycling.

[18] T. Mura, "Mechanics of Fatigue," AMD-Vol. 47, ASME, 1981.

[19] N. E. Frost, K. J. Marsh, and L. P. Pook, "Metal Fatigue," Oxford University Press, 1974.

[20] P. G. Forrest, "Fatigue of Metals," Pergamon Press, 1962.

[21] J. J. Burke, N. L. Reed, and V. Weiss, "Fatigue—An Interdisciplinary Approach," Proceedings of the 10th Sagamore Army Materials Research Conference, Syracuse University Press, 1964.

[22] S. S. Manson, "Fatigue: A Complex Subject—Some Simple Approximations," *Experimental Mechanics*, Vol. 5, No. 7, 1965, pp. 193-226.

[23] G. E. Dieter, "Mechanical Metallurgy, Second Edition," McGraw Hill Book Co., 1976, pp. 329-374.

[24] M. R. Mitchell, "Fundamentals of Modern Fatigue Analysis for Design," ASM, pp. 385-437.

[25] D. F. Socie, M. R. Mitchell, and E. M. Caulfield, "Fundamentals of Modern Fatigue Analysis," FCP Report No. 26, Fracture Control Program, University of Illinois, 1978.

[26] L. E. Tucker, R. W. Landgraf, and W. R. Brose, "Proposed Technical Report on Fatigue Properties for the SAE Handbook," SAE Technical Paper 740279, SAE, 1974.

[27] "Standard Test Method for Young's Modulus at Room Temperature," ASTM E111-61, Annual Book of ASTM Standards, Part 10, ASTM, 1980.

[28] "Static Determination of Young's Modulus of Metals at Low and Elevated Temperatures," ASTM E231-69, Annual Book of ASTM Standards, Part 10, ASTM, 1980.

[29] "Tension Testing of Metallic Materials," ASTM E8-79a, Annual Book of ASTM Standards, Part 10, ASTM, 1980.

[30] "Tensile Strain Hardening Exponent of Metallic Sheet Materials," ASTM E646-78, Annual Book of ASTM Standards, Part 10, ASTM, 1980.

[31] "Tensile Test Specimens," SAE J416, SAE Handbook, Vol. 1, SAE, 1983.

Table 3-10
Summary of Microstructural Effects on Fatigue Crack Propagation Rate (FCPR)
2024 Type Alloy [140]

Microstructural Variant	Low Humidity Air $K = 5$ ksi$\sqrt{\text{in.}}$	High Humidity Air $K = 2$ ksi$\sqrt{\text{in.}}$	Comments
Type of precipitate GP vs. S	GP zones gave lower FCPR. Disparity between FCPR of both precipitate types larger than in higher humidity.	GP zones gave lower FCPR when $K=3$ ksi$\sqrt{\text{in.}}$	With decreasing K FCPR converge for both types of precipitate. No effect of precipitate when K 3 ksi in.
Dislocation density 1% vs. 5% stretch after quenching.	1% Stretch gave lower FCPR $K=15$ ksi$\sqrt{\text{in.}}$	1% Stretch gave lower FCPR $K=15$ ksi$\sqrt{\text{in.}}$	Effect of stretch greater in T8 than in T3 temper.
Vol. % insoluble constituent a1.4 vs. 2.28	Low vol. % gave lower FCPR in T86 temper $K=15$ ksi$\sqrt{\text{in.}}$	Low vol. % gave lower FCPR in T86 temper $K=15$ ksi$\sqrt{\text{in.}}$	No effect in T31 temper. $K=15$ ksi$\sqrt{\text{in.}}$
Vol. % Mn dispersoid 1.1 vs. 2.6%	No effect.	No effect.	No tests at $K=5$ ksi$\sqrt{\text{in.}}$
Cu content 3.25 vs. 4.25%	Low Cu gave lower FCPR in both T86 and T31 tempers.	Low Cu gave lower FCPR in both T86 and T31 tempers.	Only T31 tested $K=5$ ksi$\sqrt{\text{in.}}$ Additional tests needed to confirm and explain Cu effect.

Table 3-11
Effect of Microstructural Variants on Constant Amplitude FCG Resistance of 7075 and 7050 Type High Strength Aluminum Alloys Tested in the Presence of Moisture and R = 1/3 [141]

Microstructure Variants	Near Threshold K $K=3.3$ MPa$\sqrt{m}$ (3 ksi$\sqrt{in.}$)	Intermediate K 3.3 K 13.2 MPa$\sqrt{m}$	High K $K=13.2$ MPa$\sqrt{m}$ (12 ksi$\sqrt{in.}$)	Comments
Purity (Fe, Si content)	Little effect noted in these experiments.	Small increase in FCG resistance with purity increase, particularly at high K within this regime.	High purity alloys show superior FCG resistance.	Lower purity implies higher volume fraction constituent from which microvoids initiate and coalesce under high tensile strains.
Cu content (1.0 to 2.3%)	Little or no effect apparent in this work. However, other studies covering wider Cu range (0.01-2.1%) indicate that FCG resistance degraded by Cu additions.	FCG resistance increases with Cu increase by reducing degradation by moisture. The benefit of Cu addition diminishes when either moisture effects are removed or K is decreased.	Effect of Cu addition on FCG performance vanishes as K values approach material toughness.	The effect of Cu at low K needs further verification. Nevertheless, this study suggests that any influence of Cu will be less than that of temper (see below).
Dispersoid type (Zr vs. Cr)	Little or no effect apparent in this work. Work by Filler (20) indicates that substitution of Zr improves FCG resistance in dry argon.	Increasing dispersoid volume fraction reduces FCG resistance at higher K values in low toughness alloys, while little or no effect is detectable at lower K values in this regime.	FCG resistance decreases with increase in dispersoid volume fraction, but reduction is on smaller scale than that attributed to purity (Fe, Si) differences.	The effect of dispersoid type on low K FCG resistance is not clear. Any effect expected to be less than that of temper (see below).
Temper	FCG resistance decreases with increase in degree of aging (i.e., with decreasing precipitate coherency).	Overaging increases FCG resistance by reducing degradation by environment.	FCG resistance increases with toughness increase in overaged tempers.	Temper has greatest effect on low K fatigue performance.

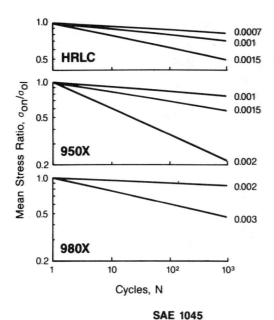

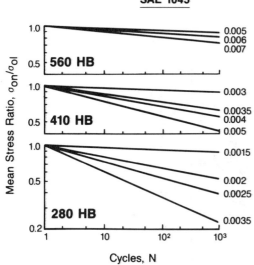

Fig. 3-70 Cycle-dependent stress relaxation of steels.

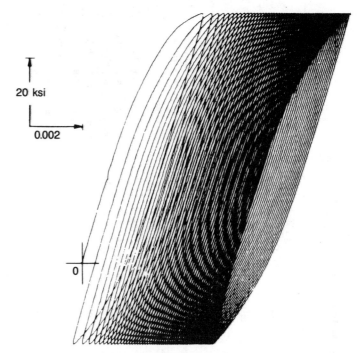

Fig. 3-71 Cycle-dependent creep of SAE 1045 steel (20 HRC) under stress cycling with a tensile mean stress.

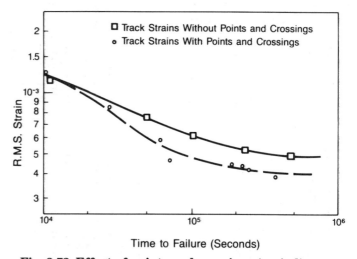

Fig. 3-72 Effect of points and crossings (periodic over strains on fatigue life.

[32] A. Esin, "A Method of Correlating Different Types of Fatigue Curves," *International Journal of Fatigue*, Vol. 2, No. 4, 1980, pp. 153-158.

[33] N. R. LaPointe, "Monotonic and Fatigue Characterizations of Metals," SAE, 1982, pp. 23-37.

[34] R. W. Landgraf, "Fundamentals of Fatigue Analysis," SAE, 1982, pp. 11-18.

[35] D. T. Raske and J. Morrow, "Mechanics of Materials in Low Cycle Fatigue Testing in Manual on Low Cycle Fatigue Testing," ASTM STP 465, ASTM, 1969, pp. 1-25.

[36] J. Morrow, "Fatigue Properties of Metals," Fatigue Design Handbook, SAE, 1968, pp. 21-30.

[37] D. F. Socie and J. Morrow, "Review of Contemporary Approaches to Fatigue Damage Analysis in: Risk and Failure Analysis for Improved Performance and Reliability," J. J. Burke and V. Weiss, Plenum Publishing Corp., 1980, pp. 141-194.

[38] R. W. Landgraf, "Effect of Mean Stress on the Fatigue Behavior of a Hard Steel," TAM Report No. 662, Department of Theoretical and Applied Mechanics, University of Illinois, 1961.

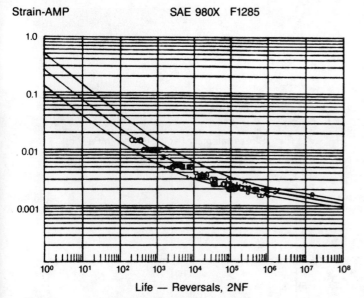

Fig. 3-73 Variability in fatigue behavior of an SAE 980X steel.

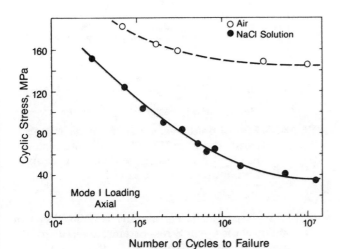

Fig. 3-74 Corrosion-fatigue behavior of 7075-T6 aluminum alloy in air and in aerated sodium chloride solution.

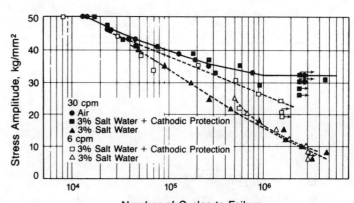

Fig. 3-75 Stress-fatigue-life plot for reversed-bending corrosion fatigue of as-rolled SM50A (0.17C-1.35Mn-0.35Si) steel.

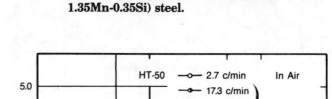

Fig. 3-76 Total strain amplitude, a, vs. number of cycles to failure, N (high strength steel).

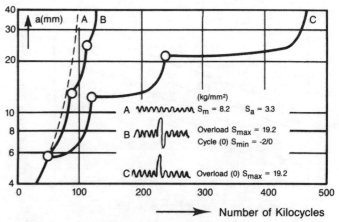

Fig. 3-77 Retardation as a result of overloads in 2024-T3 Al-alloy.

[39] R. W. Landgraf, Research Staff, Ford Motor Co., Unpublished.

[40] "Manual on Low Cycle Fatigue Testing," ASTM STP 465, ASTM, 1969.

[41] R. W. Landgraf, "Cycle Deformation Behavior of Engineering Alloys," Proceedings of Fatigue-Fundamental and Applied Aspects Seminar, Saabgarden, Remforsa, Sweden, August 1977.

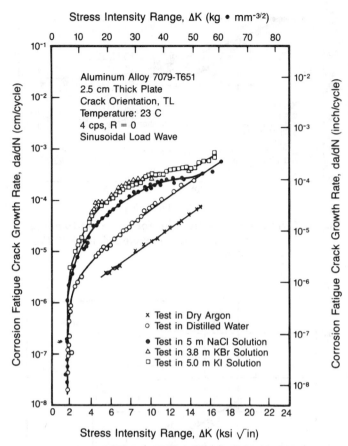

Fig. 3-78 Effect of cyclic stress intensity range on the growth rate of corrosion fatigue cracks in a high strength aluminum alloy exposed to various environments.

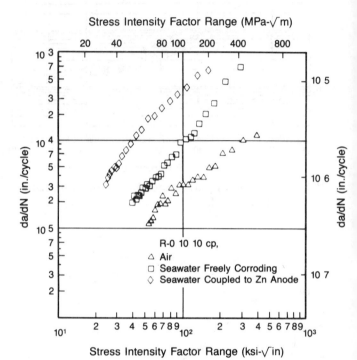

Fig. 3-80 Plot of da/dN vs. stress intensity factor range data for HY-100 steel plate tested in air and in natural sea water uncoupled and coupled to zinc anode.

[42] R. W. Landgraf, J. Morrow, and T. Endo, "Determination of the Cyclic Stress-Strain Curve," *Journal of Materials*, JMLSA, Vol. 4, No. 1, 1969, pp. 176-188.

[43] H. Mughrabi, "Materials Science and Engineering," 33, 1978, pp. 207.

[44] H. Mughrabi, K. Herz, and F. Ackerman, "Proceedings of 4th International Conference on Strength of Metals and Alloys," Nancy, France, Vol. 3, 1976, pp. 1244.

[45] S. P. Bhat and C. Laird, Scripta Met., 12, 1978, pp. 687.

[46] L. F. Coffin, "A Study of Cyclic-Thermal Stresses in a Ductile Metal," ASME Transactions, Vol. 76, ASME, 1954, pp. 931-950.

[47] S. S. Manson, "Behavior of Materials Under Conditions of Thermal Stress," National Advisory Committee on Aeronautics Tech. Note No. 2933, July 1953.

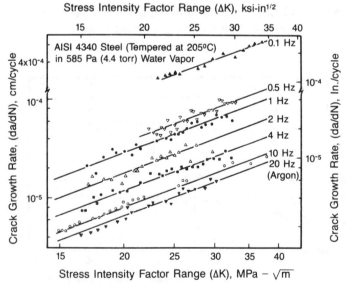

Fig. 3-79 Room temperature fatigue crack growth kinetics on AISI 4340 steel tested in dehumidified argon and in water vapor (below K_{Iscc}) at R = 0.1.

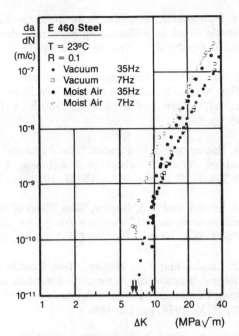

Fig. 3-81 Effect of frequency on near-threshold fatigue crack propagation in air and vacuum for E460 steel.

[48] "Constant-Amplitude Low-Cycle Fatigue Testing," ASTM E606-80, Annual Book of ASTM Standards, Part 10, ASTM, 1980, pp. 694-713.

[49] P. C. Paris, "The Fracture Mechanics Approach to Fatigue," Proceedings of the 10th Sagamore Army Materials Research Conference, Syracuse University Press, 1964, pp. 107-126.

[50] D. A. Broek, "Fundamentals of Fracture Mechanics," Noordhoff International Publishing, Leyden, The Netherlands, 1974.

[51] J. D. Landes, J. A. Begley, and G. A. Clarke, "Elastic Plastic Fracture," ASTM STP 668, ASTM, 1979.

[52] N. E. Dowling and J.A. Begley, "Fatigue Crack Growth During Gross Plasticity and the J-Integral," Mechanics of Crack Growth, ASTM STP 590, ASTM, 1976, ASTM, pp. 82-103.

[53] N. E. Dowling, "Geometry Effects and the J-Integral Approach to Elastic-Plastic Fatigue Crack Growth Rate," Cracks and Fracture, ASTM STP 601, ASTM, 1976.

[54] S. T. Rolfe and J. M. Barsom, "Fracture and Fatigue Control in Structures - Applications of Fracture Mechanics," Prentice Hall, 1977.

[55] W. G. Clark, Jr., "Fracture Mechanics in Fatigue," *Experimental Mechanics*, September 1971, pp. 421-28.

[56] S. J. Hudak, Jr., A Saxena, R. J. Bucci, and R. C. Malcolm, "Development of Standard Methods of Testing and Analyzing Fatigue Crack Growth Rate Data," Final Report by Westinghouse R & D Center to Air Force Materials Laboratory, AFML-TR-78-40, 1978.

[57] S. J. Hudak and R. J. Bucci, "Fatigue Crack Growth Measurement and Data Analysis," ASTM STP 738, ASTM, 1981.

[58] M. E. Fine and R. O. Ritchie, "Fatigue-Crack Initiation and Near Threshold Crack Growth," ASM, 1978, pp. 245-278.

[59] P. C. Paris and F. Erdogan, "A Critical Analysis of Crack Propagation Laws," *Journal of Basic Engineering*, ASME Transactions, Ser. D, 85, 4, ASME, 1963, pp. 528-534.

[60] C. Laird and G. C. Smith, "Crack Propagation in High Stress Fatigue," *Philosophical Magazine*, Vol. 7, 1962, pp. 847-857.

[61] C. Laird, "The Influence of Metallurgical Structure on the Mechanisms of Fatigue Crack Propagation," Fatigue Crack Propagation, ASTM STP 415, ASTM, 1966, pp. 131.

[62] P. J. E. Forsyth and D. A. Ryder, "Some Results of the Examination of Aluminum Alloy Specimen Fracture Surfaces," *Metallurgia*, Vol. 63, 1961, pp. 117-124.

[63] J. Backlund, A. F. Blom, and C. J. Beevers, "Fatigue Thresholds: Fundamentals and Engineering Applications," Vol. I and II, Engineering Materials Advisory Services Ltd., 1982.

[64] D. Davidson and S. Suresh, "Fatigue Crack Growth Threshold Concepts," The Metallurgical Society of AIME, 1984.

[65] R. O. Ritchie, "Near-Threshold Fatigue-Crack Propagation in Steels," Review 245, *International Metals Reviews*, No. 5 & 6, 1979, pp. 205-230.

[66] R. J. Bucci, "Development of a Proposed ASTM Standard Test Method for Near Threshold Fatigue Crack Growth Rate Measurement," ASTM STP 738, ASTM, 1981, pp. 5-28.

[67] R. Rungta, O. Deel, and N. Frey, "Fatigue-Crack Propagation Behavior of Several H-53 Helicopter Materials," Final Report to Sikorsky Aircraft Co. by Battelle Columbus Div., February, 1986.

[68] A. A. Griffith, "The Phenomena of Rupture and Flow in Solids," Phil. Trans. Roy. Soc. London, Vol. A 221, 1921, pp. 163-197.

[69] G. R. Irwin, "Fracture Dynamics," Fracturing of Metals, ASM, 1948, pp. 147-166.

[70] J. E. Srawley and W. F. Brown, "Fracture Toughness Testing Methods," Fracture Toughness Testing and Its Applications, ASTM STP 381, ASTM, 1965, pp. 133-145.

[71] W. F. Brown and J. E. Srawley, "Plane Strain Crack Toughness Testing of High Strength Metallic Materials," ASTM STP 410, ASTM, 1966.

[72] J. E. Srawley, "Plane Strain Fracture Toughness," Fracture, H. Liebowitz, Vol. 4, Academic Press, 1969, pp. 45-68.

[73] W. F. Brown, Jr., "Review of Developments in Plane Strain Fracture Toughness Testing," ASTM STP 463, ASTM, 1970.

[74] "Standard Test Method for Plane-Strain Fracture Toughness of Metallic Materials," ASTM E399, Annual Book of ASTM Standards, Vol. 03.01, ASTM, 1984, pp. 519-554.

[75] L. P. Pook and R. A. Smith, "Theoretical Background to Elastic Fracture Mechanics," Fracture Mechanics—Current Status, Future Prospects, R.A. Smith (Ed.), Pergamon Press, 1979, pp. 29-67.

[76] C. E. Feddersen, F. A. Simonen, L. E. Hulbert, and W. S. Hyler, "An Experimental and Theoretical Investigation of Plane Stress Fracture of 2024-T351 Aluminum Alloy," Battelle Memorial Institute Report, 1970.

[77] D. E. McCabe, "Fracture Toughness Evaluation by R-Curve Methods," ASTM STP 527, ASTM, 1973.

[78] D. E. McCabe, "Determination of R-Curves for Structural Materials by Using Nonlinear Mechanics Methods," Flaw Growth & Fracture, ASTM STP 631, ASTM, 1976, pp. 245-266.

[79] G. R. Irwin, "Fracture Mechanics," Metals Handbook, Vol. 8, Mechanical Testing, Ninth Edition, ASM, 1985, pp. 439-464.

[80] "Standard Practice for R-Curve Determination," ASTM E561, ASTM E399, Annual Book of ASTM Standards, Vol. 03.01, ASTM, 1984, pp. 612-631.

[81] S. R. Novak, "Resistance to Plane-Stress Fracture (R-Curve Behavior) of A572 Structural Steel," ASTM STP 591, ASTM, 1976.

[82] J. R. Rice, "A Path Independent Integral and the Approximate Analysis of Strain Concentrations by Notches and Cracks," *Journal of Applied Mechanics*, ASME, Vol. 35, June 1968, pp. 379-386.

[83] J. W. Hutchinson, "Singular Behavior at the End of a Tensile Crack in a Hardening Material," *Journal of Mech. Phys. Solids*, Vol. 16, 1968, pp. 13-31.

[84] J. R. Rice and G. F. Rosengren, "Plane Strain Deformation Near a Crack Tip in a Power Law Hardening Material," *Journal of Mech. Phys. Solids*, Vol. 16, 1968, pp. 1-12.

[85] J. A. Begley and J. D. Landes, "The J-Integral as a Fracture Criterion," Fracture Toughness, ASTM STP 514, H. T. Corten (Ed.), ASTM, 1972, pp. 1-20.

[86] J. D. Landes and J. A. Begley, "The Effect of Specimen Geometry on JIC," Fracture Toughness, ASTM STP 514, H. T. Corten (Ed.), ASTM, 1972, pp. 24-39.

[87] J. D. Landes and J. A. Begley, "Test Results from J-Integral Studies: An Attempt to Establish a JIC Testing Procedure," Fracture Analysis, ASTM STP 560, ASTM, 1974, pp. 170-186.

[88] J. D. Landes and J. A. Begley, "Recent Developments in JIC Testing," Developments in Fracture Mechanics Test Methods Standardization, W. F. Brown, Jr. and J. G. Kaufman, ASTM STP 632, ASTM, 1977, pp. 57-81.

[89] Standard Test Method for JIC , A Measure of Fracture Toughness, ASTM E813, ASTM E399, Annual Book of ASTM Standards, Vol. 03.01, ASTM, 1984, pp. 763-781.

[90] D. E. McCabe, J. D. Landes, and H. A. Ernst, "An Evaluation of the JR-Curve Method for Fracture Toughness Evaluation," Elastic Plastic Fracture, C. F. Shih and J. P. Gudas, ASTM STP 803, ASTM, Vol. 2, 1983, pp. 562-581.

[91] B. N. Leis, R. Rungta, S. Collard, and P. Skulte, "Fatigue and Fracture Properties of Steam Turbine Shaft Materials," Fracture Mechanics: Fourteenth Symposium—Vol.II: Testing and Applications, J. C. Lewis and G. Sines, ASTM STP 791, ASTM, 1983, pp. II-101-II-119.

[92] K. J. Marsh, "Full Scale Testing of Components and Structures," Fatigue Testing and Design, R. G. Bathgate, Vol. 2, Society of Environmental Engineers, 1976, pp. 22.1-22.11.

[93] J. L. Duncan, "Full Scale Testing on Highly Stressed Light Weight Structures," Fatigue Testing and Design, R. G. Bathgate, Vol. 2, Society of Environmental Engineers, 1976, pp. 23.1-23.9.

[94] W. P. McKinlay, "Dynamic Response of a Heavy Goods Vehicle Chassis," Fatigue Testing and Design, R. G. Bathgate, Vol. 2, Society of Environmental Engineers, 1976, pp. 24.1-24.17.

[95] S. Mochizuki and N. Yasuda, "The Effects of Vibrational Factors on Bending Fatigue Strength of Truck Frames," Transactions of the Society of Automotive Engineers of Japan, Vol. 1, 1970, pp. 127-134.

[96] M. R. Mitchell and R. M. Wetzel, "Cumulative Fatigue Damage Analysis of a Light Truck Frame," SAE Technical Paper 750966, SAE, 1975.

[97] "Third International Conference on Vehicle Structural Mechanics," Vol. P-83, SAE, 1979.

[98] P. Watson and S. J. Hill, "Fatigue Life Assessment of Ground Vehicle Components," ASTM STP 761, ASTM, 1982, pp. 5-27.

[99] H. R. Jaeckel, "Design Validation Testing," SAE Technical Paper 820690, P-109, SAE, 1982, pp. 153-159.

[100] O. Buxbaum, "Random Load Analysis as a Link Between Operational Stress Measurement and Fatigue Life Assessment," Service Fatigue Loads Monitoring, Simulation and Analysis, P. R. Abelkis and J. M. Potter, ASTM STP 671, ASTM, 1979, pp. 5-20.

[101] E. Gassner and W. Lipp, "Long Life Random Fatigue Behavior of Notched Specimens in Service, in Service Duplication Tests, and in Program Tests," Service Fatigue Loads Monitoring, Simulation and Analysis, P. R. Abelkis and J. M. Potter, ASTM STP 671, ASTM, 1979, pp. 222-239.

[102] J. Schjive, "The Analysis of Random Load Time Histories with Relation to Fatigue Tests and Life Calculations," NLR Report MP.201, National Aircraft and Space Laboratory, Amsterdam, 1960.

[103] N. E. Dowling, "Fatigue Failure Predictions for Complicated Stress-Strain Histories," *Journal of Materials*, Vol. 7, No. 1, 1982, pp. 71-87.

[104] G. M. Van Dijk, "Advanced Approaches to Fatigue Evaluation," NASA Special Publication 309, 1972, pp. 565-598.

[105] R. W. Landgraf, "Cyclic Stress-Strain Responses in Commercial Alloys," Work Hardening in Tension and Fatigue, A. W. Thompson, The Metallurgical Society of AIME, 1975, pp. 240-259.

[106] H. Sunwoo, M. E. Fine, M. Meshii, and D. H. Stone, "Cyclic Deformation of Pearlitic Eutectoid Rail Steel," Met. Trans., Vol. 13A, November 1982, pp. 2035-2047.

[107] H. Abdel-Raouf and A. Plumtree, "On the Steady State Cyclic Deformation of Iron", Met. Trans., Vol. 2A, April 1971, pp. 1251-1254.

[108] M. Klesnil, M. Holzmann, P. Lukas, and P. Rys, American Iron & Steel Institute, Vol. 203, 1965, pp. 47.

[109] J. T. McGrath and W. J. Bratina, Czech. *Journal of Phys.*, Vol. B19, 1969, pp. 284.

[110] A. M. Sherman, "Fatigue Properties of High Strength Low Alloy Steels," Met. Trans., Vol. 6A, May 1975, pp. 1035-1040.

[111] Y. H. Kim and M. E. Fine, "Fatigue Crack Initiation and Strain Controlled Fatigue of Some High Strength Low Alloy Steels," Met. Trans., Vol. 13A, January 1982, pp. 59-71.

[112] G. Gonzalez and C. Laird, "The Cyclic Response of Dilute Iron Alloys," Met. Trans., Vol. 14A, December 1983, pp. 2507-2515.

[113] R. W. Landgraf, "Fatigue Considerations in Use of High-Strength Sheet Steel," P-109, SAE, 1982, pp. 273-280.

[114] R. W. Landgraf, "Control of Fatigue Resistance Through Microstructure—Ferrous Alloys," ASM, 1978, pp. 439-466.

[115] P. Bearmore and R. W. Landgraf, Technical Report No. SR-74-100, Ford Motor Co., 1974.

[116] L. F. van Swam, R. M. Pelloux, and N. J. Grant, Met. Trans., Vol. 6A, 1975, pp. 45.

[117] M. R. Mitchell, "Effects of Graphite Morphology, Matrix Hardness, and Structure on the Fatigue Resistance of Gray Cast Iron," SAE Technical Paper 750198, SAE, 1975.

[118] M. R. Mitchell, "A Unified Predictive Technique for the Fatigue Resistance of Cast Ferrous-Based Metals and High Hardness Wrought Steels," SP-448, SAE, 1979, pp. 31-66.

[119] L. Molinaro and D. F. Socie, "Fatigue Behavior and Crack Development in Compacted Graphite Cast Iron," FCP Report No. 39, Fracture Control Program, University of Illinois, 1981.

[120] J. C. Radon, D. J. Burns, and P. P. Benham, "Push-Pull Low Endurance Fatigue of Cast Irons and Steels," *Journal of the American Iron and Steel Institute*, September 1966, pp. 928-935.

[121] C. Laird, "The General Cyclic Stress-Strain Response of Aluminum Alloys," Cyclic Stress-Strain and Plastic Deformation Aspects of Fatigue, L. F. Impellizzeri, ASTM STP 637, ASTM, 1977, pp. 3-35.

[122] C. Calabrese and C. Laird, *Materials Science and Engineering*, Vol. 13, 1974, pp. 141-157.

[123] C. Calabrese and C. Laird, *Materials Science and Engineering*, Vol. 13, 1974, pp. 159-174.

[124] R. E. Stoltz and R. M. Pelloux, Scripta Met., Vol. 8, 1974, pp. 269.

[125] T. H. Sanders, Jr. and J. T. Staley, "Review of Fatigue and Fracture on High Strength Aluminum Alloys," ASM, 1978, pp. 467-522.

[126] A. J. Nachtigall, ASTM STP 579, ASTM, 1975.

[127] R. O. Ritchie and J. F. Knott, "Effects of Fracture Mechanisms on Fatigue Crack Propagation," Mechanics and Mechanism of Crack Growth, Proceedings of a Conference organized by British Steel Corp., April 1973, pp. 201-225.

[128] R. O. Ritchie and J. F. Knott, "Mechanisms of Fatigue Crack Growth in Low Alloy Steel," *Acta Metallurgica*, Vol. 21, May 1973, pp. 639-648.

[129] C. E. Richards and T. C. Lindley, "The Influence of Stress Intensity and Microstructure on Fatigue Crack Propagation in Ferritic Materials," *Engineering Fracture Mechanics*, Vol. 4, 1972, pp. 951-978.

[130] M. Kanao, E. Sasaki, A. Ohta, and M. Kosuge, "Fatigue Crack Propagation Properties and Delta K Threshold for Several Structural Steel Plates," Transactions of National Research Institute for Metals, Japan, Vol. 27, No. 2, 1985, pp. 97-113.

[131] J. M. Barsom, ASME Transactions, *Journal Eng. Ind.*, Vol. 6, 1971, pp. 1190.

[132] R. O. Ritchie, ASME Transactions, Vol. 99, *Journal of Engineering Materials Technology*, 1977, pp. 195.

[133] E. J. Imhof and J. M. Barsom, "Fatigue and Corrosion Fatigue Crack Growth of 4340 Steel at Various Yield Strengths," Progress in Flaw Growth and Fracture Toughness Testing, ASTM STP 536, ASTM, 1973, pp. 182-205.

[134] B. M. Kapadia and E. J. Imhof, Jr., "Fatigue Crack Growth in Cast Irons and Cast Steels," Cast Metals for Structural and Pressure Containment Applications, MPC-11, ASME, 1979, pp. 117-151.

[135] B. M. Kapadia and E. J. Imhof, Jr., "Fatigue Crack Propagation in Electroslag Weldments," Flaw Growth & Fracture, ASTM STP 631, 1976, pp. 159.

[136] J. M. Barsom and E. J. Imhof, Jr., "Fatigue and Fracture Behavior of Carbon-Steel Rails," Rail Steels—Developments, Processes, and Use, ASTM STP 644, ASTM, 1978.

[137] R. I. Stephens et. al., "Constant and Variable Amplitude Fatigue Behavior of Eight Steels," *Journal of Testing and Evaluation*, Vol. 7, No. 2, 1979, pp. 68-81.

[138] B. R. Kirby and C. J. Beevers, "Slow Fatigue Crack Growth and Threshold Behavior in Air and Vacuum of Commercial Aluminum Alloys," Fatigue of Engineering Materials and Structures, Vol. 1, 1979, pp. 203-215.

[139] E. Zaiken and R. O. Ritchie, "On the Development of Crack Closure and the Threshold Condition for Short and Long Fatigue Cracks in 7150 Aluminum Alloy," Met. Trans., Vol. 16A, August 1985, pp. 1467-1477.

[140] W. G. Truckner, J. T. Staley, R. J. Bucci, and A. B. Thakker, "Effects of Microstructure on Fatigue Crack Growth of High Strength Aluminum Alloys," AFML-TR-76-169, Final Report by Aluminum Co. of America to Air Force Materials Laboratory, August 1976.

[141] P. E. Bretz, A. K. Vasudevan, R. J. Bucci, and R. C. Malcolm, "Effect of Microstructure on 7XXX Aluminum Alloy Fatigue Crack Growth Behavior Down to Near-Threshold Rates," Final Report by Aluminum Co. of America to Naval Air Systems Command, Contract No. N00019-79-C-0258, October 1981.

[142] G. T. Hahn and R. Simon, "A Review of Fatigue Crack Growth in High Strength Aluminum Alloys and the Relevant Metallurgical Factors," *Engineering Fracture Mechanics*, Vol. 5, 1973, pp. 523540.

[143] R. W. Landgraf, "The Resistance of Metals to Cyclic Deformation," Achievement of High Fatigue Resistance in Metals and Alloys, ASTM STP 467, ASTM, 1970, pp. 3-36.

[144] N. E. Dowling, "Notched Member Fatigue Life Predictions Combining Crack Initiation and Propagation," Fatigue of Engineering Materials and Structures, Vol. 2, 1979, pp. 129-138.

[145] D. F. Socie, "Fatigue Life Estimates for Bluntly Notched Members," *Journal of Engineering Materials and Technology*, Vol. 102, January 1980, pp. 153-158.

[146] H. Neuber, "Theory of Stress Concentration for Shear Strained Prismatical Bodies With Arbitrary Nonlinear Stress-Strain Law," *Journal of Applied Mechanics*, ASME, December 1961, pp. 544-550.

[147] T. H. Topper, R. M. Wetzel, and J. D. Morrow, "Neuber's Rule Applied to Fatigue of Notched Specimens," *Journal of Materials*, JMSLA, Vol. 4, No. 1, March 1969, pp. 200-209.

[148] R. E. Peterson, "Notch Sensitivity," Metal Fatigue, G. Sines (Ed.), McGraw-Hill Book Co., 1959, pp. 293-306.

[149] P. Watson, D. S. Hodinott, and J. P. Norman, "Periodic Overloads and Random Fatigue Behavior, Cyclic Stress-Strain Behavior Analysis, Experimentation, and Prediction," ASTM STP 519, ASTM, 1973, pp. 271-284.

[150] W. R. Brose, N. E. Dowling, and J. D. Morrow, "Effect of Periodic Large Strain Cycles on the Fatigue Behavior of Steels," SAE Technical Paper 740221, SAE, 1974.

[151] R. C. Rice and D. Broek, "Fatigue Crack Initiation Properties of Rail Steels," DOT/FRA/ORD-82/05, Final Report to Department of Transportation By Battelle Columbus Div., 1982.

[152] R. C. Rice, "Fatigue Data Analysis," Metals Handbook, Vol. 8, Mechanical Testing, Ninth Edition, ASM, 1985, pp. 695-720.

[153] R. E. Little, "Manual on Statistical Planning and Analysis for Fatigue Experiments," ASTM STP 588, ASTM, 1975.

[154] R. E. Little and J. C. Ekvall, "Statistical Analysis of Fatigue Data," ASTM STP 744, ASTM, 1981.

[155] A. Conle and R. W. Landgraf, "A Fatigue Analysis Program for Ground Vehicle Components," Society of Environmental Engineers, 1983, pp. 1-28.

[156] O. Devereux, A. J. McEvily, and R. W. Staehle, "Corrosion Fatigue: Chemistry, Mechanics and Microstructure," NACE-2, National Association of Corrosion Engineers, 1971.

[157] C. E. Jaske, J. H. Payer, and V. S. Balint, "Corrosion Fatigue of Metals in Marine Environments," MCIC-81-42, Metals and Ceramics Information Center, 1981.

[158] T. W Crooker and B. N. Leis, "Corrosion Fatigue: Mechanics, Metallurgy, Electrochemistry, and Engineering," ASTM STP 801, ASTM, 1983.

[159] R. P. Gangloff, "Embrittlement by the Localized Crack Environment," The Metallurgical Society of AIME, 1984.

[160] R. Rungta, "Predictive Capabilities in Environmentally Assisted Cracking," PVP-Vol. 99, ASME, 1985.

[161] H. L. Craig, Jr., T. W. Crooker, and D. W. Hoeppner, "Corrosion Fatigue Technology," ASTM STP 642, ASTM, 1978.

[162] D. J. Duquette, "Environmental Effect I: General Fatigue Resistance and Crack Nucleation in Metals and Alloys," ASM, 1978, pp. 335-363.

[163] K. Nishioka, K. Hirakawa, and I. Kitaura, "Low Frequency Corrosion Fatigue Strength of Steel Plate," The Sumitomo Search, No. 16, 1976, pp. 40-54.

[164] K. Endo and K. Komai, "Effects of Stress Wave Form and Cycle Frequency on Low Cycle Corrosion Fatigue," NACE-2, National Association of Corrosion Engineers, 1971, pp. 437-450.

[165] C. Laird and D. J. Duquette, "Mechanisms of Fatigue Crack Nucleation," NACE-2, National Association of Corrosion Engineers, 1971, pp. 88-117.

[166] J. Schjive and D. Broek, "Crack Propagation Tests Based on a Gust Spectrum with Variable Amplitude Loading," Aircraft Engineering, 34, 1962, pp. 314-316.

[167] E. F. J. Von Euw, R. W. Hertzberg, and R. Roberts, "Delay Effect in Fatigue Crack Propagation," Stress Analysis and Growth of Cracks, ASTM STP 513, ASTM, 1972, pp. 230-259.

[168] R. W. Hertzberg, "Deformation and Fracture Mechanics of Engineering Materials," John Wiley & Sons, 1976.

[169] J. R. Rice, "Mechanics of Crack Tip Deformation and Extension by Fatigue," Fatigue Crack Propagation, ASTM STP 415, ASTM, 1967, pp. 247-311.

[170] W. Elber, "The Significance of Fatigue Crack Closure," Damage Tolerance in Aircraft Structures, ASTM STP 486, ASTM, 1971, pp. 230-242.

[171] S. Suresh and R. O. Ritchie, "Near-Threshold Fatigue Crack Propagation: A Perspective on the Role of Crack Closure," The Metallurgical Society of AIME, 1984, pp. 227-261.

[172] R. P. Wei, "On Understanding Environment Enhanced Fatigue Crack Growth—A Fundamental Approach," ASTM STP 675, ASTM, 1979, pp. 816-840.

[173] Z. A. Foroulis, "Environment Sensitive Fracture of Engineering Materials," The Metallurgical Society of AIME, 1979.

[174] M. O. Speidel, M. J. Blackburn, T. R. Beck, and J. A. Feeney, "Corrosion Fatigue and Stress Corrosion Crack Growth in High Strength Aluminum Alloys, Magnesium Alloys, and Titanium Alloys Exposed to Aqueous Solutions," NACE-2, National Association of Corrosion Engineers, 1971, pp. 324-345.

[175] P. S. Pao, W. Wei, and R. P. Wei, "Effect of Frequency on Fatigue Crack Growth Response of AISI 4340 Steel in Water Vapour," The Metallurgical Society of AIME, 1979, pp. 565-580.

[176] D. A. Davis and E. J. Czyryca, "Corrosion Fatigue Crack Growth Characteristics of Several HY-100 Steel Weldments with Cathodic Protection," ASTM STP 801, ASTM, 1983, pp. 175-196.

[177] A. Bignonnet et. al., "Environmental and Frequency Effects on Near Threshold Fatigue Crack Propagation in a Structural Steel," The Metallurgical Society of AIME, 1984, pp. 99-113.

[178] B. F. Brown, C. T. Fujii, and E. P Dahlberg, "Methods for Studying the Solution Chemistry Within Stress Corrosion Cracks," *Journal of the Electrochemical Society*, 116, 1969, pp. 218-219.

[179] A. J. Markworth and L. R. Kahn, "A Hierarchical Model for Mass Transport Kinetics Within a Crevice Like Region," PVP-Vol. 99, ASME, 1985, pp. 143-152.

[180] A. Turnbull, "Progress in the Understanding of the Electrochemistry in Cracks," The Metallurgical Society of AIME, 1984, pp. 3-31.

[181] W. H. Hartt, J. S. Tennant, and W. C. Hooper, "Solution Chemistry Modification Within Corrosion Fatigue Cracks," ASTM STP 642, ASTM, 1978, pp. 5-18.

Addresses of Publishers

Academic Press, Orlando, FL 32887

AISI, American Iron and Steel Institute, 1000 16th Street NW, Washington, DC 20036

ASM International, 9639 Kinsman Road, Metals Park, OH 44073

ASME, American Society of Mechanical Engineers, 345 E. 47th Street, New York, NY 10017

ASTM, American Society for Testing & Materials, 1916 Race Street, Philadelphia, PA 19103

Battelle Columbus Lab, 505 King Street, Columbus, OH 43201-2693

British Steel Corp., Churchill College, Cambridge, England

Elsevier Science Publishing Co., 52 Vanderbilt Avenue, New York, NY 10017

Ford Motor Company, 20000 Rotunda Drive, Dearborn, MI 48121

MacMillan, 866 Third Avenue, New York, NY 10022

McGraw-Hill Book Co., 1221 Avenue of the Americas, New York, NY 10020

Metallurgical Society of AIME, 420 Commonwealth Drive, Warrendale, PA 15086

Metals and Ceramics Information Center, Columbus, OH

National Association of Corrosion Engineers, P.O. Box 218340, Houston, TX 77218

Oxford University Press, 200 Madison Avenue, New York, NY 10016

Pergamon Press, Maxwell House, Fairview Park, Elmsford, NY 10523

Plenum Publishing Corp., 233 Spring Street, New York, NY 10013

Prentice Hall, Route 9W, Englewood Cliffs, NJ 07632

SAE, Society of Automotive Engineers, 400 Commonwealth Drive, Warrendale, PA 15096

SAE of Japan, 10-2, Goban-cho, Chiyoda-ku, Tokyo 102, Japan

Society of Environmental Engineers (Note: Merged with Institute of Environmental Sciences to form the Institute of Environmental Sciences), 940 E. Northwest Highway, Mt. Prospect, IL 60056

Syracuse University Press, 1600 Jamesview Avenue, Syracuse, NY 13244-5160

University of Illinois, 249 Armory Building, 505 E. Armory Street, Champaign, IL 61820

John Wiley & Sons, 605 Third Avenue, New York, NY 10158

TABLE 3-3

Conversion factors

To convert from	to	multiply by	To convert from	to	multiply by	To convert from	to	multiply by
Angle			Force			Mass per unit length		
degree	rad	1.745 329 E − 02	lbf	N	4.448 222 E + 00	lb/ft	kg/m	1.488 164 E + 00
Area			kip (1000 lbf)	N	4.448 222 E + 03	lb/in.	kg/m	1.785 797 E + 01
in.²	mm²	6.451 600 E + 02	tonf	kN	8.896 443 E + 00	Mass per unit time		
in.²	cm²	6.451 600 E + 00	kgf	N	9.806 650 E + 00	lb/h	kg/s	1.259 979 E − 04
in.²	m²	6.451 600 E − 04	Force per unit length			lb/min	kg/s	7.559 873 E − 03
ft²	m²	9.290 304 E − 02	lbf/ft	N/m	1.459 390 E + 01	lb/s	kg/s	4.535 924 E − 01
Bending moment or torque			lbf/in.	N/m	1.751 268 E + 02	Mass per unit volume (includes density)		
lbf·in.	N·m	1.129 848 E − 01	Fracture toughness			g/cm³	kg/m³	1.000 000 E + 03
lbf·ft	N·m	1.355 818 E + 00	ksi $\sqrt{\text{in.}}$	MPa $\sqrt{\text{m}}$	1.098 800 E + 00	lb/ft³	g/cm³	1.601 846 E − 02
kgf·m	N·m	9.806 650 E + 00	Length			lb/ft³	kg/m³	1.601 846 E + 01
ozf·in.	N·m	7.061 552 E − 03	Å	nm	1.000 000 E − 01	lb/in.³	g/cm³	2.767 990 E + 01
Bending moment or torque per unit length			in.	m	2.540 000 E − 02	lb/in.³	kg/m³	2.767 990 E + 04
lbf·in./in.	N·m/m	4.448 222 E + 00	mil	mm	2.540 000 E + 01	Power		
lbf·ft/in.	N·m/m	5.337 866 E + 01	in.	cm	2.540 000 E + 00	Btu/s	kW	1.055 056 E + 00
Current density			ft	m	3.048 000 E − 01	Btu/min	kW	1.758 426 E − 02
A/in.²	A/mm²	1.550 003 E − 03	yd	m	9.144 000 E − 01	Btu/h	W	2.928 751 E − 01
A/ft²	A/m²	1.076 400 E + 01	mile	km	1.609 300 E + 00	erg/s	W	1.000 000 E − 07
Energy (impact, other)			Mass			ft·lbf/s	W	1.355 818 E + 00
ft·lbf	J	1.355 818 E + 00	oz	kg	2.834 952 E − 02	ft·lbf/min	W	2.259 697 E − 02
Btu	J	1.054 350 E + 03	lb	kg	4.535 924 E − 01	ft·lbf/h	W	3.766 161 E − 04
(thermochemical)			ton (short, 2000 lb)	kg	9.071 847 E + 02	hp (550 ft·lbf/s)	kW	7.456 999 E − 01
cal	J	4.184 000 E + 00	ton (short, 2000 lb)	kg x 10³(a)	9.071 847 E − 01	hp (electric)	kW	7.460 000 E − 01
(thermochemical)			ton (long, 2240 lb)	kg	1.016 047 E + 03	Power density		
kW·h	J	3.600 000 E + 06	Mass per unit area			W/in.²	W/m²	1.550 003 E + 03
W·h	J	3.600 000 E + 03	oz/in.²	kg/m²	4.395 000 E + 01	Pressure (fluid)		
Flow rate			oz/ft²	kg/m²	3.051 517 E − 01	atm (standard)	Pa	1.013 250 E + 05
ft³/h	L/min	4.719 475 E − 01	oz/yd²	kg/m²	3.390 575 E − 02	bar	Pa	1.000 000 E + 05
ft³/min	L/min	2.831 000 E + 01	lb/ft²	kg/m²	4.882 428 E + 00	in.Hg (32F)	Pa	3.386 380 E + 03
gal/h	L/min	6.309 020 E − 02				in.Hg (60F)	Pa	3.376 850 E + 03
gal/min	L/min	3.785 412 E + 00				lbf/in.² (psi)	Pa	6.894 757 E + 03
						torr (mmHg, 0C)	Pa	1.333 220 E + 02
						tonf/in.² (tsi)	MPa	1.378 951 E + 01
Stress (force per unit area)			Thermal expansion			kgf/mm²	MPa	9.806 650 E + 00
ksi	MPa	6.894 757 E + 00	in./in.·°C	m/m·K	1.000 000 E + 00	Volume ksi		
lbf/in.² (psi)	MPa	6.894 757 E − 03	in./in.·°F	m/m·K	1.800 000 E + 00	in.³	m³	1.638 706 E − 05
MN/m²	MPa	1.000 000 E + 00				ft³	m³	2.831 685 E − 02
						fluid oz	m³	2.957 353 E − 05

TABLE 3-3 (Continued)

Conversion factors

To convert from	to	multiply by	To convert from	to	multiply by	To convert from	to	multiply by
Temperature			Velocity			gal (U.S. liquid)	m³	3.785 412 E − 03
°F	°C	5/9·(°F − 32)	ft/h	m/s	8.466 667 E − 05	Volume per unit time		
°R	°K	5/9	ft/min	m/s	5.080 000 E − 03	ft³/min	m³/s	4.719 474 E − 04
Temperature interval			ft/s	m/s	3.048 000 E − 01	ft³/s	m³/s	2.831 685 E − 02
°F	°C	5/9	in./s	m/s	2.540 000 E − 02	in.³/min	m³/s	2.731 177 E − 07
			km/h	m/s	2.777 778 E − 01	Wavelength		
			mph	km/h	1,609 344 E + 00	Å	nm	1.000 000 E − 01
			Velocity of rotation					
			rev/min (rpm)	rad/s	1.047 164 E − 01			
			rev/s	rad/s	6.283 185 E + 00			

(a) kg × 10³ = 1 metric ton

Chapter 4 Effects of Processing on Fatigue Performance

4.1 Introduction
    4.1.1 Scope
    4.1.2 Relationship to Other Chapters
    4.1.3 Chapter Plan

4.2 Considerations in Evaluating the Effects of Fabrication and Processing
    4.2.1 Limitations of Standard Specimen Fatigue Data
    4.2.2 Fabrication and Processing Effects

4.3 Fatigue Performance Data on the Effects of Fabrication and Processing

4.4 Estimating Fabrication and Processing Effects on Fatigue Performance
    4.4.1 Example No. 1
    4.4.2 Example No. 2

4.5 Summary

4.6 References

Appendix 4 Supporting Data

Appendix 4A Mechanical Prestressing
    4A.1 General
    4A.2 Shot Peening
    4A.3 Surface Rolling
    4A.4 Overloading
    4A.5 Prestressing and Preloading of Holes in Sheets
    4A.6 Conclusion
    4A.7 References

Appendix 4B Heat Treatment
    4B.1 General
    4B.2 Furnace Hardening Through Heat and Quench
    4B.3 Induction and Flame Hardening: Surface Heat and Quench
    4B.4 Carburizing
    4B.5 Nitriding
    4B.6 Stress Relief
    4B.7 References

Appendix 4C Weldments
    4C.1 Linear Weldments and Surface (Notch) Effects
    4C.2 Spot Welds
    4C.3 Residual Stresses
    4C.4 Other Methods for Improvement in Weld Fatigue Performance
    4C.5 References

Appendix 4D The Effect of Machining on Residual Stresses and Fatigue Resistance
    4D.1 Introduction
    4D.2 Metal Removal Parameters
    4D.3 Residual Stresses
    4D.4 Observed Correlations
    4D.5 Summary
    4D.6 References

Appendix 4E The Effects of Forming on the Fatigue Strength of Metals
    4E.1 Introduction
    4E.2 Observation and Analysis
    4E.3 References

Appendix 4F Fatigue Design Considerations for Forgings
    4F.1 Bulk Effects
    4F.2 Surface Effects

Appendix 4G Plating and Coating
    4G.1 Effects on Fatigue Resistance
    4G.2 References

4.1 Introduction

4.1.1 *Scope*

The purpose of this chapter is to provide a link between the largely idealized test specimens of Chapter 3 and actual applications. This chapter deals only with factors related to the material. The influence of component shape and loading is dealt with in Chapter 10.

The effect of processing on performance is application specific. That is, the effect depends on how processing (1) alters local composition, (2) alters or orients local microstructure, (3) introduces long or short range self-stresses (residual stresses) due to constraint and (4) alters the surface in comparison to the reference data developed using laboratory test specimens. The effect of processing on performance could be discussed and illustrated by a series of data sets representing specific materials, typical component geometries, and a range of different processing methods used in ground vehicles as compared to laboratory test results. This presentation format would be a catalog of available data, but it would not, nor could it, cover *all* current applications. And, of course, it could not cover any new material, or new application.

A different approach has been taken here. Trends in available data have been analysed with respect to influences on microstructure, orientation, self stress effects, and surface effects. In this way the available data provides guidance for both current and future applications. In addition, the available data, when presented in this manner, provide insight for specific cases not covered by the current data, including new applications, materials, and so on.

As such this chapter is not material or application specific. It reflects the behavior of steels, cast irons, and aluminum alloys and considers both crack initiation and propagation to the extent they are affected by processing. Processing procedures used in automotive applications including fabrication procedures, thermal and thermal-mechanical procedures, and mechanical procedures are discussed in terms of their influence on the fatigue process. Factors that influence fatigue are explored in terms of surface effects, self stress effects, and chemistry and microstructure effects. Following the general discussion, a number of application specific results are presented in Appendix 4. This collection of appendices is not all inclusive, but is meant to serve as a guide for applications commonly encountered in current design.

4.1.2 Relationship to Other Chapters

The preceding chapter laid the foundation for characterizing a material's flow and fatigue behavior. This has been done primarily in terms of machined and polished, smooth specimens (a test geometry similar to a tension specimen) and various precracked specimen geometries. Throughout the previous chapter, it has been tacitly assumed that the element of material being tested and characterized represents an element of material at the critical area in, or along, the cracking path of some component. Yet, nothing has been said about the fact that the microstructure (e.g., grain size) for this specimen may differ from that in the critical areas or along the cracking path. Nor has anything been said about the influence of the surface finish on crack initiation in the test specimen as compared to that in the application of interest. Finally, the stress gradient in a component almost certainly will be different from that of a test specimen. The correlation, or lack of it, between simple specimen fatigue data and fatigue lives obtained with real components depends substantially on the extent to which processing effects have been properly accounted for in the analysis. This chapter addresses these issues.

4.1.3 Chapter Plan

The chapter plan is shown in Fig. 4-1 in the context of an evaluation of processing effects on fatigue performance of a component. Once a fabrication or processing procedure has been accepted, this chapter suggests the development of test data to characterize the processing effect in a given application. This chapter also addresses the less satisfactory alternative—using previously generated trend data to estimate processing effects. In this case the component's fatigue resistance can be estimated by looking at baseline properties for the material (as in Chapter 3) and then modifying the fatigue properties in accordance with the anticipated effect of the process. The chapter closes with two examples and an appendix that provides details on specific processing and fabrication effects on crack initiation for a range of materials.

4.2 Considerations in Evaluating the Effects of Fabrication and Processing

4.2.1 Limitations of Standard Specimen Fatigue Data

As outlined in the preceding chapter, a material's fatigue properties are normally determined using standard specimens. Standard specimens often do not reflect the conditions that exist in real components. For example, the standard smooth specimen has a lathe turned, often ground or polished

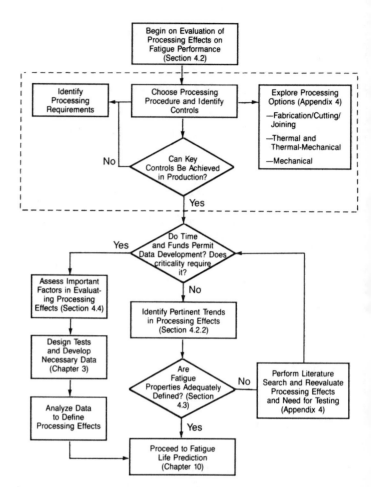

Fig. 4-1 A procedure for evaluating processing effects on fatigue performance, with references to pertinent handbook material.

surface. In contrast, in practical situations fatigue cracks often initiate on a mill-scale surface, at a sheared surface, on a plated surface or at a surface that is roughened due to plastic flow associated with significant stretching due to forming. Since fatigue crack initiation generally tends to occur at the surface, differences in surface condition may significantly alter life. Generally, the rougher the surface the shorter the life in the high cycle fatigue regime. As is shown in Fig. 4-2, other facets of processing may counter this tendency [1]. However, as a general rule, the data presented earlier in Chapter 3 (representing polished smooth specimens) provide an upper bound estimate of fatigue life or a lower bound estimate of damage in practical situations.

Standard specimens are also designed in most cases to minimize self stresses. ASTM E606 [2] discusses smooth specimen preparation with a view to minimize initial stresses in the specimens. Likewise ASTM E647 [3] warns of the effect of residual stresses and provides guidance to assess their potential influence on crack propagation data. In contrast, actual hardware may contain self stresses and related gradients due to processing, for example surface hardening methods or shot peening. (It is also possible that processing treatments, such as hard chrome plating, can develop surface microcracks.)

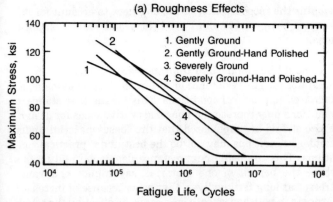

(a) Roughness Effects

S-N Curves for Flat Steel Bars (R_c -59 Hardness) Showing Effect of Grinding Severity

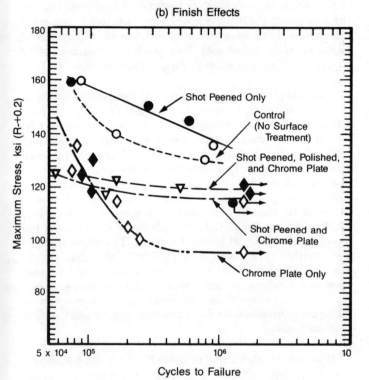

(b) Finish Effects

Fatigue Life of Axially Loaded, Round 4130 Steel Specimens Heat-Treated from 140-160 ksi

Fig. 4-2 Influence of surface roughness and finish on fatigue resistance of smooth specimens. (Note: Roughness and finish effects include some self stress effects since the available tests tend to deal with a procedure rather than isolate a roughness or finish effect [1]).

Since it is known that mean stresses influence fatigue life (as shown in Chapter 3 for smooth specimens, for example), it is logical to assume self stresses would also influence life and should be accounted for. This influence of self stress is illustrated in Fig. 4-3 for precracked specimens [4].

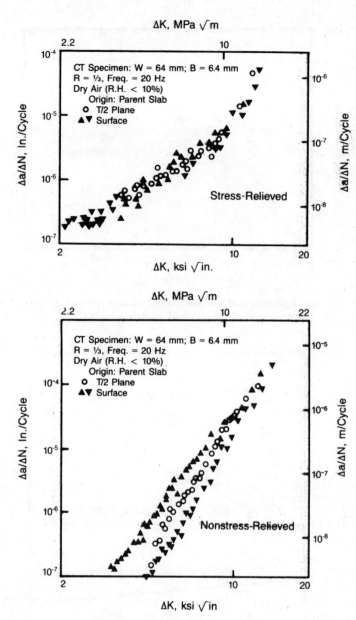

Fig. 4-3. Influence of self stresses on precracked specimens [5].

Finally, because standard specimens tend to impose minimum size requirements, it may be difficult to represent the microstructural and composition gradients that exist at critical areas and along crack paths using standard specimen sizes and geometries. The literature shows that factors like microstructure and composition influence fatigue resistance, so that processing and fabrication, that alter the microstructure (type, orientation, size) and composition (segregation, inclusions, etc.) from that in the reference data should be accounted for. Figs. 4-4 and 4-5 illustrate some aspects of the dependence of fatigue resistance on microstructure for unnotched [5] and precracked [6] specimens, respectively.

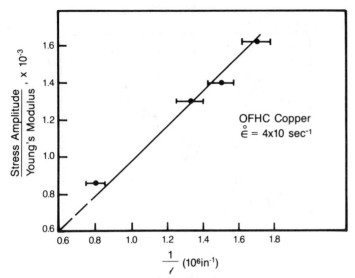

Fig. 4-4 Influence of microstructural size on the fatigue strength of unnotched specimens [6].

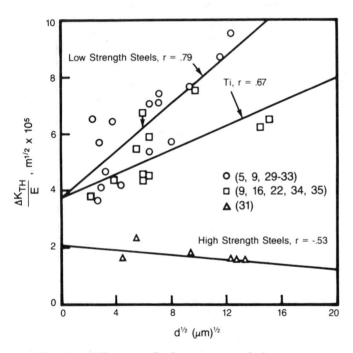

Fig. 4-5 Influence of microstructural size on near threshold behavior of precracked specimens [7].

It follows from the above discussion that standard specimens represent factors that affect fatigue in ways that may differ significantly from the conditions in real components. This does not mean that the standard specimens and related test standards discussed in Chapter 3 are ill conceived. Standard specimen designs and test conditions are typically defined so that a common basis for comparison of properties exists between materials. It is this common or standard basis for comparison that permits the ground vehicle engineer to assess which is the "best" material for his application. It remains the responsibility of the engineer to account for differences between standard conditions and service applications.

It should be emphasized that standard conditions tend to maximize fatigue resistance or minimize fatigue damage, as compared to practical conditions. This means that standard conditions may represent a nonconservative basis for design. It also should be emphasized that the "best" material under standard conditions may not be the best under practical conditions. This confusing circumstance develops in situations where the benefit of one material, say higher endurance stresses at long fatigue lives, that arise because of increased hardness in polished smooth specimens, is offset by the rough surface in the application, that reduces the life through notch sensitivity (that has been increased by the increased hardness). Care, therefore, must be taken when invoking the maxim that "if a little is good, more is better." Increased notch sensitivity, formation of quench cracks, and a range of other negative factors may develop while introducing other changes designed to enhance fatigue performance.

The final point of emphasis relates to components for which there is a significant fraction of the total life involved in both crack initiation and propagation. Attempts to choose the best material to improve long life fatigue resistance by using more refined microstructures may lead to decreases in crack growth thresholds. This circumstance is illustrated in Figs. 4-4 and 4-5. Refined grain sizes tend to increase the long life fatigue strength for smooth specimens, but they also tend to decrease the threshold for cracking in precracked specimens, at least for some materials. In such cases, the choice of the "best" material is not straight forward. To complicate matters further, processing may produce a variation in microstructure throughout a component. Figs. 4-4 and 4-5 suggest that a refined surface structure, coupled with a coarser structure in the midthickness would represent an optimum processing procedure for some lower strength steels (in some applications).

However, for high strength steels Figs. 4-4 and 4-5 suggest that an optimum processing procedure would involve maintenance of the refined surface structure throughout the thickness. However, "optimum" conditions are not always practically achievable, as in this last instance, because not all steels through-harden in heavier sections.

Results as shown in Figs. 4-2 and 4-5 suggest that the design analysis cycle can only move forward if data on the effects of fabrication and processing are available for the range of viable materials, covering the scope of fabrication and processing methods that are both practical and economically attractive. Unfortunately, this is seldom the case. Necessary data simply are not available to cover this spectrum of materials and processes. Designers are obliged to move forward in the presence of often very limited data. In some cases they can develop data using standard tests but seldom are designers afforded the luxury of data that cover all fabrication and processing effects. Accordingly, this chapter discusses fabrication and processing effects from the standpoint of why the effect exists and how one can assess its significance using either standard specimen data and/or application-specific experiments. Trends in processing effects are presented in Appendix 4 for a range of fabrication and processing procedures and materials. Throughout it is assumed that the reader is familiar with the various fabrication and processing methods which are discussed and is acquainted with the key parameters for controlling these methods. Further background on fabrication and processing methods may be found in the ASM Materials (Metals) Handbooks [7], among other sources.

4.2.2 Fabrication and Processing Effects

Tables 4-1 through 4-3 outline the effects of fabrication and thermal and mechanical processing procedures, respectively. The effects are analyzed first as physical effects and they are described in terms of their tendency to alter either the material, the stress field, or the surface, as compared to that of the standard smooth or precracked specimens as discussed in Chapter 3.

Material effects denoted (M) include cold work (increased hardness), microstructure (type, orientation and size), composition (chemistry), hot work, and electrochemical alteration. When cold work (plastic flow) is confined or occurs locally in a gradient (notch root, bent beam), the cold work also influences the stress field by creating compressive (or tensile) residual stresses that are concentrated at the surface. Surface residual stresses, which are reacted by lower, nominal stresses in the bulk elastic field, are denoted (SS). Material and cold work induced hardness and residual stresses are both key factors in fabrication, as can be seen from Table 4-1. Increased hardness and residual stresses due to cold work are also key factors in mechanical processing, as is evident in Table 4-3.

Residual stresses concentrated at the surface, SS, may also be created by thermal gradients during the quench cycle of thermal processing or by mechanical gradients due to phase transformation on the quench cycle, or plastic deformation of surface grains by shot peening or rolling. As detailed in Appendix 4, both processing techniques may cause the surface to

TABLE 4-1
ANALYSIS OF THE EFFECT OF FABRICATION ON FATIGUE PERFORMANCE OF PRODUCTION COMPONENTS*

Procedure		Physical Effect**	Relative Influence	Typical Effect on Fatigue Resistance
Form or Deep Draw		Cold Work (M, SS)	Significant	
		Orientation (M)	Significant	Variable
		Roughness (S)	Minor to Significant	
Forge Extrude		Hot Work (M)	Minor	Decrease
		Orientation (M)	Minor	(Can increase in
		End Grain (S)	Significant	direction of flow)
		Laps (S)	Significant	
Machine	Punching Cutting Turn/Mill Grinding	Roughness (S) Cold Work (SS)	Significant	Variable, but often decrease
Cast	Die	Orientation (M)		Decrease
		Size (M)		with respect
		Chemistry (M)		to Wrought
		Residuals (SS)		
		Voids (S)	Significant	
		Cold Shuts (S)		
	Sand	As for Die Roughness (S)		
Weld	Linear or Spot	Chemistry (M) Roughness (S) Residuals (SS) Geometry	Significant	Decrease

TABLE 4-1
(Continued)

Procedure		Physical Effect**	Relative Influence	Typical Effect on Fatigue Resistance
Weld (cont)	Dressed, Contoured Peened	Geometry	Significant	Increase with respect to As-Welded
		Residuals (SS)	Significant	
Powder Compact		Porosity (S)	Significant	Decrease
Plate or Annodize		Electrochemical (S, M)	Significant	Decrease with respect to Bare
		Structure (M, S)	Significant	

*As compared to laboratory fatigue test results on polished, residual stress-free samples. Applies only for intermediate to long life crack initiation conditions and near threshold to intermediate crack growth rate conditions.

** M = Material/Microstructure Change
SS = Surface Residual Stress Contributor
S = Surface Alteration

TABLE 4-2
ANALYSIS OF THE EFFECT OF THERMAL PROCESSING ON FATIGUE PERFORMANCE OF PRODUCTION COMPONENTS*

Procedure		Physical Effect**	Relative Influence	Typical Effect on Fatigue Resistance
Ferrous				
	Full Section Quench	Structure (M)	Significant	Increase (Generally)
	Anneal Temper Normalize	Structure (M)	Significant	Decrease
Surface				
	Alloying	Composition (M) Structure (M) Residual (SS)	Significant	Increase (Generally)
Aluminum				
	Full Section Age	Structure (M)	Significant	Increase (Generally)
	Surface Laser	Structure (M) Residuals (SS)	Significant	Increase (Generally)

*As compared to laboratory fatigue test results on polished, residual stress-free samples. Applies only for intermediate to long life crack initiation conditions and near threshold to intermediate crack growth rate conditions.

** M = Material/Microstructure Change
SS = Surface Residual Stress Contributor
S = Surface Alteration

TABLE 4-3
**ANALYSIS OF THE EFFECT OF MECHANICAL PROCESSING ON
FATIGUE PERFORMANCE OF PRODUCTION COMPONENTS***

Procedure		Physical Effect**	Relative Influence	Typical Effect on Fatigue Resistance
Peening		Residual (SS)	Significant	
		Cold Work (M)	Significant	Increase
		Structure (M)	Minor	(Generally)
		Roughness (S)	Minor	
Rolling		Residuals (SS)		
		Cold Work (M)	Significant	Increase
		Structure (M)		
Nominal Overload or Strain		Residuals (SS)	Significant	Increase
		Cold Work (M)		
		Roughness (S)	Minor	Decrease
Local	Prestress (Coin, Expand Etc.)	Residual (SS)		
		Coldwork (M)	Significant	Increase
		Structure (M)		

*As compared to laboratory fatigue test results on polished, residual stress-free samples. Applies only for intermediate to long life crack initiation conditions and near threshold to intermediate crack growth rate conditions.

** M = Material/Microstructure Change
SS = Surface Residual Stress Contributor
S = Surface Alteration

yield in tension. When the processing transients are completed, the surface yielding may lead to a steep gradient of compressive residual stresses, at the surface, that are reacted by much lower elastic tensile stresses in the core. Because of this gradient fatigue cracks may initiate subsurface. A complete fatigue life analysis for such a case probably should consider the behavior at the surface, as well as at several sites below the surface. Quenching from high temperatures also produces a change in the material, usually most evident as a refined structure with increased hardness (see Table 4-2).

It is very basic but noteworthy to remember that residual (self or internal) stresses are always balanced within a part. Compressive stresses are normally sought at critical areas to reduce tensile mean stress effects. In contrast, preloading into compression at critical areas may reduce the range of the local stress cycle that will enhance fatigue performance. In both cases equilibrium requires that the compressive stresses be balanced by tensile stresses. Wise design of a component and its processing/fabrication procedure will ensure that the tensile stresses are located in non-critical areas.

The final key factor, the surface, denoted (S) in Tables 4-1 through 4-3, is important for both fabrication and processing. The surface that develops in practice is much rougher than that of the polished specimen, so that surface shows up invariably as a negative factor in Tables 4-1 through 4-3. In contrast, the tables show that increased hardness (strength) and local compressive residual stresses (thermal or mechanical), and changes in material structure (that tend to refine or clean-up the microstructure) are beneficial.

Table 4-1 also introduces a number of other factors that relate to the material and the surface quality, and are procedure specific. Of these factors, those that effect the surface are critical since they represent either crack-like features or blunt notches that can quickly sharpen into cracks. Care must be taken to ensure that quality control procedures are implemented to preclude these surface defects. Otherwise, the fatigue life may be orders of magnitude less than predicted based on standard smooth specimen data. In such cases, the fatigue behavior may best be modeled by pre-cracked specimen results and analysis.

4.3 Fatigue Performance Data on the Effects of Fabrication and Processing

Tables 4-1 through 4-3 are useful in selecting fatigue performance data to characterize fabrication and processing effects. For example, consider a part that is most economically produced by forging. Table 4-1 indicates that care must be taken to account for orientation, laps and end grains, with secondary concern for hot working.

The factors in Table 4-1 may be used as a guide for designing forging dies and processing procedures. The previously mentioned forging will serve as an example product. Experience and existing data on file and in the literature suggest a loss in performance by creating a strongly oriented structure. Fig. 4-6a indicates that an order of magnitude decrease in life may occur when cracks are allowed to initiate and grow along a strongly oriented structure, as compared to across it [8]. Obviously, cutting across this elongated structure to expose end grains would cause a major life penalty, as is illustrated in Fig. 4-6b. It follows that forging dies and subsequent machining should be chosen to avoid either exposing end grains or creating macroscopic grain patterns that could be exposed accidentally in service or through subsequent reworking.

Those factors listed in Table 4-1 are equally useful in assessing how to design specimens to be cut from actual hardware to characterize fatigue performance. First critical locations and probable cracking planes are identified. Care should be taken in this step to account for stress redistribution and changes in load transfer that may activate other critical locations or redirect the cracking path. The next step makes use of Tables 4-1 through 4-3 to identify which areas of the component should be sampled to characterize properties. For the forging example, key areas to be characterized would include crack paths parallel to strongly oriented microstructure for precracked samples and areas of exposed (intersected) end grain for smooth specimens. Once sites to be characterized have been identified, the next step is to design test specimens that place the cracking direction perpendicular to the test load and provide for adequate gripping. It will often not be possible to obtain standard specimens so some creativity may be necessary. The final step is to develop data such as that presented in Chapter 3, but care should be taken to avoid those aspects of the standard test methods, like polishing, that are not representative of the actual hardware and material.

When the opportunity does not exist to develop application specific material/fabrication and processing fatigue data, the designer can make use of the factors in Tables 4-1 through 4-3 and literature data to help estimate fatigue performance data. The process of estimating performance data parallels the steps taken in developing actual test results. As outlined in the preceding paragraph, locations that represent critical areas and crack paths have to be identified and the related fatigue data must be estimated. If the factors in Tables 4-1 through 4-3 have been used in selecting the fabrication and processing methods and procedures, and if these factors have then played a role in implementing the fabrication or processing method selected, there is a good chance that the related loss of fatigue life as compared to the smooth specimen data will be minimized. The analyst should still make his best estimate of the effect of fabrication and processing for use in design.

The key question in estimating the effect of a fabrication or processing method for a given application is: how does one judge the significance of the effect relative to standard specimen data? A related question is: how well can I control fabrication or processing, and how sensitive will the fatigue life be to variations in fabrication or processing? Another related question is: how well do data for similar materials (or perhaps distinctly different materials chosen as a reference in estimating fatigue behavior) represent the material at hand for the procedure being considered?

General answers to these key questions are not possible but by categorizing fabrication and processing methods according to their physical effects, similarities between methods are apparent. Within a given generic material there are classes of material with comparable structures and strengths so that microstructural and cold work effects may be comparable. Finally, given that surface effects range from the influence of a notch to that of a crack, fracture mechanics for blunt cracks [9] could be used to estimate the role of the surface effects based on the related roughness. In all cases the estimate should be "benchmarked" with results from some comparable circumstances to ensure that the estimate is reasonable and conservative. In the absence of benchmark tests, an adequate factor of safety should be used. Here "adequate" is judged by the criticality of the component to continued safe operation. Care should also be taken to ensure that the factor of safety reflects the degree of uncertainty in the estimation process. When, because of large uncertainty or high criticality the factor of safety becomes too large, consideration should be given to developing application specific data supported by benchmark component tests.

4.4 Estimating Fabrication and Processing Effects on Fatigue Performance

Several authors have recognized the need to make estimates of fatigue performance and have assessed the sensitivity of performance to a given parameter based on literature data. Data dealing with initiation performance embrace all factors indicated in Tables 4-1 through 4-3. However, since surface is not a factor in crack propagation, the propagation performance literature focuses more on material effects and to a much lesser extent on self stress effects. It should be noted that when compressive self stresses are higher than the imposed tensile stress, there is zero stress intensity range and no growth of cracks.

The focus here is on fatigue performance. However, it should be remembered that stress-strain response, that is used in some crack initiation life analyses, may also depend on fabrication and processing. In particular, stress-strain response is sensitive to microstructural changes and accumulated strain. For ferrous alloys the effect of microstructural changes on stress-strain response is reasonably estimated by data for other materials with a similar microstructure at a comparable hardness level. The effect of accumulated strain on stress-strain behavior for ferrous metals varies. Increased hardness generally leads to elastic response at higher strain levels, but cyclic softening might negate this effect. Thus, cyclic strains approaching or exceeding the initial mechanical strain that caused the hardening often generate stress-strain response similar to the stable curve of the "virgin" material. The behavior of aluminum alloys is less patterned. Testing should be done if there is a significant change from "virgin" material due to fabrication and processing.

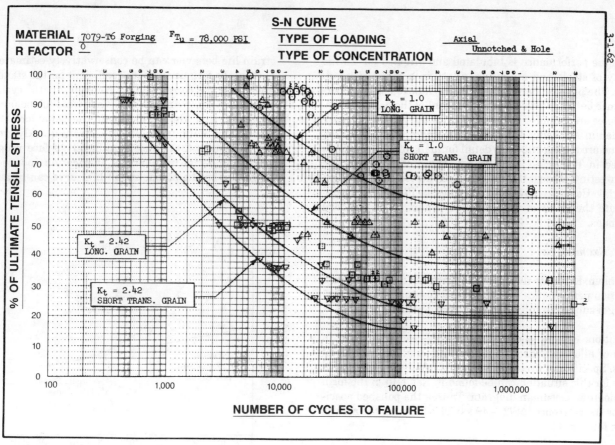

a) Initiation

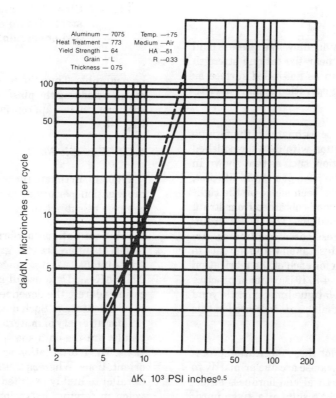

7075 Aluminum Extrusion Crack Growth Rate

b) Propagation

Fatigue performance is tabulated and graphed for a wide variety of steels and aluminum alloys in a range of handbooks. The purpose here is to illustrate how to use such performance data and cite the most relevant of the data sources. Examples are included in Refs. [7,10,11]. Additional data are included in Appendix 4 that presents and discusses fabrication and processing effects in detail for many of the conditions covered in Tables 4-1 through 4-3. These data can also be used to estimate the influence of a given procedure for a given material. Consider now two examples to illustrate how to estimate the effect of processing as compared to standard specimens.

4.4.1 Example No. 1

Problem: Estimate the effect of fabrication on fatigue crack initiation performance of a 1045 steel component (BHN 235 average) subject to fully reversed axial loading.

Solution: Fatigue performance in this case is assessed in terms of allowable stress to survive for a given number of cycles. Specific data representing the fatigue strength at 10^7 cycles for this situation can be found in Ref. [10] in the form of a modified Goodman diagram. Taking the polished specimen as the reference ($\Delta S/2 = 49$ ksi, at 10^7 cycles) the results are as follows:

as forged	0.38 × fatigue strength of polished specimen
hot rolled	0.51
turned	0.71
ground	0.88

This shows a very strong penalty on allowable stress is paid if the component incurs stresses near the fatigue strength and is left with an as forged or an as hot-rolled surface as compared to some final machining operation applied in critical areas.

Finite life fatigue behavior can be estimated if the fatigue strength data are used in conjunction with data for polished specimens for this material condition, such as that shown in Fig. 4-7. Note that Fig. 4-7 is in the form of a strain life plot, but that a stress based ordinate has been added on the right side. An estimate of stress amplitude is obtained from strain amplitude by multiplying by the modulus. The elastic strain-life curve thus can be read on stress based coordinates. Accordingly, fatigue strength estimates (taken with reference to the data in Fig. 4-7) can be shown for each of the fabrication methods, as has been done in Fig. 4-8. It is common practice (for many steels) to assume that a fatigue limit exists beyond about 2×10^6 cycles (4×10^6 reversals), as is indicated in this figure.

Differences in fatigue strength for these fabrication procedures and the smooth specimen response are due primarily to surface condition since microstructural and hardness differences are normalized by comparing results at a given hardness. Since surface does not significantly influence a "strength" coefficient, it is reasonable to assume the fatigue strength coefficient for polished specimens represents that for the other fabrication conditions. Therefore, finite life elastic strain-life behavior can be conservatively estimated by connecting the fatigue strength coefficient for the standard specimen with the corresponding point at 2×10^6 cycles for the fabrication process of interest. The stress-life behavior can be estimated from the elastic-strain-life behavior through the modulus. Finally, inelastic cycle strain quickly creates surface slip steps that cause significant surface roughness. It follows that the influence of fabrication induced surface roughness on the plastic strain life curve can be assumed to

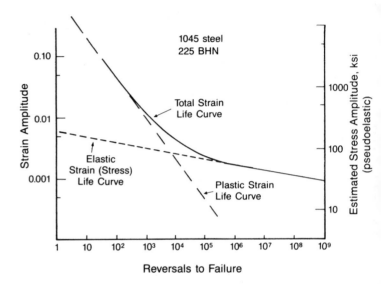

Fig. 4-7 Strain life and stress life behavior of 1045 steel with an endurance of 49 ksi from standard specimen tests.

be negligible. Thus the total strain life response can be estimated by adding the plastic response for the standard smooth specimen to the appropriate elastic response, as shown in Fig. 4-8.

4.4.2 Example No. 2

Problem: Estimate the effect of fabrication on fatigue crack propagation performance on an aluminum alloy (7075-T6) component subject to axial fatigue cycling at $R = -0.33$.

Solution: Fatigue performance in this case is assessed in terms of differences in growth rate at the same effective stress intensity factor range. For simplicity it is assumed that $\Delta K = \Delta K_{EFF}$. Data useful in assessing the influence of fabrication address the dependence of growth rate on microstructural orientation. Such data are tabulated in handbooks like [11] for a variety of materials and conditions. Typical of these results are the data shown in Fig. 4-6b for L and T orientations. The L orientation is typical of most standard specimen orientations. Whereas the T orientation is typical of cracking parallel to highly oriented microstructures such as may develop in forging, extrusion, drawing, forming, casting and some weldments. The effect of fabrication as compared to typical bar and plate products is evident in the difference in rates between L and T orientations, with the extent depending on the difference in orientation.

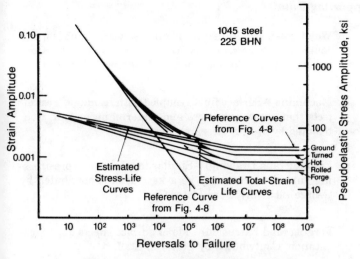

Fig. 4-8 Example of estimated fatigue behavior for various surface finishes using standard specimen data as a reference.

Fig. 4-6b shows that growth rates are almost uniformly increased. The influence of fabrication therefore is to increase the growth rate in a way that can be represented by multiplying the standard specimen growth rate data by a constant or by reducing the life calculated using standard data by the inverse of that constant. In the case of the data shown in Fig. 4-6b, this increase in growth rate is almost negligible. However, in highly oriented structures differences as great as a factor of 10 are not unusual.

4.5 Summary

The effect of processing and fabrication have been evaluated in terms of their physical effects on microstructure, surface, and internal stresses. Tables 4-1 through 4-3 present this evaluation and assess the significance of these effects as compared to standard specimen results. Modification of standard specimen data to account for processing and fabrication effects have been illustrated for smooth and precracked specimens. Modifications illustrated made use of how the physical effect carried over into the standard specimen framework, and tried to isolate the effect of one factor by normalizing the results to keep other factors constant.

The appendix that follows continues analysis of fabrication and processing effects by presenting data for selected materials and fabrication/processing methods. Analysis of the mechanism that underlies a given fabrication or processing methods effect on fatigue is presented to aid the reader in using Tables 4-1 through 4-3 in implementing these data.

4.6 References

[1] Madayag, A. F., Editor, Metal Fatigue, Wiley, 1969.

[2] "Standard Recommended Practice for Constant-Amplitude Low-Cycle Fatigue Testing", ASTM Standard E 606, ASTM.

[3] "Standard Test Method for Constant-Load-Amplitude Fatigue Crack Growth Rates Above 1 X 10E-08 Meters/Cycle" ASTM Standard E 647, ASTM.

[4] Bucci, R. J., "Effect of Residual Stress on Fatigue Crack Growth Rate Measurement. STP 743, ASTM, 1981, pp. 28-47.

[5] Abdel-Raouf, H., et. al. Temperature and Strain Rate Dependence in OFHC Copper and 304 Stainless Steel" Met. Trans, V5, 1974, pp. 267-277.

[6] Gerberich, W. W., and Moody, N. R., Review of Fatigue Fracture Topology Effects on Threshold and Growth Mechanisms, pp. 292-341. STP 675, ASTM.

[7] ASM Handbooks: Volumes 1 through 11 plus various additional sourcebooks, ASM. See also MIL-HDBK-5, Metallic Materials and Elements for Aerospace Vehicle Structures, Naval Publications and Forms Center.

[8] Rockwell International Fracture Mechanics Data Bank.

[9] Creager, M. and Paris, P. C., "Elastic Field Equations for Blunt Cracks with Reference to Stress Corrosion Cracking," *International Journal of Fracture Mechanics*, Vol. 3, 1967, pp. 247-252.

[10] Lipson, C. and Juvinall, R. C., *Handbook of Stress and Strength: Design and Materials Applications*, Macmillan Co.

[11] *Damage Tolerant Design Handbook*, MCIC-HB-01, Metals and Ceramics Information Center, Battelle Columbus Div., 1983.

Addresses of Publishers

ASM International, 9639 Kinsman Road, Metals Park, OH 44073

ASTM, American Society for Testing & Materials, 1916 Race Street, Philadelphia, PA 19103

Battelle Columbus Div., 505 King Street, Columbus, OH 43201

MacMillan, 866 Third Avenue, New York, NY 10022

Naval Publications & Forms Center, 5801 Tabor Avenue, Philadelphia, PA 19120

Appendix 4 Supporting Data

This appendix presents data on the influence of fabrication and processing on fatigue performance. Because the bulk of ground vehicle applications currently make use of an initiation approach only such data are presented. If and when the emphasis shifts to recognize a damage tolerant design/analysis approach, fatigue crack propagation data can be added as appropriate.

This appendix addresses the following topics:

Mechanical Prestressing—An increase in hardness through cold work coupled with beneficial residual stresses in parts with strain gradients (perhaps offset somewhat by surface roughness if there is significant plastic straining). Contributed by Henry Fuchs, Standford University.

Heat Treatment—A refinement of microstructure and attendant increase in hardness that is coupled with beneficial residual stresses in the case of surface treatments. Contributed by R. E. Ricklefs and W. P. Evans, Caterpillar Tractor Co.

Welding—Local influence of microstructure and surface coupled with complicated residual stress effects. Contributed by Harold Reemsnyder, Bethlehem Steel Corp.

Machining—Surface finish coupled with residual stress effects that are very process and material dependent. Contributed by Metcut Research Associates.

Forming (and Drawing)—Similar to mechanical prestressing but orientation may be a key factor. Contributed by George Libertiny, Ford Motor Co.

Forging (and Extrusion)—Major surface effects due to orientation. Contributed by Editorial Staff.

Plating (and Coatings)—Significant surface and environmentally related problems coupled with residual stress effects. Contributed by Editorial Staff.

Appendix 4A Mechanical Prestressing

Mechanical prestressing can be done by many different methods. All of them produce tensile yielding and cold work near the most highly stressed or most vulnerable surface. The springback of the elastic material that was not subjected to yielding then produces compressive stresses in the yielded regions.

The most common methods are:

Shot Peening
Surface rolling, including thread rolling
Overloading
Coining

The aim is to produce compressive self stresses in those regions (like notch roots, holes, highly stressed surfaces) that are most exposed to fatigue damage. They may also be used to overcome the damaging effects of tensile self stresses (residual stresses) that might have existed in those regions. There is an added benefit associated with a local increase in hardness that may account for a significant fraction of the life improvement.

An excellent survey of the field was given by Horger in Ref. [1].

4A.1 General

The strong effects of self stresses on fatigue strength are based on the fact that the initiation or growth of cracks is prevented by compressive self stresses and accelerated by tensile self stresses. This is related to the mechanics of crack initiation and crack growth. Without going into the complexities of nonlinear analysis and fracture mechanics of small cracks growing in non-uniform stress fields one can obtain excellent estimates of the effects of self stress by using Haigh diagrams such as Fig. 4A-1 [2]. Any point in this diagram represents a combination of mean stress and alternating stress. The stresses are nominal, but any self stress in a fatigue critical region is added algebraically to the nominal mean stress.

The diagram is constructed with straight lines as follows: From compressive yield stress at A = −500 MPa go to the cyclic yield stress at E = ±600 MPa and from the tensile yield stress at H = +500 MPa. Any combination of stresses represented by a point outside this triangle produces yielding. Self stresses go toward zero if the alternating stresses are above the lines of the triangle. For instance, a compressive self stress of 400 MPa cannot exist with an alternating stress of 200 MPa. If the alternating stress of 200 MPa is applied the self stress will readjust itself and become 330 MPa, so that the point that represents the combination falls on the line AE.

Point N = ±300 MPa represents the fully reversed smooth specimen fatigue strength for the target number of cycles of interest. In general this will be a number larger than 500,000

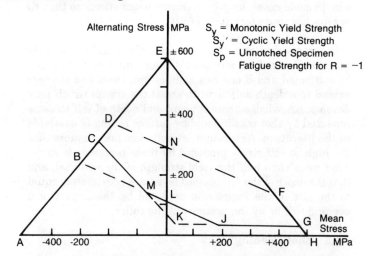

Fig. 4A-1 Construction of a Haigh Diagram for a part with a fatigue notch factor, $K_f = 2.5$, made of a material with $S_y = 500$ MPa $S_y' = 600$ MPa $S_p = 300$ MPa.

that yielding will decrease the self stress to a small value, as explained above. The inclination of the line D-N-F corresponds to the modest influence of mean stress and self stress on the fatigue strength of smooth specimens.

Point L = ±120 MPa corresponds to the fully reversed fatigue strength S_f/K_f of the notched part without any self stress. The inclined line M-L-J is parallel to D-N-F. (This is contrary to the slope of the lines drawn in some machine design textbooks, but it corresponds more closely to available experimental data.)

Line G-J-K is drawn at a constant alternating stress of 30 MPa. Depending on crack length and geometry this corresponds to some value of stress intensity range that is well below the threshold stress intensity range for all reasonable crack lengths (less than 1 cm for instance). The range of tensile stress is 60 MPa.

Line C-M-K corresponds to the same stress intensity range. For zero mean stress an alternating stress of 60 MPa will give 60 MPa range of tensile stress. For 100 MPa compressive mean stress an alternating stress of 160 MPa will give a 60 MPa range of tensile stress.

Segmented line A-C-M-J-G-H represents the Haigh diagram for the example part. Any combination of stresses below this line is expected to produce a median life longer than the target life. In the region above B-M but below C-M the stresses will produce cracks, but the cracks will not propagate [3]. One can see from this diagram that compressive self stresses between 240 and 500 MPa will permit the part to support alternating stresses up to ±300 MPa (point C) without decreasing the median life below the target. That this amount is exactly equal to the un-notched fatigue strength in this case is a coincidence. It might be somewhat less or somewhat more. It illustrates the well known fact that self stresses can, in some cases, largely overcome notch effects so that K_f approaches unity (see Refs. [4-6]).

Diagrams similar to Fig. 4A-1 can give excellent insight into the effects of self stresses if the amount of self stress can be estimated and if one can assume that these self stresses extend to a depth sufficient to arrest the cracks which may develop. Knowledge about amount and depth of self stresses produced by shot peening and by surface rolling is available in the literature. As a rough estimate one may assume that the highest self stress produced by these processes is somewhat more than half the yield strength of the material, and that the depth of the compressed layer is approximately equal to the size of the impression produced by the impact of a peening ball or by the pressure of the roller.

4A.2 Shot Peening

Shot peening is the most versatile method of producing self stresses. Small balls are thrown at high velocity against the part. They produce local yielding, evidenced by small dimples. "Springback" of the material at the surface after yielding produces the compressive stress. The process is described in detail in SAE J808a. Shot peening intensity is specified in terms of Almen numbers, according to SAE Handbook 2442 [7].

Fig. 4A-2(a) shows typical shot peening intensities as a function of section thickness. If the intensity is too high the tensile stresss in the core of the part becomes so high that failures occur there. Too low an intensity produces less strengthening than the optimum which is defined by equal probability of failures starting at the surface and below the surface. Compressive stresses extending over somewhat between 10 percent and 25 percent of the cross-section seem to approximate the optimum.

Fig. 4A-3(a) shows typical distribution of compressive stresses produced by peening.

Fig. 4A-4 shows the amount of stress produced by peening on steel as a function of the hardness.

Fig. 4A-5 shows the depth to which the compressive stress extends in steel and titanium as a function of peening intensity and material hardness.

Figs. 4A-6 and 4A-7 are examples of the results that have been obtained by shot peening. For the coil springs, peening eliminated fatigue as a failure mode. The stress limit for the peened springs was then set to avoid yielding. For the gears the permissible bending stress was increased 20 percent at 100,000 cycles and 45 percent at 10 million cycles.

4A.3 Surface Rolling

The results of this process can be comparable to those of shot peening. It can be applied to production lots of round parts with fillets, such as wheel spindles, and especially to the production of threads. Rolled threads have better fatigue strength than cut or ground threads. For exterior threads the process is often less expensive than cutting or grinding. For interior threads it is usually more expensive, but it may be justified by the increased fatigue resistance [10,11]. A survey of surface rolling and similar processes has been published by SAE (J811) [12].

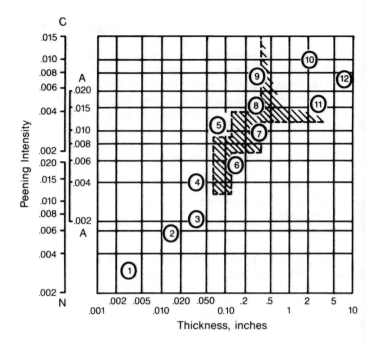

CHART POINT	PART	APPROXIMATE SIZE	ALMEN INTENSITY
1	Steel Tape	0.003 Thick	.003N
2	Steel Finger	0.015 Thick	.005N
3	Coil Spring	0.040 Wire Dia.	.007N
4	Valve Plate	0.040 Thick	.004A
5	Tuning Bracket	0.075 Thick	.010A
6	Bucket	0.120 Thick	.005A
7	Gear	32" O.D. 170 Teeth	.010A
8	Leaf Spring	0.3 Thick	.014A
9	Gear	14" O.D. 60 Teeth	.007C
10	Torsion Bar	2" O.D.	.010C
11	Crank Pin	2.7" O.D.	.015A
12	Shaft	8" O.D.	.007C

The Shaded Area Corresponds to Mil S-13165A

Fig. 4A-2 Typical shot peening intensities as a function of section thickness. [3]

4A.4 Overloading

Improved resistance to fatigue damage and static yielding can be achieved by overloading, if the service loads are predominant in one direction only, and below levels that would cause subsequent cyclic softening or yielding. The autofrettage of gun barrels is a classic example [13]. Presetting of springs is another example [14]. These processes depend on the difference in stress between different regions of the part, that is a stress or strain gradient. The most highly stressed region will yield first and, on springback, have a self stress of opposite sign. If it is yielded in tension, it will have a compressive self stress. The amount of self stress can be estimated if

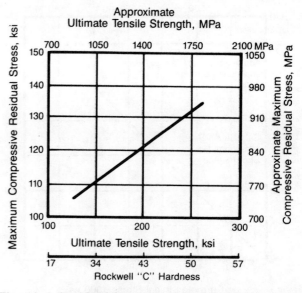

Fig. 4A-4 Residual stress produced by shot peening vs. tensile strength of steel.

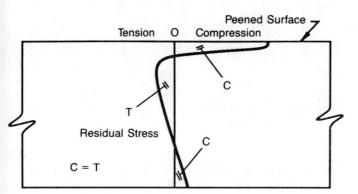

Fig. 4A-3 Distribution of stress in a shot-peened beam with no external load.

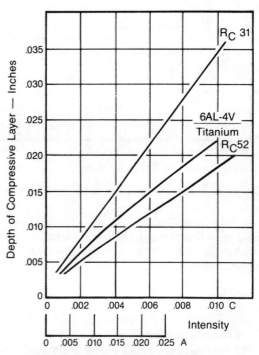

Fig. 4A-5 Depth of compressive stress vs. Almen intensity for steel and titanium.

the stress-strain curve is known [15]. Overstressing is also beneficial in arresting or slowing the growth of cracks [16,17]. This is caused by the compressively stressed zone of yielded material near the tip of the crack. In this manner, periodic overstressing may lead to extension of fatigue life. If the overload is properly specified, it can also serve to produce ductile growth or fracture during testing if cracks have grown to a size too large for continued safe operation.

4A.5 Prestressing and Preloading of Holes in Sheets

Holes in sheets present fatigue problems, especially in the aerospace industry. Many methods for prestressing them have been tried. One of the most promising is coining [18,19]. Other methods are expansion and dimple flattening [20]. Improvements in fatigue strength up to 100 percent have been

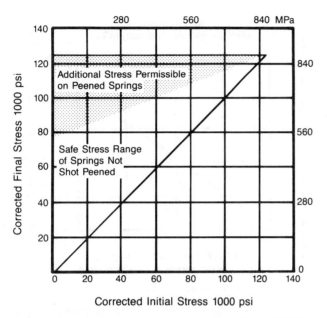

Fig. 4A-6 Safe stress ranges for coil springs (.207 inch diameter steel) [9].

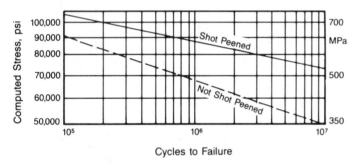

Fig. 4A-7 Fatigue chart for carburized gears.

reported. Preloading through the use of interference fit fasteners that reduce the local stress range also leads to significant improvements in life.

4A.6 Conclusion

The various methods of prestressing that have been described can serve to increase fatigue strength by 30 percent or more. They are more effective on harder materials which can retain self stress better than materials with lower yield strength. This is fortunate because the harder materials are most susceptible to fatigue strength reduction by surface imperfections and by notches. Intelligent application of mechanical prestressing thus permits the use of materials of higher yield strength.

Fig. 4A-8 shows fatigue limit plotted versus hardness from data by Nelson, Ricklefs, and Evans [5]. It shows how optimum use of compressive residual stress (self stress) provides much greater improvements for hard steels.

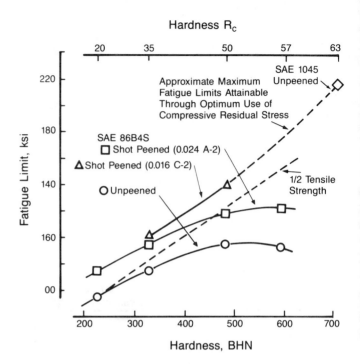

Fig. 4A-8 Fatigue limit vs. hardness and shot peening intensity.

Conversely, poor processing will do much more damage to the harder materials. With the need for lower weight and the use of higher strength materials, proper processing becomes more and more important.

4A.7 References

[1] Horger, O. J. ed. *Metals Engineering, Design second ed.* McGraw-Hill, 1965. (includes 13 sections on Fatigue Considerations Resulting from Processing and a section on Residual Stresses, each with extensive bibliography).

[2] Haigh, B. P., "The Relative Safety of Mild and High-Tensile Alloy Steels under Alternating and Pulsating Stresses", *Proc. Instn. Automobile Engineers*, (London), Vol. 24 (1929/30), pp. 320-362.

[3] Fuchs, H. O., "A Set of Fatigue Failure Criteria", *Trans ASME*, Vol. 87 Ser D, Journal of Basic Engineering (June 1965), 333-343.

[4] Harris, W. J., *Metallic Fatigue*, Pergamon, 1961, pp. 47-56.

[5] Nelson, D. V., Ricklefs, R. V., and Evans, W. P., "The Role of Residual Stresses in Increasing Long-Life Fatigue Strength of Notched Machine Members," *Achievement of High Fatigue Resistance in Metals and Alloys*, STP 467, ASTM, 1970, pp. 228-253.

[6] Almen, J. O. and Black, P. H., *Residual Stresses and Fatigue in Metals*, McGraw-Hill Book Co., 1963.

[7] SAE Handbook 2442, pp. 9.05-9.06.

[8] Provided courtesy of Metal Improvement Company.

[9] Zimmerli, F. P., "Heat Treating, Setting and Shot-Peening of Mechanical Springs," *Metal Progress*, Vol. 67, No. 6, June 1952, pp. 97-106.

[10] Dinner, H., and Felix, W., "Rolling of Screw Threads", Engr. Dig., Vol. 6, No. 12, pp. 332-333, 1945.

[11] Almen, J. O., "Fatigue Durability of Prestressed Screw Threads", Production Engineering, April 1951.

[12] Surface Rolling and Other Methods for Mechanical Prestressing of Metals, SAE J811, SAE.

[13] Fuchs, H. O., "Surface Stressing to Avoid Fatigue" in *Metal Fatigue*, G. Sines and J. L. Waisman ed., McGraw-Hill, 1959.

[14] Manual on Design and Application of Helical and Spiral Springs, SAE J795, SAE, p. 9.

[15] Gerber, T. L., and H. O. Fuchs, "Analysis of Nonpropagating Fatigue Cracks in Notched Parts with Compressive Mean Stress", *Journal of Materials*, JMLSA, Vol. 3, No. 2, June 1968, pp. 359-379.

[16] Nelson, D. V., "Review of Fatigue-Crack Growth Prediction Methods", *Experimental Mechanics*, Vol. 17, No. 2, Feb. 1977, p. 41-49.

[17] Stephens, R. I., E. C. Sheets, and G. O. Njus, "Fatigue Crack Growth and Life Predictions in Man-Ten Steel Subjected to Single and Intermittent Tensile Overloads" in Cycling Stress-Strain and Plastic Deformation Aspects of Fatigue Crack Growth, ASTM STP 637, 1977, pp. 176-191.

[18] Phillips, A., Douglas Aircraft Engineering Paper, #1068, 1961.

[19] Speakman, E. R., "Fatigue Life Improvement through Stress Coining Methods" in Achievement of High Fatigue Resistance in Metals and Alloys, STP 467, ASTM 1970, pp. 209-227.

[20] Nawwar, A. M., J. Shewchuk and D. J. Lloyd, "The Improvement of Fatigue Strength by Edge Treatment", *Experimental Mechanics*, Vol. 15, No. 5, May 1975, pp. 161-168.

Addresses of Publishers

ASTM, American Society for Testing & Materials, 1916 Race Street, Philadelphia, PA 19103

McGraw-Hill Book Co., 1221 Avenue of the Americas, New York, NY 10020

Metal Improvement Co., 10 Forest Avenue, Paramus, NJ 07542

SAE, Society of Automotive Engineers, 400 Commonwealth Drive, Warrendale, PA 15096

Stanford University, Stanford, CA 94306

Appendix 4B Heat Treatment

Increases in fatigue strength from heat treatment of steel parts are due to an increase in hardness or tensile strength and the production of compressive residual stress in the surface layer. This discussion does not differentiate between the roles of hardness and residual (or self) stresses in improving fatigue resistance. Herein, residual stress is considered to be the macrostress (rather than the microstress) that is balanced over the entire part, and can, in principle, be varied independently of the microstructure. Microstress, on the other hand, varies over microscopic regions [1], and changes with the tensile strength of a hardened steel. Cold working at lower hardness can increase microstress, however, with an increase in fatigue strength independent of the macrostress [2,3]. Specific benefits of through section heat treatments are well established for steels [4], and to a lesser extent for aluminum alloys [5].

4B.1 General

A number of reviews have addressed heat treatment residual stresses and their effects on fatigue resistance, including those of Baldwin [5,6], Almen and Black [7], and Evans [8]. The effect is not as readily demonstrated for heat treatment stresses as for those from mechanical prestressing because, with a part of specific geometry, material, and heat treatment, a unique residual stress distribution is obtained. However, residual stresses have been shown to have the equivalent effect of a mechanically applied mean stress, so the applied mean stress can be changed to simulate a residual stress effect [2]. Alternately, the geometry can be changed to vary the residual stress, as by boring the core out of an induction hardened cylinder [9].

Most data showing the effect of residual stresses in fatigue concern axial or bending loading. Residual stresses in the loading direction, which is the longitudinal direction in bars, are the most significant. A static normal stress has been shown to improve fatigue performance of bars subjected to cyclic torsional stresses [10]. Also, residual stress effects are commonly considered in contact fatigue as applied to gear teeth and bearing races.

In general, when a distinct case-and-core relationship is produced from heat treatment, the case contains compressive residual stresses and core balancing tensile stresses. A superficial surface layer may contain stresses different from the bulk of the case due to decarburization or to an effect resulting from carburizing. The transition from compression to tension stress usually occurs approximately at the case depth. The maximum compression in the case may or may not vary noticeably with case depth, but the maximum tension below the case increases as the volume of the case approaches that of the core. This effect tends to promote subsurface yielding and failure, that should be kept in mind when selecting the material and heat treatment for a part.

Residual stresses from heat treatment, as well as those from other sources, may relax under the influence of alternating stress [9,11,12]. In most instances, after some readjustment, the residual stresses do not exceed the cyclic yield strength of the material. If yielding or cyclic softening occurs, the residual stresses will tend to "wash out" and their benefit will eventually be lost.

Residual stresses and fatigue performance resulting from several heat treatment processes are considered in the following sections. No attempt is made to recommend particular processes for given applications.

4B.2 Furnace Hardening Through Heat and Quench of a Steel

The mechanism of residual stress production by furnace (through) heating and quenching has been qualitatively described in several papers, including those of Rose and Hougardy [13], Liss, et. al. [14], and Evans [8]. There are three general categories, but not all produce beneficial effects.

In the first category cooling occurs from a temperature or at a rate, that does not produce a martensite phase transformation. As the surface layer of a part is cooled, it tends to shrink due to thermal contraction but is restrained by the hot interior. If the resulting strain gradient in depth is sufficient, the thermal stress will exceed the yield strength of the surface layer. This layer then will yield in tension, accommodating the misfit. After the interior cools and contracts, the surface layer is left with a compressive stress that is balanced by tension in the core. This condition is generally beneficial in terms of improved fatigue resistance.

The second category is that of shallow-hardening in which a transformation to martensite is promoted at the surface to form a case of some depth. As the surface layer of material cools during quenching, tensile yielding occurs. When the critical temperature, M_s, is reached, this material transforms to martensite and expands. Because the strength increases, the stress from transformation is accommodated by yielding in tension of the hot, soft interior. The opposing effects of expansion of the interior core material due to transformation to upper temperature products, followed by contraction during cooling, leave the case in compression, and the core in balancing tension. This condition is also beneficial.

The last category considered is that of through-hardening. Transformation to martensite in the interior occurs at a lower temperature than that at which the core transforms in the shallow-hardening situation, and at a later time than the tranformation of the surface layer. This results in a surface layer residual stress ranging from tension with a mild quench, to substantial compression in a larger part with a severe quench (the latter result being due to high thermal compressive stresses superimposed on the tensile stresses caused by transformation). The result here is very material and process dependent with higher critical temperatures for materials with lower carbon content.

Surface compressive residual stresses tend to increase with decreasing hardenability or case depth, due to a decreasing case-to-core volume ratio or an increasing hardness gradient between case and core. Fig. 4B-1 shows the effect of a difference in hardenability on residual stress as a function of depth. The shallow-hardening SAE 1042 bar contained a higher peak compressive stress than the deeper-hardening SAE 10B39 bar.

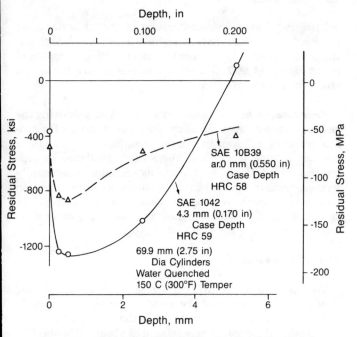

Fig. 4B-1 Comparison of longitudinal residual stresses in depth in furnace hardened steels of different hardenability [16].

The magnitude of the surface residual stresses tends to increase in compression with increasing quench severity. Finally, increasing the section size of shallow-hardening bars generally leads to greater compressive residual stresses.

Compressive residual stresses are especially effective in improving the fatigue performance of notched parts. This effect was shown by Liss, et. al. [14] with severely-notched cylindrical specimens tested in unidirectional bending. Fig. 4B-2 shows results obtained with various materials, quenches, and tempering temperatures. Essentially constant hardness was maintained except for the point indicated. The residual stresses were measured near the notch, since it was not possible to measure them at the root of the notch.

Buenneke [15] found improvements in the fatigue performance of pins when compressive residual stresses near the surface were increased. It has also been found that compressive residual stresses improve the resistance to both crack initiation and crack propagation in fatigue, with the crack propagation effect being the more pronounced. The beneficial effect of compressive residual stresses from heat treatment in

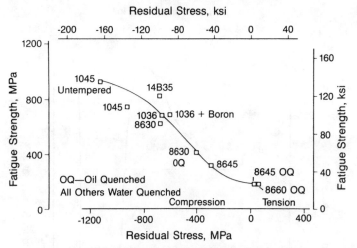

Fig. 4B-2 Effect of longitudinal residual stress on the endurance limit of selected steels—44.5 mm (1.75 in) bar diameter out of notch, 39.4 mm (1.55 in) diameter under 60 degree V-notch of 0.64 mm (0.025 in) radius. HRC 50-56 hardness, except untempered SAE 1045 at HRC 62-63 [14].

arresting surface cracks resulting from an inappropriate application of shot peening on HRC 63 hardness, reversed-bending fatigue specimens was shown by Nelson, et. al. [16]. Because of the high sub-surface residual stress from heat treatment, the fatigue performance of specimens with the weakened surface due to peening was as good as that of the as-heat-treated specimens. High hardness combined with a high compressive residual stress of 1720 MPa (250 ksi) resulted in a fatigue strength at one million cycles life of ±1590 MPa (±230 ksi). The residual and applied stresses were measured in the notch of 6.4 mm (0.25 in.) radius with a stress concentration factor of 1.5.

4B.3 Induction and Flame Hardening: Surface Heat and Quench

When a material is induction or flame hardened the surface layer is heated above the austenitizing temperature, followed by quenching. A region just below the surface layer contains a temperature gradient from the critical transformation temperature downward. Further into the core there is little heating. When the surface layer is quenched, it first contracts until plastic deformation accommodates the misfit between it and the underlying region. Upon expansion of the surface layer during transformation to martensite, along with cooling and contraction of the intermediate region, the surface layer goes into compression. A high balancing tensile stress may be concentrated in the soft intermediate region, as shown in Fig. 4B-3 [17]. The stress is shown as a function of cross-sectional area from the bar center outward, demonstrating the balance of tensile and compressive longitudinal forces.

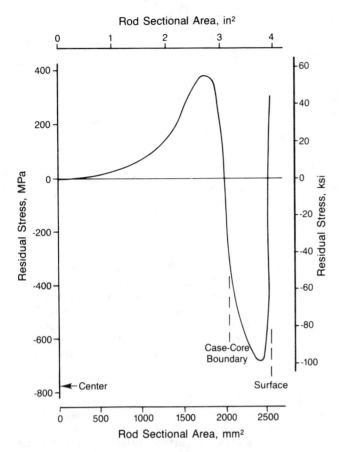

Fig. 4B-3 Longitudinal stresses in an induction hardened rod—AISI-A 4140, 57.2 mm (2.25 in) diameter, HRC 60 hardness [17].

4B.4 Carburizing

The mechanism of residual stress production by carburizing has been discussed in several papers, including those of Koistinen [18], Rose and Hougardy [13], Evans [8], and Motoyama [19]. This process depends on a M_S temperature gradient that results from the carbon gradient below the surface, with the M_S temperature being lowest at the highest carbon content near the surface. As the part cools with the quench, the surface layer transforms and expands whereas the core does not, creating a compressive, near-surface residual stress field. The maximum compression in the case may lie somewhat below the surface because of the transformation sequence, or because of retained austenite near the surface, meaning less expansion.

Bond [20], showed that deep freezing transforms some of the retained austenite near the surface, increasing the compressive residual stress in this region. Bond also showed a decrease in compressive stress in the case due to tempering. Motoyama and Horisawa [21] showed the effect of quench severity and case depth on the residual stress from carburizing. In a subsequent paper [19], features of the residual stress versus depth curves were explained by the predicted sequence of transformation.

In another paper [12], Motoyama showed the effect of carburizing residual stresses in rotating bending fatigue. A redistribution of the stresses occurred due to cycling, even at stresses below the fatigue limit. The fatigue limit was highest when the established residual stress was high in compression.

4B.5 Nitriding

Nitriding differs from carburizing in that it is normally preceded by a furnace hardening heat treatment. A high initial temper is usually given so that the nitriding operation at about 540°C (1,000°F) does not subsequently lower the strength of the core.

Richmond, et. al. [22] described a method for predicting the residual stresses from nitriding. They assumed that the concentration of nitrogen in solution was kept low enough to prevent nitride formation. The stress calculated by this method is relatively low. Actually, nitrides are usually formed, resulting in an increased volume expansion and an increased compressive residual stress after slow cooling. The case, that contains the compressive stress, is usually shallower for nitriding than for carburizing.

4B.6 Stress Relief

The mechanism operating in stress relief depends on the heat-treated structure of the steel. If the steel is initially untempered, stresses will relax due to the tempering reaction [23]. They will be completely relaxed at about 425°C (800°F), as was shown in Fig. 4B-1. However, if the material has previously been tempered at a high temperature or if the material is a low-strength structural steel that contains welding stresses, stress relaxation will proceed due to changes in elastic modulus and yield strength, and creep deformation [24,25]. In this situation, depending on the magnitude of the applied stress or residual stress, the relaxation may start at a higher temperature and be completely relaxed at a temperature of 600°C (1110°F) or more.

Residual stresses occur in iron castings due to plastic deformation accompanying non-uniform cooling and resistance of the mold to contraction. These stresses are relieved by slow heating to a temperature of 595-650°C (1,100-1,200°F), which is below the temperature of transformation to austenite. A lower temperature may be used to prevent significant loss in tensile strength. After holding at temperature for 1-2 hours per 25 mm (1 in.) of casting thickness, the part is slowly cooled so new residual stresses are not introduced. Lower stress relieving temperatures require longer heating times [26].

4B.7 References

[1] Evans, W. P. and Littmann, W. E., "'Macro' and 'Micro' Residual Stresses," *SAE Journal*, Vol. 71, March 1963, pp. 118-121. See also "Definitions for Macrostrain and Microstrain," SAE J932, SAE Handbook Volume 1, SAE.

[2] Evans, W. P. and Millan, J. F., "Effect of Microstrains and Particle Size on the Fatigue Properties of Steel," SAE Technical Paper 793B, January 1964.

[3] Evans, W. P., Ricklefs, R. E., and Millan, J. F., "X-Ray and Fatigue Studies of Hardened and Cold-Worked Steels," Local Atomic Arrangements Studied by X-Ray Diffraction, Cohen, J. B. and Hilliard, J. E., Gordon and Breach, Science Publications, Inc., 1966, pp. 351-377.

[4] Lipson, C. and Juvinal, R. C., Handbook of Stress and Strength, Macmillan, 1963.

[5] Private communication, R. J. Bucci (ALCOA) with B. N. Leis, 1987.

[6] Baldwin, W. M., Jr., "Residual Stresses," ASTM Proceedings, Vol. 49, 1949, ASTM, pp. 539-583.

[7] Almen, J. O. and Black, P. H., "Residual Stresses and Fatigue in Metals," McGraw-Hill Book Co., 1963.

[8] Evans, E. B., "Residual Stresses: Types and Sources," Residual Stress Measurement by X-Ray Diffraction," SAE J784a, SAE, 1971, pp. 3-11.

[9] Hayama, Toru, "Effect of Residual Stress on Fatigue Strength of Induction-Hardened Steel," *Bulletin of the Japan Society of Mechanical Engineers*, Vol. 18, No. 125, November 1975, pp. 1194-1200.

[10] Sines, George, "Failure of Materials Under Combined Repeated Stresses with Superimposed Static Stresses," NACA Technical Note 3495, November 1955.

[11] Larsson, L. E. and Spiegelberg, Per, "The Gradual Change of Residual Stresses During Bending Fatigue of Induction-Hardened Steel," *Scandinavian Journal of Metallurgy*, Vol. 2, 1973, pp. 19-23.

[12] Motoyama, Moritaro, "Residual Stress and Fatigue Strength of Carburized and Quenched Steel," SAE Technical Paper 760716, 1976.

[13] Rose, A. and Hougardy, H. P., "Transformation Characteristics and Hardenability of Carburizing Steels," Symposium: Transformation and Hardenability in Steels, Climax Molybdenum Co., February 1967, pp. 155-166.

[14] Liss, R. B., Massieon, C. G., and McKloskey, A. S., "The Development of Heat Treat Stresses and Their Effect on Fatigue Strength of Hardened Steels," SAE Technical Paper 650517, 1965.

[15] Buenneke, R. W., "Applications of Residual Stress Measurements," SAE Technical Paper 720243, 1972.

[16] Nelson, D. V., Ricklefs, R. E., and Evans, W. P., "The Role of Residual Stresses in Increasing Long-Life Fatigue Strength of Notched Machine Members," Achievement of High Fatigue Resistance in Metals and Alloys, ASTM STP 467, ASTM, 1970, pp. 228-253.

[17] Hanslip, R. E., "Residual Stresses in Surface-Hardened Oil Field Pump Rods," SESA Proceedings, Vol. 10, No. 1, 1952, pp. 97-112.

[18] Koistinen, D. P., "The Distribution of Residual Stresses in Carburized Cases and Their Origin," ASM Transactions, Vol. 50, 1958, pp. 227-241.

[19] Motoyama, Moritaro, "The Effect of Carburizing Variables on Residual Stresses in Hardened Chromium Steel," SAE Technical Paper 750050, 1975.

[20] Bond, W. B., "X-Ray Diffraction Studies of the Residual Stress Level in an AMS 6260 (SAE 9310) Carburized Case," *Norelco Reporter*, Vol. 13, No. 1, January-March 1966, pp. 18-20, 35.

[21] Motoyama, Moritaro and Horisawa, Hiroshi, "Residual Stress Measurements in Case-Hardened Steels," SAE Technical Paper 710281, 1971.

[22] Richmond, O., Leslie, W. C., and Wriedt, H. A., "Theory of Residual Stresses Due to Chemical Concentration Gradients," ASM Transactions, Vol. 57, ASM, 1964, pp. 294-300.

[23] Brown, R. L. and Cohen, Morris, "Stress Relaxation of Hardened Steel," *Metal Progress*, Vol. 81, No. 2, February 1962, pp. 66-71.

[24] Ritter, J. C. and McPherson, R., "Anisothermal Stress Relaxation in a Carbon-Manganese Steel," *Journal of The American Iron and Steel Institute*, Vol. 208, Part 10, October 1970, pp. 935-941.

[25] Ritter, J. C. and McPherson, R., "Stress Relaxation in Steel," *Welding Research Abroad*, Vol. 18, April 1972, pp. 20-31 (from Australian Welding Research, October 1971).

[26] Schaum, J. H., "Stress Relief of Gray Cast Iron," AFS Transactions, Vol. 56, 1948, pp. 265-278.

Addresses of Publishers

AFS, American Foundrymen's Society, Golf & Wolf Roads, Des Plaines, IL 60016

AISI, American Iron & Steel Institute, 1000 16th Street, N.W., Washington, DC 20036

ASM International, 9639 Kinsman Road, Metals Park, OH 44073

ASTM, American Society for Testing & Materials, 1916 Race Street, Philadelphia, PA 19103

Climax Molybdenum Co., AMAX, Inc., 1600 Huron Parkway, Ann Arbor, MI 48106

Caterpillar Tractor Co., 100 N.E. Adams Street, Peoria, IL 61629

Gordon & Breach Science Publications, Inc., P.O. Box 789, Cooper Station, New York, NY 10276

McGraw-Hill Book Co., 1221 Avenue of the Americas, New York, NY 10020

SAE, Society of Automotive Engineers, 400 Commonwealth Drive, Warrendale, PA 15096

SESA, Society for Experimental Stress Analysis (Note: Renamed Society for Experimental Mechanics (Stress Analysis)), Seren School Street, Bethel, CT 06801

Appendix 4C Weldments

Fabrication by welding is widely utilized in the ground vehicle industry to achieve economy and reduce dead weight. However, the fatigue resistance of weldments generally does not increase proportionally with an increase in base metal tensile strength. The trends of fatigue data for carbon and constructional alloy steels, plain material, transverse butt welds with reinforcement intact, and continuous non-load carrying longitudinal fillet welds,* are shown in Figs. 4C-1 and 4C-2 for fatigue lives of 1×10^5 and 2×10^6 cycles, respectively. The welded constructional alloy steel demonstrates an advantage over welded carbon steel when either the stress ratio, R, is greater than 0.25 to 0.50, or the expected life is short. However, at long lives, or in the case of alternating stresses ($R < 0$), the welded constructional alloy and carbon steels are similar in behavior. The sensitivity of alloy steels to notches is greater than in carbon steels and increases with increases in life. For example, Figs. 4C-1 and 4C-2 show that the ratio of the fatigue strength at two million cycles to the fatigue strength at 1×10^5 cycles is greater for the welded carbon steels than for the welded constructional alloy steels.

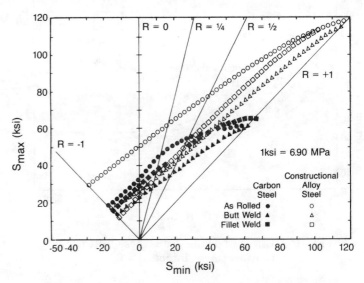

Fig. 4C-2 Constant life diagram $-N = 2 \times 10^6$ cycles.

Presence of slag or scale on the surfaces to be welded may lead to fusion. Porosity is caused by gas entrapped during solidification of the weld metal, excessive moisture in the electrode coating, or disturbance of the arc shield by drafts. Slag inclusions from the electrode coating are one of the most common weld discontinuities encountered. This can be caused by imperfect cleaning of the weld between successive passes.

The importance of mechanical notching on the fatigue resistance of weldments is illustrated by the fact that removal of butt-weld reinforcement raises the fatigue resistance to that of the base metal [1]. Also, increasing the height of reinforcement can significantly reduce the fatigue strength of transverse butt welds at two million cycles [1]. The geometric notch severity of the weld reinforcement accounts for the frequently observed phenomenon that the fatigue strength of transverse butt welds with reinforcement intact (or of transverse fillet welds) is insensitive to tensile strength. This effect is demonstrated in Fig. 4C-3, where test results for hot-rolled carbon steels, high-strength low-alloy structural steels, and constructional alloy steels, and data bands for additional tests on carbon steels, quenched and tempered carbon steels, and constructional alloy steels are plotted [1,2]. Although the tensile strengths of the steels shown in Fig. 4C-3 varied from 58 to 148 ksi (400 to 1020 MPa), there are no significant differences among the fatigue strengths of the various steels.

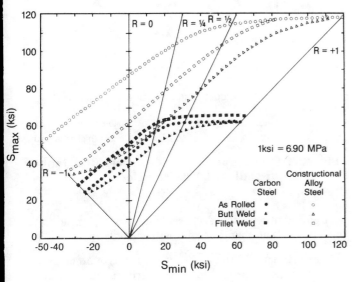

Fig. 4C-1 Constant life diagram $-N = 10^5$ cycles.

4C.1 Linear Weldments and Surface (Notch) Effects

A weldment generally contains both external and internal notches. Notches include changes in section due to reinforcement or weld geometry, surface ripples, undercuts, and lack of penetration. In addition, welds may be subject to such internal heterogeneities as shrinkage cracks, lack of fusion, porosity, and inclusions.

Shrinkage cracks are caused by excessive restraint exerted by adjacent material or shrinkage of the weld zone on cooling.

Lack of penetration lowers the fatigue strength of transverse welds significantly but has relatively little effect on longitudinal welds. Porosity and slag inclusions decrease fatigue resistance in proportion to the decrease in effective weld area of transverse welds. Microstructural changes due to severe quenching concomitant with the sudden extinguishing

---

*The fatigue strengths of transverse and intermittent longitudinal fillet welds are about one-half of those for continuous longitu-

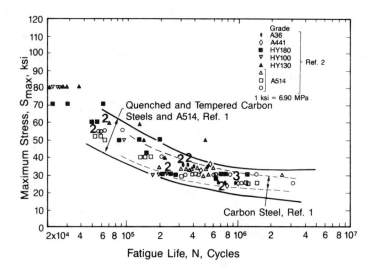

Fig. 4C-3 Effect of grade, transverse butt welds, reinforcement intact, R = 0.

of the welding arc may initiate a fatigue crack at the point of change of electrode. Severe quenching also results from stray flashes and weld splatter. Such stress raisers may be reduced or eliminated by control of the welding procedure.

4C.2 Spot Welds

Resistance spot welding is widely used to join metal sheets. The fatigue strength at one million cycles of single-lap spot-welded sheets of as-rolled carbon steel [3-8], cold-rolled (50% reduction) carbon steel [8], heat-treated carbon steel [8] and high-strength low-alloy (HSLA) steel [3-8] are shown in Fig. 4C-4. Fatigue strength is presented as both axial stress range, $\Delta\sigma$, (load range divided by cross-sectional area of sheet) and shear stress range, $\Delta\tau$, (load range divided by circular area of spot weld nugget), and is plotted versus the ratio of the nugget diameter to sheet thickness. The data include two stress ratios, R = 0 and R = −1, and both single spot welds and double spot welds (two welds on a line perpendicular to the load).

Although the sheet tensile strengths varied from 46 to 120 ksi (317 to 828 MPa), there were no significant differences in long life fatigue resistance either in terms of $\Delta\sigma$ or $\Delta\tau$ among the various grades. However, at lives less than 1×10^5 cycles, the fatigue strength, in terms of $\Delta\sigma$ or $\Delta\tau$, increases with sheet tensile strength [4,5,7]. Also, carbon steel joints develop cyclic strengths that are a greater proportion of their static strength, P/P_u, at one million cycles than HSLA spot welded joints.

There is no difference between joints with one or two spots on line, as shown in Fig. 4C-4, but the fatigue strength per

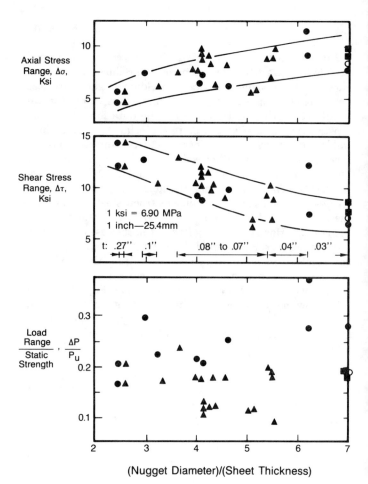

Fig. 4C-4 Single lape spot welded sheet steel, fatigue strength at 1,000,000 cycles.

spot of a joint with two spot welds in a row (parallel to the load) is less than that for a joint with a single spot weld [3,9].

The variation of fatigue strength in terms of $\Delta\sigma$ or $\Delta\tau$ with sheet thickness appears to be primarily a function of the bending due to the eccentricity concomitant with a single lap test specimen. For a given nominal stress range $\Delta\sigma$ or $\Delta\tau$, the critical local stress at the spot weld nugget increases with sheet thickness. Therefore, the fatigue strength based on nominal stress range would decrease with an increase in sheet thickness.

At lives greater than 1×10^4 cycles, fatigue cracks initiate at the edge of the nugget in the sheet heat affected zone* (HAZ) and propagate through and across the sheet. On the other hand, at lives less than 1×10^3 cycles, the nugget fails in a manner similar to that of a static failure [4].

---

*The heat-affected zone is that portion of the base metal that has not been melted but in which the mechanical properties and microstructure have been changed by welding.

4C.3 Residual Stresses

The heat-affected zone of the base metal is heated by the welding procedure to a temperature above the lower critical temperature (the temperature at which a steel changes phase), and the rate of cooling, governed by the conduction of heat away from the fusion zone, will produce metallurgical transformations similar to those produced by deliberate heat treatment.

Residual stresses due to welding are formed as a result of the differential in heating and cooling rates at various locations in the material. In addition, due to these thermal gradients, some of the material will be elastically deformed while other regions are plastically deformed. The interaction between these regions results in residual or internal stresses after cooling. These stresses may be quite large and will be tensile in the vicinity of the weld, where their magnitude is approximately equal to the yield strength of the weld metal, as shown in Fig. 4C-5.

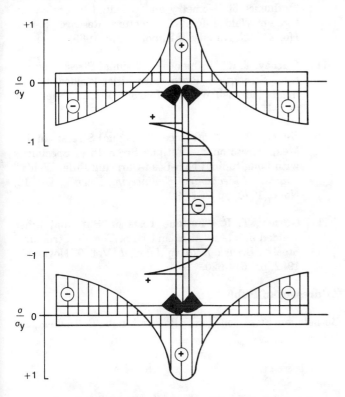

Fig. 4C-5 Residual stresses in welded beam.

Thermal stress relief can increase the fatigue strength of weldments at lives greater than one million cycles provided that the stress ratio is less than or equal to zero and notches are present in the weldment. For example, the fatigue strength of transverse butt welds (with reinforcement intact) at two million cycles has been increased by 12 percent and 24 to 33 percent for stress ratios of 0 and -1 respectively through thermal stress relief [10-17]. Such improvement has also been shown for longitudinal, non-load carrying fillet welds (e.g., attachments, gussets, etc.) where the increases in fatigue strength at two million cycles due to thermal stress relief for stress ratios of 0, -1, and -4 were, respectively, 15, 57, and 168 percent [15]. On the other hand, thermal stress relief has little or no effect on the fatigue resistance of longitudinal butt welds, transverse butt welds with the reinforcement removed, or longitudinal butt welds terminating well beyond the highly stressed region [10,13].

4C.4 Other Methods for Improvement in Weld Fatigue Performance

The introduction of compressive residual surface stresses at stress raisers can increase the fatigue resistance of weldments. For example, shot peening of non-load-carrying fillet-welded carbon steel and butt-welded constructional alloy steel has increased the fatigue strengths at two million cycles by 20 to 40 percent [2]. The efficacy of shot-peening for fatigue resistance is strongly influenced by shot size, arc height, and percent coverage. For example, the 20 to 40 percent improvement in fatigue strength for quenched and tempered constructional alloy steel was achieved by peening at an arc height of 0.006 to 0.012 (Almen C strip) [1]. Therefore, for an improvement in fatigue resistance to be significant and repeatable, shot peening must be closely controlled.

Hammer peening has been observed to improve the two million cycle fatigue strength of carbon steel butt welds by 15 to 25 percent and that of fillet welds by 20 to 50 percent [11]. In contrast, hammer peening of a quenched and tempered carbon steel was observed to reduce the long-life fatigue strength by 9 percent [1]. In general, hammer peening should not be considered the equivalent of a carefully controlled shot-peening program in the fabrication of cyclically loaded, welded elements.

In summary, good detail practice does much to improve the fatigue resistance of a welded joint subjected to repeated loads. Sections should be changed gradually to reduce the stress concentration effect. Joints or details with a large variation in stiffness should be avoided. High restraint in localized zones may cause high secondary stresses that are not considered in the design calculations. Some "Do's" and "Don'ts" for design and fabrication of welded joints are listed below.

DO'S

1. Change sections gradually.
2. Machine or otherwise dress weld at critical locations to obtain smoothness.
3. Use butt joints instead of lap joints.
4. Extend cover plates on girders well beyond theoretical cut-offs.
5. Streamline fillet welds.
6. Give preference to a structure that will not collapse after fatigue failure of a detail.
7. Locate joints where fatigue conditions are not severe.
8. Use welding procedures that will eliminate gas pockets, slag inclusions, etc.
9. Avoid undercutting, cracks, spatters and other stress raisers.

DON'TS

1. Don't use joints with large variation in stiffness.
2. Don't introduce high restraint in details.
3. Don't use intermittent welding.
4. Don't permit striking of an arc outside of weld area.
5. Don't overweld.

4C.5 References

[1] Reemsnyder, H. S., "Fatigue Properties of Welded Roller-Quenched and Tempered (RQ) Steels", Paper No. presented at SAE Earthmoving Industry Conference, Peoria, April 1979.

[2] Reemsnyder, H. S., "The Development and Application of Fatigue Data for Structural Steel Weldments", Fatigue Testing of Weldments, ASTM STP 648, 1978, pp. 3-21.

[3] Welter, G., and Choquet, A., "Fatigue Tests of Spot Welds in Cor-Ten and Mild Steel", Welding Journal, Vol. 33, March 1954, pp. 134-s - 140-s.

[4] Pollard, B., "Spot Welding Characteristics of HSLA Steel for Automotive Applications", Welding Journal, Vol. 53, August 1974, pp. 343-s - 350-s.

[5] Kan, Y-R, "Fatigue Resistance of Spot Welds - An Analytical Study", Metals Engineering Quarterly, Vol. 16, November 1976, pp. 26-36.

[6] Overbeeke, J. L., "Fatigue of Spot Welded Lap Joints", Metal Construction, Vol. 8, May 1976, pp. 212-215.

[7] Cappelli, P. G., Castagna, M., and Ferrero, P., "Fatigue Strength of Spot-Welded Joints of HSLA Steels", FIAT, November 1976.

[8] Chandel, R. S., and Garber, S., "Mechanical Aspects of Spot-Welded Joints in Heat-Treated Low-Carbon Mild-Steel Sheets", Metals Technology, January 1977, pp. 37-44.

[9] Overbeeke, J. L., and Draisma, J., "Fatigue Characterization of Heavy-Duty Spot-Welded Lap Joints", Metal Construction and British Welding Journal, Vol. 6, July 1974, pp. 213-219.

[10] Munse, W. H., and Grover, L., "Fatigue of Welded Steel Structures", Welding Research Council, New York, 1964.

[11] Gurney, T. R., "Fatigue of Welded Structures," Cambridge University Press, London, 1968.

[12] Reemsnyder, H. S., "Some Significant Parameters in the Fatigue Properties of Weld Joints", Welding Journal, Vol. 48, May 1969, pp. 213-s - 220-s.

[13] Baren, M. R., and Hurlebaus, R. P., "The Fatigue Properties of a Welded Low Alloy Steel", Welding Journal, Vol. 50, May 1971, pp. 207-s - 212-s.

[14] Selby, K. A., Stallmeyer, J. E., and Munse, W. H., "Influence of Geometry and Residual Stress on Fatigue of Welded Joints", Structural Research Series, No. 297, University of Illinois, June 1965.

[15] Gurney, T. R., "Influence of Residual Stresses on Fatigue Strength of Plates with Fillet Welded Attachments", British Welding Journal, Vol. 7, June 1960, pp. 415-431.

[16] Gurney, T. R., "Influence of Residual Stresses and of Mean Stress on the Fatigue Strength of Specimens with Longitudinal Non-Load-Carrying Fillet Welds," *Journal of Mechanical Engineering Science*, Vol. 12, No. 6, 1970, pp. 381-390.

[17] Gurney, T. R., "Fatigue Tests on Butt and Fillet Welded Joints in Mild and High Tensile Structural Steels", *British Welding Journal*, Vol. 9, November 1962, pp. 614-620.

Addresses of Publishers

Bethelehem Steel Co., Homer Research Lab., Bethlehem, PA 18016

Appendix 4D The Effect of Machining on Residual Stresses and Fatigue Resistance

4D.1 Introduction

The effects of different machining and finishing methods have been measured in terms of a variety of mechanical properties [1-3]. The greatest and most consistent effects have been observed with fatigue properties. The influence of surface and residual stress effects, caused by machining, is greatest for conditions where there is limited cyclic plasticity. Concern for machining on fatigue performance, therefore, is focused at lives beyond the transition fatigue life.

Fig. 4D-1 is a summary of the high cycle fatigue response of Inconel 718 for various machining operations. Inconel 718 is a nickel-base alloy used at high temperatures that has been extensively studied in regard to machining. The response shown in Fig. 4D-1 is attributable to different metal removal methods and ranges of parameters of several of these methods. Note that fatigue strength varies by about a factor of 3 as a result of machining/finishing differences.

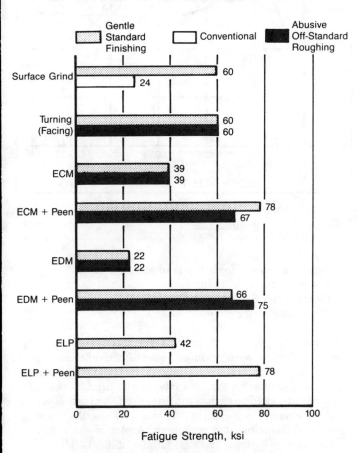

Fig. 4D-1 Summary of high cycle fatigue behavior of Inconel 718, solution treated and aged ($R_c 44$).

Fig. 4D-2 is a similar summary of intermediate life fatigue behavior of the same material. The type and magnitude of variations indicated in Figs. 4D-1 and 4D-2 have been observed on other nickel-base alloys as well as on aluminum, cobalt and titanium base materials and steels. The high cycle fatigue data summarized in Fig. 4D-1 is the result of cantilever bending tests run in fully reversed bending. The intermediate life fatigue data were obtained in the four-point bending, strain controlled test mode. All tests were conducted at room temperature unless otherwise noted.

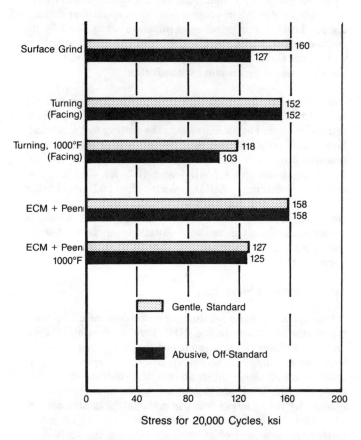

Fig. 4D-2 Summary of low cycle fatigue behavior of Inconel 718, solution treated and aged ($R_c 44$).

In order to explain observed fatigue differences, a detailed metallurgical study was conducted on both machined coupons and failed test specimens. In some cases, reduced fatigue strength values were related to microcracking in the surface that, in turn, could be traced to a particular metal removal operation. A small bias in fatigue behavior could frequently be attributed to surface finish differences. However, the influence of surface finish was studied at some length and the conclusion was reached that surface finish is much less significant in determining fatigue strength than has been traditionally accepted. It also was evident that different degrees of cold working at the surface occasionally influenced fatigue strength. The evidence suggests that the major factor responsible for most of the fatigue strength differences was the surface residual stress. Many correlations between residual stress and fatigue behavior have been identified. A summary of typical data observed in this area are summarized in the following paragraphs.

4D.2 Metal Removal Parameters

Variations in the grinding process include those conditions that produce low level compressive stresses and no detectable surface damage. Such parameters consist of low grinding wheel speeds (2,000 ft./min.), the use of grinding oils as fluids, and a low rate of infeed progression. More abusive or damaging combinations or grinding parameters are typified by increased wheel speeds, increased rates of infeed and the use of water-base fluids as coolants. Still greater surface damage is usually created by grinding without fluid.

Variations in the effects of turning and milling are usually related to the sharpness of the tool. When tool sharpness is lost, as measured by an increase in the cutting tool wearland, larger zones of plastic deformation and higher residual compressive stresses in the machined surface typically result. The data shown in Figs. 4D-1 and 4D-2 for electrical discharge machining (EDM), electro-chemical machining (ECM), chemical machining or chemical milling (CHM) and electropolishing (ELP) are for one or two different sets of parameters. Except in unusual situations, surface integrity variations due to differences in nontraditional machining parameters are small.

4D.3 Residual Stresses

Examples of typical residual stress profiles resulting from machining are shown in Fig. 4D-3. This illustration is based on data obtained from a quenched and tempered steel finished by grinding. These results are quite typical of those associated with the grinding process. Note that the stresses at the outer fibers (at zero depth) are very low in magnitude, essentially zero. Deeper into the material, peak stresses are identified. The peak stress for the abusive grinding operation in this example occured at 0.002 in. beneath the surface.

Referring again to Fig. 4D-1, the peak stresses associated with gentle, conventional and abusive grinding as labeled on that illustration would be −25 ksi, +90 ksi and +105 ksi, respectively. It has been found experimentally that a positive correlation exists between such peak residual stresses, when plotted against the resulting endurance limit or fatigue strength of a material.

4D.4 Observed Correlations

Extensive work has been done on the grinding of high strength steels to identify surface integrity effects. Many alloy modifications of AISI 4340 at several different hardness levels have been evaluated [1]. The range of parameters used was as follows:

Grinding Wheel Abrasive: Aluminum Oxide
Grinding Wheel Speed: 2,000 to 6,000 surface feet per minute

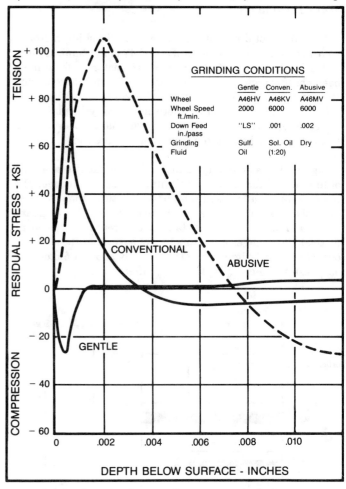

Fig. 4D-3 Residual stress in AISI 4340: surface grinding.

Grinding Wheel Hardness: H to M
Infeed: 0.002 to 0.003 in./pass
Fluid: Water Soluble, Oil, Dry
Wheel Dressing Procedure: Coarse, Fine

Ultimately, fifteen groups of fatigue specimens, ten specimens per group, were produced using fifteen different sets of grinding parameters. Each group of specimens was run in fatigue to determine the 10^7 cycle fatigue limit. Separately, coupons produced by each of the fifteen different sets of grinding parameters were evaluated to determine the residual stress profile associated with each. A plot of the resulting peak residual stress versus fatigue data is shown in Fig. 4D-4. Peak residual stresses varied from about 160 ksi in tension to about 35 ksi in compression. The associated fatigue strengths ranged from 45 ksi to an average of about 120 ksi. This observation supports the contention that at least a 3:1 variation in fatigue strength may be observed in steel as a result of variations in surface finishing parameters. As indicated earlier, this is not an isolated case; similar behavior has been observed on several other types of steels. This example is reported, however, since it summarizes a very complete example of the effects of grinding on the surface integrity of steels.

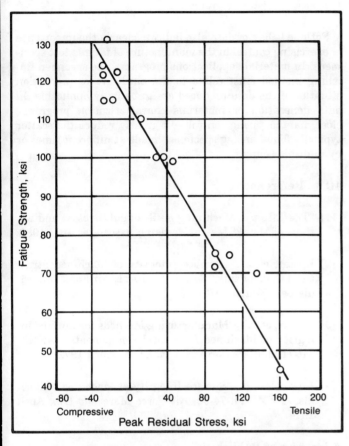

Fig. 4D-4 Fatigue strength vs. peak residual surface stress in ground AISI 4340 (R_c50).

Fig. 4D-5 is a compilation of similar data obtained on 17-4 PH (which is a precipitation hardening stainless steel) [1]. In this work, the 17-4 PH alloy had been solution treated and precipitation hardened to the fully aged or H900 condition. A consistent trend similar to that shown in Fig. 4D-4 can be observed. The single data points related to EDM and CHM are reasonably consistent with the results obtained from ground samples. The data points for milled surfaces, however, appear to deviate somewhat from the fatigue strength trend suggested by grinding. This is not usually the case, but in this situation it may have been related to two simultaneous effects. On one hand, the milling operation probably produced compressive stresses in the surface, which increased in magnitude as the tool became dull (compare Δ1 and Δ2 in Fig. 4D-5). On the other hand, with this alloy, the localized surface heating (related particularly to the dull cutting tool) may have served to significantly over-age the surface layer, thereby reducing its inherent fatigue resistance. Superposition of the two effects resulted in lower fatigue strength values than would be predicted from residual stress data alone.

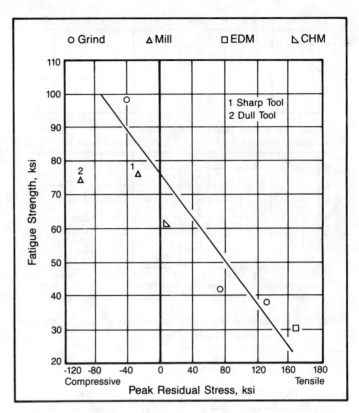

Fig. 4D-5 Fatigue strength vs. peak residual stress in 17-4 PH, H900 condition.

Fatigue strength versus residual stress magnitude summaries for a number of nickel base alloys are shown in Fig. 4D-6. Data for ground, milled, turned, EDM, ECM and CHM surfaces are indicated on these plots [1]. Note that there is a reasonably consistent trend evidenced by all of these materials, irrespective of the metal removal method used to produce the various surfaces being studied.

Fig. 4D-7 summarizes data available on two different titanium alloys [1]. Again, the fatigue behavior correlates well with the measured peak residual stress.

4D.5 Summary

A positive correlation has been found to exist between peak surface residual stress and fatigue behavior. This correlation has been found, not only among variations in a single process such as grinding, but also transcends the methods of metal removal involved, and can be shown to exist within a given material for a wide variety of metal removal processes. Generally, compressive residual stresses tend to increase fatigue strength when compared to that associated with the nominally stress free condition as a baseline. The higher the level of these stresses, the greater the influence that is usually exerted by them.

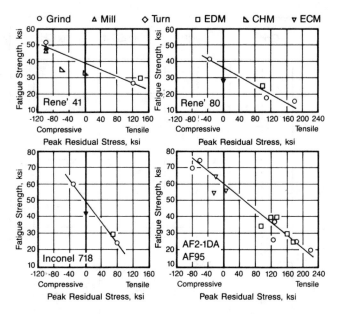

Fig. 4D-6 Surface integrity summaries for a number of nickel-base alloys.

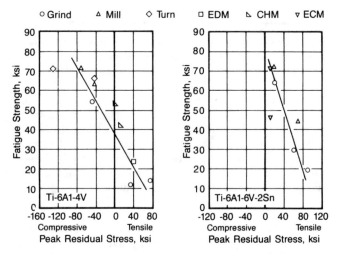

Fig. 4D-7 Fatigue strength vs. peak residual stress in Ti-6Al-4V and Ti-6Al-6V-2Sn.

In seeking to identify the possible relationship between surface residual stresses and fatigue strength, it was found that the peak stress in the surface profile (not necessarily at the surface) rather than the stresses at the surface provided the most consistent correlation.

Fatigue behavior data also indicate clearly the importance of exercising control in the manufacture of fatigue test specimens. In materials applications programs, the specimen finishing method must relate to the hardware or production situation if the data produced are to be truly applicable. In data generation or materials characterization programs, close manufacturing control will greatly reduce the scatter typically found among specimens made at different times or by multiple sources.

4D.6 References

[1] "The Effect of Machining on Residual Stresses and Fatigue of Metal," Metcut Research Associates, Inc., 1982.

[2] Koster, et. al., "Surface Integrity of Machined Structural Components," AFML-TR-70-11, Air Force Materials Lab, March 1970.

[3] Koster, et. al., "Manufacturing Methods for Surface Integrity of Machined Structural Components," AFML-TR-71-258, Air Force MaterialsLab, April 1972.

[4] Koster, et. al., "Surface Integrity of Machined Materials," AFML-TR-74-60, Air Force Materials Lab, April 1974.

Addresses of Publishers

Air Force Materials Lab, Wright Patterson AFB, OH 45433

Metcut Research Associates, 3980 Rosslyn Drive, Cincinnati, OH 45209

Appendix 4E The Effects of Forming on the Fatigue Strength of Metals

4E.1 Introduction

The fatigue life of cold formed machined parts is affected by the change in hardness, the surface roughness and the magnitude of residual stresses and strains generated during the cold forming operation. Consequently, relying on the properties of the virgin material could lead to erroneous conclusions.

In order to predict the fatigue life of a cold-formed part, the stresses and strains generated by the cold forming must be known in all directions. The method of cold forming operation affects the fatigue life of a part only because each method generates different strains (and related hardness), surface roughness, and residual stresses. By knowing how these variables influence life, all types of cold forming (stamping, stretching, straightening, bending, cold rolling, etc.) can be treated together.

4E.2 Observation and Analysis

The first step in the fatigue analysis of a cold-formed part, is the determination of the magnitudes of the principal strains caused by the cold-forming operation. The strain values at various locations can be established by etching small circles on the surface of the blank prior to forming. After forming, the circles change to ellipses. The length of the two axes of the ellipses and the thicknesses of the material at the center of the ellipses define the state of strain at the locations the circles are etched. (Suitable equipment to etch accurate small circles on the surface of blanks is commercially available.) A number of commercially available computer programs also exist to determine the strain due to various cold-forming operations.

The second step in the analysis is the determination of hardness and residual stresses generated by the cold-forming operation. If the cold formed part is to be subjected to large cyclic stresses/strains (short life fatigue) a knowledge of the residual stresses and hardness is not overly important, since the large cyclic strains will remove the residual stresses almost immediately and cyclically soften the material, unless the virgin metal is very brittle. If the cold formed part is to be subjected to long life fatigue conditions, a knowledge of residual stresses is essential, as is knowledge of the surface condition and hardness. Various techniques are availble to determine the hardness and the magnitude of residual stresses (discussed in other chapters). Small prestrains don't appreciably alter the surface finish. However, large prestrains cause significant roughness that significantly reduce fatigue life. None of the available techniques to assess residual stresses are simple; it is a complex but possible task.

Once the strain generated by cold-forming and the corresponding hardness and residual stresses are known, the fatigue life (or at least the life to crack initiation at a given point) can be estimated if the fatigue properties of the virgin material are known for the corresponding stress/strain state. Unfortunately, no systematic work has yet been carried out to establish the fatigue properties of metals when they are pre-strained by cold forming to various values and, at the same time, subjected to various magnitudes of residual stresses.

The lack of information on the basic fatigue properties of the materials in complex stress/strain states and the difficulty in carrying out a crack propagation analysis on a non-homogeneous material subject to a non-homogeneous cyclic stress/strain field usually necessitates the follow-up fatigue testing of the actual part. The results of the analysis, using available data, can only be considered as estimates, helping the designer to reduce the number of trial and error steps needed to finalize a design. Nevertheless, such analyses could significantly reduce the number of expensive full scale fatigue tests on actual parts.

The following paragraphs summarize approaches to estimate forming effects based on the results of a series of studies [1-9].

The short life (lives less than the transition fatigue life) fatigue characteristics of cold formed parts subjected to in-plane stretching will be considered first. A number of authors have investigated this problem [1-9]. The results for uniaxial, stretched cold-forming indicated that large tensile prestrains decrease strain controlled fatigue lives (Fig. 4E-1) and produce no effect on stress (load) controlled fatigue (Fig. 4E-2). No valid test results exist to establish the effect of large compressive pre-strain on strain controlled fatigue behavior, but it is generally considered to be advantageous.

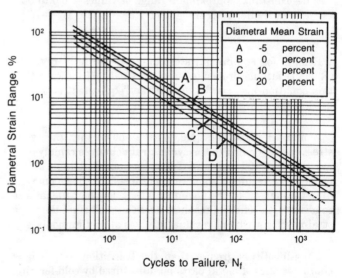

Fig. 4E-1 The effect of pre-strain on low cycle fatigue [4].

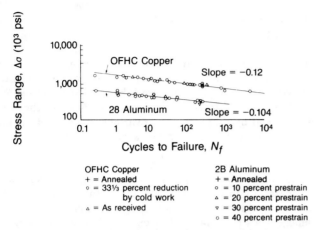

Fig. 4E-2 Relation between stress range at fracture and number of cycles to failure for OFHC copper and 2S aluminum alloy in various conditions of prestrain [6].

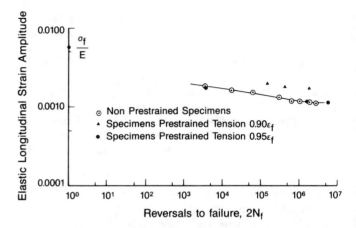

Fig. 4E-3 The effect of prestrain on the fatigue resistance of Maxiform 50 steel [5].

Results of strain controlled fatigue testing on prestrained material can be described approximately by the following equation.

$$N = \left[\frac{e_f - ae_m}{e}\right]^{\frac{1}{n + be_m}}$$

where

N = number of cycles to failure
e_f = true fracture strain
e_m = effective pre-strain
Δe = effective cyclic strain range
n = slope of the strain-life curve when $e_m = 0$ (approximate range of values = 0.5 to 0.7)
a = constant (approximate range of values = 0.9 to 1.0)
b = constant (approximate range of values = 0.0 to 0.2)
e = equivalent strain = $0.71\left[(e_1 - e_2)^2 + (e_2 - e_3)^2 + (e_3 - e_1)^2\right]^{0.5}$

Results for equibiaxial stretching on the fully reversed (R = −1) fatigue properties of some steels [7] indicate that this equation also can be used if the "equivalent strain" is used in the above calculations.

A number of authors [4,6,7] have shown that the cyclic stress corresponding to the cyclic strain changes rapidly during the first 100 cycles of a test conducted at high strain ranges. The results of some relevant tests are shown on Figs. 4E-3 through 4E-5. The net effect is that, for short life fatigue conditions, the effect of residual stress is small since the large strain cycles wash-out residual stresses almost immediately.

Long-life (lives greater than the transition life) fatigue characteristics of metal parts manufactured by cold-forming can be significantly affected. Results indicate that when the residual stresses generated during the stretching process are

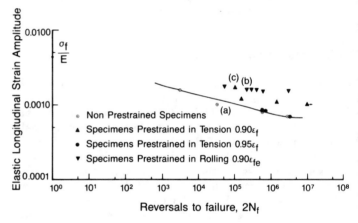

Notes (a) Some Cyclic Creep During Initial Portion of Test
(b) Sequence c
(c) Sequence e, Failed From the "Outside-In"

Fig. 4E-4 The effect of prestrain on the fatigue resistance of SAE 1010 steel [5].

small, pre-strains of up to 90 percent of the true fracture strain will increase fatigue life. The available data are insufficient to quantify this effect mathematically. It is clear, however, that the smaller the cyclic strain range, and the larger the pre-strain, the greater the improvement in fatigue life, as illustrated in Figs. 4E-6 through 4E-8.

A group of investigators attempted to correlate the improvement in fatigue life to the Brinell hardness number of the pre-strained metal [5]. Data for various types of steel, together with a trend line for the data, are given in Fig. 4E-9. The general acceptability of this method is not yet established, but it can be used as a first estimate of the fatigue life until better methods are developed.

As with short life fatigue conditions, the stress range typically changes at the beginning of the cycling. The number of cycles it takes to stabilize the strain amplitude depends on

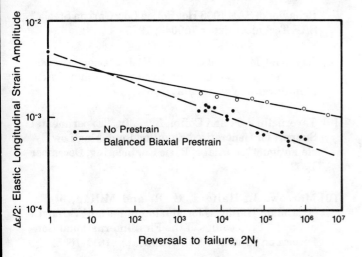

Fig. 4E-5 Elastic strain-life lines for 0 and 40% balanced biaxial prestrain SAE 1008 HRLC steel [8].

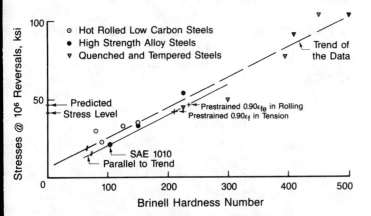

Fig. 4E-6 Endurance stress versus Brinell hardness number for several steels [5].

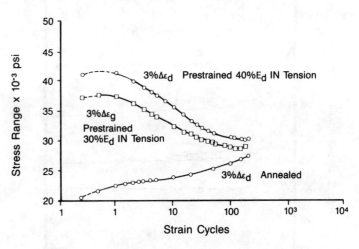

Fig. 4E-7 Commercially pure aluminum [9].

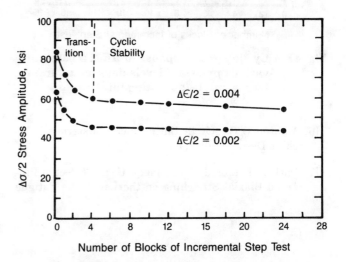

Fig. 4E-8 Cyclic stress response during incremental step test after 40% balanced biaxial pre-strain showing cyclic softening transition region and cyclically stable region, SAE 1008 HRLC steel [8].

the pre-strain and cyclic strain range. That is, for higher pre-strains and higher cyclic strain amplitudes, the change in stress amplitude will be more rapid.

There is little information available on the combined effect of pre-strain generated by cold forming and residual stress in the long life fatigue range. Since the effect of pre-strain is usually advantageous in long life fatigue, a conservative lower limit estimation of fatigue life can be made by neglecting the effect of pre-strain and considering only the effect of residual stress on fatigue. In such analyses the residual stress can be considered as an addition to the mean stress. A somewhat less conservative approach is to estimate the effect of pre-strain using the Brinell hardness number (as described earlier).

4E.3 References

[1] Brown, L. M., "An Investigation into Some Effects of Cold Drawing on Strength and Endurance of Mild Steel," Ship Builders in Scotland, Vol. 71, 1928.

[2] Forrest, P. G., "Fatigue of Metals," Pergamon Press, 1962.

[3] Libertiny, G. Z., "The Problem of Life Prediction of Stamped Structural Parts Subjected to Cyclic Loads," ASME Technical Paper 74-WA/DE-18, ASME, 1974.

[4] Libertiny, G. Z., "Effect of Hydrostatic Pressure on the Short Life Fatigue Property of an Alloy Steel," International Journal of Mechanical Engineers, Vol. 182, Part 36, 1967-68.

[5] Libertiny, G. Z., Topper, T. H. and Leis, B. N., "The Effect of Large Pre-Strains on Fatigue," Experimental Mechanics, February 1977.

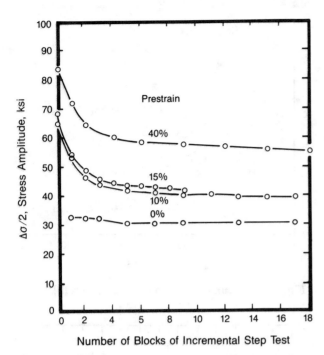

Fig. 4E-9 Cyclic stress response at $\Delta E_t/2 = 0.004$ for various prestrain levels determined from incremental step testing [8].

[6] Manson, S. S., "Cyclic Life of Ductile Materials," Machine Design, July 1960.

[7] Parker, T. E. and Montgomery, G. L., "Effect of Balanced Biaxial Stretching on the Low Cycle Fatigue Behavior of SAE 1008 Hot Rolled Low Carbon Steels," SAE Technical Paper 750048, 1975.

[8] Raymond, M. H. and Coffin, L. F., Jr., "Geometric and Hysteresis Effects in Strain-Cycled Aluminum," Acta Metallurgica, Vol. 11, July 1963.

[9] Tavernelli, J. F. and Coffin, L. F., Jr., "Experimental Support for Generalized Equation Predicting Low Cycle Fatigue," Journal of Basic Engineering, December 1962.

[10] Morrow, J., Halford, G. R. and Millan, J. F., "Optimum Hardness for Maximum Fatigue Strength of Steel." Proceedings of the First International Conference on Fracture, Vol. 3, 1965, pp. 1611-1635.

Addresses of Publishers

ASME, American Society of Mechanical Engineers, 345 E. 47th Street, New York, NY 10017

Ford Motor Co., 20000 Rotunda Drive, Dearborn, MI 48121

Pergamon Press, Maxwell House, Fairview Park, Elmsford, NY 10523

SAE, Society of Automotive Engineers, 400 Commonwealth Drive, Warrendale, PA 15096

Appendix 4F Fatigue Design Considerations for Forgings

4F.1 Bulk Effects

Fatigue data generated from tests on uniform section bars, while useful for comparing material properties, should be used with caution when properties of forgings are being considerd.

Forgings are generally non-isotropic in structure and, as a result, fatigue properties tend to vary with location and direction. This follows from the orientation of any non-metallic inclusions (which, for example, tend to be heavier at the flash line) and also from any residual texture or preferred orientation within the structure.

In addition, it is difficult to obtain uniform work and hence uniform structure in forgings. Even flat "pancake" forgings show a wide range in structure throughout. Since fatigue is a structure sensitive property, fatigue properties tend to be dependent on location and orientation in that forging. Die locked or unworked areas cannot be expected to reflect the properties of the billet since preheating a forging will change the structure and properties, whether subsequently worked or not. Surface regions are frequently influenced by friction effects and die chill, heavily worked areas may experience kinetic heating from working and in extreme cases shear bands can form. Finally, forgings with varied sections may respond differently to heat treatment which may influence fatigue properties.

For fully-machined forgings, the bulk effects mentioned above shall be considered in a fatigue analysis.

4F.2 Surface Effects

In addition to bulk effects, forgings subjected to fatigue environments are significantly influenced by surface condition. Many forgings are used in service with some or all of the as-forged surface remaining. It follows that the fatigue life is reduced by features such as forging laps, folds, cracks and other imperfections which can act as crack initiation sites. Surface metallurgical features can also be important. For example, steels may show decarburization and other alloys exhibit changes in chemistry and phase distribution resulting from the heating and cooling cycles associated with forging and heat treatment.

The full effects of all these changes can only be evaluated by fatigue tests on actual components.

Appendix 4G Plating and Coating

4G.1 Effects on Fatigue Resistance

Typically, metallic components are coated for one of two reasons. The first may be to enhance the esthetic appearance of the component and the second is to provide protection to the component in a hostile environment.

The designer should be aware that, in most cases, the fatigue resistance of load bearing components is reduced. The exception is the polymeric type coatings (enamels, PVC, etc.).

Plating can be detrimental to fatigue strength because the plate material may be in a state of tension. This is true for most nickel plating and chrome plating. The damage is particularly great on parts made of hard steel. Cohen [1] reports a decrease in fatigue limit from 720 MPa without plating, to 380 MPa after plating for steel of hardness 52 HRC.

One alternative is to shot peen the parts before plating. Federal Specification QQ-C-320 [2] requires shotpeening before plating for all steel parts designed for long fatigue life.

Electrochemically applied films tend to be brittle and, when load is applied to the structure or component, the surface film may crack or craze. The net effect is to produce points of stress concentration in the base metal surface. In one case electrolysis nickel reduced the tensile strength of a material from 350 to 270 MPa [3].

Anodic surface treatments (anodizing) produce a ceramic like structure that has poor tensile and torsional properties and result in the same stress concentration problem as electrochemically applied films.

Obviously, the designer should carefully consider the ramifications of utilizing a surface treatment or coating in view of the loss of strength and fatigue life reduction which can be anticipated.

4G.2 References

[1] Cohen, B., "Effect of Shot-Peening prior to Chromium Plating on the Fatigue Strength of High Strength Steel:" *Wright Air Development Center Tech. Note 57-178*, June 1957.

[2] Federal Specification QQ-C-320.

[3] Metals Handbook, Vol. 5, 9th Edition, ASM.

Addresses of Publishers

ASM International, 9639 Kinsman Road, Metals Park, OH 44073

Chapter 5 Service History Determination

5.1 Introduction
   5.1.1 Scope
   5.1.2 Relationship to Other Chapters
   5.1.3 Chapter Plan

5.2 Measurement Objectives
   5.2.1 Setting Measurement Objectives
   5.2.2 Customer Environment Sampling
   5.2.3 Test Site Sampling

5.3 Service History Measurement
   5.3.1 Types of Data
   5.3.2 Measurement Transducers
   5.3.3 Special Problems

5.4 Data Acquisition Techniques
   5.4.1 Signal Conditioning
      5.4.1.1 Shielding Practices
      5.4.1.2 Suppression of Automotive Transients
      5.4.1.3 Preventing Ground Noise
   5.4.2 Signal Transmission
      5.4.2.1 Wires and Cables
         5.4.2.1.1 Magnetic Noise
         5.4.2.1.2 Static Noise
         5.4.2.1.3 Crosstalk Noise
         5.4.2.1.4 Ground-Loop Noise
         5.4.2.1.5 Common Mode Noise
      5.4.2.2 Telemetry
   5.4.3 Digital Techniques
      5.4.3.1 General Principles
      5.4.3.2 Computer Facilities

5.5 Signal Recording, Storage and Display
   5.5.1 Oscilloscopes
   5.5.2 Strip Chart Recorders
   5.5.3 X-Y Recorders
   5.5.4 Tape Recorders
   5.5.5 On-Board Data Collectors

5.6 Data Reduction Techniques
   5.6.1 Filtering
      5.6.1.1 Sample Rate Definition
      5.6.1.2 Aliasing Problems
      5.6.1.3 Prevention of Aliasing
   5.6.2 Spectrum Analysis
      5.6.2.1 Time Compression Analyzers
      5.6.2.2 FFT Analyzers

5.7 Data Evaluation Techniques
   5.7.1 Statistical Summarizations (Spectrum Analysis)
   5.7.2 Amplitude Based Summarization
      5.7.2.1 Peak-Valley Sequencing
      5.7.2.2 Cycle Counting Procedures
         5.7.2.2.1 Level Crossing Counting
         5.7.2.2.2 Peak Counting
         5.7.2.2.3 Simple Range Counting
         5.7.2.2.4 Rainflow Counting and Re-

5.8 References

Appendix 5A Sample Programs for Cycle Counting

Appendix 5B Examples of Cycle Counting

5.1 Introduction

From the viewpoint of product durability, a designer's objective is to specify a product that achieves some defined reliability level when placed in the hands of customers. Ideally this requires the designer to:

(a) know what the customers are doing with the product

(b) know how product test procedures compare with customer usage

(c) have access to analysis tools that can accurately predict the effects of changes to the product, customer usage, or test procedures.

5.1.1 *Scope*

The purpose of this chapter is to discuss the measurement and processing of service loads that arise either due to customer usage or during product testing.

Because much of the chapter is concerned with concepts that are often hardware related, where many of the devices are evolving as rapidly as the computer industry, the reader should be aware that the material presented here is only a point of reference for a specific instance in time. The reader is urged to extend his search to encompass the state of the art of his time.

5.1.2 *Relationship to Other Chapters*

The techniques described in this chapter for the measurement and processing of service loads relate primarily to strain measurement and flaw detection (Chapter 7) and to making fatigue life predictions (Chapter 10). In fact, service history determinations are made, in most cases, to identify potentially critical sites in a structure and to make predictions concerning that structure's resistance to fatigue damage.

This chapter is, in many ways, the experimental equivalent of Chapter 6, that deals with the analytical estimation of vehicle "loads" that will arise during normal service. The experimental approach has been used much longer and is still by far the most common; but the analytical techniques for the estimation of vehicle loads are becoming more and more sophisticated and widespread in their usage.

The output from a vehicle loads evaluation is often the input for a numerical analysis (Chapter 8) to convert "far-field" forces into local estimates of stress and strain or a laboratory evaluation (Chapter 9). If local strains are measured directly, the cycle counting techniques described in this chapter can serve as a direct input to fatigue life prediction, as described in Chapter 10.

5.1.3 *Chapter Plan*

This chapter is subdivided into a series of separate but related topics that pertain to service history determination. Section 5.2 reviews measurement objectives, because the approach that is used to experimentally determine the service environment should evolve directly from a clearly stated objective.

Section 5.3 provides an overview of service history measurement approaches. Sections 5.4 through 5.7 delve into the specifics of service history measurement, ranging from data acquisition and transmission (5.4) to recording and storage (5.5), to data reduction (5.6) and evaluation (5.7). References are cited in Section 5.8.

5.2 Measurement Objectives

5.2.1 *Setting Measurement Objectives*

A wide variety of engineering load analysis situations exist. Lund and Donaldson [1] have provided useful guidelines on the subject of load variable selection. When planning the load data acquisition exercise it is important to determine which signals, or variables, should be monitored. Other constraints, such as time and facilities, may prevent investigation of some of the selected variables, but it will usually be possible to establish a minimum set of variables from a clear statement of the objectives. The objective, for example, may be to duplicate service or proving ground inputs on a vehicle, and thus, it may be necessary to record triaxial force and moment inputs at each front end spindle. If, however, the only available vehicle test rig has only four vertical actuators, the data acquisition problem could be simplified to those directly useful for test input. In such cases it would be advisable to acquire extra channels in order to check the validity of simplifying assumptions.

5.2.2 *Customer Environment Sampling*

Recent advances in recording technology are allowing increasingly detailed samplings of actual customer service usage of a product. Because of the expense of the monitoring equipment and supporting data interpretation functions, the first products to benefit from continuous monitoring tended to be high cost products, such as aircraft. Bridges, off-shore platforms, large cranes, and some types of buildings are more recent additions to the list. The ground vehicle industry has not undertaken extensive direct loads monitoring programs, due to the numbers of components involved. The more common practice in the ground vehicle industry is to perform selected samplings, after initial screening of customer groups, as described by Smith and Stornant [2]. As a result, most customer loads monitoring is done for the purpose of revising test or proving ground design/durability schedules. In this situation the number of signals, and amounts of data to be measured are governed more by time and quantity constraints, than in short term data acquisition exercises such as durability, circuit or rig testing.

5.2.3 *Test Site Sampling*

One of the major sources of component load information in the ground vehicle industry is durability or duty cycle testing of prototypes and production components. The 'events' built into such tests often consist of extreme case samples of expected customer usage. Because the location and test event duration are known, and environments can be controlled, it is easier to collect more signals, in better formats, than would be the case in customer environment sampling. This type of testing usually yields load information that is useful for both design and for laboratory component testing. Most of the requirements for load information monitoring that the typical engineer will encounter involves test site sampling. The following sections elaborate on the methodology and instrumentation used for service history determination.

5.3 Service History Measurement

5.3.1 *Types of Data*

Service history measurement is an important part of the design process both in the initial design stages and the final development stages. Service loads are needed in the initial design stage so the component can be properly sized for the conditions it will encounter. Service loads are also very valuable for accelerated laboratory testing of components or complete structures. Loads that are predicted to do little or no damage can be eliminated, and the remainder applied at an accelerated rate to test the durability of a component or structure. The advantage of simulated service loads is that they can be applied to many components that perform the same function but which vary somewhat in design. For instance, if the service loads for a rear axle housing are known, the stresses for a similar design can often be calculated, or a laboratory fatigue test performed without repeating the data acquisition exercise.

Service loading variables that are calculated from measured strains include force, moment, torque, and acceleration. The load variable which is "measured" in a given case depends on the component being tested and on an analysis of the forces that act upon the component. The analysis of loads to be measured is very important because it prevents wasted effort in measuring loads of no value and insures an understanding of the component's behavior. For instance, a rod with ball joints at both ends may be used to hold the rear axle of a truck upright. Since there are ball joints at both ends, only an axial force can be carried by the rod since the ball joints will not transmit a moment or torque. With this background information, it is a relatively simple matter to strain gage the rod for measuring axial force and to obtain a service history as discussed in Section 5.3.2. A cantilever beam, front engine support is another example. In this case moments could easily be converted to loads at the end of the support. Still other examples are the measurement of load on a driveshaft to determine torque and the measurement of accelerations on a fuel tank to estimate the loads on fuel tank brackets.

After the prototype parts have been designed and constructed, service strain histories are generally needed. The components can be coated with a brittle lacquer that will crack when the underlying material is deformed. Cracks in the lacquer indicate areas of maximum strain and the direction of these strains. Strain gages can then be attached to the component in these critical areas; each one being oriented in the direction of maximum strain (i.e. normal to the crack pattern). The component can then be subjected to its expected customer usage or durability test and a service strain history can be recorded. From this strain history, cumulative damage methods can be used to estimate the component's life. To the extent possible, strain histories should include bolt-up, residual, static, and dynamic strains.

Strain histories are only useful for the particular component design under test. If, for example, the strain gage is located in the corner of the frame of a vehicle and the design is changed because of insufficient fatigue life, the original strain history will no longer be useful. Load histories, in contrast to strain histories, are collected where the component interfaces with the structure. As a result, small changes in design details do not normally affect a service load history. In addition, load histories are most useful during the initial stages of prototype testing and strain histories are most useful during final testing.

The service history, whether strain or load, should be evaluated for peak values, frequencies and critical vibration speeds and by a cycle counting method such as level crossing or rainflow for cumulative damage analysis.

5.3.2 Measurement Transducers

Most transducers use a metallic foil resistive-type strain gage attached to an elastic structure to measure force, torque, moment and acceleration. Examples of some currently available strain gages are shown in Fig. 5-1. There are many manufacturers of off-the-shelf transducers including tension-compression load cells, multi-axis load cells, torque transducers, strain gage accelerometers, clevis pins, bolts or studs. Also available are special purpose transducers such as an automobile torque wheel, aircraft propeller torque sensor and dynamometer trunnion torque sensor. In addition, many firms also make custom transducers for a customer's specific application.

In some instances it may be desirable, or even necessary, to make the load carrying structure a transducer. This can be done by strain gaging an elastic member that will be deformed by the load being measured. A calibration curve must be developed that provides the desired transformation between strain and load. A tension load cell can be made from a piece of round stock with four strain gages attached to it as

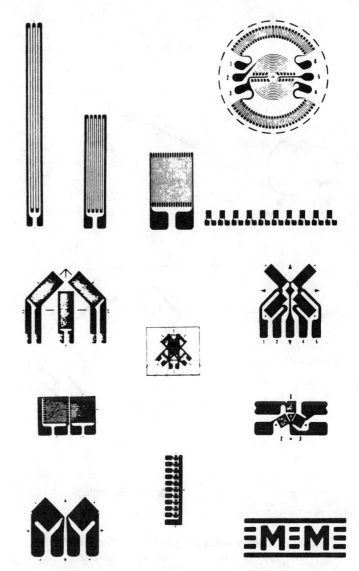

Fig. 5-1 Examples of currently available strain gage designs.

shown in Fig. 5-2. Two gages should be placed in the axial direction and two in the transverse direction. This gage arrangement can be used to provide temperature compensation and to eliminate bending strains (in the plane of the gages).

Transducers can also be made to measure compression, moment and torque. Gage locations and descriptions can be found in Refs. [1] and [3]. The importance of accurate and traceable calibrations cannot be over emphasized. Of particular concern are the effects of transducer overload or other types of detrimental usage which can lead to erroneous readings. The engineer should be cognizant of misuse and check the sensors with regular calibrations.

5.3.3 Special Problems

There are several special problems to be considered when designing transducers. Most of these problems are interre-

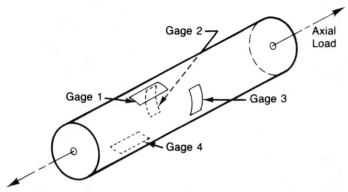

Gages 2 And 3 Will Provide A Measure Of μ Poisson's Ratio: Value Will Be (−) And Will Add To Axial Strain Giving 2 (1 + μ) Shown Below.

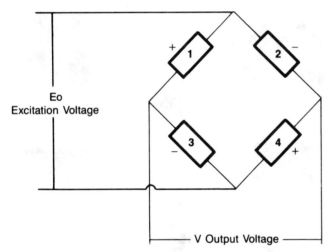

Two Axial Gages And Two Circumferential Gages Providing Temperature Compensation And A Bridge Constant Of 2 (1 + μ).

Fig. 5-2 Example of a simple load cell made from four strain gages and a round bar.

lated and require a compromise to satisfy several conditions. The problems include sensitivity, linearity, hysteresis, cross talk and calibration.

The problem of sensitivity involves making the load cell small enough to produce sufficient strain output for accurate results, while making the load cell large enough to provide for an overload factor beyond the normal range. Most commercial transducers provide a 50 percent overload factor; so if the transducer is overloaded by 50 percent or less the operation of the transducer will not be affected adversely, nor will the zero balance shift. When designing (or selecting) a load cell the size of the overload factor should be based on the accuracy of the estimate of the loads to be measured. To achieve maximum sensitivity without overloading, several things can be done. First, full Wheatstone bridges can be used for all transducers, since they amplify the strain output. Second, a high strength material can be used to allow a higher strain output without overloading. Finally, bending strains can be measured by use of a beam member (because higher strains will result for the same load).

Linearity is defined as the maximum deviation of the calibration curve from a straight line drawn between the no-load and rated-load outputs, expressed as a percentage of the rated output. Linearity checks generally apply to increasing loads only. Linearity values between 0.1 percent and 0.2 percent of full scale are common for commercial transducers. In most cases, linearity can be maintained simply by keeping the stress levels on the transducer material well within the elastic range.

Hysteresis can be described as the maximum difference between load cell output readings for the same load magnitude; one reading obtained by increasing the load from zero and the other by decreasing the load from rated load. Linearity and hysteresis are shown graphically in Fig. 5-3. Hysteresis is usually measured at half-rated output and expressed as a percent of rated output. Values of 0.1 percent to 0.2 percent for hysteresis are common. Hysteresis can be minimized by keeping the maximum stress on the transducer material at a low percentage of its yield strength. It is common practice to limit stresses to 30 percent of yield strength.

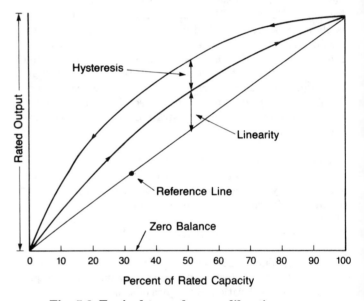

Fig. 5-3 Typical transducer calibration curve.

Cross talk can be a problem with multi-axis transducers if a force or moment in one direction produces a reading in another direction. The transducer can be designed to minimize these effects by using the correct bridge arrangement to measure the desired load. For example, a bridge arrangement for bending moment can cancel any effects of axial load. Two bending moments could be measured on the same element by placing the gages for one bending moment on the neutral axis of the other bending moment. Proper transducer design is the key to eliminating cross talk. To separate forces on two or more axes, the structure must be designed and the gages placed properly to isolate the different loads. Shielding practices, as described in Section 5.4.1.1, are an important consideration in the transducer design.

All transducers should be calibrated to determine the strain versus load response and to check linearity and hysteresis. The transducer is only as accurate as the method of calibration so an appropriate calibration method is required, based on the desired accuracy. More detailed information on the design and manufacturing of transducers is available from strain gage and transducer manufacturers.

5.4 Data Acquisition Techniques

5.4.1 Signal Conditioning

Conversion of the strain gage change of resistance to an electrical signal is termed signal conditioning. (Ref. [4] provides a helpful review of basic strain-gage circuitry.) The strain gage or gages are normally wired in a Wheatstone bridge input circuit, and excitation voltage applied across opposite ends of this bridge, while the output signal is taken from the remaining bridge ends (see Fig. 5-2). This results in an output signal that is at zero voltage level with no strain and a positive or negative signal with corresponding gage resistance changes. Since various gage sizes and resistances require different excitation voltages, the conditioner should include a well regulated and filtered bridge excitation power supply (with a span control to adjust this voltage level). To provide for variations in gage and gage lead wire resistances, the conditioner should also include an input balance circuit that applies limited resistive shunting to achieve a zero bridge output voltage with zero strain input. A calibration circuit should also be included to resistively unbalance the measuring bridge by switching a calibration resistor across a selected input bridge arm. The resultant step change in output voltage will then be the equivalent to the calculated gage microstrain level or to an observed percentage of a known applied force.

It should be noted that both DC and AC carrier excitation supplies may be used in signal conditioning, with the latter requiring additional controls to compensate for input reactance and circuitry to provide a detectable output signal. Buffer amplification to isolate the bridge output and amplify the millivolt signal to a higher level is normally included as the final element in the conditioner, and variable attenuation and low pass filtering may also be added.

Although the conditioner balance circuit may accommodate a single gage input with long extension lead wires, measurement errors can be introduced when ambient temperature changes cause variations in two-lead extensions to the shorter gage runs. The three-lead extension method represents a practical way of running longer extension wires without this problem. By running two wires from one end of the strain gage, with one going to the completion arm and the other returning as the output lead, both bridge arms then have the same changes in lead wire resistance to produce balance and lead wire temperature compensation. The three-lead method also reduces remote calibration errors when the measuring gage is not directly shunt calibrated at its terminals, since the output lead resistance is in series with the calibration resistor and not the measuring gage, as with the

less than three percent, resistance of each extension wire run should be less than one-half percent of the gage resistance with two-lead extensions and less than one percent with three-lead hookups for gages with a gage factor of 2.00, where the change in strain, , that is measured is defined by

$$\Delta\epsilon = \left(\frac{\Delta R}{R}\right)\left(\frac{1}{G_f}\right) \qquad (5\text{-}1)$$

In Eq. 5-1, R represents the unstrained gage resistance, R represents the change in resistance due to the strain excursion, and G_f is the gage factor. Constant current excitation and four-lead extensions (which provide full supply and signal isolation) can be used when high accuracy, single gage data are required.

5.4.1.1 Shielding Practices

Instrumentation and thermocouple cables should be properly shielded, grounded and physically separated from power circuits to minimize noise.

The three most common types of cable noise are normal mode noise, common mode noise and cross talk. Normal mode noise is the noise that develops between signal wires as a result of

(a) electrostatic induction due to distributed capacitance between the signal wires and the surroundings

(b) electromagnetic induction due to magnetic fields linking with the signal wires

(c) junction or thermal potentials due to improper connections

Common mode noise is the noise that can develop between each signal wire and ground as a result of

(a) incorrect location of signal ground

(b) multiple grounding of the shield

Cross talk, in this context, is the noise caused by AC and pulse type signals in adjacent circuits.

Noise in instrumentation and thermocouple cables can be reduced in the following ways:

(a) Electrostatic normal mode noise can best be minimized by shielding.

(b) Electromagnetic normal mode noise can effectively be reduced by twisting signal conductor pairs, and by physically separating the power and instrumentation cables.

(c) Common mode noise caused by ground potential differences can best be minimized by grounding a shield at only one point and by the correct location of that ground point.

(d) Cross talk can best be reduced by using cables with twisted pair conductors and by using individually insulated shields over each pair.

There are a variety of specific techniques that can be used to improve shielding:

(a) Aluminum or copper backed tape, 100% coverage, with a copper drain wire (tinned copper should be used with aluminum) is a more effective electrostatic shield than a braided or wrapped wire armor.

(b) At a minimum, low-level analog signal cables should be made up of twisted pairs with an overall shield.

(c) Digital signal cables and pulse type circuits should be wired with individually twisted and shielded pairs. In multipair cables, the shields should be isolated and an overall shield should also be provided.

(d) Except where specific reasons dictate otherwise, cable shields should be electrically continuous. When two lengths of shielded cables are connected together at a terminal block, a point on the terminal block should be used for connecting the shields.

(e) The shield of each cable should be covered with an insulating tape or jacket in order to prevent stray and multiple grounds to the shield.

With respect to grounding, the following recommendations apply:

(a) All shields should be grounded.

(b) Shields should be grounded at only one point.

(c) Digital signal circuits should not be grounded at any point external to the power supply.

(d) The shields on grounded as well as ungrounded thermocouple circuits should be grounded at the thermocouple well.

(e) Multipair cables used with thermocouples must have individual isolated shields so that each shield may be maintained at the particular couple ground potential.

(f) The low or negative potential side of a signal pair should be grounded at the same point where the shield is grounded. Where a common power supply is used, the low side of each signal and its shield should be grounded at the power supply.

5.4.1.2 *Suppression of Automotive Transients*

The designer of electronic circuits for automotive applications has to insure reliable circuit operation in a severe transient environment. The transients on the automobile power supply range from the severe, high energy, transients generated by the alternator/regulator system to the low-level "noise" generated by the ignition system and various accessories (motors, radios, transceivers, etc.). Transients are also coupled to the input and output terminals of automotive electronics by magnetic and capacitive coupling in the wiring harness, as well as conductive coupling in common conductor circuits (especially the chassis "ground"). Steady state overvoltages may be applied by the circuit power supply due to failure of the voltage regulator or the use of 24 V battery "jump" starts. The circuits must also be designed against the possibility of the battery being connected in reverse polarity. Circuits which drive inductive loads must be protected against the transients resulting from the energy stored in the field of the inductor. These transients can be defined from the load inductance and load current. Table 5-1 summarizes the automotive power supply transients as documented by the Society of Automotive Engineers (SAE).

5.4.1.3 *Preventing Ground Noise [7] [5]*

The "ground loop" is one of the most common sources of error in vibration measurement systems. Spurious or unwanted signals result from "ground" currents flowing in the common or shield paths between the transducer and signal conditioner and interfere with the true vibration signal. These currents are caused by connections made at multiple grounding points of differing potential and create voltage drops in the common or shield paths. Such a voltage appears as an error, E, and is added to the vibration signal seen by the conditioner (Fig. 5-4).

Mutual linkages to surrounding electrical equipment may also produce magnetically-induced error signals. Fig. 5-4 illustrates this common problem, often encountered in using standard high impedance piezoelectric accelerometers. In these devices, the outer case is often tied to the signal common or shield. When the transducer is mounted to a grounded structure, such as a motor bearing housing, gear box, etc., the potential for two grounds is created. In most cases the common side of the signal conditioner would also be grounded, thereby completing the "loop."

One solution to this problem is to break the ground connection at the signal conditioner (Fig. 5-5). In practice, this is often not possible and the connection must be broken at the transducer by using an "isolated" or ground-insulated design (Fig. 5-6). There are three methods of isolation in general use. For the reasons listed below, only method I and III are recommended for general use:

Table 5-1 Typical Automotive Power Supply Transient Summary.

Length of Transient	Cause	Energy Capability / Voltage Amplitude	Possible Frequency of Application
Steady State	Failed Voltage Regulator	∞ / +18V	Infrequent
5 Minutes	Booster Starts with 24V Battery	∞ / ±24V	Infrequent
4.5 ms to 0.1s	Load Dump — i.e. Disconnection of Battery While at High Charging Rates	≥10J / ≤125V	Infrequent
≤0.32s	Inductive Load Switching Transient	<1 J / −300V to +80V	Often
≤0.20s	Alternator Field Decay	<1 J / −100V to −40V	Each Turn-Off
90 ms	Ignition Pulse, Battery Disconnected	<0.5 J / ≤75V	≤500 Hz Several Times in Vehicle Life
1 ms	Mutual Coupling in Harness*	<1 J / ≤200V	Often
15 μs	Ignition Pulse, Normal	<0.001 J / 3V	≤500 Hz Continuous
	Accessory Noise	≤1.5V	50 Hz to 10kHz
	Transceiver Feedback	≈20mV	R.F.

*These transients may be present on any wire in the vehicle.

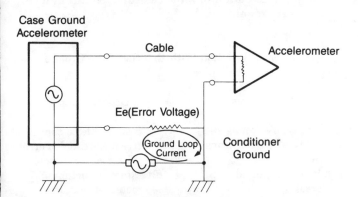

Fig. 5-4 Typical installation showing two ground points and potential "ground loop" path.

I. Separate Isolated Mounting Stud (See Fig. 5-7)

Advantages

(1) Eliminates ground loops.
(2) Adapts to standard case grounded units.
(3) May allow axis orientation (Fig. 5-7b).

Disadvantages

(1) Easily lost - unless bonded in place.
(2) Easily damaged or shorted out.
(3) Adds some mass.
(4) Reduces mounted resonance of transducers.
(5) Adds small cost to transucer.
(6) May have temperature limit.

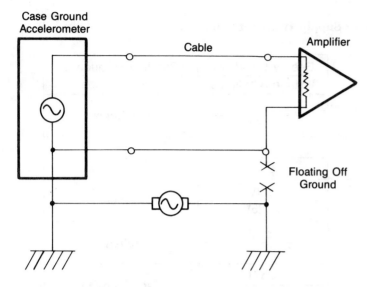

Fig. 5-5 Breaking the ground connection at the signal conditioning amplifier will effectively break the "ground loop" circuit but is often not convenient.

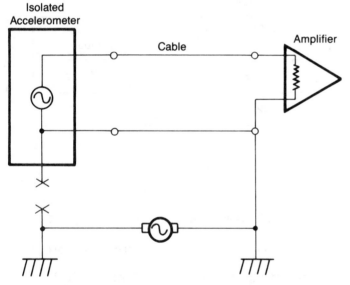

Fig. 5-6 Most effective method of breaking "ground loop" is to isolate the accelerometer off ground.

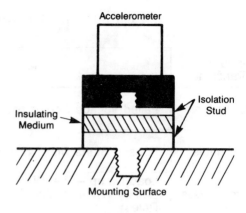

(a) Stud Mounted

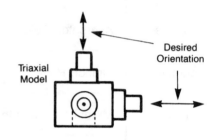

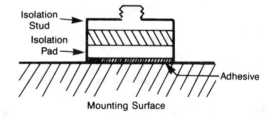

(b) Adhesive Mounted

Fig. 5-7 Isolation of an accelerometer through mounting.

II. Internal Isolation With External Case Ground (See Fig. 5-8)

Advantages

(1) Not easily damaged.
(2) Integral part of transducer.
(3) Reduces ground loop error.

Disadvantages

(1) Susceptible to capacitive noise coupled error (Fig. 5-8b). See note.
(2) Minor reduction of mounted resonance.

106

NOTE: This is a serious problem with internally isolated units unless internal electrostatic shielding is provided around the high impedance crystal or amplifier circuits. In effect, the ground loop is still present but with a capacitive link. (See Fig. 5-8b.)

III. External Permanent Isolation (See Fig. 5-9)

Advantages	Disadvantages
(1) Eliminates ground loop error.	(1) Minor reduction in mounted resonance.
(2) Not easily damaged.	
(3) Integral part of transducer.	

Figs. 5-7a and b show a separate isolated stud. Care must be taken to avoid damaging exposed insulation. In some cases, a permanently installed stud can also provide a convenient means for orienting the sensitive axis of a triaxial accelerometer by cementing the stud in the desired orientation. The stud may also be cemented to the accelerometer to permit use in other locations.

Fig. 5-8a illustrates a typical internal isolation technique for a piezoelectric accelerometer in which the element and connector are insulated from the case.

Fig. 5-8b illustrates the capacitive error coupling from the case that may be driven by a high-level error signal to a high impedance input circuit. This error can also occur with built-in impedance-matching amplifiers.

Fig. 5-9a shows that permanent ground isolation provides an electrostatic shield and effectively prevents "ground loops." In addition, protected construction avoids damage to thin film insulation from external wrench tightening.

Fig. 5-9b illustrates the electrostatic shielding effect of permanent construction.

Fig. 5-10 illustrates the potential for multiple "loops" when outer cases of co-axial connectors are grounded.

Outer casings of connectors that are part of the shield or common path should be wrapped in insulating material to avoid mutual contact or contact with ground points. This is a commonly overlooked source of ground signal errors (as shown in Fig. 5-10).

5.4.2 Signal Transmission

Signal transmission involves the sending of the conditioned signal to some form of display, recording, or processing device in a manner that preserves the required signal information without distortion or alteration of the signal, and may involve one or more of the following methods.

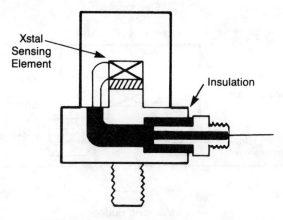

(a) Accelerometer Insulated from Case

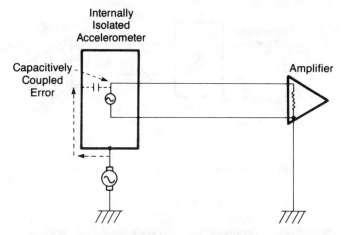

(b) Illustration of Capacitive Error Coupling That Can Develop from the Case

Fig. 5-8 Internal isolation with an external case ground.

5.4.2.1 Wires and Cables

Wires and cables range from short, two-wire runs from a signal conditioner to a recorder, to long distance, multiconductor cable runs from a test cell to a computer location. Wired signal transmission is a fundamental transmission form that is subject to several primary forms of noise; specifically magnetic, static, crosstalk, ground-loop, and common mode noise. Techniques for controlling each of these types of noise are reviewed in the following paragraphs.

5.4.2.1.1 Magnetic Noise

Motors, transformers, and power lines create magnetic fields that inductively couple to wires located near them and generate a 60 Hz noise voltage proportional to the electrical current involved and inversely proportional to the distance

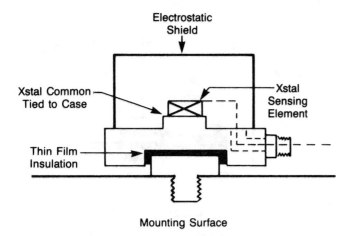

(a) Accelerometer with Permanent Ground Isolation

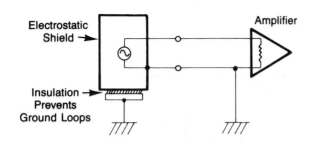

(b) Illustration of Electrostatic Shielding Effects

Fig. 5-9 External permanent isolation.

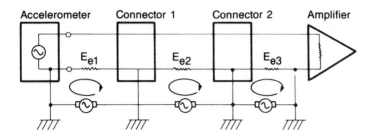

Fig. 5-10 Multiple ground loops may develop when outer cases of coaxial connectors are grounded.

from them. The opening or closing of circuits containing inductances such as solenoids or relay coils can produce very high rates of change in the electro-magnetic fields that result in high amplitude interference voltage spikes appearing on the signal. Twisting of the two signal wires will minimize this problem since it effectively breaks the total signal feed loop into many single loops that have adjacent and opposing, induced voltages that cancel one another.

5.4.2.1.2 Static Noise

Power lines and voltage sources establish electrostatic fields that capacitively couple to develop an unwanted AC of capacitance from the source to the signal wire. Metal shielding wrapped around the signal wires and brought to the ground serves to intercept this noise by capacitively coupling it to ground and leaving the signal wires with a small value of leakage capacitance to the source and a correspondingly reduced static noise signal.

5.4.2.1.3 Crosstalk Noise

An interfering signal on a data channel that originates from an adjacent channel is termed crosstalk. This may be a problem when different AC signals are carried on separate wires that are wrapped together creating significant, distributed capacitance between the wires. Physical isolation or shielding of individual signal pairs is the accepted way to control this noise.

5.4.2.1.4 Ground-Loop Noise

Two different grounds existing in the same signal circuit and at different ground potentials result in current flow that creates ground-loop noise. Use of one ground point at the source and elimination of all other "unintended" grounds will eliminate this problem.

5.4.2.1.5 Common Mode Noise

An unwanted signal that appears at both terminals simultaneously with the same phase and amplitude is termed common mode noise. Although it usually originates from electromagnetic or electrostatic fields, common mode noise is separately identified since it is minimized by using amplifiers specially designed to handle this problem. Floating input differential amplifiers are available with high common mode rejection characteristics to accept the data signal while rejecting common mode noise appearing on symmetrical signal lines.

5.4.2.2 Telemetry

Although defined as the measurement and transmission of data from remote sources (by many different means), telemetry usually employs radio frequency (RF) signal transmission using either a radio link or a conducting link. A radio link is valuable in vehicle field test work since small, lightweight signal conditioning and RF transmitting equipment requiring minimal power can be conveniently mounted on a test vehicle. More vibration sensitive or larger receiving and recording equipment with higher power requirements can be located in a chase vehicle or at a stationary receiving station. Many system configurations are available using frequency or time division multiplexing to permit several channels of data to be carried on one RF carrier. Frequency-division multiplexing is most commonly used for vehicle test work since it offers good frequency response and flexible operation with standardized Inter Range Instrumentation Group (IRIG) subcarrier channels designated to give compatibility between different manufacturer's equipment. The normally used FM/FM system employs subcarrier channels that include voltage controlled oscillators (VCO's) that are frequency modulated by the conditioned signals and mixed to collectively frequency modulate the higher carrier frequency. RF transmission of the carrier frequency to the receiver antenna conveys the composite signal into the receiver where it is first demodulated to remove the carrier and is then fed to a group of band pass filters that recover each subcarrier signal from the mixed signal. Individual filter outputs are then demodulated to recover the original conditioned signals that first modulated the subcarrier VCO's. Since five watts RF power is required to cover a one mile radius in the 200 mhz VHF telemetry band, a broadcasting license is required (in some areas) to operate this equipment. Unlicensed equipment with power under 100 mW is available, but low power limits the coverage range and, normally, the data channels to only special applications. Pulse code modulation (PCM) lightmeter systems (that are a recent development), will probably see increased usage in field data acquisition applications. Fig. 5-11 shows the elements of a PCM system including the multiplexing, modulation, and magnetic tape recording components.

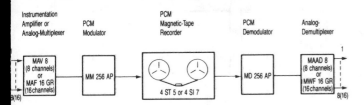

Fig. 5-11 Data acquisition with a pulse code modulation system.

5.4.3 Digital Techniques

5.4.3.1 General Principles

Analog to digital (A/D) converters are normally involved in signal transmission and when electronic processing of the data signals is required. Although analog transmission methods may be used, it is difficult to eliminate the inherent signal degradation associated with analog equipment. Consequently, when significant processing and exchange of data is required, it is preferable to convert the analog signal to an equivalent time sequence of digital words, that then represent the sequential point by point makeup of the analog signal. In this digital form the information may be repeatedly processed and transmitted without degradation since accuracy is a function of the number of bits used in computation.

Commercial A/D converters are usually part of an integrated system that includes a multiplexer to sequentially connect each of a group of analog signals to the A/D converter which generates a digital output representing a point amplitude value. Such systems usually include buffer amplifiers to provide programmable gain isolation, as well as sample and hold circuits to permit faster multiplexing rates of the analog input signals and input anti-aliasing filters to prevent alias. Programming and logic control completes the system to determine multiplexer sequence and rate, sample and hold timing, and A/D conversion.

Actual A/D conversion is normally performed using a successive approximation technique that employs a comparator referenced to a program controlled digital register and digital to analog decoder. The program begins conversion by clearing the input register and setting a zero reference voltage to establish the sign of the signal voltage. It then enters a value of one in the most significant bit of the D/A decoder as the comparator determines which voltage is larger. If the D/A converter voltage is larger than the input voltage, it removes the one—or leaves it in, if smaller—and moves to the next bit position. This programmed procedure continues until the least significant bit is resolved and the digital equivalent of the analog input signal is then available in the storage register. This process, including multiplex switching and timing procedures, may be performed in 10 microseconds with conventional 12-bit converters operating at throughput rates of 100,000 samples per second.

The digital time resolution of an analog signal is basically proportional to the A/D conversion rate. Since data handling economics favor minimal conversion rates, the determination of proper A/D conversion rates is important. A per channel conversion rate of 5 times the highest significant frequency component in the analog data signal is desirable to define nonrecurrent waveforms, although higher accuracy measurement of a transient peak amplitude such as that obtained from a field test vehicle may require twice this rate. Low pass filters with cutoff frequencies just above the highest frequency of interest may be used to treat the signal going to the A/D converter. These anti-aliasing filters can also be used to reduce errors from higher frequency noise on the data signal.

5.4.3.2 Computer Facilities

Digital data computers have allowed increased sophistication in all areas of service load data acquisition, storage, processing, and display. Fig. 5-12 shows the elements of a typical digital computer data acquisition facility.

The first key ingredient in any such computer facility is the ability to digitize signals at the appropriate rates. The second necessity is an adequate capability to either store the values for later reference or process them into other forms without overloading the computer. Most other hardware criteria are of secondary (but still considerable) importance.

Present trends are to digitize most load signals and to apply the necessary analysis methods to disk files of the digitized voltage values. Storage on disk facilitates editing, (that may include scaling, point removal, merging, superimposing, etc.), as well as display and analysis of the signals. Unfortunately, the extensive use of these capabilities has resulted in a greatly increased demand for random access computer storage. The advent of laser disk technology may aleviate the problem, but at the present time, the data acquisition engineer must determine optimal sample rates for the signal dig-

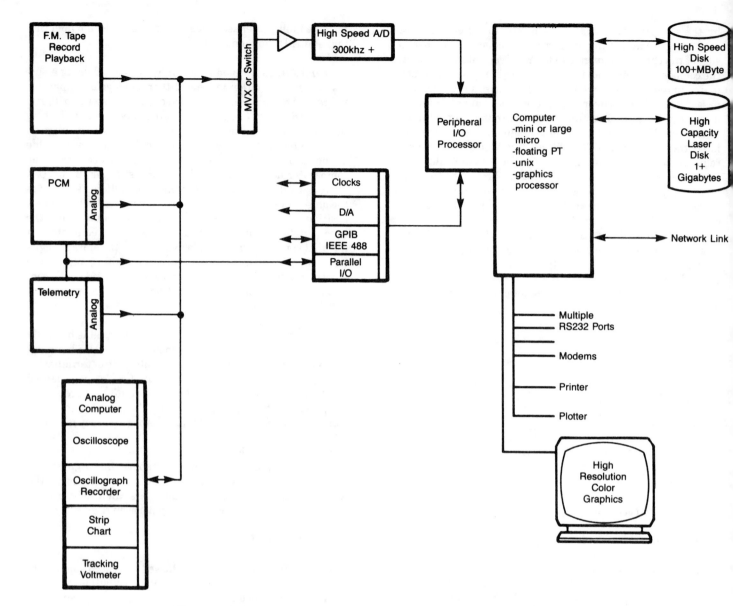

Fig. 5-12 Data acquisition computer facility—state of the art (1986).

itization process. When the data storage bottleneck has been overcome, the next difficulty is often computer processing speed. The computer time needed to condense sequences or create statistical summaries becomes significant when the number of voltage readings exceeds 100,000. (At a 1,000 Hz sampling rate this represents only 100 seconds of data on a signal channel.) Processing algorithms generally require much more time than the initial digitization routines. This almost always forces the data acquisition engineer to consider the fastest computer available.

The final phase of this data acquisition process is display. Present technology, although expensive for higher speed systems, allows graphical display of the signals on a terminal screen. Such a facility is vital for a quick overview of the complete histories or detailed analyses of specific events.

The relative importance of each of the major features; data acquisition, storage, processing and display depend on the specific application. Because of the variety of tradeoffs between computer features and price, and the inherent dynamic nature of the computer industry, a reader interested in constructing a data acquisition system would be well advised to seek the advice of experts in the data acquisition field, or at least the services of computer knowledgeable personnel.

5.5 Signal Recording, Storage and Display

In addition to digital computer systems there are a number of different recording and storage devices commonly used for field data acquisition. The characteristics and limitations of oscilloscopes, strip chart recorders, X-Y recorders, oscillographs, tape recorders, histogram recorders, and untended data collectors are described in the following paragraphs.

5.5.1 Oscilloscopes

Oscilloscopes can be used as indicators of both static and dynamic strain gage signals but are normally used with associated excitation and measurement instruments to provide better resolution than permitted by direct physical measurement of the cathode ray tube (CRT) waveform. The oscilloscope can accurately display higher frequency strain gage signal information than other instruments, and may include electronic storage capability and camera units to record such information. This also makes the scope helpful when judging signal quality, frequency content, and noise.

5.5.2 Strip Chart Recorders

These devices are usually limited to fixed location laboratory use, and can provide large chart widths for full scale resolution of DC to 1 Hz input signals with chart speeds ranging from a few centimeters per hour to over ten centimeters per second. Since they use a servo system to drive the recorder pen, high amplifier gain can provide input sensitivities of several millivolts full scale with high accuracy and signal noise immunity. Application of strain gage signals is provided by plug-in or separate excitation and balance units that feed directly into the millivolt recorder inputs. Although multipen units that overlap to cover the complete chart width are available, they can be troublesome to use when precise time correlations of such data are required.

Light beam oscillographs are used to record a wide range of frequency signals ranging from DC to over 20 kHz. Direct write oscillographs are usually limited to an upper frequency of about 100 Hz. Oscillographs are useful for prompt visual comparison and measurement of a signal relative to its amplitude and phase. This is particularly useful when conducting shorter term vehicle or component tests, since more detailed graphical analysis of such recorded data is quite time consuming. A direct write system should include low mass ink pens and heated tip stylus assemblies that write on prescaled paper—specially heat sensitized in the case of thermal writing techniques. A variety of paper speeds are available ranging up to roughly 20 cm/sec and a timing generator and marking pen may be included to aid in time/frequency measurement. Plug-in signal conditioners are often used to accommodate many different signal sources including strain gages, so that up to sixteen channels or more of data may be recorded with a reasonably portable package.

Light beam oscillographs employ galvanometers that provide a wide range of sensitivities and natural frequencies to adapt the channels to specific response needs. The galvanometers have mirrors attached to their moving coil suspensions, that are located in a magnet block assembly. This assembly positions the mirrors to reflect the light sources through an optical system that focuses and optically amplifies the signal deflection of the galvanometer movements. Using photographic paper to receive these light images, frequencies as high as 20 kHz may be recorded with chart speeds up to 500 cm/sec. Recording paper and photodevelopment methods should be selected according to writing and paper speed requirements. A near immediate image display can be produced at lower paper speeds on direct print paper. Trace identifiers make it possible to overlap one channel on another, but good signal resolution and trace visibility may restrict the number of channels to one per 30 mm of paper width when transient signals having higher frequency components are involved. Separate signal input systems are normally used with optical oscillographs since optical system space requirements, photographic paper handling needs, timing and control electronics and galvanometer mounting block provisions make an isolated package more desirable.

5.5.3 X-Y Recorders

X-Y recorders are similar to strip chart recorders in recording characteristics since they both employ servo pen drive systems. However, X-Y recorders are useful in certain strain gage related work to directly display one measured effect on another such as applied force measured with a strain gage load cell vs. the resultant response strain of a tested structural element. Since individual sheets of graph paper are used to record such runs, various pen colors may be used to generate a group of related plots on the same paper to allow for a direct visual comparison. Time bases may also be included to provide output level vs. time ranging from seconds to minutes for one full scale sweep.

5.5.4 Tape Recorders

Data acquisition tape recorders are used to record electrical data signals on magnetic tape for immediate or later playback to oscillographs, analyzers, or computers. They are most useful when extensive reduction of long-term field test data are required. They permit storage of the original data signals on tape, while allowing repeated playback of these signals at recorded speed or other speeds that provide a time base expansion or compression. Instrumented tape recorders range from basic cassette units to automated multichannel instruments with high operating flexibility and 2 MHz bandwidth capabilities. Although several recording methods, ranging from simple direct to multiplexed digital may be used, the IRIG intermediate band FM operating mode is commonly used with strain gage signals since it can record DC to 625 Hz data at a tape speed of 1-7/8 inches/second with a good signal to noise ratio. This gives 6.4 hours of recording time on a 3,600 foot reel of tape. This ability to record signals well beyond the required frequency range of interest, is a major advantage, but because of the necessity to subsequently digitize the signals, much of the workload lies in the digitization procedures that will follow. Another advantage for the data analyst is that the digitization process can be repeated if, for example, it was found that the sample rate was not set high enough. This is not the case of techniques such as PCM, which have a predetermined sample rate. The advantage of PCM systems lie in the fact that the data have already been digitized and the analysing computers hardware and tasks are thus considerably simplified. The decision to use analog (FM) or digital (PCM) type techniques is application dependent and potential users should seek expert advice if such a selection must be made. It should be noted that recorders of any manufacturer, built to IRIG specifications for electronics and mechanical configuration, can handle tapes made on any other IRIG recorder with assured performance levels. This is important to users with separate acquisition and reduction recorders and those exchanging tapes with other facilities.

Since analog data are normally converted to digital form somewhere in the data reduction process, digital recorders that include the capability for both A/D input conversion and direct recording of digitized information are very useful for field test data acquisition. Digital to analog conversion output can then aid in verifying proper A/D conversion, but care must be taken to assure that digitizing rates are adequate to resolve the data and that noise (error) components are not part of the digital representation.

As a special form of recorder, the write-only, incremental, digital cassette tape recorder is useful for long duration, low frequency data logging applications when continuous data is not required. By gating the input A/D converter to operate only when a significant event occurs, such as an amplitude exceeding a given threshold or by using an interval clock pulse, the A/D conversion is performed at a rate determined by the input clock and fed to a first in, first out (FIFO) memory buffer. The actual recording is then controlled by the record clock that uses a stepping motor to incrementally advance the tape while serially recording from buffer to tape. By operating the recorder under input data control, only the most meaningful data are recorded and power requirements are minimized to permit long total operating time. A representative recorder of this type records at 100 bits per second with a density of 615 bits per inch to yield a 2.2×10^6 bit capacity with standard cassettes. Caution: good procedural techniques must be developed to minimize transient noise on recorded data.

5.5.5 *On-Board Data Collectors*

Data recorders are available that can be placed on board a vehicle and left unattended for long periods of time [6]. Such systems can be very useful in situations where the conditions or duration of the field test do not allow continuous operator involvement. However, successful data collection using untended data recorders requires extreme care during test preparation and initial installation. In particular, specific rules should be followed in shielding and grounding, strain gage selection, power supply connections, and mounting of the recorder.

Some of the major problems in using such recorders can be attributed to the following:

(a) Poor emf environment.

(b) Poor mechanical environment.

(c) Inadequate sampling rates.

(d) Inadequate digital storage space on the device.

(e) Inappropriate software.

The problems are basically caused by the on-board environment not being suitable for a data acquisition computer exercise and because of size or power restrictions, inadequate computer facilities for the job required. It is important that the user ensures that the acquisition requirements are matched to the features of the particular on-board data monitoring unit.

The decision to use an on-board signal summarizing device or a more conventional tape recording system may also depend on the expense of the data acquisition exercise itself. When it is easy to repeat the run several times, an onboard device can be applied more reliably. If the test run is a one time affair it is probably better to record as much information as possible in a "raw" form. For example, if it was found that the sampling or voltage digitization rate was too slow or too fast, it is simpler to replay a tape recorded signal than to perform a new test.

As mentioned before, the advances of the computer and data storage electronics industry are allowing more and more features, previously available only on laboratory computers, to be down loaded onto smaller on board devices. It is foreseeable that the engineers of the future will not need to be concerned as greatly with the above tradeoffs. This trend is of course counter balanced by the desire to record increased amounts of information.

5.6 Data Reduction Techniques

Once the necessary data have been acquired and stored, it is often of interest to reduce the data into a more usable form before completing a detailed evaluation. This section addresses filtering methods, digital data reduction issues, spectrum analysis tools and computer techniques for data reduction.

5.6.1 *Filtering*

Filtering represents a basic data reduction form since it may be used to isolate primary interest signals from noise or unimportant masking frequencies. There are four types of commonly used filters identified as low pass, high pass, band pass, and band reject. The low pass filter rejects signals above a "cutoff" frequency and passes frequencies from DC to cutoff, while the high pass performs in an opposite manner to reject or attenuate the frequencies below a cutoff (cutoff being defined as the frequency at which the filter output amplitude drops 3 dB relative to its passband level, while continuing to increase attenuation with greater frequency changes). It is useful to recall that a dB, or decibel, is a term used to express the relative magnitude of two acoustic or electrical signals. Since a dB is calculated based on the common logarithm of the ratio of the two levels, a 3 dB drop would correspond to about a 50% drop in filter output amplitude. Band pass filters employ two cutoff frequencies to define high and low frequencies between which signal frequencies can pass without attenuation. Band reject filters conversely attenuate or reject the frequencies between the high and low cutoff points and pass frequencies both above and below this region.

Rolloff rate expressed in dB/octave is an important measure of filter effectiveness and relates to the number of poles or reactive elements in the filter. Each pole additively provides a 6 dB/octave rolloff rate so that a "softer" 2-pole filter has a 12 dB/octave rolloff and a sharper cutoff, 6-pole filter has a 36 dB/octave rolloff. Since an octave represents a 2:1 ratio of frequencies, this means that a 4-pole, low pass filter with a 20 Hz cutoff frequency will attenuate a 40 Hz input signal 24 dB or to 6% of its original value in accordance with the 24 dB/octave rolloff rate of this filter.

It should be noted that phase shift caused by filtering must be considered when mechanical phase relationships are to be analyzed from various strain gage or transducer locations. Normally a filter will add a 45° phase shift per pole at the cutoff frequency, but phase characteristics at other frequencies depend on the configuration of the filter. For example, a Butterworth design will give a nearly flat amplitude response, while a Bessel design will provide a nearly linear phase response. However, since the signals to be compared for phase relationships will normally be of the same frequency, valid phase comparisons can be made by applying identical filtering to such signals, while allowing for phase shift effects when analyzing composite waveforms.

Where the frequency makeup of a signal is required, the signal may be fed to parallel filters adjusted to individual pass bands to provide a breakdown of a composite input signal. Its makeup may then be described in terms of "approximate" frequencies and their respective amplitudes. The values should be considered approximate since filter bandwidth adjustment and rolloff regions overlapping adjacent filter passbands may introduce resolution errors. Care must also be taken when paralleling filters to a common source, since impedance mismatches may introduce errors. This is true even when using some buffered active filters, although it is a primary concern with passive filters.

The following paragraphs present a discussion of some fundamental concepts concerning filtering with respect to the acquisition and analysis of random field data for use in the assessment of fatigue damage, and in simulation testing and component testing. The relationship between the rate at which a field time history is sampled for digital processing, and the accuracy with which the peak values of data are measured is addressed. The phenomena of aliasing and techniques for minimizing its effects are also described.

5.6.1.1 Sample Rate Definition

The required sample rate versus data frequency for acceptable accuracy has been a source of lengthy discussions among data analysts. However, a great deal of the discussion stems from the fact that there are several uses for the results of analyzed data. For example, those interested in the frequency properties have developed sampling criteria which provide them with adequate results. Those who are interested in the material's fatigue behavior have developed their sampling criteria. Most of the sampling criteria for these cases are quite different. However, there are some common issues with respect to the phenomena of aliasing, which must be dealt with regardless of the end use of the data.

To illustrate these differences and common phenomena, consider first, a simple sinusoidal function. This function represents the simplest form of a periodic phenomenon, one that repeats itself over time, is continuous, and is completely predictable. A graphical representation of a continuous sine function is shown in Fig. 5-13.

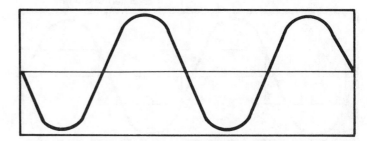

Fig. 5-13 Continuous sine wave.

This simple function can be represented in the discrete digital domain by a set of samples or equally spaced "snapshots" in time. This may be observed in Fig. 5-14. Obviously, the more samples per cycle of the sine which are recorded, the more representative the sample set is of the original continuous wave.

What is the minimum number of samples per cycle required to have any information about the sine wave? Information scientists Nyquist and Shannon investigated this question and found that there must be at least two equally spaced samples per cycle to determine the periodicity of a periodic waveform. This may be observed graphically in Fig. 5-15.

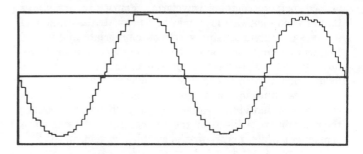

Fig. 5-14 Sampled sine wave.

A comparison of Figs. 5-15a and b shows that, although periodicity is determined with this theoretical minimum sample rate, there is a possibility that the actual amplitude of the sinusoid may not be accurately determined. This is due to the fact that the sample timing may not exactly coincide

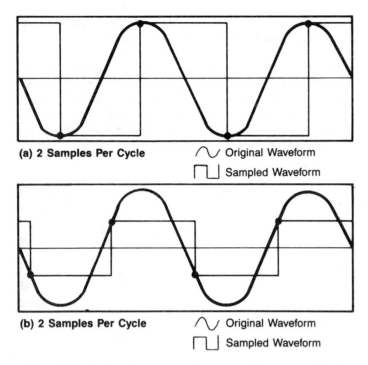

Fig. 5-15 At least two samples per cycle are required to determine the periodicity of a periodic waveform.

with the continuous waveform peaks. It is this fundamental problem of amplitude determination which varies in importance between frequency and fatigue analysis and also affects their sampling criteria.

A frequency equal to one half the sample rate is commonly known as the Nyquist frequency. (Nyquist Frequency = F_N = 1/2* Sample Rate)

The above discussion of the sampling criterion may be extended to more complex waveforms. Fourier theory states that any complex waveform may be represented as the sum of a series of individual sine waves of appropriate and varied frequency, amplitude, and phase. Complex random waveforms have Fourier attributes as well as amplitude and phase modulation applied to the sinusoid components. So, to apply the above sinusoidal analysis to a complex wave which is comprised of individual sinusoids, it is sufficient to consider the sampling rate effects on the highest frequency sine wave component which makes up the complex waveform.

Frequency analysts have developed techniques for averaging of sampled signals which will give a high confidence, statistical measure of the amplitude of a time series. However, these averaging techniques discard the time sequence of the amplitudes and provide only a root mean square (RMS) value of the amplitude.

Fatigue analysts, on the other hand, have a different set of requirements for data analysis. They have determined that fatigue phenomena are not dependent upon average amplitude but upon the peak amplitudes and the sequence of peak stress amplitudes which a material or component experiences. It is, therefore, a requirement for fatigue related analyses of sampled time series to preserve both an accurate measure of each peak and valley, as well as the sequence of the peaks and valleys in the sample set. In the case of a multiaxial fatigue analysis, the phase relationship between multiple channels must also be maintained.

From the above simple analysis, it is clear that two extremes in sampling criteria may be considered: The Nyquist criterion, that determined periodicity but falls short on amplitude, and the "brute force" approach (a near infinite number of samples) which generally is not practical. These conflicting approaches must be reconciled, because the sampling rate has a dramatic effect on the requirements of a digital acquisition and processing system.

Donaldson [7] has identified some parameters that help define the error in peak levels of sampled waveforms. The error in peak resolution of a sinusoid is a function of the ratio of data frequency to sample frequency.

$$Pk_{err} = 2 \sin^2 (\pi f_d / 2 f_s) \qquad (5\text{-}2)$$

where:

Pk_{err} = percent error on peaks/100

f_d = data frequency

f_s = sampling frequency

This relationship is expressed graphically in Fig. 5-16. From the graphical representation, it is obvious that in order to achieve 1 percent accuracy in peak values, the data frequency must be no greater than 5 percent of the sample frequency, or in other terms, 20 samples per cycle of the highest frequency must be taken. For a fatigue waveform which contains underlying frequencies up to 100 hz, the sample rate for 1 percent accuracy is 2000 points per second. In multichannel analysis and control systems, this accuracy requires high system throughputs and massive data storage capabilities. The problem is compounded by the common requirement for multiple channels of data and data record lengths of 1 hour or more.

From Fig. 5-16, it can be observed that at a sampling rate of 4 points per cycle of the highest frequency (for a normalized data frequency of 1/4 or 25%), the error in peaks can be 30 percent. The 4 point per cycle sampling criterion is typically used in simulation test systems as the reasonable engineering tradeoff for throughput and storage considerations, but, it sometimes results in accuracy problems in fatigue analysis. The main situation where this is a problem is the one in which the highest frequency events are relatively large in amplitude. This is a relatively common occurrence, since the highest loads are often due to impact or other transient loads. If the sampling rate is not high enough these transient loads (or strains) will likely be underestimated.

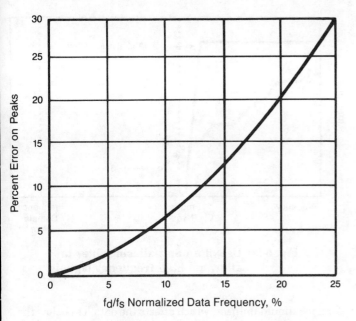

Fig. 5-16 Probable error in peaks of a history based on the ratio of data frequency sampling frequency.

5.6.1.2 Aliasing Problems

If the sample rate is less than twice the maximum data frequency ($f_s < 2*f_d$) the phenomena of aliasing occurs.

Fig. 5-17 graphically illustrates this condition. Note that when the sinusoid is sampled at a rate of less than two points per cycle, that the reconstructed waveform appears to be at a different and lower frequency. This alias frequency is in error with respect to the original waveform, not only in frequency, but also in terms of the number of peaks and valleys which are of concern in a fatigue analysis. This phenomena extends to complex waveforms comprised of several sinusoidal components. Alias frequencies will always appear in a reconstructed waveform if frequencies exist in the original continuous waveform which are greater than one half the sample rate, or greater than the Nyquist frequency.

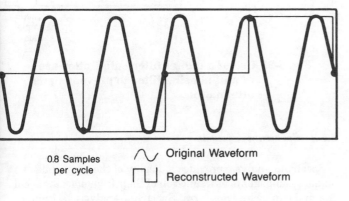

Fig. 5-17 0.8 samples per cycle.

The actual frequency of the alias will be equal to the sample rate, less the data frequency that is greater than one-half the sample rate ($f_{alias} = f_s - f_d$ if $f_d > 1/2f_s$). Aliasing will "fold," or mirror, frequencies that occur above the Nyquist frequency around the Nyquist frequency to frequencies below this frequency. This is illustrated in Fig. 5-18.

Area 1 represents the original spectrum of data to be sampled. Area 2 represents the spectral content in the original data which extends to a frequency greater than the Nyquist frequency. When the waveform is sampled, the phenomena of aliasing will cause the frequencies above the Nyquist frequency to fold about the Nyquist frequency as depicted by Area 3 and this area will appear below the Nyquist frequency as spectral energy in Area 4. This spectral energy appearing below the Nyquist frequency will be added to the original spectral energy as shown in Area 5. The data frequencies below the region of alias frequencies are untouched and completely valid, and the frequencies which have been "aliased into" are incorrect. A sampled waveform that contains alias frequencies has both an erroneous frequency content and incorrect peak and valley relationships.

5.6.1.3 Prevention of Aliasing

The solution to the problem of aliasing is to perform data sampling under the conditions decribed in Fig. 5-19. The sample rate should be chosen to be high enough to put the Nyquist frequency above all frequencies in the original signal. Note, this frequency must be above all frequencies in the original signal, not just above the frequencies of interest.

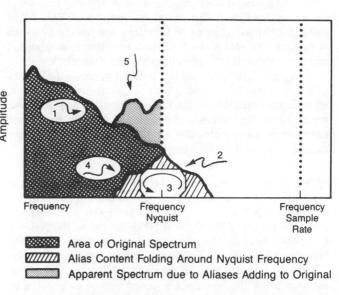

Fig. 5-18 An illustration of aliasing effects on a frequency domain plot of a spectrum.

There are some practical limitations to performing sampling as illustrated in Fig. 5-19. Often this approach requires high sample rates which may require great amounts of digital mass storage if the data are to be retained as a sampled

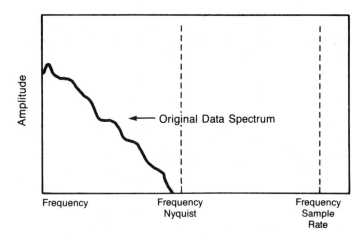

Fig. 5-19 Selection of sample rates sufficiently high to prevent aliasing.

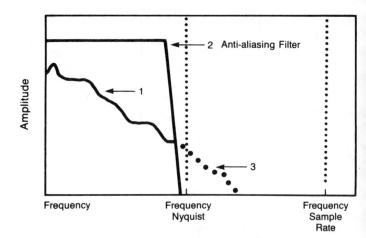

Fig. 5-20 Use of an anti-aliasing filter to alternate high frequency signals.

time history. If the data are to be cycle counted or otherwise reduced immediately upon sampling, this technique is practical, given the hardware system is capable of the required sample rates. It is important that the instrumentation systems used to condition and sample the analog signal be noise free, as noise added to the signal is as subject to aliasing as is the original signal. Often electrical noise is of such a high frequency that it is deemed "out of the range of interest." However, the higher frequency noise will be subject to aliasing, and the valid lower frequencies may be affected.

A more practical technique to prevent aliasing may be to use an analog filter before the signal is sampled. This filter, called an Anti-Aliasing or A^2 filter, is a low pass filter which is designed to attenuate frequencies in the analog signal which are above the Nyquist frequency as shown in Fig. 5-20. The original data spectrum is identified by arrow 1. The application of the Anti-Aliasing filter, characterized by arrow 2, attenuates or eliminates the frequencies of the original spectrum (arrow 3) which extend beyond the Nyquist frequency. Since these frequencies are eliminated by the filter, they cannot be aliased in the sampling operation.

There are also some practical considerations regarding the use of Anti-Aliasing filters. First, physically realizable filters do not have infinitely sharp cutoff characteristics. Thus, the cutoff frequency of the filter must be set below the Nyquist frequency in order to insure that the signals above the Nyquist frequency are sufficiently attenuated not to cause significant aliasing. It is important to realize that some useful data below the Nyquist frequency may be attenuated by the filter. To avoid distortion of the original data, the filter used must exhibit linear phasing.

Typical Anti-Aliasing filters are of 6 to 8 poles (orders) and exhibit attenuation rates in the cutoff band of 36-48 dB per octave attenuation. A Bessel filter is often used because of its linear phase characteristics, however, a Butterworth filter is also used in some cases because its passband can be designed for a maximum flatness, which affects the data less below the Nyquist frequency.

A technique often employed to deal with the effect of signals attenuated by the passband attenuation of the antialiasing filter is to choose sample rate, and anti-aliasing cutoff filter frequencies such that both are above the frequency range of interest. Then, after the signal is sampled, a digital filter with a much sharper cutoff characteristic can be applied to the data to band limit the data to the frequency range of interest. This is illustrated in Fig. 5-21 by arrow 4.

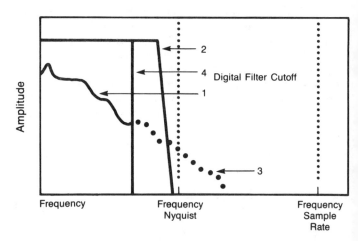

Fig. 5-21 Use of a digital filter cutoff after use of anti-aliasing filter to prevent signal attenuation.

5.6.2 Spectrum Analysis

Spectrum analysis provides a display of the amplitude of a signal's components as a function of their frequency to reveal the makeup of its waveform. Spectrum analysis techniques are often used in vibration studies, but seldom used in detailed fatigue analyses, since history accountability (strain sequence effects) are lost.

At times both vibration and fatigue analysis are of concern because specific frequencies may be causing unexpected fatigue damage and the engineer must find and eliminate the fatigue inducing vibrations. In most ground vehicle cases, however, the two fields are treated by differing design groups. Because vehicle or component resonances cannot usually be eliminated entirely, but only tuned to frequencies that cause no ill effects, it is important that the data acquisition engineer be cognizant of the tools provided by frequency or spectrum analysis techniques.

There are several differing methods of analyzing or extracting the significant frequencies that are present in a signal. One common method is to place a constant amplitude dynamic load input into some part of the structure (a wheel for example) and observe the resonance at specific points of the structure. By "sweeping" or varying the frequency of the input one can usually find specific frequencies that cause very high response resonances in other sections of the vehicle. In this manner it is possible to assess the sensitivity of vehicle air volume resonance to specific large amplitude "tire hop" wheel input conditions, such as wash board gravel roads.

There are a number of ways of deriving component sensitivity to these types of excitation. Constant amplitude frequency sweeping, mentioned above is one. Another method is to use random "white noise" (a signal with many frequencies intermixed) and compute the most dominant component response frequencies. The effects of impulse inputs, such as from a hammer, etc., can also be used to determine component response.

The general objective is to operate the component such that its response to all service inputs does not cause excessive vibrations; excessive both in terms of vehicle function and component fatigue durability.

This study of the "frequency" effects of vibrations is generally termed spectrum analysis.* A variety of hardware and software exists to analyze vehicular behavior in such frequency terms.

5.6.2.1 Time Compression Analyzers

Real time spectrum analyzers offer strong performance capabilities for vibration analysis work and especially those applications involving low frequencies and random characteristics. The time compression, real-time spectrum analyzer usually includes a cathode ray tube (CRT) display with electronic cursor to permit high accuracy determination of both frequency and amplitude values that may be annotated on the CRT.

Analyzers of this type have controls that automatically adjust A/D conversion rate and resolution to complement the selected frequency range. Many operating features are also available for calibration, spectrum averaging, X-Y recorder output control (assuring optimized measurement), and display of analyzed signals.

5.6.2.2 FFT Analyzers

Another form of real time spectrum analyzer termed the Fast Fourier Transform (FFT) analyzer uses fast Fourier techniques to digitally process analog input signals with a computational algorithm that determines the Fourier coefficients and generates a graphical display of the spectrum using all digital processing. These analyzers may be in portable form and include a CRT display and a wide range of operating modes that are controlled by a microprocessor to determine given modes of operation. An FFT system has linearity and dynamic performance advantages over other systems and is available in multichannel input to permit the inclusion of other analytical functions such as cross correlation, cross spectrum, coherence functions and transfer functions to further increase the power of such an instrument for vibration analysis work.

5.7 Data Evaluation Techniques

Fatigue problems, because they encompass large numbers of load events, are usually exercises in information compression and summarization. When the engineer is presented with a signal, such as on the left of Fig. 5-22, for example, it becomes very difficult to "process" such a signal visually and extract the critical features that cause the fatigue damage.

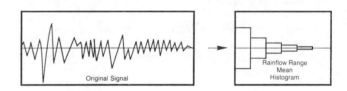

Fig. 5-22 Schematic representation of the process of summarizing a signal for fatigue life prediction.

Some form of compression or summarization must be applied before the data can be transformed into a life prediction. Although it is possible nowadays to use the complete signal to model the fatigue process, it is common to derive a number of different secondary variables that accentuate particular features of the signal and simplify the actual fatigue damage computations. Some of the more popular techniques are described in this section.

*For a list of computer programs that can be used by anyone, the reader is referred to Ref. [7].

Most of the techniques involve cycle counting methods that process the random signals into an equivalent set or blocks of constant amplitude cycles. Many of these methods arose from the need for data reduction in the years before the application of digital computers. The increase in computerized processing has allowed greater flexibility in selection of data reduction techniques, and allowed the retention of more data in non-reduced form. The advent of full scale component or vehicle testing using realistic service load histories has rejuvenated the use of frequency based techniques for fatigue purposes. The methods are required for the correct laboratory testing of structures that have components sensitive to the frequency of load inputs; such as shock absorber gromets, etc., as described in Refs. 8-11.

In order to achieve correct frequency and amplitude response it becomes necessary to test using a realistic load history; preferably the measured service signal itself, and to compensate in the test control computer for test setup and vehicle dynamic effects such that the component(s) of interest are loaded with the desired service spectrum. The trend, therefore, is towards the retention of service history data in more or less its original signal form, and to use the frequency and cycle count summarizing methods to subsequently characterize the signal for various purposes; e.g. the rainflow histogram of Fig. 5-22 shows the various equal amplitude blocks of cycles sorted by range as they have been extracted from the original signal.

The following sections describe the various summarization techniques that are commonly in use.

5.7.1 *Statistical Summarizations (Spectrum Analysis)*

When the data are reduced statistically, the following properties are usually defined:

(1) Distribution function of instantaneous values (or just peaks and valleys)

(2) σ, the standard deviation

(3) N_0, the number of zero (or mean) level crossings with positive slope

(4) λ, the clipping ratio (or crest factor), that is, the largest deviation from the mean divided by the standard deviation

(5) ϕ, the irregularity factor, equal to N_0/N_p where N_p is the number of peaks. When $\phi = 1.0$, the loading crosses the zero (or mean) load between every peak and valley, whereas $\phi < 1.0$ indicates that the random process contains higher frequency-lower amplitude loadings.

Power spectra density (PSD) analysis methods are useful in the definition of load spectra where information on the frequency of the loading process is needed. This is particularly important in calculations of the response of linear loading systems. PSD analysis is a procedure in which Fourier analysis is combined with probability and statistics to describe random processes. In order to use this technique, the process should be stationary (its statistical properties should not change with time); it is also simplest if the instantaneous values are normally distributed.

A random process is defined by the PSD function in the frequency domain:

$$\phi(\omega) = \lim_{t \to \infty} \frac{1}{T\pi} \left(\int_0^T y(t) e^{-i\omega t} dt \right)^2 \quad (5\text{-}3)$$

where

ω = radians/second or circular frequency,
y = a random process variable, such as the loading history
t = time,
T = time, upper limit of integral,
e = the base of natural logarithms, and
$i = \sqrt{-1}$.

If the random process is normally distributed, useful statistical properties can be directly obtained from the PSD function in terms of the moments M about the zero frequency axis, as follows:

$$M_n = \int_0^\infty \omega n \phi(\omega) \, d\omega \quad (5\text{-}4)$$

From this it is possible to derive the statistical properties N_y, the number of level y crossings with positive slope; N_0, the zero crossings; N_p, the peak frequency; and h the average range:

$$N_y = (1/2)(M_2/M_0)^{1/2} e^{-(y^2/2M_0)} \quad (5\text{-}5)$$

$$N_0 = (1/2)(M_2/M_0)^{1/2} \quad (5\text{-}6)$$

$$N_p = (1/2)(M_4/M_2)^{1/2} \quad (5\text{-}7)$$

$$h = M_2 (2/M_4)^{1/2} \quad (5\text{-}8)$$

The average amplitude of the loads is usually expressed as the root mean square of the loads or strains

$$Y_{RMS} = \sqrt{1/t \int_{t_1}^{t+T} y^2 \, dt} \quad (5\text{-}9)$$

While statistical analysis can be used to characterize loading histories, it cannot be used to accurately estimate fatigue lives. A series of tests reported in Ref. [12] will be used as an example. The strain-time history from a strain-gaged bogie of a tank car traveling over continuously welded track was recorded. Periodic overstrains occurred when the vehicle passed over points and crossings in the track. A second loading history was prepared by eliminating the periodic high loads. The power spectral densities of both load histories is shown in Fig. 5-23.

The average frequency for both loading histories was 3 Hz. The PSD for these two signals are nearly identical and one would expect the fatigue life of specimens tested with both histories to be the same if the PSD was related to fatigue life. Results from fatigue tests utilizing these two histories are shown in Fig. 5-24. The fatigue life was reduced by a factor of 10 when the periodic overstrains were included. If the root mean square of strain or PSD was the controlling fatigue parameter the fatigue lives for both histories would have been the same.

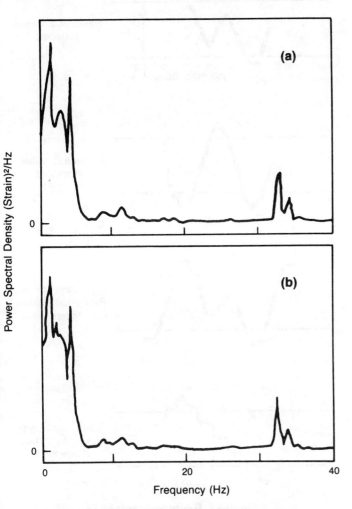

Fig. 5-23 PSD's of testing signals: (a) track strains only; (b) track strains with point and crossing strains.

Statistical analysis can be used to characterize the average behavior of a loading history. As shown above, fatigue damage can be caused by a small number of large events, which will not be the average behavior but will affect the fatigue life. For this reason, fatigue life estimates based on PSD and RMS are not reliable and should not be used.

5.7.2 Amplitude Based Summarization

Most of the methods used for summarizing signals for subsequent direct fatigue life prediction computations do not use the time variable of the signal, i.e. for metal components in ground vehicle applications the speed of loading does not usually influence fatigue life. Of primary importance are amplitude and sequence. When a signal has been digitized, one of the first compression techniques applied is to throw away

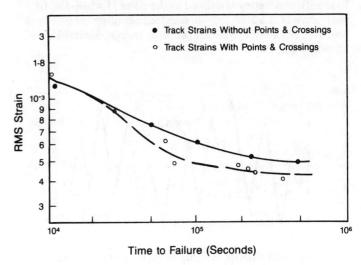

Fig. 5-24 RMS (strain) vs. time to failure.

all samples between adjacent peaks and valleys as shown in Fig. 5-25.

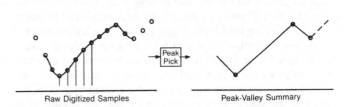

Fig. 5-25 Compression of a digitized signal.

Although this yields a major reduction in the amount of storage required to save the history, it also inherently restricts the type of information that may be obtained from subsequent calculations. The frequency analysis techniques, for example, cannot be applied to peak and valley summaries of signals; nor can multi channel or multiaxially loaded components be analyzed. This peak-valley extraction process is a prerequisite for, or is inherent in, all the subsequent cycle counting methods that are needed for computing the fatigue damage expended by the component history.

5.7.2.1 Peak-Valley Sequencing

The first step in data reduction for a detailed fatigue analysis is to determine the sequence of peaks and valleys in the loading history. A peak is defined as the point at which the first derivative of the load time history changes from positive to negative. A valley is defined as the point at which the first derivative of the load time history changes from negative to positive. A valley is also known as a trough. Examples are shown in Fig. 5-26.

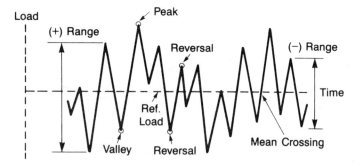

Fig. 5-26 Basic fatigue loading parameters.

A hysteresis threshold, or dead band, is usually employed when processing field loading histories. (A signal existing at some level is assumed to have changed only when it exceeds the dead band limits in either direction.) It is used to eliminate small peaks and valleys (i.e. noise) while retaining the overall sequence of loading [13]. An example is shown in Fig. 5-27.

Although analog devices are available to determine peaks and valleys, a digital computer is usually employed because it can process the load history and perform the fatigue analysis. A sample computer program for picking peaks and valleys is included in Appendix 5A. One of the problems encountered in processing load histories to determine peaks and valleys is to determine the appropriate sampling rate. In dynamic studies, it is necessary to sample the signal at a frequency at least two times the maximum frequency of interest. For fatigue analysis, the sampling rate must be higher in order to ensure that the actual peak value is measured (see Section 5.6.1.1). Twenty times the maximum frequency of interest is often used. However the reader is cautioned that impact and other transient loadings may occur at much higher frequencies than normal operating loads.

For some data acquisition purposes the peak/valley sequences, with or without a time variable, are the objectives of the acquisition exercise [13]. Multi channel component histories with phase critical inter channel behavior are an example. Presently most acquisition exercises apply additional history compression techniques to further reduce the information load of the designer/analyst.

5.7.2.2 Cycle Counting Procedures

Various counting methods have been developed to reduce cyclic time histories to some simple form of cycle count for analysis and testing purposes. Cycle counts can be made for

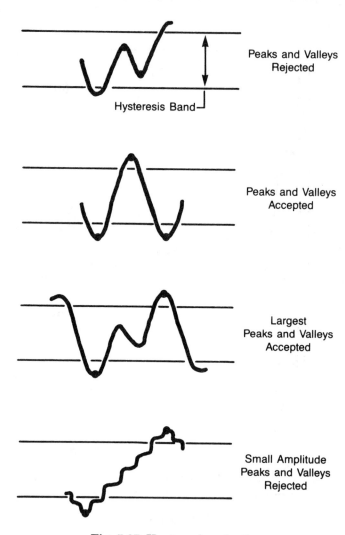

Fig. 5-27 Hysteresis rejection.

time histories of force, stress, strain, torque, acceleration, deflection, or other loading parameters of interest. The three basic counting methods used over the years have been level crossing count, peak count, and range count. Numerous variations of each method have evolved [14], mainly to eliminate higher frequency-lower amplitude cycles or to match some preconceived theory of fatigue damage accumulation or crack propagation. All of these methods basically count one parameter and eliminate all knowledge of the loading sequence. Since two independent parameters are needed to define a cycle, assumptions have to be made about the other parameter, and these methods are thus sometimes found to be inadequate. As a result, two-parameter counting methods, such as the range-pair-range, rain-flow and racetrack methods, have been developed. These methods reflect some aspects of loading sequence and are extensively used in conjunction with local stress-strain notch fatigue damage calculation schemes. Rainflow counting is perhaps the most widely accepted method for the identification of fatigue critical events and is useful when pursuing a basic understanding of material behavior (such as memory effects).

The following paragraphs include a description of the cycle counting procedures in common use. This presentation is derived from the current ASTM standard on the subject [15]. Note with caution that the definition of a cycle varies with the method of cycle counting.

5.7.2.2.1 *Level Crossing Counting*

Results of a level crossing count are shown in Fig. 5-28a. One count is recorded each time the positive sloped portion of the load exceeds a preset level above the reference load, and each time the negative sloped portion of the load exceeds a preset level below the reference load. Reference load crossings are counted on the positive sloped portion of the loading history. It makes no difference whether positive or negative slope crossings are counted. The distinction is made only to reduce the total number of events by a factor of two.

In practice, restrictions on the level crossing counts are often specified to eliminate small amplitude variations that can give rise to a large number of counts. This may be accomplished by filtering small load excursions prior to cycle counting. A second method is to make no counts at the reference load and to specify that only one count be made between successive crossings of a secondary lower level associated with each level above the reference load, or a secondary higher level associated with each level below the reference load. Fig. 5-28b illustrates this second method. A variation of the second method is to use the same secondary level for all counting levels above the reference load, and another for all levels below the reference load. In this case the levels are generally not evenly spaced.

The most damaging cycle count for fatigue analysis is derived from the level crossing count by first constructing the largest possible cycle, followed by the second largest, etc., until all level crossings are used. Reversal points are assumed to occur halfway between levels. This process is illustrated by Fig. 5-28c. Note that once this most damaging cycle count is obtained, the cycles could be applied in any desired order, and this order could have a secondary effect on the amount of damage. Other methods of deriving a cycle count from the level crossings count could be used.

5.7.2.2.2 *Peak Counting*

Peak counting identifies the occurrence of a relative maximum or minimum load value. Peaks above the reference load level are counted, and valleys below the reference load level

valleys are usually reported separately. A variation of this method is to count all peaks and valleys without regard to the reference load.

To eliminate small amplitude loadings, mean crossing peak counting is often used. Instead of counting all peaks and valleys, only the largest peak or valley between two successive mean crossings is counted as shown in Fig. 5-29b.

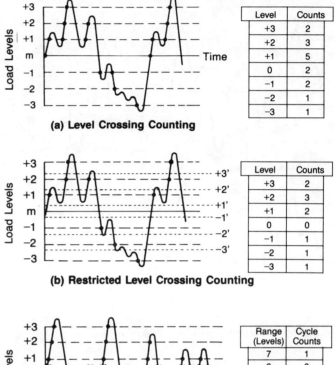

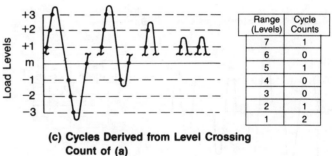

Fig. 5-28 Level crossing counting example.

The most damaging cycle count for fatigue analysis is derived from the peak count by first constructing the largest possible cycle, using the highest peak and lowest valley, followed by the second largest cycle, etc., until all peak counts are used. This process is illustrated by Fig. 5-29c. Note that once this most damaging cycle count is obtained, the cycles could be applied in any desired order, and this order could have a secondary effect on the amount of damage. Alternate methods of deriving a cycle count, such as randomly selecting pairs or peaks and valleys, are sometimes used.

5.7.2.2.3 *Simple Range Counting*

For this method, a range is defined as the difference between two successive reversals, the range being positive when a valley is followed by a peak and negative when a peak is followed by a valley. The method is illustrated in Fig. 5-30. Positive ranges, negative ranges, or both may be counted with this method. If only positive or only negative ranges are counted, then each is counted as one cycle. If both positive and negative ranges are counted, then each is counted as one-half cycle. Ranges smaller than a chosen value are usually eliminated before counting.

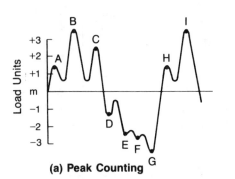

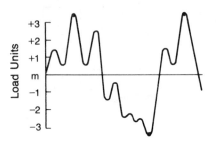

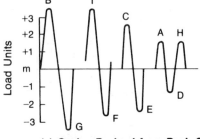

Fig. 5-29 Peak counting example.

When the mean value of each range is also counted, the method is called simple range-mean counting. An example is provided in Fig. 5-30.

5.7.2.2.4 Rainflow Counting and Related Methods

A number of different terms have been employed in the literature to designate cycle counting methods that are similar to the rainflow method. These include range-pair counting [16,17], the Hayes method [18], the original rainflow method [19-21], range-pair-range counting [22], ordered overall range counting [23], racetrack counting [24], and hysteresis loop counting [25]. If the load history begins and ends with its maximum peak, or with its minimum valley, all of these give identical counts. In other cases, the counts are

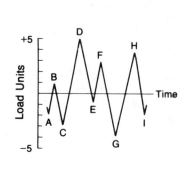

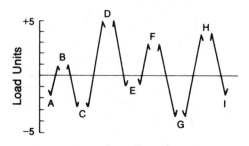

Fig. 5-30 Simple range counting example—both positive and negative ranges counted.

similar, but not generally identical. Three methods in this class are defined here: range-pair counting, rainflow counting, and a simplified method for repeating histories.

The various methods similar to the rainflow method may be used to obtain cycles and the mean value of each cycle; they are referred to as two parameter methods. When the mean value is ignored, they are one parameter methods, as are simple range counting, peak counting, etc.

For details of the specific cycle counting procedures, see Appendix 5B. Examples of programs for picking peaks and valleys and rainflow counting are provided in Appendix 5A.

5.8 References

[1] Beckwith, T. G. and Buck, N. L., "Mechanical Measurements," Addison-Wesley Publishing Co., Inc., 1961, pp. 287-295.

[2] Smith, K. V. and Stornant, R. F., "Cumulative Damage Approach to Durability Route Design," SAE Technical Paper 791033, SAE, 1979.

[3] Perry, C. C. and Lissner, H. R., "The Strain Gage Primer," McGraw-Hill, 1955, pp. 198-236.

[4] Dally, J. W. and Riley, W. F., *Experimental Stress Analysis*, McGraw-Hill, 1978, pp. 217-272.

[5] Judd, J. E., "Preventing Ground Noise in Vibration Measurement Systems," *Test*, October/November 1980, pp. 12-16.

[6] Anon., "Accessories Bulletin No. 207," Electro General Corp., 1985.

[7] Digital Signal Processing Committee, "Programs for Digital Signal Processing," IEEE Press, New York, John Wiley & Sons, New York, New York, 1979.

[8] Cryer, B. W., Nawrocki, P. E., and Lund, R. A., "A Road Simulation System for Heavy Duty Vehicles," SAE Paper 760361, 1976.

[9] Jacoby, G., "Load Simulation of Wheel Axle and Complete Vehicle Load," Paper presented at Symposium on Experimental Simulation Tests of Motor Vehicles and Their Elements, Warsaw, Polant, October 1-2, 1979.

[10] Lund, R. A., "Multiple Channel Environmental Simulation Techniques," Paper presented at Symposium on Experimental Simulation Tests of Motor Vehicles and Their Elements, Warsaw, Poland, October 1-2, 1979.

[11] Donaldson, K., "Field Data Classification and Analysis Techniques," SAE Technical Paper 820685, SAE, 1982.

[12] *Fatigue Under Complex Loading:Analysis and Experiments*, SAE, 1977.

[13] Conle, A., Oxland, T. R., Wurtz, D., and Topper, T. H., "Tracking Time in Service Histories for Multiaxis Fatigue Problems," ASTM, 1987.

[14] Conle, A. and Topper, T. H., "Fatigue Service Histories:Techniques for Data Collection and History Reconstruction," SAE Technical Paper 820093, 1982.

[15] "Standard Practices for Cycle Counting in Fatigue Analysis," ASTM Standard E1049, ASTM, 1985.

[16] Anon., "The Strain Range Counter," VTO/M/416, Vickers-Armstrongs Ltd. (now British Aircraft Corp. Ltd.), April 1955.

[17] Burns, A., "Fatigue Loadings in Flight:Loads in the Tailpipe and Fin of a Varsity," Aeronautical Research Council Technical Report C.P. 256, London, 1956.

[18] Hayes, J. E., "Fatigue Analysis and Fail-Safe Design," *Analysis and Design of Flight Vehicle Structures*, E. F. Bruhn, Editor, 1965, pp. C13-1 to C13-42.

[19] Matsuishi, M. and Endo, T., "Fatigue of Metals Subjected to Varying Stress," Japan Society of Mechanical Engineers, March 1968. See also T. Endo, et. al., "Damage Evaluation of Metals for Random or Varying Loading," *Proc. of the 1974 Symposium on Mechanical Behavior of Materials*, Vol. 1, Society of Materials Science, Japan, 1974, pp. 371-380.

[20] Anzai, H. and Endo, T., "On-Site Indication of Fatigue Damage Under Complex Loading," *International Journal of Fatigue*, Vol. 1, No. 1, 1979, pp. 49-57.

[21] Endo, T. and Anzai, H., "Redefined Rainflow Algorithm: P/V Difference Method," *Society of Materials Science*, Japan, Vol. 30, No. 328, 1981, pp. 89-93.

[22] van Dijk, G. M., "Statistical Load Data Processing," Sixth ICAF Symposium, May 1971. See also NLR MP 71007U, National Aerospace Laboratory, Amsterdam, 1971.

[23] Fuchs, H. O., et al., "Shortcuts in Cumulative Damage Analysis," *Fatigue Under Complex Loading: Analyses and Experiments*, R. M. Wetzel, Editor, SAE, 1977, pp. 145-162.

[24] Nelson, D. V. and Fuchs, H. O., "Predictions of Cumulative Fatigue Damage Using Condensed Load Histories," *Fatigue Under Complex Loading: Analyses and Experiments*, R. M. Wetzel, Editor, SAE, 1977, pp. 163-187.

[25] Richards, F. D., LaPointe, N. R. and Wetzel, R. M., "A Cycle Counting Algorithm for Fatigue Damage Analysis," SAE Technical Paper 740278, SAE, 1974.

Addresses of Publishers

Addison-Wesley Publishing Co., Inc., 1 Jacob Way, Reading, MA 01867

ASTM, American Society for Testing & Materials, 1916 Race Street, Philadelphia, PA 19103

British Aircraft Corp., Technical Office, Weybridge, Surrey, England

Electro General Corp., Minnetonka, MN 55343

McGraw-Hill Book Co., 1221 Avenue of the Americas, New York, NY 10020

SAE, Society of Automotive Engineers, 400 Commonwealth Drive, Warrendale, PA 15096

Tristate Offset Co., Cincinnati, OH

Appendix 5A

Sample Programs for Cycle Counting

The following two sections include program listings for picking peaks and valleys and rainflow cycle counting.

5A.1 Program to Pick Peaks and Valleys

```
C
C       CHECKS 1ST PT FOR INTEGER OR REAL, THEN ASSUMES REST SAME CHARACTER*80 INP
C       GET THE ENVIRONMENT FILE
        OPEN (1, FILE='ppick.env', STATUS='OLD', ERR=50)
        REWIND (1)
        READ (1, *, ERR=60) IFILT
        READ (1, *, ERR=60) NSTART
        READ (1, *, ERR=60) NEND
        CLOSE (1)
        NOUT=0
        N=1
        IF (NSTART.NE.1) THEN
C          GO TO THE DATA
           NM=NSTART-1
           CO 8 I=1, NM
           READ (5, 10, END=100) INP
           N=N+1
      8    CONTINUE
        ENDIF
C
C       WE ARE AT THE FIRST REQUIRED PT
        READ (5, 10, END=100) INP
     10 FORMAT (A80)
        READ (INP, *, ERR=500) I1
C       IF THIS READ WORKED, 1ST PT IS INTEGER
        IOLD=I1
        WRITE (6, *) IOLD
        NOUT=NOUT+1
     12 N=N+1
        READ (5, *, ERR=100, END=100) INOW
        IF (INOW.EQ.IOLD) GO TO 12
C       SET SLOPE
        IUP=-1
        IF (INOW.GT.IOLD) IUP=1
C       INITIAL 1 PTS READ, SLOPE SET. READY TO RUN
C
     20 CONTINUE
C       GET NEXT PT
        N=N+1
C       CHECK FOR END OF REQUIRED DATA
        IF (N.GT.NEND) GO TO 50
        READ (5, *, ERR=50, END=50) INEXT
        IF (INEXT.EQ.INOW) GO TO 20
        IF (INEXT.LT.INOW) GO TO 40
C       NO?, THEN ITS BIGGER AND A +VE RAMP. CHECK IF RAMP IS ALREADY +VE IF (IUP.EQ.1) THEN
C          NOT A REVERSAL, DISCARD INOW & USE INEXT AS NEW POTENTIAL REV
           INOW=INEXT
           GO TO 20
        ENDIF
C       IUP IS -VE, A REV HAS OCCURED, IS IT BIG ENOUGH?
        IF ((INEXT-INOW).LT.IFILT) THEN
C          THIS REV IS TOO SMALL, IGNORE IT
           GO TO 20
        ENDIF
```

```
C         YES, ITS BIGGER THAN IFILT & HAS DIRECTION CHANGE. INOW WAS A REVERSAL.
          NOUT=NOUT+1
          WRITE (6, *) INOW
          IOLD=INOW
          INOW=INEXT
          IUP=1
          GO TO 20
C
   40     CONTINUE
C         INEXT IS GOING DOWN -VE. CHECK PREVIOUS RAMP DIRECTION
          IF (IUP.EQ.-1) THEN
C            NO DEAL, SAME DIREC.
             INOW-INEXT
             GO TO 20
          ENDIF
C         CHANGE IN DIREC. OCCURRED, IS IT BIG ENOUGH?
          IF ((INOW-INEXT).LT.IFILT) THEN
C            TOO SMALL
             GO TO 20
          ENDIF
C         YES, BIG ENOUGH
          NOUT=NOUT+1
          WRITE (6, *) INOW
          IOLD=INOW
          INOW=INEXT
          IUP=-1
          GO TO 20
C
C         END OF FILE ENCOUNTERED
   50     CONTINUE
          NOUT=NOUT+1
          WRITE (6, *) INOW
          WRITE (6, 55) NOUT
   55     FORMAT ('TOTAL REVS=', I10)
          STOP
C
   60     WRITE (6, 65)
   65     FORMAT ('***ERROR READING ppick.env FILE***')
          STOP
C
  100     WRITE (6, 105)
  105     FORMAT ('***NOT ENOUGH DATA IN FILE***')
          STOP
  500     CONTINUE
          WRITE (6, 505)
  505     FORMAT ('***REAL NOS, REWRITE ppick.f***')
          STOP
          END
```

Auxillary file 'ppick.env'

```
400      window value that causes filtering in ppick
1        starting value for picker
120000   end value for picker
```

5A.2 Program for Rainflow Cycle Counting

```fortran
C         MAINLINE TO ACTIVATE RAINFLOW S/R. compile using Fortran 77
C
C         This routine is supplied courtesy of members of
C         the SAE Fatigue Design & Evaluation Committee,
C         as a sample only, and is not warrantied in any
C         way. The user is expected to ensure that the
C         routines are correct.

          CHARACTER*1 INP1(80)
          CHARACTER*80 INP
          INTEGER UNI
C
C         RATHER THAN PRESCAN THE DATA FILE IN ORDER TO
C            FIND THE MAXIMA, READ THEM FROM A USER CREATED
C            FILE.
          OPEN(1,FILE='rain.env',STATUS='OLD',ERR=50)
          REWIND(1)
          READ(1,*,ERR=60)IVMAX
          READ(1,*,ERR=60)IVMIN
          CLOSE(1)
C         NOW TRANSFER THESE VALUES TO THE RAINFLOW SUBR.
          CALL RAINCUT(0,IVMAX,IVMIN,IER)
          IF(IER.NE.0)STOP
C
C         OK, FETCH THE FIRST TWO PTS, MAKE SURE THEY
C            ARE A HALF CYCLE. THE 1/2 CYCLE SHOULD BE
C            BIG ENOUGH TO SPAN 1/32 OF THE TOTAL RANGE
C            DEFINED BY IVMAX AND IVMIN
          UNI=5
        3 N=1
          read(UNI,*)I
        4 CONTINUE
          IV1=I
          IF(IV1.GT.IVMAX .OR.IV1.LT.IVMIN)THEN
             WRITE(6,12)IV1,N
               STOP
          ENDIF
        7 N=2
          read(UNI,*)I
        8 CONTINUE
          IV2=I
          IF(IV2.GT.IVMAX .OR.IV2.LT.IVMIN)THEN
             WRITE(6,12)IV2,N
               STOP
          ENDIF
          CALL RAINCUT(-1,IV1,IV2,IER)
          IF(IER.NE.0)STOP
C
C         RUN SECTION, THE 1ST TWO PTS HAVE BEEN ESTABLISHED,
C            GET THE REST OF THE DATA
          NSKIP=0
          NOR=0
          IVOLD=IV2
       10 CONTINUE
          N=N+1
          READ(5,*,END=20,ERR=30)IV
          IF(IV.GT.IVMAX .OR.IV.LT.IVMIN)THEN
             WRITE(6,12)IV,N
```

```fortran
   12     FORMAT(' ERROR , OUT OF RG: V=',I6,',N=',I10)
          NOR=NOR+1
          GO TO 10
        ENDIF
C       CHECK FOR EQUAL TO LAST PT. NOTE THAT
C         DIFFERENCE HERE MAY STILL IMPLY THE SAME VALUE
C         IN RAINCUT BECAUSE OF THE WINDOWING OR DEAD BAND
C         EFFECT OF THE 32 BINS.
        IF(IV.EQ.IVOLD)THEN
          NSKIP=NSKIP+1
          GO TO 10
        ENDIF
        CALL RAINCUT(1,IV,IV2,IER)
        IF(IER.EQ.O)THEN
          IVOLD=IV
          GO TO 10
        ENDIF
C       NO?, THEN SOME FORM OF ERROR
        IF(IER.EQ.1)THEN
          NSKIP=NSKIP+10
          GO TO 10
        ENDIF
C       OTHER ERROR, EXIT
        STOP
C       END OF DATA
   20   CONTINUE
C       CALL THE S/R TO PRINT OUT THE DATA
        CALL RAINCUT(100,IV,IV2,IER)
        WRITE(6,23)N,NOR,NSKIP
   23   FORMAT(' **** NPTS=',I8,',NO. OUT OF RG=',I8,',NO. SKIPPED=',I8)
        STOP
C
C       READ ERROR
   30   CONTINUE
C       IGNORE ERROR, MAY BE TEXT COMMENT INSERT
        N=N-1
        GO TO 10
C
   40   WRITE(6,45)N
   45   FORMAT('NO DATA?')
        STOP
C
   50   WRITE(6,55)
   55   FORMAT('***ERROR, CANNOT OPEN rain.env FILE***')
        STOP
   60   WRITE(6,65)
   65   FORMAT('***READ ERR, rain.env FILE***')
        STOP
        END

        SUBROUTINE RAINCUT(IFUNC,IVT,IVT2,IER)
C       This routine is supplied courtesy of members of
C         the SAE Fatigue Design & Evaluation Committee,
C         as a sample only, and is not warrantied in any
C         way. The user is expected to ensure that the
C         routines are correct.
C
C================================================================
```

```
C         For further information the user is refered to the paper by:
C         Downing, S. D. and Socie, D. F., "Simplified Rainflow Counting
C         Algorithms," International Journal of Fatigue, Vol.4, No. 1,
C         January 1982, pp.31-40.
C
C         THE VARIABLE IFUNC IS USED TO CONTROL THE OPERATION OF THIS S/R:
C         IFUNC= 0 : INITILIZE CALL: IVT, IVT2 ARE ACTUALLY IMAX,IMIN
C                    USED FOR MATRIX LIMITS
C              =-1 : IVT,IVT2 FIRST TWO HISTORY POINTS.
C              = 1 : NORMAL OPERATION DATA TO BE COUNTED IS IN IVT
C              =100: INPUT FINI, PRINT RESULTS
C
C         CHANGED TO USE INTEGER ONLY VOLTAGES MARCH 86, A.C.
C         CHANGE TO BIN NO. AFTER RAINFLOW DETECTED.
C
C         RAINFLOW ALGORITHM TO COUNT A HISTORY BY S. DOWNING
S         STATEMENTS 1-9 ARE RAINFLOW COUNTING RULES.
C         REARRANGEMENT OF DATA IS NOT NECESSARY.
C
          INTEGER MAT,E,SROW,SCOL,P,X,Y,SUM,CNT,SLOPE,ALOW,BLOW
          DIMENSION MAT(32,32),E(200),SROW(32),
        & SCOL(32)
          LOGICAL BUG
C         IN ORDER TO PRESERVE VARS BETWEEN CALLS, PLACE IN COMMON:
          COMMON/RAINF/MAT,E,SROW,SCOL,NA,NB,DA,DB,ALOW,BLOW,
        & CNT,SUM,IREV,J,N,SLOPE,ISTART
C
          BUG=.FALSE.
C             = USED FOR DEBUG
          IER =0
          IF(IFUNC.EQ.1)GO TO 1
          IF(IFUNC.EQ.100)GO TO 6
          IF(IFUNC.EQ.-1)GO TO 75
          IF(IFUNC.EQ.0)GO TO 2000
          IER=1
          WRITE(6,70)IFUNC
       70 FORMAT("ERROR IN RAINCUT:IFUNC=",I6)
          RETURN

     2000 CONTINUE
          IMAX=IVT
          IMIN=IVT2
C         INITILIZE THE VARIABLES
C         WE DIVIDE INTO A & B HERE TO ALLOW FOR FUTURE VARIABLE DIM. OF
C         STORAGE IN THE RAINFLOW MATRIX
          NA=32
          NB=32

C         ASSUME IMIN AT CENTER OF BIN 1, IMAX AT C OF BIN 32
          DA=FLOAT(IMAX-IMIN)/(NA-1)
          ALLOW=IFIX(IMIN-DA*0.5)
          DB=FLOAT(IMAX-IMIN)/(NB-1)
          BLOW=IFIX(IMIN-DB*0.5)
C         ALOW, & BLOW ARE THE LOWER BOUNDS OF THE MATRIX
C         CNT COUNT THE REMAINDER IN P.D. LIST, SUM COUNTS STUFF THAT
C         COMES OFF THE STACK
          IF(BUG)WRITE(6,2003)DA,ALOW,DB,BLOW
     2003 FORMAT('INTERVAL DA=',E10.3,'ALLOW=1,I6)
        &  'INTERVAL DB=',E10.3,'BLOW=',I6)
          CNT=0
          SUM=0
```

```
            IREV=0
            J=0
            N=2
            DO 69 KA=1,NA
            SROW(KA)=0
            SCOL(KA)=0
            DO 69 KB=1,NB
            MAT(KA,KB)=0
         69 CONTINUE
            ISTART=1
            RETURN
C           END OF MAX MIN INITILIZE SECTION
C
C
C           SET THE FIRST TWO PEAKS
         75 CONTINUE
            E(1)=IVT
            E(2)=IVT2
            IF(E(1).EQ.E(2))THEN
              IER=1
              WRITE(6,2004)IVT,IVT2
       2004   FORMAT("RAINCUT ERRO: 1ST & 2ND PTS=",2I8)
            ENDIF
       2005 SLOPE=1
            IF(E(1).GT.E(2))SLOPE=-1
            RETURN
C------------------------END OF INIT SECTION
C
C
C           NORMAL CALL: DATA INPUT. COUNT CYCLES
C       FIRST VECTOR SHOULD NOW BE IN
          1 CONTINUE
            P=IVT
            N=N+1
            SLOPE=-SLOPE
            E(N)=P
            IF(BUG)WRITE(6,3003)N,P
       3003 FORMAT('PT.',I6,'=',I6)

C       P IS A VECTOR
          2 IF(N.LT.ISTART+1)RETURN
            X=SLOPE*(E(N)-E(N-1))
            IF(X.LE.O)GO TO 200
            IF(N.LT.ISTART+2)RETURN
            Y=SLOPE*(E(N-2)-E(N-1))
C       THE STATEMENT BELOW IS RAINFLOW RULE 3
            IF(X.LT.Y)RETURN
            IF(X.EQ.Y.AND.ISTART.EQ.(N-2))RETURN
            IF(X.GT.Y.AND.ISTART.EQ.(N-2))GO TO 4
            IF(X.GE.Y.AND.ISTART.NE.(N-2))GO TO 5
          4 ISTART=ISTART+1
            RETURN
C
C           LOOP HAS BEEN CLOSED, COUNT IT
C           Y IS THE RANGE
          5 CONTINUE
            SUM=SUM+1
C*            MEAN=(E(N-1)+E(N-2))/2
C
            KA=IFIX(FLOAT(E(N-1)-ALOW)/DA)+1
```

```
C         GENERATE MATRIX TEST FOR TENSION OR COMPRESSION
C            ENSURE COMP. TIP IS THE ROW.
             IF(KA.GT.KB)THEN
                IT=KA
                KA=KB
                KB=IT
             ENDIF
C         ENTER THE REV INTO MAX MIN MATRIX, WHERE MAX IS COL, MIN IN ROW
             IF(BUG)WRITE(6,4200)SUM,E(N-1),E(N-2),KA,KB
      4200   FORMAT("LOOP COUNTED, ",I8," 1ST PT=",I6," 2ND PT=",I6,
           & "KA,KB=",2I4)
             MAT(KA,KB)=MAT(KA,KB)+1
             IF(BUG)WRITE(6,4202)KA,KB,MAT(KA,KB)
      4202   FORMAT("MAT(",I2,",",I2,")=",I7)
             N=N-2
             E(N)=E(N+2)
             GO TO 2
C
C---------------END OF DATA INPUT. EMPTY PUSH DOWN LIST
         6   J=J+1
             IF(J.GT.ISTART)GO TO 999
             N=N+1
             SLOPE=-SLOPE
             E(N)=E(J)
         7   IF(N.LT.ISTART+1)GO TO 6
             X=SLOPE*(E(N)-E(N-1))
             IF(X.LE.O)GO TO 300
             IF(N.LT.ISTART+2)GO TO 6
             Y=SLOPE*(E(N-2)-E(N-1))
C         STATEMENT BELOW IS RAINFLOW RULE 8
             IF(X.LT.Y)GO TO 6
         9   CONTINUE

C*              RANGE=Y
             CNT=CNT+1
C*              XMEAN=(E(N-1)+E(N-2))/2
             KA=IFIX(FLOAT(E(N-1)-ALOW)/DA)+1
             KB=IFIX(FLOAT(E(N-2)-BLOW)/DB)+1
C         GENERATE MATRIX TEST FOR TENSION OR COMPRESSION
             IF(BUG)WRITE(6,4206)KA,KB,E(N-1),E(N-2)
      4206   FORMAT(' KA,KB,E(N-1),E(N-2)=',4I6)
C         CHECK FOR MIN ON THE ROW
             IF(KA.GT.KB)THEN
                IT=KA
                KA=KB
                KB=IT
             ENDIF
C         ENTER THE REV INTO MAX MIN MATRIX, WHERE MAX IS COL, MIN IN ROW
             MAT(KA,KB)=MAT(KA,KB)+1
             IF(BUG)WRITE(6,4600)CNT,E(N-1),E(N-2),KA,KB,MAT(KA,KB)
      4600   FORMAT(" P.D.LIST LOOP ",I8," 1ST PT=",I4," 2ND PT=",I4,
           & "MAT(",I2,",",I2,")= ",I7)
             N=N-2
             E(N)=E(N+2)
             GO TO 7
       200   N=N-1
             E(N)=E(N+1)
             SLOPE=-SLOPE
             GO TO 2
```

```
      300  N=N-1
           E(N)=E(N+1)
           SLOPE=-SLOPE
           GO TO 7
C
C          COUNTING DONE, OUTPUT DATA
      999  IREV=CNT+SUM
C
C          DATA TO TAPE
C*           WRITE(6,9100)
C*    9100   FORMAT("DO YOU WISH RG-MN-OCCUR DATA WRITTEN TO TAPE?")
C*           CALL YESNO(IS)
C*           IF(IS.EQ.0)RETURN
C          YES, GET NON-ZERO MAT ENTRIES, SCALE & OUTPUT.
C          ASSUME CENTER OF INTERVAL AS REPRESENTATIVE.
           WRITE(6,9110)
      9110 FORMAT('DRAINFLOW RANGE-MEAN-CYCLES')
           XMIN=ALOW+DA/2.
           DO 9150 I=1,NA
           DO 9150 J=I,NB
           JVT=MAT(I,J)
           IF(JVT.EQ.0)GO TO 9150
           X1=FLOAT(J-1)*DB+XMIN
           X2=FLOAT(I-1)*DA+XMIN
           RG=ABS(X1-X2)
           XMEAN=(X1+X2)/2.
           IF(RG.EQ.0.)GO TO 9150
           WRITE(6,9130)RG,XMEAN,JVT
C*           WRITE(21,9130)RG,XMEAN,JVT
      9130 FORMAT(2X,G10.3,2X,G10.3,2X,I8)
      9150 CONTINUE
           RG=0.
           XMEAN=0.
           JVT=0
           WRITE(6,9130*RG,XMEAN,JVT
C*           WRITE(21,9130)RG,XMEAN,JVT
C
C
C      PRINT OUT THE INFORMATION FROM RAINFLOW, FIRST THE CYCLES
           WRITE(6,900)IREV
      9000 FORMAT('OTHE TOTAL CYCLES ARE',I6)
C _____
C          exit here normally.
           if(.not.BUG)return
C _____
C      NOW PRINT OUT THE INFORMATION IN THE MATRIX
           ID=NA
           XT=ALOW
           WRITE(6,9001)XT
      9001 FORMAT('OTHE VALUE OF ALOW IS ',F10.4)
           WRITE(6,9020)
      9020 FORMAT('OFIRST HALF MATRIX INFORMATION')
           WRITE(6,9003)XT
           DO 77 KA=1,NA
           WRITE(6,9002)(MAT(KA,KB),KB=1,16)
           XT=XT+DA
           WRITE(6,9003)XT
      9003 FORMAT(3X,F10.0)
       77  CONTINUE
      9002 FORMAT(3X,16(I5))
```

```
      9030  FORMAT('OTHE SECOND HALF OF THE MATRIX IS')
            XT=ALOW
            WRITE(6,9011)XT
            DO 78 KA=1,NA
            WRITE(6,9010)(MAT(KA,KB),KB=17,NB)
            XT=XT+DA
            WRITE(6,9011)XT
      9011  FORMAT(3X,F10.0)
        78  CONTINUE
      9010  FORMAT(3X,16(I5))
C
            RETURN
            END
```

Appendix 5B

Examples of Cycle Counting

The sections which follow provide examples of the cycle counting methods referred to in Section 5.7.2.2. Note that these examples illustrate (1) simple range counting, (2) range-pair counting, (3) rainflow counting, and (4) simplified rainflow counting for repeating histories, which are the methods that can be used as two parameter methods.

5B.1 Range-Pair Counting

The range-pair method counts a range as a cycle if it can be paired with a subsequent loading in the opposite direction. Rules for this method are as follows:

Let X denote the range under consideration; and Y, the previous range adjacent to X.

Step 1: Read the next peak or valley. If out of data, go to Step 5.

Step 2: If there are less than three points, go to Step 1. Form ranges X and Y using the three most recent peaks and valleys that have not been discarded.

Step 3: Compare the absolute values of ranges X and Y.

    (a) If X Y, go to Step 1.
    (b) If X Y, go to Step 4.

Step 4: Count range Y as one cycle and discard the peak and valley of Y; go to Step 2.

Step 5: The remaining cycles, if any, are counted by starting at the end of the sequence and counting backwards. If a single range remains, it may be counted as a half or full cycle.

The load history in Fig. 5B-1 is replotted as Fig. 5B-2a and is used to illustrate the process. Details of the cycle counting are as follows:

1. Y = [A−B]; X = [B−C]; and X Y. Count [A−B] as one cycle and discard points A and B. (See Fig. 5B-2b. Note that a cycle is formed by pairing range A−B and a portion of range B−C.)

2. Y = [C−D]; X = [D−E]; and X Y.

3. Y = [D−E]; X = [E−F]; and X Y.

4. Y = [E−F]; X = [F−G]; and X Y. Count [E−F] as one cycle and discard points E and F. (See Fig. 5B-2c.)

5. Y = [C−D]; X = [D−G]; and X Y. Count [C−D] as one cycle and discard points C and D. (See Fig. 5B-2d.)

6. Y = [G−H]; X = [H−I]; and X Y. Go to the end and count backwards.

7. Y = [H−I]; X = [G−H]; and X Y. Count [H−I] as one cycle and discard points H and I. (See Fig. 5B-2e.)

8. End of counting. See the table in Fig. 5B-2 for a summary of the cycles counted in this example, and see Appendix 5B.2 for this cycle count in the form of a range-mean matrix.

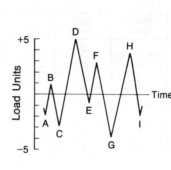

Range (Units)	Cycle Counts	Events
10	0	
9	0	
8	1.0	C-D, G-H
7	0.5	F-G
6	1.0	D-E, H-I
5	0	
4	1.0	B-C, E-F
3	0.5	A-B
2	0	
1	0	

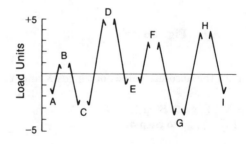

Fig. 5B-1

5B.2 Rainflow Counting

Rules for rainflow counting are as follows:

Let X denote the range under consideration; Y, the previous range adjacent to X; and S, the starting point in the history.

Step 1: Read the next peak or valley. If out of data, go to Step 6.

Step 2: If there are less than three points, go to Step 1. Form ranges X and Y using the three most recent peaks and valleys that have not been discarded.

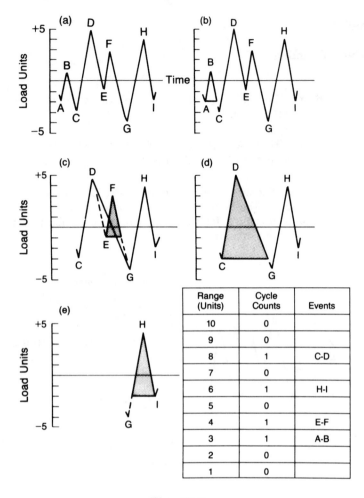

Fig. 5B-2

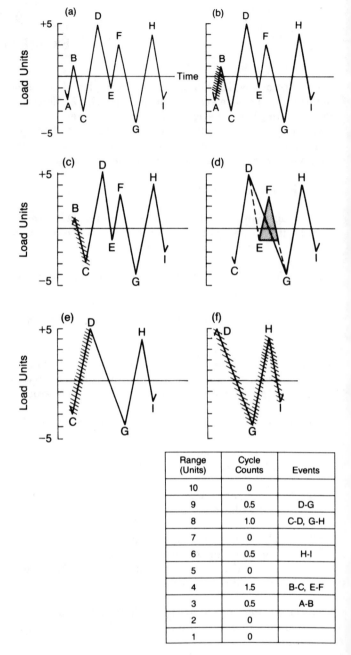

Fig. 5B-3

Step 3: Compare the absolute values of ranges X and Y.

    (a) If X Y, go to Step 1.
    (b) If X Y, go to Step 4.

Step 4: If range Y contains the starting point S, go to Step 5; otherwise count range Y as one cycle; discard the peak and valley of Y; and go to Step 2.

Step 5: Count range Y as one-half cycle (or reversal); discard the first point (peak or valley) in range Y; move the starting point to the second point in range Y; and go to Step 2.

Step 6: Count each range that has not been previously counted as one-half cycle.

The load history of Fig. 5B-1 is replotted as Fig. 5B-3a and is used to illustrate the process. Details of the cycle counting are as follows:

1. S = A; Y = [A−B]; X = [B−C]; X Y. Y contains S, i.e., point A. Count [A−B] as one-half cycle and discard point A; S = B. (See Fig. 5B-3b.)

2. Y = [B−C]; X = [C−D]; X Y. Y contains S, i.e., point B. Count [B−C] as one-half cycle and discard point B; S = C. (See Fig. 5B-3c.)

3. Y = [C−D]; X = [D−E]; X Y.

134

4. Y = [D−E]; X = [E−F]; X < Y.

5. Y = [E−F]; X = [F−G]; X ≥ Y. Count [E−F] as one cycle and discard points E and F. (See Fig. 5B-3d. Note that a cycle is formed by pairing range E−F and a portion of range F−G.)

6. Y = [C−D]; X = [D−G]; X ≥ Y. Y contains S, i.e., point C. Count [C−D] as one-half cycle and discard point C. S = D. (See Fig. 5B-3e.)

7. Y = [D−G]; X = [G−H]; X < Y.

8. Y = [G−H]; X = [H−I]; X < Y. End of data.

9. Count [D−G] as one-half cycle, [G−H] as one-half cycle, and [H−I] as one-half cycle. (See Fig. 5B-3f.)

10. End of counting. See the table in Fig. 5B-3 for a summary of the cycles counted in this example, and see Table 5B.2 for this cycle count in the form of a range-mean matrix.

5B.3 Simplified Rainflow Counting for Repeating Histories

It may be desirable to assume that a typical segment of a load history is repeatedly applied. In this case, once either the maximum peak or minimum valley is reached for the first time, the range-pair count is identical for each subsequent repetition of the history. The rainflow count is also identical for each subsequent repetition of the history, and for these subsequent repetitions, the rainflow count is the same as the range-pair count. Such a repeating history count contains no half cycles, only full cycles, and each cycle can be associated with a closed stress-strain hysteresis loop [1-4]. Rules for obtaining such a repeating history cycle count, called "simplified rainflow counting for repeating histories," are as follows:

Let X denote range under consideration; and Y, previous range adjacent to X.

Step 1: Arrange the history to start with either the maximum peak or the minimum valley. (More complex procedures are available that eliminate this requirement; see Ref. [4].)

Step 2: Read the next peak or valley. If out of data, STOP.

Step 3: If there are less than three points, go to Step 2. Form ranges X and Y using the three most recent peaks and valleys that have not been discarded.

Step 4: Compare the absolute values of ranges X and Y.

(a) If X < Y, go to Step 2.

(b) If X ≥ Y, go to Step 5.

Step 5: Count range Y as one cycle; discard the peak and valley of Y; and go to Step 3.

The loading history of Fig. 5B-1 is plotted as a repeating load history in Fig. 5B-4a and is used to illustrate the process.

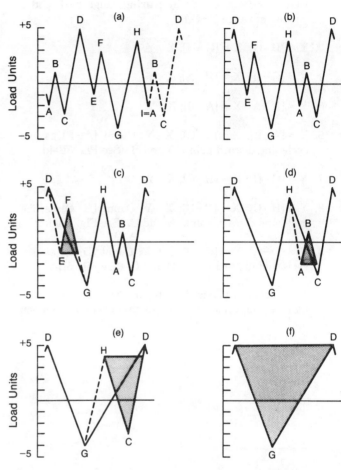

Range (Units)	Cycle Counts	Events
10	0	
9	1	D-G
8	0	
7	1	H-C
6	0	
5	0	
4	1	E-F
3	1	A-B
2	0	
1	0	

Fig. 5B-4

Rearraing the history to start with the maximum peak gives Fig. 5B-4b, reversal points A, B, and C being moved to the end of the history.

Details of the cycle counting are as follows:

1. Y = [D−E]; X = [E−F]; X < Y.

2. Y = [E−F]; X = [F−G]; X > Y. Count [E−F] as one cycle and discard points E and F. (See Fig. 5B-4c. Note that a cycle is formed by pairing range E−F and a portion of range F−G.)

3. Y = [D−G]; X = [G−H]; X < Y.

4. Y = [G−H]; X = [H−A]; X < Y.

5. Y = [H−A]; X = [A−B]; X < Y.

6. Y = [A−B]; X = [B−C]; X > Y. Count [A−B] as one cycle and discard points A and B. (See Fig. 5B-4d.)

7. Y = [G−H]; X = [H−C]; X < Y.

8. Y = [H−C]; X = [C−D]; X > Y. Count [H−C] as one cycle and discard points H and C. (See Fig. 5B-4e.)

9. Y = [D−G]; X = [G−D]; X = Y. Count [D−G] as one cycle and discard points D and G. (See Fig. 5B-4f.)

10. End of counting. See the table in Fig. 5B-4 for a summary of the cycles counted in this example, and see Table 5B.3 for this cycle count in the form of a range-mean matrix.

5B.4 References

[1] Matsuishi, M. and Endo, T., "Fatigue of Metals Subjected to Varying Stress," Paper presented to Japan Society of Mechanical Engineers, Fukuoka, Japan, March 1968. See also T. Endo, et al., "Damage Evaluation of Metals for Random or Varying Loading," *Proc. of the 1974 Symposium on Mechanical Behavior of Materials*, Vol. 1, the Society of Materials Science, Japan, 1974, pp. 371-380.

[2] Richards, F. D., LaPointe, N. R., and Wetzel, R. M., "A Cycle Counting Algorithm for Fatigue Damage Analysis," Automotive Engineering Congress, Society of Automotive Engineers, Paper No. 740278, Detroit, MI, February 1974.

[3] Dowling, N. E., "Fatigue Failure Predictions for Complicated Stress-Strain Histories," *Journal of Materials*, ASTM, Vol. 7, No. 1, March 1972, pp. 71-87.

[4] Downing, S. D. and Socie, D. F., "Simplified Rainflow Counting Algorithms," *Int. Jnl. of Fatigue*, Vol. 4, No. 1, January 1982, pp. 31-40.

Chapter 6 Vehicle Simulation

6.1 Introduction
   6.1.1 Scope
   6.1.2 Relationship to Other Chapters
   6.1.3 Chapter Plan

6.2 Suspension Design Cycle
   6.2.1 Selection of Front or Rear Suspension
   6.2.2 Suspension Geometry Optimization
   6.2.3 Kinematic Modeling of Front or Rear Suspensions
   6.2.4 Dynamic Modeling of Front or Rear Suspensions
   6.2.5 The Full Vehicle Model

6.3 Simulation Types

6.4 Full Vehicle Model Simulations
   6.4.1 Suspension Abuse Test—Lab Simulations
   6.4.2 Full Vehicle Model Road-Load Simulation

6.5 Summary/Remarks

6.6 References

6.1 Introduction

6.1.1 *Scope*

The following material is intended to illustrate the application of computer simulations as an aid in the design of passenger car suspensions. The method is used to determine early road-load data for suspension components. This early road-load data can be used in conjunction with finite element and fatigue analysis methods to estimate the fatigue lives of suspension components *before* the first prototypes are made. This particular component package is used only as an example. The analysis methods are applicable to other vehicular designs requiring dynamic considerations.

6.1.2 *Relationship to Other Chapters*

The computer simulation techniques described in this chapter are, in many ways, an analytical equivalent to the experimental techniques described in Chapter 4. The analytical techniques are much more recent in their evolution than experimental procedures, and for that reason, they are not as widely used or accepted in the ground vehicle industry. However, great advances in the sophisticated and widespread usage of these computer simulation techniques are expected in the near future. The output from computer simulations (road-loads) can be used as direct inputs to computer stress algorithms as described in Chapter 8. If the computer simulation package automatically computes local stress and strains, it may be possible to use this information as direct input to a fatigue life prediction as described in Chapter 10.

6.1.3 *Chapter Plan*

This chapter is divided into three brief subsections. Section 6.2 describes the new suspension design cycle and shows where in the cycle the technique can be applied. Section 6.3 describes the types of simulations that are commonly per-

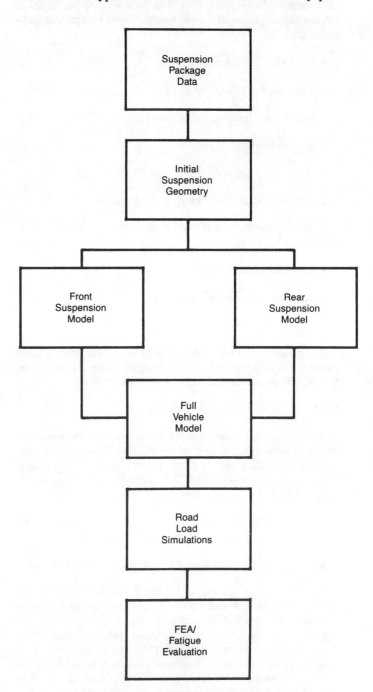

Fig. 6-1 The design cycle including vehicle simulation techniques.

formed. Section 6.4 provides two examples of computer simulations that utilize the technique. Fig. 6-1 shows the process in block form.

6.2 Suspension Design Cycle

Product planners use market survey indicators to decide what type of vehicle they want to manufacture. Following this decision, full vehicle, as well as front and rear suspension design assumptions are established. Some typical assumptions are:

- full vehicle dimensions
- suspension package space
- suspension technical objectives
- cost and weight objectives
- other

6.2.1 Selection of Front or Rear Suspension

Front or rear suspension selection is accomplished by considering, at a minimum:

- package space available
- suspension technical objectives
- cost and weight objectives

The suspension design that is the best compromise for the above criteria is selected for vehicle application. At this point, the initial suspension layouts have been created.

6.2.2 Suspension Geometry Optimization

Most vehicle manufacturers have suspension geometry analysis software that is used to optimize geometry. Geometry characteristics such as camber, caster, and toe-in are optimized using these programs. The initial suspension layouts are updated to reflect geometry optimization.

The suspension geometry points along with a typical short-long arm (SLA) suspension are shown in Fig. 6-2.

6.2.3 Kinematic Modeling of Front or Rear Suspensions

By using programs such as ADAMS, DADS, SDRC, and others, kinematic models can be constructed for the front and rear suspensions. The basic input for the models is the geometry points as taken from the suspension layouts. Various joint types particular to each program provide for suspension connectivity. The suspension models should be validated by comparing the geometry of the layout with the geometry as generated by model jounce/rebound travel simulation.

6.2.4 Dynamic Modeling of Front or Rear Suspensions

The same programs used above can be used to construct dynamic suspension models. These models can be used to study suspension load distribution and used as the front and rear suspension subsystems of a full vehicle model. In addition to the geometry points, adding the following component data to the kinematic model will create a dynamic model:

- mass
- inertia
- c.g.
- springing members
- bushings

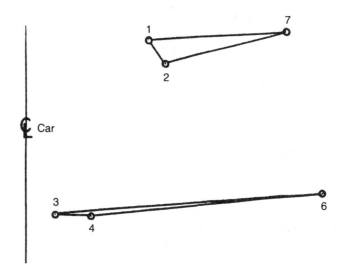

(a) Front Suspension Geometry Points

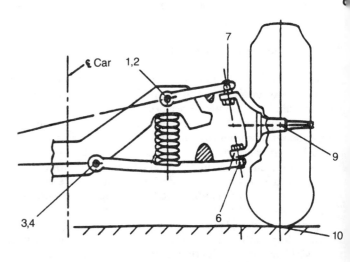

(b) Short Long Arm (SLA) Front Suspension

Fig. 6-2 Example suspension geometry.

- shock/struts

- jounce/rebound bumper

Fig. 6-3 shows typical force members added to form a dynamic suspension model. Validation of the dynamic model should be done. Checking suspension rates, jounce/rebound bumpers, and shock forces can validate the models functional characteristics.

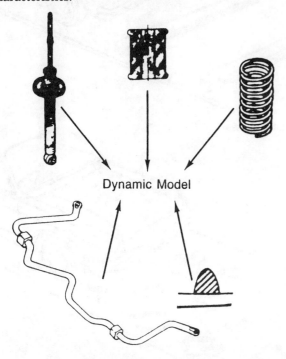

Fig. 6-3 Typical force members that comprise a dynamic suspension model. (Part 1)

6.2.5 The Full Vehicle Model

Much of the work for creating the full vehicle model has already been done. The front and rear suspension dynamic models are the major components of the full vehicle model. In addition, a body and tires must be included. Defining the body part requires the mass, c.g., and inertia data be input to the model. Tires must also be added using a suitable tire model. Data typical for tire model definition are: vertical and lateral stiffness, static loaded radius, free radius, rolling resistance, etc. Combining: (1) front and rear suspension, (2) body, and (3) tires complete the full vehicle model.

Fig. 6-4 shows a typical full vehicle model.

6.3 Simulation Types

Various types of simulations can now be performed on the full vehicle model. Suspension load data can be obtained for

Mechanical problem	Domain	Type of boundary-value problem
torsion of prismatic bars		elliptic type equation with constant coefficients
torsion of non-prismatic, axially-symmetric bars		elliptic type equation with variable coefficients
bending of plates of arbitrary shape		biharmonic equasion
		variational approach
stationary heat flow in a mechanical device		elliptic type equation in a multiconnected domain
non-stationary heat flow		parabolic type equation
plane stress analysis, stress concentration		set of elliptic type equations
elasto-plastic torsion of prismatic bars		nonlinear elliptic type equation
large deflections of membranes		nonlinear elliptic type equation
large deformations of membrane shells		set of nonlinear equations of variable type ell.-hyp.

Fig. 6-3 Typical force members that comprise a dynamic suspension model. (Part 2)

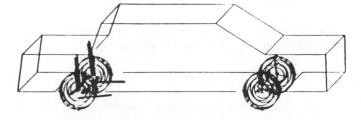

Fig. 6-4 Full vehicle model.

lab simulations or for road profiles that model test track durability events. Cobblestone, chuckhole, J-turns, etc., are typical of durability events that generate high suspension loading and considered to cause considerable damage to suspension components. The following two examples illustrate both types of simulations.

6.4 Full Vehicle Model Simulations

6.4.1 *Suspension Abuse Test—Lab Simulations*

Suspension abuse tests [1] were simulated, that included (1) curb impact, (2) chuckholes, (3) railroad ties, (4) rough roads, (5) wheel drops. Fig. 6-5 shows these events.

Laboratory tests were conducted that simulated these vehicle durability type events. An ADAMS model was constructed for the vehicle that used a simple spring and camper to model the tires. The simulation results for the lower ball joint forces along with the strut forces for each test are given in Fig. 6-6. The results of the ADAMS simulation were in good agreement with those measured on the actual experimental simulation.

6.4.2 *Full Vehicle Model Road-Load Simulation*

A full vehicle ADAMS model using ADAMS/tire software to model tires was used to simulate the chuckhole and railroad tie durability events. Fig. 6-7 shows these events (Ref. 2).

The results for the vertical forces of the front left and right lower ball joints are given in Figs. 6-8 and 6-9.

Although there are no experimental data available for comparison, the loads based on experience appear to be reasonable. Fig. 6-10 shows the relationship in time for the vertical and longitudinal force at the left lower ball joint. It is seen that the maximum value occurs at different instants of time.

6.5 Summary/Remarks

In summary, the method flows as follows:

- generate a background suspension layout from assumptions
- using background layout along with suspension technical objectives, select the suspension type
- optimize the suspension geometry
- create the suspension kinematic model
- create the dynamic model

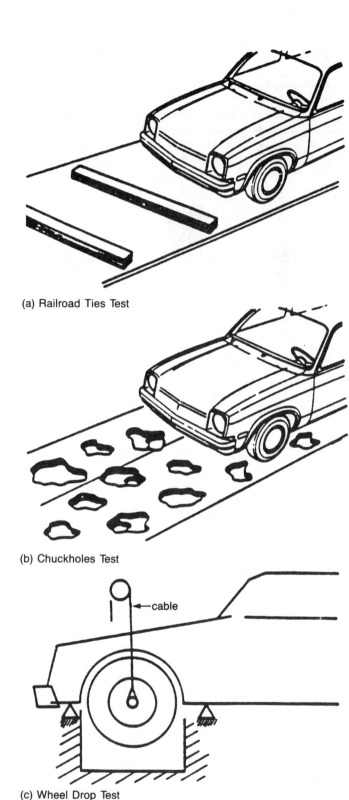

(a) Railroad Ties Test

(b) Chuckholes Test

(c) Wheel Drop Test

Fig. 6-5 Suspension abuse tests. (Part 1)

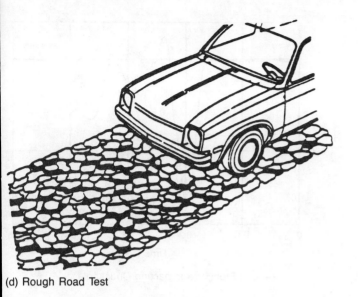

(d) Rough Road Test

Fig. 6-5 Suspension abuse tests. (Part 2)

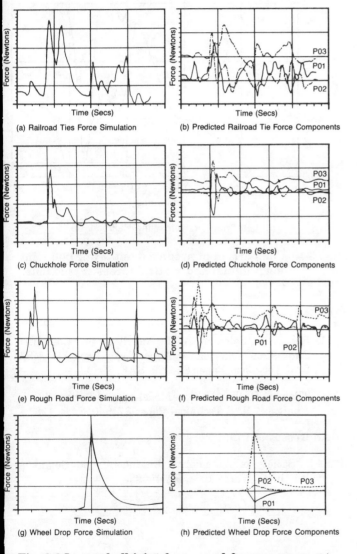

(a) Railroad Ties Force Simulation
(b) Predicted Railroad Tie Force Components
(c) Chuckhole Force Simulation
(d) Predicted Chuckhole Force Components
(e) Rough Road Force Simulation
(f) Predicted Rough Road Force Components
(g) Wheel Drop Force Simulation
(h) Predicted Wheel Drop Force Components

Fig. 6-6 Lower ball joint forces and force components.

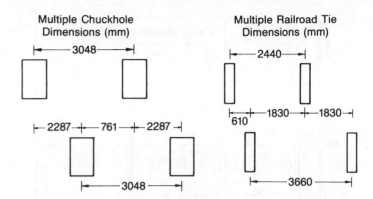

Fig. 6-7 Chuckhole and railroad tie durability events.

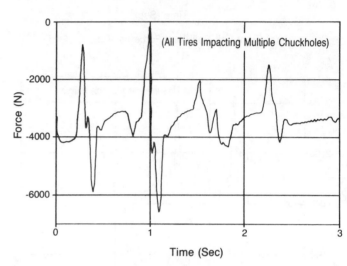

(a) Front-Left Ball Joint Force vs. Time

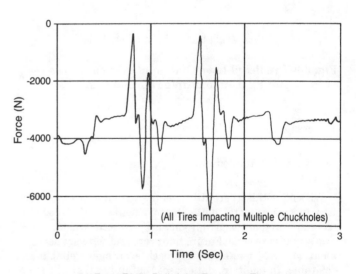

(b) Front-Right Ball Joint Force vs. Time

Fig. 6-8 Predicted lower ball joint vertical forces resulting from multiple railroad ties.

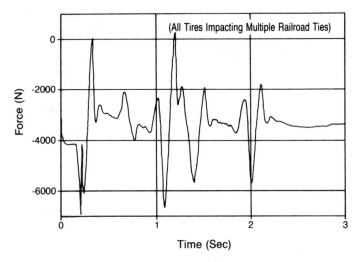

(a) Front-Left Ball Joint Force vs. Time

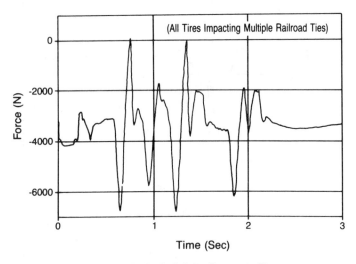

(b) Front-Right Ball Joint Force vs. Time

Fig. 6-9 Predicted lower ball joint vertical forces resulting from multiple railroad ties.

- create the full vehicle model
- run the simulated event

With all of this, there still remains a great need for correlation to experimental data. Improvements to the modeling technique and/or the actual software will occur when correlation studies are done. Future hardware and software development will only make this approach even more reliable and desirable. In addition, data bases containing typical component data would be very valuable.

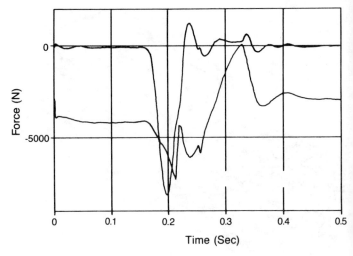

(a) Left Front Tire Impacting Chuckhole

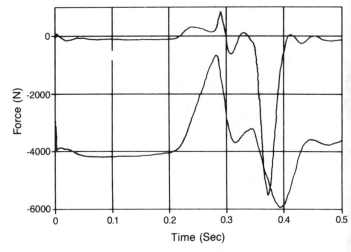

(b) Left Front Tire Impacting Railroad Tie

Fig. 6-10 Comparison of predicted left-lower ball joint longitudinal and vertical forces versus time.

6.6 References

[1] Ral, N. S., Solomon, A. R., Angell, J. C., "Computer Simulation of Suspension Abuse Tests Using ADAMS," SAE Technical Paper 820079, SAE, 1982.

[2] Thomas, D. W., "Vehicle Modeling and Service Loads Analysis," SAE Technical Paper 871940, SAE, 1987.

Addresses of Publishers

SAE, Society of Automotive Engineers, 400 Commonwealth Drive, Warrendale, PA 15096

Chapter 7 Strain Measurement and Flaw Detection

7.1 Introduction
    7.1.1 Scope
    7.1.2 Relationship to Other Chapters
    7.1.3 Chapter Plan

7.2 Techniques to Define Fatigue Critical Location
    7.2.1 Brittle Coatings
    7.2.2 Strain Gages
    7.2.3 Photoelasticity
    7.2.4 Holographic and Speckle Interferometry
    7.2.5 Thermoelasticity

7.3 Flaw Detection Techniques
    7.3.1 Ultrasonic
    7.3.2 Dye Penetrant
    7.3.3 Magnetic Particle
    7.3.4 Radiography
    7.3.5 Eddy Current
    7.3.6 Holography
    7.3.7 Acoustic Emission
    7.3.8 Potential Drop
    7.3.9 Gel Electrode

7.4 Residual "Self-Stress" Determination
    7.4.1 X-ray Diffraction Method
    7.4.2 Sectioning
    7.4.3 Blind Hole Drilling
    7.4.4 Ultrasonic Method
    7.4.5 Barkhausen Noise
    7.4.6 Other Methods

7.5 Advantages and Limitations of Various Techniques

7.6 References

7.1 Introduction

7.1.1 Scope

This chapter first summarizes experimental techniques widely used for identifying regions of concentrated stress or strain, where fatigue cracks generally have the greatest likelihood of forming, and for quantifying the stress or strains in those regions. Next, techniques for detecting, locating, and determining the size of material flaws are described. Such flaws may be present from the time of fabrication (e.g., cast or welded components) or develop as a result of fatigue loadings. Finally, since residual stresses can have a strong influence on fatigue performance, experimental techniques for determining such stresses are also summarized.

7.1.2 Relationship to Other Chapters

The experimental techniques concerned with defining fatigue critical locations relate primarily to service history determinations (Chapter 5) and to making life predictions (Chapter 10) since knowledge of stresses and strains likely to be encountered in service is of crucial importance in estimating component life.

Flaw detection techniques are of interest to determination of material properties (Chapter 3) and to effects of processing on fatigue performance (Chapter 4) since determination of fatigue crack initiation and crack growth properties depends on our ability to detect and monitor cracks, and since defects that may result from processing can have a marked influence on the fatigue behavior of components. Flaw detection techniques also relate to life prediction (Chapter 10) and to failure analysis (Chapter 11) since knowledge of initial flaw size is needed for predicting life spent in crack growth, and since the causes of failure are often linked to the presence of material defects.

The techniques for determination of residual or "self" stresses are related primarily to the same chapters as the techniques for flaw detection. Such stresses are produced by various processing procedures, and their presence can affect the determination of fatigue crack initiation and crack growth properties. Life predictions should ideally take them into account, and the cause of failure is often related to their presence in components.

7.1.3 Chapter Plan

This chapter provides a description of the main features of each of the experimental techniques commonly employed in strain measurement and flaw detection. In the space available here, it is not possible to discuss all of the important details involved in the implementation of a given technique. Instead, references are cited that are readily obtainable by a reader and which consider such details as well as theory behind a technique, when relevant. Finally, a comparison of the relative advantages and limitations of those techniques in a certain category (e.g., flaw detection) is given in Section 7.5.

When it actually comes to applying a particular technique, the technical literature and advice of companies that furnish the equipment and supplies associated with the technique are needed. A number of companies offer such services. For some of the well-established techniques (e.g., brittle coatings, strain gages), a few companies dominate the market. These companies are generally well-known throughout industry by virtue of the widespread use of their products, educational seminars and advertising. For some of the newer techniques (e.g., crack detection and sizing by modern non-destructive methods), numerous companies are involved, but to attempt to identify and list all of them would be a difficult task and one which could easily miss certain companies. Moreover, by the time a list was published and read at some later time, it could well be out-of-date. In this case, companies can generally be found by referring to industrial and trade directories, by their advertisements in periodicals read by design and test engineers, and by word-of-mouth from those who are utilizing a particular technique. Finally, it should be noted that some references in this chapter cite the literature of certain companies, but this is not intended to be an endorsement or any indication of favoritism.

7.2 Techniques to Define Fatigue Critical Location

Experimental stress analysis is a necessary and important stage in most product design evaluations and may be applied to models or prototypes of components and machines. A two-step procedure in common use is to first apply a "whole field" technique like brittle coating to qualitatively study the strain distribution over the surface of a part, followed by a "point" technique like strain-gaging to quantitatively determine strains at fatigue critical locations. Powerful optical methods such as photoelasticity, holography, and thermoelasticity are other whole field techniques that may be used. Application of all of these techniques in fatigue loading situations requires more careful control of experimental parameters than for static analyses, especially if precise quantitative data are required.

7.2.1 *Brittle Coatings*

Resin based or ceramic brittle coatings, bonded to the surface of a part, develop crack patterns when subjected to loading. The primary use of such coatings is to locate regions of tensile stress concentration and to determine principal strain directions for subsequent strain-gage placement. For resin-based coatings, the following steps are needed for preparation: (a) the surface to be investigated is cleaned and degreased; (b) a reflective coating is applied; (c) the brittle coating is sprayed on by pressure cans or air-spray guns; and (d) the coating is dried at several degrees above the test temperature. If quantitative results are expected, calibration bars should be coated simultaneously and in an identical fashion to the test part. The threshold strain required to cause the coating to crack must be determined from the calibration bars, generally loaded in a cantilever fixture. Typically, the threshold strain is on the order of 500 micrometers per meter. The component is loaded in as short a time as practicable (15 seconds to one minute) and then unloaded. The time to load is noted and a creep correction applied from available charts (ceramic coatings are not affected by creep at ambient temperatures). The line around the boundary of the cracked region is called an isoentatic, that encloses an area where the strain equals or exceeds the threshold value. Crack lines are called isostatics and are perpendicular to the maximum principal stress direction if the secondary principal stress is small. (A random crack pattern is observed under equi-biaxial tension conditions.) Crack patterns may be studied under oblique lighting or colored with a red dye for photographic purposes. Great care in calibration and interpretation is needed if quantitative results are expected, since the technique is, for example, quite sensitive to temperature, humidity, and variations in coating thickness. A more detailed description of brittle coating methods may be found in Refs. [1-3].

7.2.2 *Strain Gages*

The bonded electrical resistance strain gage is the most widely used method of experimental stress analysis due to its versatility and precision of measurement. In these gages, a thin conductor adhesively bonded to a surface subjected to strain undergoes dimensional changes that, in turn, cause changes in its resistance and in the resistivity of the material. Common types of gages available are metallic wire, foil, semiconductor, and weldable. Metallic foil gages have found the most widespread application and can be used to measure strains as small as one microinch per inch.

Commonly used metallic gage materials for room temperature use are constantan, isoelastic, and karma alloys, each having different drift, hysteresis, nonlinearity, and resistance to cyclic strain. Epoxy, phenolic, polyimide, and polyamide resins are used as backing with occasional glass fiber reinforcement. Careful and precise installation and alignment of gages and lead wires are needed to ensure reliable results. The adhesive cement selected to bond the gage to the surface can also affect gage performance. Other factors influencing strain gage usage are temperature sensitivity and transverse sensitivity. Self-temperature compensated gages, compensating "dummy" gage circuit arrangements, and correct lead wire configurations can reduce thermally induced apparent strain. Formulas for correcting for the influence of both transverse and temperature sensitivities are also available. (Often the influence of transverse sensitivity can be neglected since it is generally on the order of one percent.) For semiconductor strain-gages (silicon P or N types), the resistance changes are nonlinearly related to strain although their properties are suited for use in miniature accelerometers and pressure transducers. They are roughly two orders of magnitude more strain sensitive than metallic foil gages but are more fragile and temperature sensitive and have much lower operating strain limits.

Two electrical circuits most commonly used in strain gage work are the potentiometer and the Wheatstone bridge, that convert the resistance change to a voltage output for data acquisition equipment. The potentiometer circuit measures total potential drop across the gage and is usually suitable for dynamic measurements. The Wheatstone bridge has found widespread use, being employed in both the null-balanced and unbalanced modes for static measurements and in the unbalanced mode for dynamic measurements. Excitation signals may be either direct or alternating current depending on available circuitry. Twisted lead wires must be used whenever electrical noise due to stray magnetic fields is expected. Cables with grounded shields may be used to reduce electrostatic noise. Strain measurements on rotating components may be achieved by slip rings or telemetry. Here, shielded wires are a must and if a Wheatstone bridge circuit is to be used, the entire bridge should be connected on the rotating member before the bridge output is transmitted through slip rings or by telemetry. (See Section 5.4.1.1 for further details on shielding.)

Strain gages are available in a variety of configurations, materials, and sizes from a number of manufacturers. Special configurations are available for constructing torque, shear-strain, or diaphragm (pressure) type transducers. Long-term transducer installations require high quality materials and installation for adequate sensitivity, low drift and hysteresis, and resistance to moisture. Gages are available in rosette configurations for determination of principal strain magnitudes and directions. Detailed information on electrical strain gages may be found in Refs. [1, 4-8].

For gages subjected to cyclic strain, constantan-alloy polimide-encapsulated gages last 10 to 20 percent of the fatigue life of structural alloys while non-encapsulated gages last only a few percent of the life [9]. Significant zero shift errors (of 100 microinches per inch) usually occur as the grid alloy hardens or softens and its specific resistance changes. Further improvement in fatigue performance can be obtained by using karma or isoelastic alloy fully encapsulated glass-reinforced epoxy-phenolic carrier gages with factory installed lead wires.

Finally, it should be noted that special care must be exercised when strain gages are used on materials that are poor heat conductors, such as plastics, since local heating effects can disturb readings.

7.2.3 Photoelasticity

Photoelastic techniques utilize the phenomenon of a change in optical properties of certain transparent materials subjected to load in order to obtain stress distributions and magnitudes in models and on components of complex geometry. These materials are called birefringent since they develop different indices of refraction along the two principal stress directions, resulting in destructive interference or reinforcement of light emerging from them. Photoelasticity may be utilized as a whole field technique in two forms. In transmission photoelasticity, a small two-dimensional model (i.e., a thin slice in a state of plane stress) is placed in a transmission polariscope for evaluation, while in reflection photoelasticity, a full-size component is covered with a coating of photoelastic material and illuminated, viewed and analyzed from a distance using a reflection polariscope.

In a plane transmission polariscope, light (monochromatic or white) is polarized in a plane parallel to a polarizer's axis, transmitted through a stressed planar bi-refringent model, and the resulting light polarized in a plane parallel to an analyzer's axis. (See Fig. 7-1.) The resulting view shows isochromatics (lines of constant maximum shear stress) and isoclinics (lines of constant principal stress direction). The number of isochromatic fringes are dependent on stress level, material stress-optical coefficient, and model thickness. Dark field patterns (no light while unstressed) are obtained if the polarizer and analyzer axes are perpendicular, that allows integral fringe orders to be measured. Light field patterns (no dark areas while unstressed) are obtained if the polarizer and analyzer axes are parallel, from which one-half fringe orders may be measured (0.5, 1.5, etc.). When monochromatic light is used, isochromatics cannot necessarily be distinguished from isoclinics. However, when white light is used, the colorful isochromatics can be easily distinguished from the black isoclinics.

In a circular polariscope, two quarter-wave plates (producing a quarter wavelength phase difference) are oriented with their axes at ±45 degree angles relative to the polarizer axis and are placed on either side of the model. (See Fig. 7-2.) The resulting isochromatic fringe pattern is not a function of the principal stress direction so that isoclinics are eliminated. Since the isochromatics give the difference in principal stresses (maximum shear stress), stress separation methods are available, such as through solution of the La Place equation, to determine the sum of the principal stresses, thus allowing determination of the individual principal stresses.

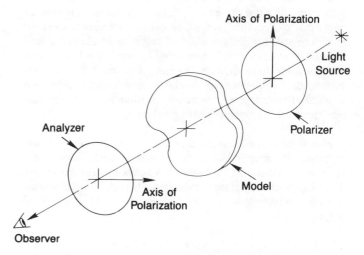

Fig. 7-1 Schematic of plane polariscope.

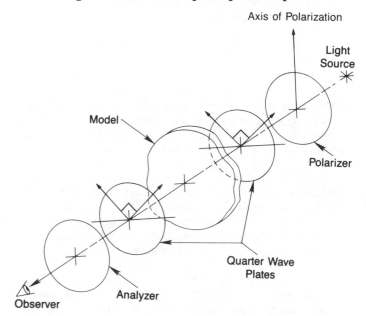

Fig. 7-2 Schematic of circular polariscope.

Compensation methods can be employed in order to determine fractional fringe orders for higher precision. In the Tardy method of compensation, the analyzer is rotated through an angle until destructive interference is obtained; the polarizer and analyzer axes are already aligned with the maximum principal stress directions at the point of interest. A simple wedge or a Babinet Soleil Compensator has an active birefringent element that can be adjusted so that zero relative retardation is achieved at the point of interest and the fringe order read from the adjustment apparatus. Greater precision can also be obtained by fringe multiplication that is achieved by placing partial mirrors on either side of the photoelastic model.

Stress distributions determined with photoelastic models represent, with few exceptions, those in geometrically similar parts made of metals (and other nearly isotropic materials) since stresses are independent of modulus of elasticity and Poisson's ratio (in simply connected bodies when body forces are absent or uniform, as in the case of gravity). Stress magnitudes determined with photoelastic models can also be used to estimate those in geometrically similar parts of different size through appropriate scaling relations.

Photoelastic model materials need to have linear elastic stress-strain properties, possess optical and mechanical isotropy, resist creep, and have a low temperature sensitivity. The classes of materials available with such characteristics include polycarbonates, epoxy resins, polyesters, and polyurethanes.

In three-dimensional photoelasticity, a model of a component is loaded statically and heated to a critical temperature where secondary molecular bonds in the photoelastic material break down and the load is carried by primary bonds. After elevated temperature deformation has stabilized, the loads are removed and the model is cooled slowly to avoid the introduction of temperature gradients and associated thermal stresses. Upon cooling the secondary bonds re-form and lock or "freeze" the stress pattern into the model. Slices are then carefully machined from the model at regions of interest and analyzed by two-dimensional methods. Individual principal stresses (three in this case) may be obtained by analytical techniques such as the shear difference method.

In reflection photoelasticity, a component or prototype is covered with a thin photoelastic coating and observed under loading through a reflection polariscope, depicted in Fig. 7-3. Thin sheets of photoelastic material may be bonded with an aluminized reflective adhesive to flat surfaces. Otherwise, a sheet cast from liquid photoelastic material may be shaped to a curved surface while it is soft, allowed to harden, and then bonded with reflective adhesive in a manner that eliminates air bubbles. A reflection polariscope may be used in a number of different configurations and with a variety of options (e.g., digital read-out) to analyze regions of interest. It is conventional in reflection photoelasticity to express results in terms of strains rather than stresses, and calibration bars are used to relate strain levels to isochromatic fringe orders (or fractions thereof). Dynamic evaluations of cyclically loaded components can be made with the aid of a strobe light.

The maximum strain sensitivity of the above photoelastic techniques is approximately 10 microinch per inch, when methods are employed to determine fractional fringe orders. When used for qualitative evaluations, regions of stress concentration are immediately revealed by the presence of closely packed fringes. A more detailed description of the

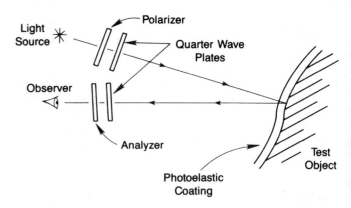

Fig. 7-3 Schematic of reflection polariscope.

techniques and available equipment may be found in Refs. [1, 4, 10-12].

7.2.4 Holographic and Speckle Interferometry

The primary purpose of holographic interferometry is to reveal microscopic changes in the geometry of test components undergoing mechanical or thermal loading. This encompasses determination of deflections and vibration behavior (e.g., resonant frequencies, vibration modes) based on observation of out-of-plane deformations (i.e., normal to the surface of a component) as well as strains derived from in-plane displacements. A holographic image will show displacements over the surface of a component as small as a fraction of the wavelength of light, from which strains on the order of 1 to 10 microinch per inch can be resolved depending on the "gage length" selected. Even higher sensitivity can be obtained by electronic processing of the holographic data, if needed.

The essence of the holographic interferometric technique is as follows. A laser beam is split into an object beam that illuminates the component being evaluated and a reference beam that proceeds directly to a photosensitive plate. The object beam reflected from the component interacts with the reference beam on the plate. With no load on the component, the plate is exposed to the object and reference beams and a hologram developed. With current commercially available holography camera systems, exposure and development is done in a matter of seconds using a plate made of a thermoplastic recording medium and a low power, continuous wave laser for illumination. During the exposure period, it is necessary that the test set-up remain dimensionally stable, that generally requires use of a vibration isolation table. When the component is then loaded, either statically or dynamically, resulting deformations of the component cause changes in the path length of the object beam, as depicted in Fig. 7-4. These changes vary from point-to-point on the surface, and manifest themselves as a pattern of interference fringes on the hologram. The spacing between each fringe corresponds to a fixed amount of displacement, that is known precisely from the laser wavelength and the angles of illumination and viewing of the component. The hologram can then be viewed in real time through a TV camera and video monitor to determine how changes in subsequent loading affect the deformation or, in the case of static loading, the fringe pattern can be stored on videotape or digitized or a hard copy made from the video monitor for later quantitative evaluation. An alternate approach is to first expose a hologram plate to the object and reference beams, then load the component statically, then re-expose the plate, and finally develop it. The resulting hologram permanently records the interference fringe pattern. This "double exposure" technique was conventional in the earlier days of holographic interferometry when it was necessary to use photographic plates for producing holograms, which required lengthy processing time in a dark room. While the technique can still be used if desired, the availability of thermoplastic recording plates, that can be quickly erased and re-used hundreds of times, makes the real time viewing technique more convenient in most situations.

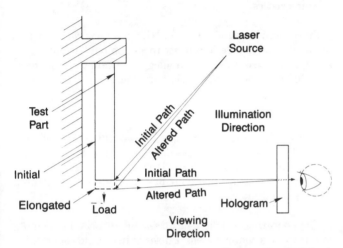

Fig. 7-4 Schematic of change in path length of object beam in holographic interferometric determination of deformation.

When the illumination and viewing directions are both set up to be approximately normal to the surface of a component, the interference fringe patterns will respond to out-of-plane displacements, and this is the configuration used most frequently for vibration visualization and analysis, as well as deflection of beams, plates, membranes, etc. If the illumination angle grazes a component (i.e., is close to parallel to its surface), then the fringe patterns will respond in large part to in-plane displacements, although if any significant out-of-plane displacements are also present, they will also influence patterns. However, methods have been developed for separating out the influences of out-of-plane deformation as well as any rigid body motions that might occur during load application.

Normally, any dynamic excitation for use with holographic interferometry is done with sinusoidal loading, rather than impulses or wideband signals such as engine noise. If necessary, the structural behavior of a test component in an environment with "dirty," in-situ excitation can be investigated using a pulsed laser to rapidly expose a hologram and thereby "freeze the action." Pulsed lasers provide exposure times on the order of nanoseconds, but are relatively expensive and specialized devices.

For identifying fatigue critical locations and quantifying strains there, two basic approaches can be employed with holographic interferometry. It is most convenient optically to use holography in a configuration where illumination and viewing directions of the object beam are approximately normal to the surface of a test component. As noted before, out-of-plane deformations will be revealed in this configuration. Regions of stress concentration will often produce significant corresponding out-of-plane displacements. The appearance of regions of closely packed interference fringes would thus indicate, at least qualitatively, regions of stress concentration. However, the out-of-plane configuration would not be convenient for trying to quantitatively relate observed out-of-plane displacements to in-plane displacements and strains. The other approach is to use a holography set-up with grazing illumination, that is more directly responsive to in-plane displacements. The optical configuration is not quite as simple as that for out-of-plane measurements, but the additional complexity is very modest. In this situation, regions of stress concentration will be revealed with little ambiguity, and strains can be determined from displacements measured over selected "gage lengths." Depending on the particular component and loading situation, it may be necessary to use several holographic images (e.g., different angles of illumination) to determine strains if out-of-plane deformations are significant relative to in-plane ones. In any case, all images can be essentially captured simultaneously.

Details concerning the theory behind and applications of holographic interferometry can be found in Refs. [13-18]. Owing to the rapid advances being made in use of holographic techniques in structural analysis, it is advisable to contact those companies supplying holographic equipment and developing the technology for widespread usage of the techniques, in order to obtain the most up-to-date information as well as the assistance needed to ascertain the suitability of utilizing holographic interferometry in a given application.

Another laser-based technique is speckle interferometry. The surface of a diffusely reflecting object illuminated by laser light will have a granular appearance of randomly distributed, twinkling light and dark speckles. In holographic interferometry, speckle is considered somewhat of a nuisance ("optical noise") that reduces the clarity of interference fringes. However, various techniques have been developed to utilize speckle patterns to determine in-plane displacement and strain. Optical set-ups required for speckle-based measurements are simpler and requirements for mechanical stability less stringent than those for holographic interferometry; however, the sensitivity of the speckle method is generally less than that of holographic interferometry.

As an example of one speckle technique, a test component is illuminated by laser light and imaged by a lens onto a photosensitive plate, that is double exposed with the component being deformed between exposures. The resulting developed plate is called a "specklegram," that can be optically filtered to yield a fringe pattern which portrays displacement derivatives, from which strains can be obtained. In general, the speckle technique allows in-plane and out-of-plane displacements to be determined independently.

The sensitivity of the technique varies depending on the particular optical configuration and method used, but is typically about an order of magnitude less sensitive than holographic interferometry. Since speckle techniques continue to evolve for use in strain determination, it is again advisable to contact those companies offering optical/laser products and technology for use in experimental mechanics to determine the applicability of speckle in a particular test situation. Details concerning various speckle techniques may also be found in Refs. [13, 14, 18-23].

7.2.5 *Thermoelasticity*

This technique is based on the principle that an object undergoing an adiabatic, elastic change in stress will also experience a small temperature change, that can be detected and monitored in terms of the infra-red radiation emitted from its surface. The change in temperature at a given point is proportional to the sum of the principal stresses. Although the theoretical principles of thermoelasticity were described by Lord Kelvin in the middle of the nineteenth century, only recently has instrumentation been developed with adequate sensitivity to allow the thermoelastic effect to be utilized for quantitative stress determination.

Implementation of the technique requires that a test specimen or structure be subjected to cyclic loading of sufficient frequency (on the order of 5 to 10 Hz) to ensure adiabatic behavior. (Broadband random loading has also been used.) The test object is scanned from a distance to produce a map of infra-red radiation, utilizing commercially available equipment consisting of a scanning head and detector unit, an analog signal processing unit and a digital control, storage and display unit. The equipment may be used in the lab or in the field.

The output of whole-field thermoelastic scanning shows regions of stress concentration and can be calibrated to give quantitative results using, for example, an orthogonal pair of strain gages on the test object to give the sum of principal stresses. Stress resolution is claimed to be on the order of 1 MPa for steels and 0.35 MPa for aluminum alloys, or, equivalently, about 5 micrometers per meter of strain, in laboratory environments. Details concerning the technique can be found in Refs. [24-27].

7.3 Flaw Detection Techniques

From the viewpoint of satisfactory fatigue life, critical regions of components should be free of those kinds of defects that can act as sites for premature crack formation. Over the years, a number of non-destructive testing (NDT) methods have been developed for flaw detection and sizing with widely varying capabilities, sophistication and ease of use. Recently, improvements in data and image processing techniques have allowed more detailed descriptions of flaw size, shape and location to be obtained, have reduced the skill needed to interpret results, and have led to automated inspection in some cases.

There are four basic methods that are the established workhorses of industry. These are ultrasonic, dye penetrant, magnetic particle, and radiography. Other methods such as holography, eddy current, acoustic emission, potential drop and gel electrode have been used to a lesser extent in the past but are receiving increased interest and application. Each of the above methods has specific applications where it excels. The selection of the most appropriate method for an application may be based on one or more of several factors such as (a) part configuration or geometry, (b) location of a critical area, (c) surface conditions, (d) sensitivity required, (e) type of material (ferrous or non-ferrous), (f) inspection rate, and (g) cost. The methods are discussed briefly below in terms of their basic principles of operation and some of their pertinent characteristics.

Since significant advances in NDT methods continue to be made, it is advisable to contact those companies and organizations involved in NDT technology to obtain the most up-to-date information on available techniques and equipment.

7.3.1 *Ultrasonic*

Ultrasonic inspection became a practical NDT method in the 1940's. Since then, it has developed into one of the more sophisticated testing methods. It is used for locating surface and internal defects in both ferrous and non-ferrous metals as well as composites.

The operating principle of ultrasonic inspection is similar in nature to a sonar system. Pulses of inaudible sound waves with frequencies centered at between 1 to 25 MHz are directed into a material and a portion of this sound energy is reflected back when it strikes material of differing acoustic reflectivity such as a crack, void, phase change or different material as in cast iron or composites. The reflected sound is transformed into an electrical signal, amplified, and displayed on a cathode ray tube (CRT) or processed by computer. In general, the more severe defects will reflect more sound energy and result in signals of greater amplitude.

Ultrasonic waves are introduced into a part via piezoelectric or electromagnetic acoustic transducers. When an electrical pulse is applied to a transducer, it mechanically deforms and vibrates, generating ultrasonic waves. Conversely, when the reflected sound strikes a transducer, an electrical signal is produced. In the often used pulse-echo mode of operation, transducers are pulsed in short bursts (typically 500 to 2000 pulses per second) and the reflected energy (echo) is received in between pulses. The amplitude of an echo depends on a number of factors such as defect size and orientation relative to propagating waves and the acoustic impedance of the defect/material interface. Strongest echoes are generated by discontinuities (e.g., voids and cracks) with large differences in acoustic impedance relative to the material in which they reside and whose areas are normal to the direction of propagation.

Transducers are typically coupled to test specimens or components through a layer of water, oil or grease. This is necessary because holding a transducer against a specimen without such a couplant can cause a small air gap that reflects most of the ultrasonic energy due to the high acoustic impedance (mismatch) at an air/material interface.

Ultrasonic testing is well suited to production inspection. The electronic nature of the technique and equipment facilitates automation. Commercially available automated scanning systems can provide precise information on the location, size and nature of defects, including colored video maps of defects. Details concerning ultrasonic testing and inspection can be found in Refs. [28, 29].

It should also be noted that combined laser-ultrasonic inspection techniques have recently been developed and are commercially available. For instance, in scanning laser acoustic microscopy, sound waves are transmitted through a part that has gold foil placed on the other side. The foil distorts to provide an "image" of the emerging sound. The pattern of foil wrinkling reveals the presence of internal defects and is made visible through laser illumination. The pattern of light reflected from the foil is displayed on a video monitor. This technique has excellent sensitivity and resolution and has been used, for example, to inspect ceramic components and small welds.

7.3.2 *Dye Penetrant*

Dye penetrant inspection also became a prominent NDT method in the 1940's. Since then it has continued to expand in use and is currently recognized as one of the more sensitive methods. It is employed extensively, for example, in the aerospace industry for locating minute surface cracks in critical parts made of aluminum, titanium, austenitic stainless steel, ceramics, etc.

This technique is based on a simple principle and is essentially a three-step process. The first step consists of cleaning the surface of a part and then coating it with a liquid containing a dye. Next, the excess penetrant is removed from the surface. Finally, a developer is applied to the surface of the part. Penetrant that had been drawn into surface cracks by capillary action is drawn out of the crack by the blotter action of the developer. The dye in the penetrant discolors the developer to pin-point the presence of a defect.

Liquid penetrants are specially formulated for good penetrating capabilities in conjunction with a soluble dye that will exhibit a high contrast relative to the developer. The formulation is also designed to be compatible with a specific means of removing the excess penetrant from the surface of a part. Removal is usually accomplished by employing either water wash, post emulsification, or the solvent method.

The dye may be of the visible or fluorescent variety. The visible type typically produces a dark red indication against the white background of the developer when viewed under white light. The fluorescent type, on the other hand, must be viewed under black light, that causes the flaw indication to fluoresce a bright yellow-green against a dull purple background. Developers consist of highly absorbent white particles and come in both the wet and dry forms.

The penetrant method is used predominantly for overall surface inspection of components. Geometrically simple as well as complex shapes can be readily accommodated by dipping or spraying. The method has been and continues to be used on a production basis. However, it is incapable of detecting internal defects or of determining depth of surface flaws. Details concerning use of the dye penetrant method can be found in Ref. [28].

7.3.3 *Magnetic Particle*

The magnetic particle method is one of the older NDT methods. It gained considerable prominence during World War II and has continued to play a significant role in the ensuing years. It is extensively used for locating surface and near surface defects in ferromagnetic materials (iron and numerous steel alloys, with the exception of austenitic stainless steels).

Its principle of operation is based on magnetizing the part undergoing test while magnetic particles are applied to its surface. Within a part, magnetic discontinuities such as cracks or seams that intersect the magnetic line of force result in leakage fields emerging at the surface. These leakage fields attract and hold finely divided ferromagnetic particles that are applied to the surface of the part. The resulting indications are apparent under visual examination and pinpoint the location and general extent of the discontinuity.

Magnetizing procedures may include the passage of high amperage current directly through a part, for circular magnetization, or placing the part inside of a current carrying coil (or solenoid) for longitudinal magnetization. Two processing steps are usually required for the full disclosure of all randomly oriented discontinuities. Complex shapes may require additional steps.

Magnetic particles can be applied in either wet or dry form (powder). In the wet form, particles are suspended in a liquid medium such as water or oil, that allows full coverage of even complex shapes to be achieved easily and quickly. The particles are available in various colors and are selected on the basis of the best contrast with the surface of a part.

The magnetic particle method has a distinct advantage over other NDT methods when overall inspection is required on ferromagnetic parts. Visibly discernible indications are produced almost instantaneously and even very irregularly shaped parts can be inspected on a production basis at a relatively low cost. It does, however, require equipment to magnetize a part, and, of course, cannot reveal "deep" internal defects. Further details on the technique can be found in Ref. [28].

7.3.4 *Radiography*

Radiography is the oldest of the NDT methods. It got its start shortly after X-rays were discovered by Roentgen in

1895. X-rays are used extensively in both the medical and industrial fields for probing into otherwise hidden internal areas. While it has excellent capabilities for locating deep internal defects, it can also be used to detect surface cracks.

Radiographic inspection is based on the differential absorption of penetrating radiation. Dense material will absorb more radiation than less dense material. In practice, radiographic film is placed on one side of a part while the other side is exposed to a beam of X-rays. Those areas of the part exhibiting the least absorption such as cracks or voids will result in the corresponding area of the film being exposed to the greatest degree. Developing the film will produce a shadow image of the part with various shades of gray in accordance with the degree of exposure. Processed film is examined with the aid of a viewer that illuminates the transparency from the rear with highly diffused white light. As in conventional photography, the film must be properly exposed if it is to provide the required degree of contrast between an image of a defect and the surrounding area.

Radiography is highly directional in that the defect to be detected must have an appreciable depth parallel to the X-ray beam. Thus, small laminar defects aligned normal to the beam can go undetected. On the other hand, defects such as slag inclusions or porosity are generally easily spotted since they are often close to round in cross section. The directionality problem can be overcome with multiple exposures with the part or beam re-oriented between each exposure.

Fluoroscopy is another form of radiographic inspection. Unlike the film process, it produces a "real time" image of the part while it is being exposed to the X-ray beam. The image can be viewed directly on a fluorescent screen or remotely on a video monitor with the aid of additional electronic equipment. The method offers greater productivity at reduced cost relative to film but its sensitivity (resolution and contrast) has been somewhat less. However, recent advances in digital image enhancement have provided images comparable to or better than the best film images.

Computerized tomography, widely used in medicine, is being adopted increasingly in industrial radiography. Conventional X-ray processes provide a two-dimensional image that superposes data on defects from different depths. In tomography, information obtained from different X-ray scans at different orientations is processed by computer to give images of selected cross sections of a part. The reconstructed images can be manipulated to furnish detailed information on particular defects. Automated inspection systems are available that compare data from inspections with pre-set standards of defect acceptability for a given part.

The radiation associated with the radiographic method is hazardous to humans. Consequently, strict safety precautions are required to protect operators and other personnel. Shielded rooms or enclosures are required to confine the radiation. The conventional radiographic method requires knowledgeable personnel skilled in the art, although this requirement is being relaxed somewhat with the advent of automated systems.

Details concerning this method may be found in Ref. [28].

7.3.5 Eddy Current

The basic principle of electromagnetic induction dates back many years but, like other NDT methods, eddy current inspection was not widely used until the 1940's. Advances in the basic technology together with those in the electronics field have greatly enhanced the general use of the method. It has a broad spectrum of applications including alloy sorting, hardness testing, conductivity measurements, and flaw detection.

When a conductive material is subjected to a rapidly changing electromagnetic field, eddy currents are generated in the material that oppose the changes in the magnetic field. The presence of a defect interrupts the normal eddy current path or pattern resulting in a basic impedance change within the driving or sensing coil. These changes are detected, amplified, and displayed on a meter or cathode ray tube (CRT). In general, the magnitude of the resulting signal will increase with the size of the defect. The electromagnetic field is derived by passing a high frequency (25 to 125 KHz) current through a coil. The coil can encircle the part or be built into a probe that can be placed near the surface of a part. Changes in the eddy current field can be sensed through either the exciting coil or a separate sensing coil.

Encircling coils are usually employed for inspecting simple shapes such as tubular products that permit the use of a close fitting coil. Thus, long lengths can be easily and quickly fed through the coil for inspection purposes. Probe type coils, on the other hand, are used for scanning the surface of a part or inspecting discrete areas of complex shapes. Some care must be exercised to avoid changes in probe to surface distance (lift-off effect) since erroneous signals can result.

While the eddy current method can be used to detect surface cracks in both ferrous and non-ferrous metals, superior results are generally obtained on non-ferrous metals. It is readily adaptable to automatic inspection of simply shaped parts. Complex shapes are a problem unless only discrete areas are to be inspected by means of a probe; however, the use of robotics could help overcome that difficulty. The eddy current method is limited to detection of surface and near surface defects and to use on electrically conductive materials, thus ruling out its application to plastics, ceramics, etc., that are gaining increased usage in structure. Details concerning the eddy current technique may be found in Refs. [28-30].

7.3.6 Holography

Holographic interferometry has been used in a wide variety of applications for detecting cracks, voids, debonds and delaminations. It can be applied to metals, composites, plastics, etc. As used in non-destructive testing, the holographic technique consists of loading a component and observing the surface deformation pattern as manifested by interference fringes. The presence of structural defects causes anomalies in surface deformation and thus in the interference pattern, as depicted in Fig. 7-5. These anomalies can be readily identified by comparison to patterns in defect-free components, but in many cases, such a comparison is unnecessary since the anomalies stand out so clearly.

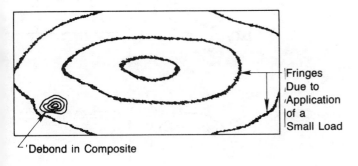

Fig. 7-5 Examples of anamolous deformation patterns caused by defects as observed using holographic interferometry.

Loadings may be applied by direct mechanical stressing, pressure or vacuum stressing, vibrational excitation, impulse loading, or thermal stressing. Low levels of stress can be used due to the high sensitivity of holographic interferometry.

This technique provides a whole field visual display of defect indications, is applicable to complicated shapes, requires no special surface preparations, and provides results in a matter of minutes using automatic, instant holographic camera systems. In most situations, only qualitative interpretation of fringe patterns is needed, although quantitative evaluations of patterns can also be performed if desired.

To utilize this technique requires mechanical stability of the optical-test set-up; however, various approaches have been developed to permit its use in a production environment. For instance, the technique is used routinely in the inspection of tires by a number of manufacturers. Details concerning non-destructive testing by holographic interferometry may be found in Refs. [13, 28, 31].

A variation of the above technique is acoustical holography. Coherent (uniformly aligned, single frequency) ultrasonic acoustic waves are utilized in a manner similar to coherent (laser) light in optical holography to produce a pattern of interference fringes. Different methods have been developed to generate acoustic holograms. The ripple-tank method, illustrated in Fig. 7-6, is an example. A hologram is produced by interference of reference and object beams impinging on a liquid surface. The ripple pattern (analogous to an optical interference pattern) is illuminated by laser light and the resulting reflected image viewed on a video monitor or electronically processed by other means. This technique, being a hybrid ultrasonic-laser method, is able to locate and size deep internal defects. Details concerning it may be found in Ref. [28].

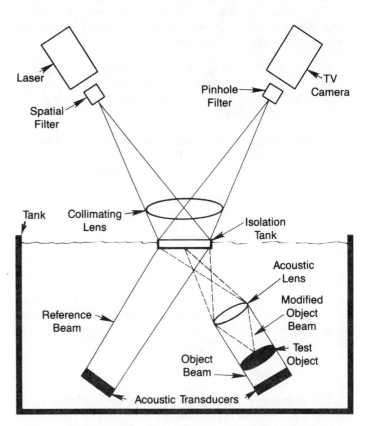

Fig. 7-6 Schematic of ripple-tank acoustic holography set-up.

7.3.7 Acoustic Emission

Acoustic emission testing is based on the phenomenon that "sound bursts" are emitted from materials when they are stressed. An example would be the sound emitted from ice on a pond, just before one falls through the ice. The major difference is the frequency of the emitted sound. Ice, as it cracks, emits sound that is audible to the human ear. Sound emitted from materials is, for the most part, inaudible in the crack initiation and growth phases, and remains mostly within the material as stress waves.

Specialized piezoelectric transducers have been developed that respond to the minute surface deformations caused by the stress waves from crack growth and/or localized plastic deformation around defects.

Transducer output is conditioned and fed to the electronics of an acoustic emission system where it may be recorded and observed during a test. The recordings may be reviewed after test completion for a more definitive evaluation.

Using "source location" techniques, similar to seismological methods, in which several transducers are "listening" to the material being tested, locations of the initial crack may be estimated even when the location is internal. Interpretation of the data displayed is improved with experience; a skilled experimenter or operator can differentiate "real" from "false" indications. These skills are supplemented through the use of computerized data evaluation found in current acoustic emission test equipment.

Extraneous "background noise" generated by loading fixtures, rubbing of structural fasteners and so forth can be eliminated by available data processing procedures to minimize false indications.

Growth of cracks can be monitored by the acoustic emission technique using empirical calibrations. The rate of acoustic energy bursts increases with the rate of crack growth.

The technique has been applied to both ferrous and nonferrous alloys as well as composites and ceramics. Typical industrial uses include monitoring pressure vessels and piping for accumulation of fatigue damage. Aerospace structures, both on the ground and in flight, have also been successfully monitored for crack detection and growth. Finally, acoustic emission can provide a quick volumetric examination of large structural components, allowing for more effective utilization of techniques for flaw sizing such as radiography and ultrasonics, if the need is indicated. Details concerning the technique can be found in Refs. [28, 32-34].

7.3.8 *Potential Drop*

The presence of a crack or defect in an electrical field causes a disturbance of the field that can be used to determine flaw size and to monitor crack growth. In its simplest form, the potential drop method involves passing a constant current through a lab specimen or part and accurately measuring the change in voltage across the crack plane. As a crack grows, the potential drop increases due to a reduction in the uncracked area. Crack size may be determined by empirical calibration between the potential drop associated with cracks of known size or by analytical solutions for the perturbing effect of a flaw in an electrical field.

Most applications of the technique have utilized direct current fields, although more recently, alternating current fields have been used as well. As depicted in Fig. 7-7, high frequency alternating currents are concentrated in the skin of a test specimen, facilitating the detection and monitoring of small surface cracks.

The potential drop technique has been able to monitor successfully the growth of surface breaking cracks in lab specimens with initial sizes as small as about 0.1 mm, even when

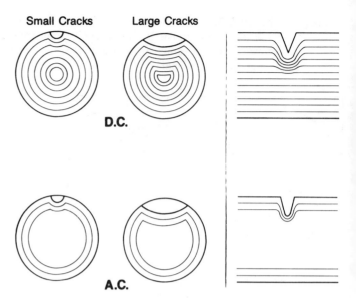

Fig. 7-7 Illustration of direct and alternating current fields as used in the potential drop method.

direct current was used. The technique has also been utilized with good results to monitor the growth of surface cracks emanating from welds, providing data on both surface crack length and depth for several semi-elliptical cracks that merged during growth.

Portable instrumentation is commercially available for applying the technique either in the lab or the field. Details concerning the theory behind and implementation of the potential drop technique may be found in Ref. [35].

7.3.9 *Gel Electrode*

Oxide films on the surface of specimens and parts rupture locally over regions where fatigue microcracks have formed in the material underneath. This observation led to development of a technique in which a small probe consisting of a potassium iodide, borax and starch gel is brought in contact with an oxide film and a potential applied between the specimen and a cathode in the probe for a brief period of time (on the order of 10^{-2} sec.). Current flows preferentially to regions of the oxide film where microcracks have formed, and the potassium iodide releases iodine ions that react with the starch to form a black adsorption complex at the interface between the probe and oxide film, such that an image of microcracks is imprinted in the gel.

Surface fatigue cracks as small as 10 microns can be detected. The flow of charge during imaging is proportional to crack length and provides an alternate way of determining length, detecting cracks as small as 100 microns. In addition, the gel electrode technique has been able to detect and monitor surface deformation and damage preceding microcrack formation, within the first 1% of specimen fatigue life.

A limitation of the technique for inspection purposes is that specimens or parts must be anodized (in the case of aluminum alloys) to produce an oxide film that is somewhat thicker than that occurring naturally. To date, this method has been used primarily in laboratory studies. Details concerning it may be found in Ref. [36].

7.4 Residual "Self Stress" Determination

Almost every part or assembly contains some compressive or tensile stresses that were "locked" into the material during its fabrication. These "residual stresses" are produced by many processing operations, e.g., hardening methods, machining, etc. If parts are misaligned during assembly or force fitted, additional residual stresses are added. The relieving of undesirable residual stresses by heat treatment is common practice but may not always be feasible or may compromise material properties. Often residual stresses are intentionally put into a part by means such as shot peening or induction hardening when it is desirable to have beneficial high compressive stresses on and below the surface.

Residual stress distributions in a part may be simple or complex, depending on many factors. For example, a part may contain residual stresses caused by forging, welding and machining, producing large stress variations in the material. Even steeper residual stress gradients may exist in such a part near the surface as a result of grinding or shot peening. These residual stresses are additive to the applied loads in service, as depicted in Fig. 7-8. Residual stress measurements are needed to check the adequacy of processes to accomplish residual stress control, such as (1) stress relieving by heat treatment or stretching, (2) development of beneficial compressive surface stresses by shot-peening or careful grinding, (3) avoidance of tension on or beneath the surface resulting from abusive machining, (4) modification of design of forming dies if tensile residual stresses appear in the formed part, and (5) monitoring of welding procedures to minimize residual tension.

Residual stresses may also develop in service, particularly at stress concentrations, as a result of an irregular loading history. The stresses can change magnitude and even sign during the course of the loading. Moreover, initial residual stresses from processing can be altered (e.g., relaxed) by service loadings.

There are many techniques for measurement of residual stress that can be considered theoretically possible, but only a few techniques have been reduced to practice to the point where equipment is available for industrial use. The following description is limited to these methods; each has advantages and disadvantages. With any method, accuracy and reliability are prime considerations but such factors as complexity, size, weight, and use in "field conditions" are often of concern. Initial equipment cost and cost of time per data point are very important. The ability to measure stress gradients along as well as beneath the surface is desirable. The nondestructive versus destructive nature of a technique is a major consideration. The possible complexity of residual stress patterns in practical structures suggests that no single

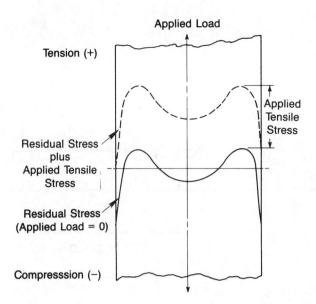

Fig. 7-8 Hypothetical stress distribution in a bar with shot-peened surface.

method can meet every test requirement; sometimes two or more methods are useful in conjunction. Stress conditions deep within the interior of a complex part are impossible to measure non-destructively by any presently available method.

7.4.1 X-ray Diffraction Method

X-ray diffraction technology was used as early as 1912 for the determination of crystal structures. Crystallography utilizes the basic concept of a "unit cell" that is the smallest group of repeating atoms of which all crystals are composed. Various families of planes may be drawn through the corners of the unit cells that form a crystallite. Such planes are separated by an interplanar spacing, d.

The angle of diffraction of an X-ray beam is a measure of the interplanar spacing. The diffraction angle, 2 , is related to the interplanar spacing and the wavelength of the radiation by Bragg's law, i.e., $2d \sin \theta = \lambda$. (See Fig. 7-9.) The X-ray wavelength, λ, is of the order of magnitude of the interplanar spacing, d, approximately one or two Angstroms. A departure of the interplanar spacing from the unstressed value represents a stress in the material and is the essence of the X-ray diffraction method for measurement of residual stress.

In accordance with the Poisson effect, the interplanar spacing will (e.g., under tension) elongate in the direction of the applied stress and contract in the transverse direction. In the classical treatment, it is assumed that the stress normal to the surface is zero; i.e., the only possible stresses are biaxial in the plane of the surface. The X-ray beam is usually directed at the surface from at least two different "tilt" angles (ψ), generally perpendicular to the surface and at an angle of 45 degrees to the surface as depicted in Fig. 7-10. If the material is stressed in tension, the X-ray beam normal to the surface will be diffracted by an angle (2θ) that corresponds to

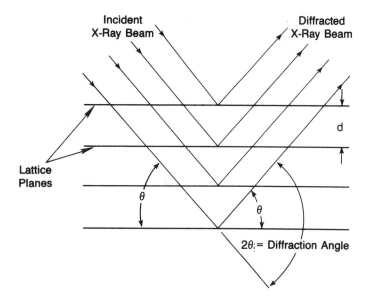

Fig. 7-9 X-ray beam diffraction in accordance with Bragg's law.

the slightly reduced interplanar spacing and the beam incident at 45 degrees will be diffracted by a slightly different angle corresponding to the increased spacing in the direction of the tensile stress component. The residual stress is a function of the change in the two interplanar spacings relative to the unstressed condition.

By differentiation of Bragg's law, it is found that a difference in diffraction angles at the two ψ angles is a measure of the change of the interplanar spacing. The equation for determining stress then is simply a constant times the shift in the diffraction angle.

The X-ray diffraction (XRD) method of residual stress determination is non-destructive; only the X-ray beam and no part of the instrument touches the specimen. The beam penetrates the material to less than 0.025 mm. For this reason, the stresses measured are essentially "surface" stresses. By electro-polishing the metal for layer removal, one can obtain stress gradients beneath the surface, as illustrated in Fig. 7-11. If a small X-ray beam is used, such stress gradients can be obtained by electro-polishing only a small area of the surface. Although frequently larger, the X-ray beam can be as small as about 1 mm in diameter. Thus, with the XRD technique, a high degree of resolution is attainable both normal to and across the surface.

There are three basic types of instruments for use in XRD residual stress determinations: film cameras, diffractometers, and position sensitive detectors.

Stress measurements have been made for many years by photographic film methods using the "back reflection" film camera. The diffraction angle shift associated with stressed material is measured by the displacement of lines produced by the diffracted X-ray beam on a film plate located to receive the radiation. In practice, the film method is found to be

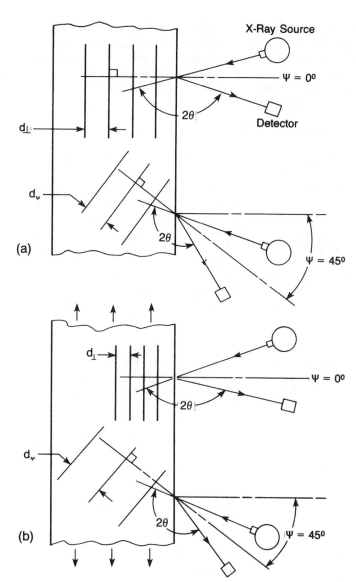

Fig. 7-10 X-ray beam angles.

difficult largely due to the problem of making accurate measurements from the developed film plates. The procedure is very time consuming, requiring several man-hours per data point and accuracy is relatively low.

Despite disadvantages, the film method has offered a degree of portability not found in many XRD devices used in the past. The X-ray camera head is quite small and can be used in geometric situations that have been inaccessible to other instruments.

A number of companies have manufactured instruments used in crystallography and chemical analysis in university and industrial laboratories. While some of this work is done with back-reflection cameras, most is done with electronic scanning diffractometers. Many of these instruments were not initially designed to permit the making of residual stress measurements but some have been modified to meet this requirement. Accurate measurements can be made with such instruments but they are limited to small parts of a few inches or less in any dimension. In all cases, the part has to be brought to the diffractometer, thus severely restricting its use in many practical industrial applications. The time required to obtain data depends upon the equipment and the precision requirements. Typically, a stress data point might require about 20 minutes.

The limitations of the laboratory diffractometers described above led engineers to develop instruments which are relatively easy to use and can accommodate large test parts. Also, the need to take diffractometers to the part (e.g., an aircraft wing) was considered. Equipment was developed with reduced size and weight, varying degrees of portability, automated data processing, and display of printed or graphical results. However, such diffractometers had to make at least two sensor scans, which limited their speed in data acquisition. On the order of 10 to 20 minutes was required per data point. Thus there was a desire to develop a device that could substantially reduce the time required to make a stress measurement.

In the 1960's, General Motors developed an industrial diffractometer ("Fastress") that uses dual X-ray tubes and a dual set of diffraction sensors. The result is a machine that can produce data at a speed of about one minute per test point. It is well-suited for use on steel alloys, although some users report good results with aluminum. The high speed produces low cost per data point and thus, extensive examinations are relatively affordable. The Fastress has the additional advantage of very small X-ray beam sizes, as low as 1 mm diameter, that permits examination of complex shapes and areas with sharp stress gradients such as in the vicinity of notches and across welds.

There has been a growing interest in field applications for measurements of residual stress in oil pipelines, large tanks, off-shore drilling platforms, ships and aircraft. Because of their weight and size, all of the diffractometers described above range from no portability to a limited degree of portability. The industrial units can be used on or near production lines or for tests on large structures in a somewhat sheltered environment such as a test laboratory or aircraft hangar.

In 1977, Northwestern University demonstrated a "Portable X-ray Analyzer for Residual Stresses (PARS)" now offered commercially. This system incorporates several new features departing from conventional diffractometer design. A fixed radiation sensor eliminates the need for mechanical scanning. The sensor is a position sensitive detector (PSD): a gas-filled tube containing a graphite coated wire. X-rays impinging on the tube cause localized ionization of the gas, producing electrical pulses that permit measurement of intensity and position to provide a diffraction profile. The light weight, high-voltage power supply weighs only a fraction of conventional transformer/voltage regulation systems. The miniature, air-cooled X-ray tube is small in size and free from the complexity of water cooling. The design incorporates a very lightweight goniometer head, easily held in place by one operator. The goniometer is "tilted" to two ψ angles. A data point can be obtained in less than 20 seconds, utilizing a microprocessor-based computer which is part of the electronics package.

The University of Denver Research Institute has also developed a portable rapid stress measuring instrument (Porta/Rapid/Stress). A solid-state radiation sensor is used and configured with respect to the X-ray tube in a fashion that corresponds to an all-electric counterpart of the back-reflection camera technique. This instrument is offered with a very small portable goniometer head that can be positioned into complex structural situations inaccessible to other diffractometers. Test data are obtained with this unit in less than one minute per stress measurement. The design incorporates a solid-state Position Sensitive Scintillation Detector (PSSD) that permits the diffracted beam peaks to be measured without mechanical motion of the sensor. The principle is based on the conversion of the diffracted X-rays to visible light by a scintillation material that is coated on the receiving face of two bundles of fiber optics. These bundles are located on opposite sides of the incident X-ray beam so as to intercept the cone of the diffracted beam. The light signals are transmitted by the fiber optics to the detector electronics section where the signals are amplified and the intensity and position are measured using a linear photodiode array. This instrument may be employed as a "single exposure" device without need for ψ angle changes for maximum speed in obtaining data. However, if desired, the instrument may also be used for multiple ψ angle tests for higher levels of accuracy.

The precision of position sensitive detector instruments has been asserted to be ±1 ksi (7 MPa), compared to ±5 ksi (35 MPa) for film devices. The precision of diffractometers falls somewhere between those values. The cost of these instruments is quite high and the skilled nature of the XRD technique has led to the existence of a number of commercial laboratories which perform both surface and depth profiling measurements of residual stress.

There are a number of important factors that must be considered when using XRD techniques for residual stress measurements. For example, the surface of the part being tested must be clean to bare metal, i.e., free of paint, plating, and corrosion. This cleaning must be done by chemical or electrochemical means but care must be taken so that any chemical cleaning does not etch away a layer of the surface containing stresses of interest. The cleaning must not be accomplished by any mechanical effort such as grinding or sanding the surface. Such physical working of the surface will completely distort the results since even light sanding of a metal surface can impose large stresses in a thin layer of the material. Manufacturing plants often use a glass-bead method of cleaning production parts after or during processing. This procedure must be strictly avoided in preparing a part for XRD tests because the effect is like a light shot peening which can eaily place a thin, highly compressive layer on the surface, completely obscuring the original stress conditions. Surface roughness is also a consideration. If heavy machining leaves marks readily visible to the eye, it is likely that XRD measurements may encounter difficulty.

The material tested should not have large grain size since this can be difficult or impossible to analyze by the XRD method. Castings and directionally solidified alloys are often questionable. A preliminary test with the diffractometer will usually reveal the presence of this problem. Preferred orientation found in rolled sheet and plate can pose equally difficult problems. The entire concept of XRD requires that the crystallites of which the material is composed be random in their distribution and small in comparison with the area of the surface irradiated by the X-ray beam.

As stated previously, all XRD stress measurements are "surface measurements" since the X-ray beam penetrates the material to less than 0.025 mm. This fact is an advantage in many cases since the region of stress measurement is quite precisely defined. If stress gradients are very steep within the thin layer penetrated by the beam, corrections can be applied to the data to compensate for the fact that the diffracted beam is influenced by this gradient. The XRD method can be used to obtain stresses within a material but in such a case, the exterior material must be removed by mechanical or chemical means to reveal the inner surface that can then be measured by the XRD technique. Obviously, the stresses measured at the exposed surface must be considered additive to any stresses released in the material removal process. Furthermore, residual stresses caused by machining cuts used to remove layers must, in turn, be removed by an etching process.

Finally, it must be recognized that one cannot investigate some problems such as very sharp corners, the interior of deep holes, or inside bends where other structure may interfere with the X-ray beams. There are thus many interesting situations that simply cannot be tested with XRD or probably any other technique. However, in difficult geometries, it may be feasible to cut apart the test item in order to gain access to a region of interest. An example is that of an engine crankshaft where the region of most interest may be on the inside fillet between the ground bearing surface and the main forging. (See Fig. 7-11.) If surface grinding stresses in this region are the desired objective, one can cut sample crankshafts apart into easily tested segments without significantly affecting the grinding residual stresses, and the production grinding methods can thus be evaluated.

Further details concerning the theory and implementation of XRD techniques can be found in Refs. [37-47].

7.4.2 Sectioning

When residual stress distributions are of interest deep within a material, it is necessary literally to cut the structure apart. Methods of residual stress determination that are inherently destructive may be listed as "sectioning" or "mechanical" methods. The principal methods are known as "layer removal," "beam dissection," and "boring and turning." The hole-drilling technique is also a mechanical method but is often referred to as "semi-destructive" and is discussed in the next section. The cost of performing residual stress measurements by mechanical methods is not very great as far as instrumentation is concerned, such equipment being pri-

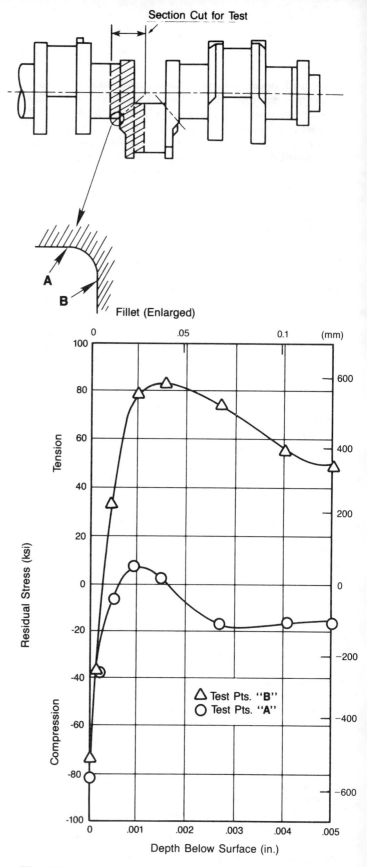

Fig. 7-11 Example of measurements of grinding residual stresses near a fillet on a crankshaft. Section shown shaded was removed to permit XRD access to fillet region. Layer removal by localized electrolytic polishing of small area at fillet. X-ray diameter: .045 in. (1.14 mm).

marily for strain or deflection measurements; however, a well equipped machine shop is required for working with the test specimens. All of the methods are very time consuming.

The "layer removal" method involves the removal of thin layers of the surface of a test beam. To the extent that stressed layers are removed, the curvature of the beam is changed. This change is assumed to result from a uniformly applied bending moment resulting from the release of compressive or tensile forces in the removed layer. The force is calculated as that required to produce the bending moment corresponding to the measured deflection. The stress is determined as the average over the cross section of the layer removed. Owing to the averaging effect, layer removal must be accomplished in very small increments if steep stress gradients are to be measured. Layer removal by thin increments results in small changes in beam deflection with the need for highly accurate deflection measurements. Layers on the order of 0.01 mm may be removed if steep stress gradients are anticipated. Thicker layers can be removed if test results indicate that the stress gradients are "mild." Layer removal is sometimes accomplished by machining. Since machining stresses can extend to a depth of as much as 0.1 mm, the thickness of the layer increment must be substantially greater than the depth of any expected stresses from this source. Chemical or electrochemical layer removal is preferable if precise work, especially on thin layers, is undertaken. If thick layers are to be removed, machining followed by etching to remove the machining-stressed layer is feasible. The layer removal method can be used on beams of almost any size; however, thin beams have less stiffness and thus give more curvature for a given amount of stressed layer removed. To test light surface machining effects, e.g. a beam of about 2 mm by 10 mm by 100 mm is a practical size.

The "beam dissection" (sometimes called "parting-out") method is used to obtain residual stresses deep within parts of theoretically any size and shape. The method involves cutting away sections of material in the region of interest. Each section is cut into smaller and smaller elements. Each element is, in effect, a beam and may be analyzed in a manner similar to the layer removal method. When dissecting a structure to produce the beam, one must consider the "parting stresses" that result from the cutting-apart process. If the region of interest contains stresses associated with adjacent material, the fact of dissecting will separate the beam from this stress influence. In such a case, strain gage or mechanical deflection measurements prior to and after dissection will reveal additional "parting" stresses that must be added to those observed in the removed beam. The beam must be removed by a dissection process that does not in itself alter the stresses in the beam. This usually involves careful machining processes that may be followed by etching to remove the machining-induced stresses.

An illustration of the beam dissection method is shown in Fig. 7-12. If stresses through the entire cylinder are of interest, a beam section (ABCD) is removed as shown in Fig. 7-12a. Parting stresses would be associated with the amount of curvature of the removed beam as shown in Fig. 7-12b. The stress is calculated assuming a linear distribution across the beam and must be added to the stresses determined by subsequent dissection. A series of "primary cuts" is made in the sequence shown in Fig. 7-12c with deflection measurements and stress calculations for each cut. As the region of steep stress gradient is encountered, the remaining beam (EBCF) is subjected to successive layer removals of thinner-dimension "secondary cuts" progressing from the original outside cylindrical surface layer inward. The result of the calculations of stress from the curvature of each cut is plotted in Fig. 7-12e where parting stresses are also shown. It should be noted that the results in this example might have been obtained using the boring and turning method described next. However, the technique shown in Fig. 7-12 could have been applied to other shapes and is not limited to the cylinder shown for simplicity in this illustration.

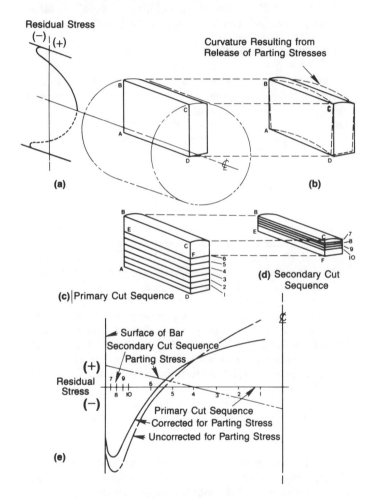

Fig. 7-12 Illustration of beam dissection method applied to a residual stress distribution in a round bar.

The "boring and turning" method is used on cylindrical shapes. The test part is usually two or more diameters in length and at least 25 mm in diameter. Strain gages are mounted in sets of tangential and axial pairs at the midsection. They are mounted on the bore if stresses on the outside surface are desired and on the exterior of the cylinder if inside surface stresses are to be investigated. The boring or turning takes place on the surface opposite to the strain gage location. The change in strain is measured with each layer removed and residual stress in that layer determined from available formulas.

For the most detailed and accurate measurements, the boring and turning methods may be combined. Layer removal is accomplished by machining or etching, or a combination of these methods, depending upon the thickness of the layer to be removed. The boring and turning method is difficult and time consuming; however, with careful procedures, good results can be obtained for the stress distribution throughout virtually the entire part.

Further details on the above techniques as well as other sectioning methods can be found in Refs. [48-51].

7.4.3 Blind Hole Drilling

The introduction of a very small hole into a part containing residual stresses relaxes such stresses locally, causing a change in surface strains in the immediate vicinity of the hole. The state of biaxial residual stress at and just below the surface can be determined from measurements of the changes in strain in three radial directions as the hole is produced.

In practice, the hole drilling technique begins with installation of a three-element rosette type strain gage whose center is located directly over the point of the desired stress measurement. A precision alignment guide for drilling which provides accurate location at the center of the rosette is then installed. (See Fig. 7-13.) A hole is then drilled. It has been shown that the strains in the immediate vicinity of the hole are fully relieved when the hole depth is about equal to its diameter, which is typically 1.5 to 3 mm. The change in strain around the hole due to relaxation is measured by the rosette and read out on a portable strain indicator. Stress is calculated from formulas that relate the measured strains, modulus of elasticity, Poisson's ratio and constants that are a function of the material and the hole/gage geometry. Alternatively, empirical relations between stress and measured strain may be used, that can be developed by experimental calibration for a given material subjected to known applied stresses.

The strain gage rosette is designed with three gages located radially from the center of the hole at angles related to each other in such a way that both the magnitude and direction of the principal stresses may be obtained from the strain measurements. The technique is most accurate when the stress distribution is uniform along the depth of the hole. However, methods do exist for determining the presence and approximate magnitude of stress gradients.

A significant concern with this technique has been the possibility of large induced stresses as a result of drilling the hole with low speed drills (operating at about 1000 rpm). Of most concern are materials that work-harden readily such as certain stainless steels. To overcome this problem, various methods for producing the hole have been used such as chemical milling and air abrasive jets. A simpler method using a high speed air turbine dental drill has produced test results indicating virtual elimination of induced stresses in several materials. This drill is readily used in conjunction with commercially available milling guides.

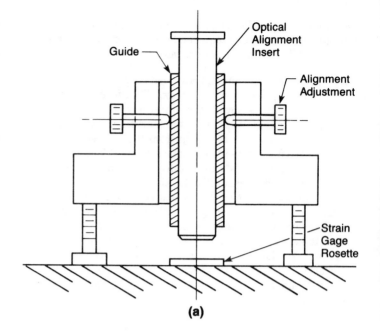

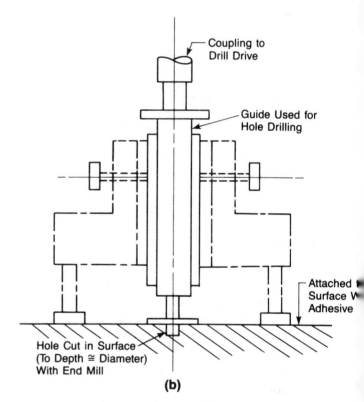

Fig. 7-13 Illustration of (a) accurate alignment and (b) drilling of hole in the blind hold drilling method.

The hole drilling method is time consuming compared with XRD but it lends itself well to use in field conditions. The equipment is of relatively small size and weight and very portable. The cost of the equipment is about one-tenth of a typical XRD machine.

Since the technique requires that metal be removed, it is not truly non-destructive. However, on larger parts, the amount of metal removal may be insignificant and the hole may, in some cases, be repaired after the test by grinding. Consequently, the method has been referred to as "semi-destructive."

Details on the hole drilling method and its implementation may be found in Refs. [52-58].

A combined hole drilling-laser holographic technique may also be used to determine residual stresses. The region of interest on a part is illuminated with a low power laser beam and a hologram made on a reusable thermoplastic recording plate (using an instant holocamera). Then a hole is drilled, producing surface displacements around the hole as the residual stresses are released locally. The displacements, in turn, cause optical interference fringes to appear on the hologram, which are viewed in real time on a video monitor. By counting the number of fringes that exist between two diametrically opposite points surrounding the hole, the magnitude and direction of residual stresses can be obtained by substituting the count into formulas that relate the count to the residual stresses that existed in the region before the hole was introduced.

This modification of the conventional hole drilling technique produces results quickly because it eliminates the need to install strain rosettes and a precision milling guide at each location. It can also be applied to regions inaccessible to rosettes and milling guides. However, the technique is not portable; a part to be evaluated must be placed on a vibration isolation table so that a hologram can be made. Further information concerning this technique can be found in Ref. [59].

7.4.4 Ultrasonic Methods

Measurement of applied and residual stresses by ultrasonic techniques is based on the approximately linear change in speed of sound in metal with stress, a weak but measurable effect, often termed the acoustoelastic effect. Acoustic waves are launched into specimens and their time-of-flight or other velocity-related parameters measured.

Three types of waves are employed in acoustoelastic stress measurements.

1. Longitudinal waves, in which particle motion is parallel to the direction of wave propagation, are used for determining bulk stresses (e.g., average stress through the thickness of a plate at a given location). The velocity of longitudinal waves is proportional to the sum of principal stresses.

2. Shear waves, in which particle motion is transverse to the direction of wave propagation, are also used to measure bulk stresses. The velocity difference between two orthogonally polarized shear waves is proportional to the difference of principal stresses (an acoustical birefringence effect similar to that in photoelasticity). Stress- velocity relations vary from metal-to-metal and are determined by calibration, typically with uniaxial tension specimens.

3. Rayleigh waves (or surface acoustic waves) travel along and just below the surface of specimens, penetrating to a depth of approximately one wavelength, that is typically a millimeter or so at a frequency of 3 MHz in aluminum and steel alloys. This type of wave is a combination of shear and longitudinal waves that produces particle motion similar to ocean waves rolling onto a beach. Since Rayleigh waves have a limited depth of penetration, that can be varied by changing ultrasonic frequency, it has the potential to provide information on stress profiles below the surface. On the other hand, bulk waves travelling through stress gradients give averaged values of stress. For example, in the case of a bulk wave traversing the depth of a beam in bending and encountering equal tension and compression, there would be no net change in velocity.

Bulk wave transducers have been developed that can scan specimens with acoustic beam diameters on the order of several millimeters. Longitudinal waves are generally transmitted into specimens through a liquid couplant. Shear waves are produced by transducers brought into contact with a specimen, or by mode conversion of oblique longitudinal waves, or by non-contacting electromagnetic-ultrasound transducers. Rayleigh waves are generally produced by wedge transducers held in contact with the surface of a specimen. Commercially available transducers and accompanying instrumentation are portable and use standard electronic components.

Application of acoustoelastic techniques to determination of stresses caused by applied loadings has generally produced good results. For instance, bolt stresses are readily measured with available equipment. However, measurement of residual stresses has produced mixed results, because other characteristics of processed metals also affect ultrasound velocity and birefringence. Variations in hardness, crystallographic texture, dislocation density, composition and phase can all have an effect, in some cases overwhelming that due to stress. Many of those processes that generate residual stresses (e.g., induction hardening, shot peening or welding) also produce the competing "material influences" that cannot be easily separated from residual stress effects. In certain cases where the competing effects have apparently been small, good results have been obtained in residual stress determination (e.g., interference fit of a plug in a hole, hydrostatic extrusion of an aluminum rod). Further development will be needed before widespread industrial applications of ultrasonic methods to residual stress determination can be made. The non-destructive nature of such methods, rapid speed of data acquisition, portability and moderate cost of equipment, and potential for automation is an appealing combination if the above problems can be overcome. Details concerning ultrasonic methods may be found in Refs. [29, 60-67].

7.4.5 Barkhausen Noise

Ferromagnetic materials contain magnetic domains in which magnetic moments of atoms are aligned. Such domains are generally comparable to or smaller than grain size and are separated by domain walls. During cycling of an applied magnetic field, a magnetization curve is produced, as depicted in Fig. 7-14.

Abrupt magnetic reorientations occur and magnetic domain walls move under the influence of a changing applied magnetic field. It has been found that a magnetic field applied smoothly to a specimen does not magnetize it smoothly but instead in abrupt, uneven increments called Barkhausen jumps. (See Fig. 7-14.) These jumps are caused by discontinuous movements of the mobile magnetic boundaries between adjacent domains. Such movements result in electrical "noise" and are influenced by the stress existing in a material. Consequently, measurement of such noise can be empirically related to stress in a manner useful for residual stress determination.

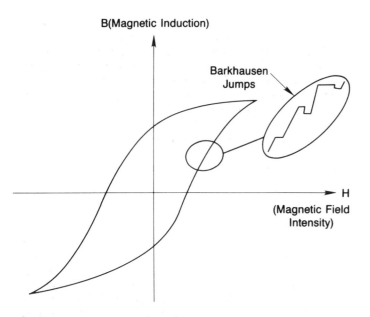

Fig. 7-14 Magnetization curve.

Instrumentation to implement this technique is small, light and portable, and lends itself to field applications. A briefcase-size electronics package contains the components for generation of the magnetic field and for analyzing the received Barkhausen noise to give a readout that is related to the stress in a material. The electronics package is connected to a hand-held probe that applies the magnetic field to a specimen and also senses the corresponding noise. However, it is necessary to calibrate the probe with relation to the test material. Calibration in "Barkhausen units" may be accomplished by flexing a beam to which a probe is attached and measuring the deflection from which stress can be calculated. Probes may be specially designed to fit complex shapes.

The Barkhausen technique appears to have a limited range of stress over which useful results can be obtained, on the order of ±280 MPa (±40 ksi) in steels. At larger magnitudes of tensile or compressive stress, a curve of Barkhausen noise vs. stress tends to become flat. The technique is responsive to stress over a shallow depth, of about 1 to 2 mm (0.04 to 0.08 in.).

Applications to measurement of residual stresses have included shot peened helicopter rotor blade spars, weldments, auto frettaged gun tubes, anti-friction bearings and gas turbine components. Use of the technique could be worthwhile in making preliminary evaluations of components, followed up as necessary by more detailed residual stress determinations using XRD or other methods. Details on the Barkhausen technique can be found in Refs. [68-71].

7.4.6 Other Methods

Some techniques are presently available that give only qualitative or semi-quantitative information. Since such methods may be useful in conjunction with other test procedures, two are mentioned below.

(a) The concept of stressed layers and resulting curvature as utilized in the sectioning methods discussed in Section 7.4.2 provide a method of measurement of residual stress produced as a result of shot peening operations. In order to enable the specification of shot peening procedures which are intended to provide both a desired intensity and depth of the compressive layer, test samples called "Almen strips" are exposed on one side to a shot-peening procedure corresponding to a particular manufacturing process. The peened side develops a convex curvature due to the surface compressive stress. Measurement of this curvature or "Almen intensity" can be related to residual stress and penetration depth by empirical procedures.

(b) Brittle or photoelastic coatings may be applied to a region of interest to obtain a general idea of the nature and magnitude of residual stresses. After the coating has been applied, small holes are drilled in the surface. As in the strain gage hole drilling method, residual stresses are redistributed in the vicinity of the hole, producing cracks in brittle coatings or isochromatic fringes in photoelastic coatings. Careful observation of the patterns can provide useful qualitative information but reliable, sufficiently accurate quantitative results are difficult to achieve.

7.5 Advantages and Limitations of Various Techniques

A comparison of *relative* advantages and limitations of the techniques for defining fatigue critical locations is provided

Table 7-1 Comparison of Techniques to Define Fatigue Critical Locations

Technique	Max. Strain Sensitivity (microm/in)	Whole Field?	Provides Strain Gradient?	Time Involved in Use	Eqpt. Cost	Primary Advantages	Principal Limitations
Brittle Coating	500	Yes	Yes, qualitatively	Moderate (primarily application time)	Low	Shows regions of stress concentration and aids placement of strain gages.	Very sensitive to environment (temp., humidity). Reliable quantitative results cannot be expected.
Strain Gages	5	No	No	Moderate (primarily installation time)	Low to Moderate	Well-suited to both static and dynamic measurements. Rosettes give directions and magnitudes of principal stresses. Excellent portability for use in the lab or field.	Point-by-point measurement only.
Photoelastic Coatings	10 (with fringe interpolation procedures)	Yes	Yes	Moderate (primarily application time)	Moderate	Reveals overall strain distribution for both static and dynamic loading.	Regions of interest must be accessible to light.
2-D Photoelastic Models	10 (with fringe interpolation or multiplication techniques)	Yes	Yes	Moderate (primarily specimen preparation time)	Moderate	Shows regions of stress concentration quickly.	Limited to evaluations of "plane stress" model geometries.
3-D Photoelastic Models	10 (with fringe interpolation or multiplication techniques)	Yes	Yes	High (specimen preparation, loading, and generation of results)	Moderate	Allows evaluation of surface and interior stresses.	Very time consuming. Largely being replaced by finite element techniques.
Holographic Interferometry	1-10 (depending on processing of data)	Yes	Yes	Low (assuming optical set-up ready to be used)	Moderate to High	Shows regions of stress concentration quickly. Distribution and magnitude of strain can be obtained.	Requires mechanical stability of test set-up unless pulsed later used. Field use for measurement of dynamic strain not possible yet.
Speckle Interferometry	10-100 (depending on processing of data)	Yes	Yes	Low (assuming optical set-up ready to be used)	Moderate	Shows regions of stress concentration quickly. Specklegram gives strain magnitude and distribution. Mechanical stability requirements less stringent than for holography.	Field use for measurement of dynamic strain not possible yet. Not as well developed for industrial application as holography.

Table 7-1 Comparison of Techniques to Define Fatigue Critical Locations (Cont)

Technique	Max. Strain Sensitivity (microm/in)	Whole Field?	Provides Strain Gradient?	Time Involved in Use	Eqpt. Cost	Primary Advantages	Principal Limitations
Thermoelasticity	5-10	Yes	Yes	Low (assuming cyclic loading fixture ready for use)	High	Portable, can be used in the lab or field. Shows regions of stress concentration quickly.	Only gives sum of principal stresses. Requires cyclic loading of components. Requires calibration to obtain quantitative data.

Table 7-2 Comparison of Flaw Detection Techniques

Technique	Detection of Internal Defects?	Need to Apply Loading?	Eqpt. Cost	Materials Which Can Be Inspected	Need to Apply Elec./Magnetic Field?	Remarks
Ultrasonic	Yes	No	Moderate to High	Metals, Composites, etc.	No	Automated systems available for giving location and size of defects.
Dye Penetrant	No	No	Low	Metals, Composites, etc.	No	Provides quick way of detecting surface defects.
Magnetic Particle	No (unless near surface)	No	Low to Moderate	Ferrous Metals	Yes	Provides quick way of detecting surface defects.
Eddy Current	No (unless near surface)	No	Moderate	Conducting Metals	Yes	Works best for nonferrous metals.
Radiography	Yes	No	Moderate to High	Metals, Composites, etc.	No	Automated systems available for giving location and size of defects.
Optical Holography	Yes (if cause surface deformation)	Yes (but very low levels)	Moderate to High	Metals, Composites, etc.	No	Provides quick way of detecting defects.
Acoustic Holography	Yes	No	Moderate to High	Metals, Composites, etc.	No	Automated systems available for giving location and size of defects.
Acoustic Emission	Yes	Yes	Moderate to High	Metals, Composites, etc.	No	Able to detect presence of defects and monitor damage under fatigue loading.
Potential Drop	No	No	Moderate	Conducting Metals	Yes	Systems available for monitoring crack size during growth.
Gel Electrode	No	No	Low	Conducting Metals	Yes	Requires oxide film of suitable thickness on test part.

Table 7-3 Results of Flaw Detection Program[72]

Method	Simulated Structure	Crack Location	Matl.	Number of Personnel Tested	Flaw Size Range (in.)	Fraction of Finds
Ultrasonic (shear wave)	wing spar	fastener holes in web and cap	Al	121	0.03—0.05 0.08—0.11 0.15—0.18 0.23—0.28	0.32 0.43 0.53 0.55
Eddy Current (bolt hole)	wing spar	fastener holes in web and cap	Al	133	0.03—0.05 0.08—0.11 0.15—0.18 0.23—0.28	0.56 0.66 0.72 0.76
Radiography	skin and stringer panel	fastener holes in sandwiched shims	Al	84	0.25—0.29 0.31—0.37 0.39—0.45 0.47—0.55	0.79 0.95 0.98 0.98
Fluorescent Penetrant	riveted panel	fastener holes	Ti	37	0.03—0.05 0.07—0.10 0.12—0.15	0.65 0.96 0.97
Magnetic Particle	clevis/link rod assembly	fatigue cracks at clevis roots and holes; EDM notches at rod thread roots and mid-section	Steel	29	0.03—0.05 0.07—0.10 0.12—0.15	0.28 0.57 0.75

Table 7-4 Comparison of Residual Stress Measurement Techniques

Technique	Nondestructive?	Reliability	Bulk Stress?	Stress Gradient?	Eqpt. Cost	Time Involved In Use	Portable?	Remarks
Sectioning	No	Good	Yes	Yes	Low (assuming machine shop available)	High	No	Only technique which can give complete stress distribution.
Hole Drilling	Semi	Good	No	Yes	Moderate	Moderate	Yes	Well-developed but restricted to near surface stresses.
XRD	Yes	Good	No	No (except by layer removal)	High	Moderate	Yes	Well-developed but restricted to surface stresses.
Ultrasonic	Yes	Unproven	Potential but Unproven	Potential but Unproven	Moderate	Low to Moderate	Yes	Still under development.
Barkhausen Noise	Yes	Fair	No	No	Moderate	Low to Moderate	Yes	Restricted to ferromagnetic materials. Restricted stress range.

in Table 7-1. Moderate cost is taken to mean on the order of $10K, low cost $1K, and high cost $100K. Moderate time is taken to mean several hours, low a matter of minutes, and high, days. Maximum strain sensitivity denotes the smallest increment of strain that can generally be resolved under laboratory conditions and should not be confused with experimental uncertainty, that can vary considerably depending on the particular test conditions. Suitability of a technique for a given test situation may depend on factors other than those considered in Table 7-1, and it is thus advisable to consult with vendors and users of a technique for more detailed information before selecting a technique(s).

A comparison of the flaw detection techniques is summarized in Table 7-2. No comments on the defect sizes which can be detected by a given technique are given because those sizes will depend on numerous factors such as the component geometry, location and nature of defects, sophistication of instrumentation used, and so forth.

A program was recently conducted by the U.S. Air Force to evaluate the proficiency of personnel using different NDT techniques in the detection of cracks that developed in simulated structural components subjected to laboratory fatigue loading. A summary of the results is shown in Table 7-3. No general conclusions regarding the efficacy of one technique vs. another should be drawn from these results; however, they do shed some light on practical ranges of crack sizes that were detected with a given frequency.

A comparison of the residual stress measurement techniques is given in Table 7-4. The precision of a given technique depends on the difficulty of the test situation, the care with which measurements are made, etc. In general, it appears that the techniques all have the potential to determine residual stresses to within about ± 35 to 70 MPa (± 5 to 10 ksi). X-ray diffraction offers the best spatial resolution of stress, on the order of 1 mm^2 (0.002 in^2), while the other techniques are at least an order of magnitude higher in that regard.

7.6 References

[1] Dally, J. W. and Riley, W. F., *Experimental Stress Analysis*, McGraw-Hill Book Co., New York, NY, 1965.

[2] Stern, F. B., "Brittle Coatings," Chapter V, *Manual on Experimental Stress Analysis*, 3rd Edition, SESA, 1978.

[3] "Brittle Coating for Stress Analysis," Bulletin S-109-D, Measurements Group, Inc., 1978.

[4] Dove, R. C. and Adams, P. H., *Experimental Stress Analysis and Motion Measurement*, C. E. Merrill Publishing Co., 1964.

[5] Perry, C. C. and Lissner, H. R., *The Strain Gage Primer*, McGraw-Hill Book Co., 1962.

[6] Weymouth, L. J., Starr, J. E., and Dorsey, J., "Bonded Resistance Strain Gages," Chapter II, *Manual on Experimental Stress Analysis*, 3rd Edition, SESA, 1978.

[7] Tech Notes TN-505 ("Strain Gage Selection"), TN-504 ("Strain Gage Temperature Effects"), TN-507 ("Wheatstone Bridge Nonlinearity"), TN-502 ("Strain Gage Excitation Levels"), TN-501 ("Noise Control in Measurements"), TN-509 ("Transverse Sensitivity Errors"), TN-511 ("Strain Gage Misalignment Errors"), TN-512 ("Plane Shear Measurements") and Tech Tips TT-603 ("Bondable Terminals"), TT-607 ("Installation and Protection in Field Environments"), TT-605 ("High Elongation Strain Measurements"), Measurements Group, Inc., copyright dates between 1979 and 1983, depending on the particular publication.

[8] Perry, C. C., "The Strain Gage Revisited," *Experimental Mechanics*, Vol. 24, No. 4, December 1984, pp. 286-299.

[9] Dowling, N. E., "Performance of Metal-Foil Strain Gages During Large Cyclic Strains," *Experimental Mechanics*, Vol. 17, No. 5, May 1977, pp. 193-197.

[10] Post, D., "Photoelasticity," Chapter IV, *Manual on Experimental Stress Analysis*, 3rd Edition, SESA, 1978.

[11] "Photoelastic Stress Analysis—Techniques and Products," Bulletin SFC-300, Photoelastic Division, Measurements Group, Inc., 1980.

[12] Tech Notes TN-702 ("Introduction to Stress Analysis by the Photostress Method"), TN-707 ("Model Making for Three-Dimensional Photoelastic Stress Analysis"), TN-701 ("Calibration of Photoelastic Coatings"), TN-704 ("How to Select Photoelastic Coatings"), TN-705 ("Calibration of Photoelastic Plastics for Two- and Three-Dimensional Model Analysis"), and TN-706 ("Corrections to Photoelastic Coating Fringe-Order Measurements"), Measurements Group, Inc., copyright dates between 1977 and 1982, depending on the particular publication.

[13] Vest, C. M., *Holographic Interferometry*, John Wiley & Sons, Inc., 1979.

[14] Taylor, C. E., "Holography," Chapter VII, *Manual on Experimental Stress Analysis*, 3rd Edition, SESA, 1978.

[15] Sollid, J., "Holography Applied to Structural Components," *Optical Engineering*, Vol. 14, No. 5, September-October 1975, pp. 460-469.

[16] Goldberg, J. L., "A Method of Three-Dimensional Strain Measurement on Non-Ideal Objects Using Holographic Interferometry," *Experimental Mechanics*, Vol. 23, No. 1, March 1983, pp. 59-73.

[17] "Design and Testing with Holography," Technical Note, Newport Corp., 1984.

[18] Luxmoore, A. R., *Optical Transducers and Techniques in Engineering Measurements*, Applied Science Publishers, 1983.

[19] Hung, Y. Y. and Hovanesian, J. D., "Full-Field Surface-Strain and Displacement Analysis of Three-Dimensional Objects by Speckle Interferometry," *Experimental Mechanics*, Vol. 12, No. 10, October 1972, pp. 454-460.

[20] Hung, Y. Y., Rowlands, R. E., and Daniel, I. M., "A New Speckle-Shearing Interferometer: A Full-Field Strain Gage," *Applied Optics*, Vol. 14, 1975, pp. 618-622.

[21] Yamaguchi, I., "Simplified Laser-Speckle Strain Gage," *Optical Engineering*, Vol. 21, No. 3, 1982, pp. 436-440.

[22] Chiang, F. P. and Kim, C. C., "Some Optical Techniques of Displacement and Strain Measurements on Metal Surfaces," *Journal of Metals*, May 1983, pp. 49-54.

[23] Asundi, A. and Chiang, F. P., "Theory and Applications of the White Light Speckle Method for Strain Analysis," *Optical Engineering*, Vol. 21, No. 4, 1982, pp. 570-580.

[24] Belgen, M. H., "Structural Stress Measurements with an Infra-Red Radiometer," ISA Transactions, Vol. 6, No. 12, January 1967, ISA, pp. 49-53.

[25] Stanley, P. and Chan, W. K. "Quantitative Stress Analysis by Means of the Thermoelastic Effect," *Proceedings of the Fifth International Congress on Experimental Mechanics*, June 1984.

[26] Oliver, D. E. and Webber, J. M. B., "Absolute Calibration of the SPATE Technique for Non-Contacting Stress Measurement," *Proceedings of the Fifth International Congress on Experimental Mechanics*, June 1984.

[27] Abstracts of Papers, *First International Conference, Stress Analysis by Thermoelastic Techniques*, Sira, Ltd., November 1984.

[28] *Nondestructive Inspection and Quality Control*, Vol. 11, *Metals Handbook*, 8th Edition, ASM International, 1976.

[29] *Review of Progress in Quantitative Nondestructive Evaluation*, D. O. Thompson and D. E. Chimenti, Vols. 1 through 4, Plenum Publishing Corp., 1981-84.

[30] *Eddy Current Characterization of Materials and Structures*, ASTM STP 722, ASTM, 1981.

[31] Vest, C., "Holographic NDE: Status and Future," Report PB81-207409, National Bureau of Standards, May 1981.

[32] Hayward, G. P., "Introduction to NDT," *Inspectors' Handbook Series*, ASQC, 1978.

[33] Tatro, C. A., "Design Criteria for Acoustic Emission Experimentation", *Acoustic Emission*, ASTM STP 505, ASTM, 1972, pp. 84-99.

[34] Geberich, W. W. and Hartbower, C. E., "Some Observations on Stress Wave Emission as a Measure of Crack Growth," *International Journal of Fracture*, Vol. 3, No. 3, September 1967, pp. 185-192.

[35] *Advances in Crack Length Measurement*, C. J. Beevers, Engineering Materials Advisory Services, Ltd., 1982.

[36] Baxter, W. J., "Oxide Films: Quantitative Sensors of Metal Fatigue," Research Publication GMR-3958, General Motors Research Laboratories, January 1982.

[37] *Residual Stress Measurement by X-Ray Diffraction*, SAE J784a, SAE, 1971.

[38] *Methods of Residual Stress Management*, SAE J936, SAE, 1965.

[39] Ruud, C. O., "X-Ray Analysis and Advances in Portable Field Instrumentation," *Journal of Metals*, Vol. 31, No. 6, June 1979, pp. 10-15.

[40] James, M. R. and Cohen, J. B., "PARS—A Portable X-Ray Analyzer for Residual Stress," *Journal of Testing and Evaluation*, Vol. 6, No. 2, 1978, p. 91.

[41] Bolstad, D. A. and Quist, W. E., "The Use of a Portable X-Ray Unit for Measuring Stresses in Al, Ti and Fe Alloys," *Advances in X-Ray Analysis*, Vol. 8, Plenum Publishing Corp., 1965.

[42] Mitchell, C. M., "A Dual Detector Diffractometer for Measurement of Residual Stress," *Advances in X-Ray Analysis*, Vol. 20, Plenum Publishing Corp., 1977.

[43] Ruud, C. O. and Farmer, G. D., "Residual Stress Measurement by X-Rays: Errors, Limitations and Applications," *Nondestructive Evaluation of Materials*, Plenum Publishing Corp., 1979.

[44] Larson, R. C., "The X-Ray Diffraction Measurement of Residual Stress in Aluminum Alloys," *Advances in X-Ray Analysis*, Vol. 7, Plenum Publishing Corp., 1964.

[45] Cullity, B. D., "Some Problems in X-Ray Stress Measurements," *Advances in X-Ray Analysis*, Vol. 20, Plenum Publishing Corp., 1977.

[46] Klug, H. P. and Alexander, L. E., *X-Ray Diffraction Procedures*, 2nd Edition, J. Wiley & Sons, 1974.

[47] Cullity, B. D., *Elements of X-Ray Diffraction*, 2nd Edition, Addison-Wesley Publishing Co., Inc., 1978.

[48] Barrett, C. S., "A Critical Review of Various Methods of Residual Stress Measurement," *Proc. of the Society for Experimental Stress Analysis*, Vol. 2, No. 1, 1944, pp. 147-156.

[49] Sachs, G. and Epsey, G., "The Measurement of Residual Stresses in Metal," *The Iron Age*, September 18 and 25, 1941, pp. 63-71 and 36-42, respectively.

[50] Fuchs, H. O. and Mattson, R. L., "Measurement of Residual Stresses in Torsion Bar Springs," SESA, Vol. 4, No. 1, 1946, pp. 64-73.

[51] "Evaluation of Methods for Measurement of Residual Stress," SAE HS- 147, SAE, 1957.

[52] Rendler, N. J. and Vigness, I., "Hole-Drilling Strain-Gage Method of Measuring Residual Stresses," *Experimental Mechanics*, Vol. 6, No. 12, December 1966, pp. 577-586.

[53] "Standard Method for Determining Residual Stresses by the Hole-Drilling Strain Gage Method," ASTM E837-81, ASTM, 1981.

[54] "Measurements of Residual Stresses by the Hole-Drilling Strain-Gage Method," Technical Note TN-503-1, Measurements Group, Inc., 1985.

[55] Sandifer, J. P. and Bowie, G. E., "Residual Stress by Blind-Hole Method With Off-Center Hole," *Experimental Mechanics*, Vol. 18, No. 5, May 1978, pp. 173-179.

[56] Kabiri, M., "Measurement of Residual Stresses by the Hole-Drilling Method: Influences of Transverse Sensitivity of the Gages and Relieved-Strain Coefficients," *Experimental Mechanics*, Vol. 24, No. 3, September 1984, pp. 252-256.

[57] Bathgate, R. C., "Measurement of Non-Uniform Bi-Axial Residual Stresses by the Hole Drilling Method," *Strain Journal of BSSM*, Vol. 4, No. 2, 1968, pp. 20-29.

[58] Schajer, G. S., "Application of Finite Element Calculations to Residual Stress Measurements," *Journal of Engineering Materials and Technology*, ASME Transactions, Vol. 103, April 1981, pp. 157-163.

[59] Design and Testing Handbook, Electro-Optical Product Directory No. 300, Newport Corp., 1986.

[60] Crecraft, D., "The Measurement of Applied and Residual Stresses in Metals," *Journal of Sound and Vibration*, Vol. 5, No. 1, 1967, pp. 173- 192.

[61] Noronha, P. J., Chapman, J. R. and West, J. J., "Residual Stress Measurement and Analysis Using Ultrasonic Techniques," *Journal of Testing and Evaluation*, Vol. 1, No. 3, May 1973, pp. 209-214.

[62] Blessing, G. V., Hsu, N. N. and Proctor, T. W., "Ultrasonic-Shear-Wave Measurement," *Experimental Mechanics*, Vol. 24, No. 3, September 1984, pp. 218-222.

[63] Goebbels, K., and Hirsekorn, S., "A New Ultrasonic Method for Stress Determination in Textured Materials," *NDT International*, Vol. 17, No. 6, 1984, pp. 337-341.

[64] Allen, D. R., and Sayers, C. M., "The Measurement of Residual Stress in Textured Steel Using an Ultrasonic Velocity Combinations Technique," *Ultrasonics*, Vol. 22, No. 4, 1984, pp. 179-188.

[65] Szabo, T. L., "Obtaining Subsurface Profiles from Surface Acoustic Wave Velocity Dispersion," *Journal of Applied Physics*, Vol. 46, No. 4, 1975, pp. 1448-1454.

[66] Clark, A. V., and Mignona, R. B., "Use of Off-Axis SH-Waves to Map Out Three-Dimensional Stresses in Orthotropic Plates," *Ultrasonics*, Vol. 22, No. 5, 1984, pp. 205-214.

[67] Hirao, M., Kyukawa, M., Sotani, Y., and Fukuoka, H., "Rayleigh Wave Propagation in a Solid With a Cold-Worked Surface Layer," *Journal of Nondestructive Evaluation*, Vol. 2, No. 1, 1981, pp. 43-49.

[68] Gardner, C. G., Matzkanin, G. A. and Davidson, D. L., "The Influence of Mechanical Stress on Magnetization Processes and Barkhausen Jumps in Ferromagnetic Materials," *International Journal of Nondestructive Testing*, Vol. 3, 1971, pp. 131-169.

[69] Sundström, O. and Törrönen, K., "The Use of Barkhausen Noise Analysis in Nondestructive Testing," *Materials Evaluation*, February 1979, pp. 51- 56.

[70] Chait, R., "Residual Stress Pattern in a High Hardness Laminar Composite Steel Weldment," *Proceedings of a Workshop of Nondestructive Evaluation on Residual Stress*, NTIAC-76-2, 1975, pp. 237-246.

[71] Barton, J. R., Kusenberger, F. N., Beissner, R. E. and Matzkanin, G. A., "Advanced Quantitative Magnetic Nondestructive Evaluation Methods, Theory and Experiment," *Nondestructive Evaluation of Materials*, J. J. Burke and V. Weiss, Plenum Publishing Corp., 1979, pp. 461-463.

[72] Sproat, W. H. and Rowe, W. J., "Ensuring Aircraft Structural Integrity Through Nondestructive Evaluation," *Proceedings 1984 Aircraft Structural Integrity Conference*, Warner Robbins Air Logistics Center, Macon, Georgia, Nov. 27-29, 1984.

Addresses of Publishers

Addison-Wesley Publishing Co., Inc., 1 Jacob Way, Reading, MA 01867

Applied Science Publications, Inc., P.O. Box 5399, Grand Central Station, NY 10163

ASM International, 9639 Kinsman Road, Metals Park, OH 44073

ASME, American Society of Mechanical Engineers, 345 E. 47th Street, New York, NY 10017

ASQC, American Society for Quality Control, 230 W. Wells Street, Suite 7000, Milwaukee, WI 53203

ASTM, American Society for Testing & Materials, 1916 Race Street, Philadelphia, PA 19103

Engineering Materials Advisory Services, Ltd., West Midlands, England

General Motors Research Laboratories, Warren, MI 48090

McGraw-Hill Book Co., 1221 Avenue of the Americas, New York, NY 10020

Measurements Group, Inc., Raleigh, NC

C. E. Merrill Publishing Co., 1300 Alum Creek Drive, Columbus, OH 43216

Newport Corp., Fountain Valley, CA

Plenum Publishing Corp., 233 Spring Street, New York, NY 10013

SAE, Society of Automotive Engineers, 400 Commonwealth Drive, Warrendale, PA 15096

SESA, Society for Experimental Stress Analysis, Westport, CT

Sira, Ltd., Southill, Chislehurst, Kent, England

Warner Robins Air Logistics Center, Macon, GA

John Wiley & Sons, 605 Third Avenue, New York, NY 10158

Chapter 8 Numerical Analysis Methods

8.1 Introduction
  8.1.1 Scope
  8.1.2 Relationship to Other Chapters
  8.1.3 Chapter Plan

8.2 Closed Form Methods
  8.2.1 Background
  8.2.2 Nominal Stress Estimation
    8.2.2.1 Strength of Materials
    8.2.2.2 Theory of Elasticity
    8.2.2.3 Energy Methods
  8.2.3 Local Stress Estimation
    8.2.3.1 The Handbook Approach
  8.2.4 Linear Elastic Fracture Mechanics
  8.2.5 Compensation for Plasticity Effects
  8.2.6 Topical Bibliography

8.3 The Finite Element Method
  8.3.1 Background
  8.3.2 Theoretical Basis
    8.3.2.1 The Element Stiffness Matrix
    8.3.2.2 Structural Stiffness Matrix
    8.3.2.3 Isoparametric Elements
    8.3.2.4 Modelling Considerations
  8.3.3 Elastic-Plastic Analysis
    8.3.3.1 Background
    8.3.3.2 Incremental or Flow Theory of Plasticity
    8.3.3.3 Deformation Theory of Plasticity
    8.3.3.4 Numerical Analysis Procedures
  8.3.4 General Applications and Examples
    8.3.4.1 Engine Block Analysis
    8.3.4.2 SAE Keyhole Specimen
    8.3.4.3 Uniformly Loaded Sheet with a Central Crack
  8.3.5 Application to Linear Elastic Fracture Mechanics
    8.3.5.1 The Stress Intensity Factor
    8.3.5.2 Calculation of SIF Using FE Methods
  8.3.6 Available Software
  8.3.7 Topical Bibliography

8.4 The Finite Difference Method
  8.4.1 Background
    8.4.1.1 Formulations and Types of Boundary Value and Initial Value Problems
    8.4.1.2 The Basic Idea of the FDM
    8.4.1.3 Advantages and Limitations
  8.4.2 Theoretical Basis
    8.4.2.1 The Classical FDM
    8.4.2.2 The Generalized FDM
  8.4.3 The GFDM Approach to Nonlinear Problems
  8.4.4 General Applications and Examples
    8.4.4.1 Deflection of a Beam
    8.4.4.2 Shear Stress in a Prismatic Bar Subjected to Torsion
  8.4.5 Application to Linear Elastic Fracture
  8.4.6 Available Software
  8.4.7 Summary

  8.4.8 Topical Bibliography

8.5 The Boundary Integral Equation and Boundary Element Methods
  8.5.1 Background
  8.5.2 Theoretical Basis
    8.5.2.1 Governing Integral Equation for Elastostatics
    8.5.2.2 Boundary Point
    8.5.2.3 Boundary Elements
    8.5.2.4 Internal Results
    8.5.2.5 Higher Order Elements
    8.5.2.6 Non-Smooth Boundaries
  8.5.3 Other Considerations
  8.5.4 Summary
  8.5.5 Topical Bibliography

8.6 Comparison of the Methods
  8.6.1 Historical Background
  8.6.2 The Finite Element Method
  8.6.3 The Finite Difference Method
    8.6.3.1 Derivative Approximation
    8.6.3.2 Variational Formulation
  8.6.4 The Boundary Integral Equation Method

8.7 References

8.1 Introduction

Stress analysis is a necessary step in assessing the adequacy or safety of a component or structure for its intended use. The stress analysis serves as a transfer function from the loads characterization to the fatigue life prediction.

The analysis requires, as input, a definition of the loads or displacements to which the structure will be subjected. The loads may be dynamic or quasi static. In either case, the stress analyst usually assumes the loads are known with certainty (although this is often an incorrect assumption). The stress calculations possess as great a degree of uncertainty as the loads, if not substantially greater. A factor of 2 difference in the alternating component of stress can result in an order of magnitude difference in the subsequently predicted fatigue lives. Therefore, some engineering judgement must be applied in balancing the investment in man hours and computational resources against accuracy and precision.

8.1.1 Scope

The purpose of this chapter is to review the commonly used numerical analysis procedures for estimating stresses in a component subjected to significant operating loads. An attempt has been made to provide sufficient detail to allow the reader to gain an elementary understanding of each of these analysis procedures. Numerous references have been cited throughout to guide the reader to more detailed discussions in areas of particular interest.

8.1.2 Relationship to Other Chapters

Numerical analysis procedures described in this chapter for the estimation of component stresses relate primarily to the chapters on Material Properties (Chapter 3), Service History Determination (Chapter 5) and Fatigue Life Prediction (Chapter 10). Loads which are measured according to techniques described in Chapter 5 are used as inputs to a numerical analysis. Stresses and strains which are computed then serve as inputs to a fatigue life prediction analysis. Basic material properties to be used in the life prediction are reviewed in Chapter 3.

8.1.3 Chapter Plan

This chapter is divided into four major topic areas. Section 8.2 provides a brief review of the closed-form, "hand" stress analysis procedures that have been used for many years and which are still quite useful in certain situations. Section 8.3 addresses the finite element method, that is used commonly today. Sections 8.4 and 8.5 address the other less widely used finite difference and boundary element methods, respectively. For the finite element and finite difference sections the following topics are reviewed:

- Background
- Theoretical Basis
- Elasto-Plastic Analysis
- General Applications and Examples
- Application to Fracture Mechanics Analyses
- Available Software
- Topical Bibliography

The boundary element section covers only the first two and last topics in this list. Section 8.6 closes this chapter with a brief comparison of the three numerical approximation stress analysis procedures. Please be aware that there are some differences in notation within different sections of this chapter, resulting from the fact that different authors prepared those sections. Matrix notation is used in the section describing the finite element method, while somewhat more complicated notation [192] is used in the section describing the finite difference method.

8.2 Closed-Form Methods

8.2.1 Background

It is standard practice to analyze complex structures subjected to complex loadings by numerical techniques. This may be done with a large variety of commercially available computer analysis programs. In fact, it often is more expedient to analyze simple problems with simple computer models than to perform the analysis by hand. This is especially true with the availability of automatic mesh generation, integrated design and analysis programs, and downsized analysis programs that operate on personal computers.

Nevertheless, many categories of problems do not have to be analyzed by computer modeling to obtain solutions that are exact, or sufficiently accurate for engineering purposes. These solutions may be obtained relatively easily and quickly.

There are many situations in which hand calculations are appropriate. For example, global models based on the stiffness properties of the structural elements are useful for estimating the distribution of loads throughout a structure. However, if the actual structural elements are irregular in cross section, or the constructional joints are composed of an assembly of many members, the numerical model may not predict correct stresses. The loads predicted by the model can be used in hand calculations to analyze specific locations of the structure in detail.

Another situation demanding hand calculations arises where solutions are sought to a problem with several variations. If a large, expensive model is required to obtain only one solution, rerunning the problem with a multitude of minor variations may be prohibitively expensive. However, if the problem can be simplified to its salient features and an approximate solution generated by hand as an algebraic formulation, then the sensitivity of the problem to variations in the significant parameters can be explored more easily and inexpensively. Then, if desired, a detailed computer analysis of an optimized configuration may be performed.

Finally, it is surprisingly easy to commit modeling errors with any numerical analysis program. Most errors occur in specifying boundary conditions or units. Geometry errors may occur if plots of the model are not available. Sometimes the program user's manual is vague or poorly written, leading to inadvertent misuse of the program options. Also, it is not unusual for analysis programs to contain subtle computational errors, even if they reside in the public domain and have supposedly been verified. Errors due to any of these potential causes may not prevent the analysis from executing and they may not be apparent at a first glance of the printed output. Therefore, a "sanity check" on any computer-generated solution should be standard practice. This may be accomplished by performing hand calculations of some simplified or idealized aspect of the problem to verify that the results are at least of the correct sign and order of magnitude.

8.2.2 Nominal Stress Estimation

Nominal stresses are the gross tension and bending stresses calculated through a cross section, neglecting the effect of local stress raisers such as notches or cracks. The nominal stress is required to assess the likelihood of static failure. It is also used with stress concentration factors to estimate local stresses and the probable location(s) of fatigue failure. There are three commonly used ways to estimate nominal stresses: strength of materials, energy methods, and the theory of elasticity. Entire books have been written about them, so only brief descriptions are given here.

8.2.2.1 Strength of Materials

The strength of materials approach is essentially a "static" analysis procedure applied to deformable or elastic bodies. The equations of statics ($\Sigma F = 0$, $\Sigma M = 0$) are combined with equations of consistent deformation, based on the geometry of the structure ($\delta = F/k$) to obtain the forces in some component of the structure. The component develops internal stresses that equilibrate the forces and which are assumed to be proportional to the deformation. This approach is readily applied to the calculation of nominal stresses in structural elements such as beams and shafts.

Since forces, deformations and stresses are linearly related, superposition of simple support or loading conditions can be invoked to obtain the solutions for more complex configurations. The application of superposition and strength of materials is limited to problems in the elastic range of material properties, and for which deflections and rotations in the structure are very small compared to the dimensions of the structure. The strength of materials approach is not adequate for calculating buckling loads, or local stresses at a hole or notch.

Solutions to a huge variety of practical stress analysis problems have been obtained using strength of materials, and are found in many references. Thus, by knowing a few basic rules like $\sigma = Mc/I$ and having a couple of references available, the nominal stresses in many structural components can be estimated easily and with acceptable accuracy.

8.2.2.2 Theory of Elasticity

The theory of elasticity replaces the physical reasoning of strength of materials with mathematics. Equilibrium on an element of material is expressed as a distribution of stresses on its surfaces. The behavior of the linear, perfectly elastic material is described by relationships between stress and strain. Strains, that are defined by displacements, are restricted by the compatibility equations to correspond to continuous, physically meaningful deformations. Applied displacements or forces constitute boundary conditions on the resulting complex system of differential equations.

Methods of solution include direct integration, selection of a stress function that satisfies the differential equations, numerical approximations and experimentation. Since the mathematics of elasticity are rather rigorous, it is unlikely that a typical stress analysis would be based on such a method, out of time and expediency considerations. Nevertheless, a number of important problems in stress analysis were initially solved by the theory of elasticity, such as the stress concentration around a hole.

8.2.2.3 Energy Methods

An alternative approach to characterizing the state of stress of a structure is to make use of the fact that for an elastic body, the work performed on the structure by external forces is equal to the change of potential or strain energy in the structure. A specific way of expressing this generalization is by Castigliano's theorem, which states that the partial derivative of the strain energy of a structure with respect to a statically independent force gives the component deflection of the location of the structure at which the force is applied, Thus it becomes possible to

ture due to any loading condition. Redundant support loads can be calculated by equating the partial derivative with zero. It is a highly useful technique for obtaining member loads, deflections and stresses.

A second method, known as the principle of virtual work, may be useful in evaluating buckling loads.

8.2.3 Local Stress Estimation

The nominal stresses can be considered accurate only in regions where the geometry is uniform, and away from locations of applied forces. Transitions in geometry and applications of loads perturb uniform or smoothly varying stress fields, such that the nominal stresses represent only an average. In the vicinity of the transition in geometry or applied force, the stress may be several times the nominal stress in magnitude. Of course, the maximum stress is the yield stress, not including strain hardening. Locally, however, strains exist that correspond to these very high elastically calculated stresses. The (elastic) stress gradients are also typically very steep.

Experience has shown that these areas of locally high stress and strain are where fatigue failures most often occur. Typical stress raisers include bolt holes, threads, fillets, welds, reentrant corners, notches, corrosion pits, damaged surfaces, etc. The sharper the disturbance in geometry, or the smaller the area over which a force is applied, the greater the local stresses.

The strength of materials approach is not adequate for calculating local stresses at the roots of these stress raisers. For that matter, a coarsely-discretized finite element model is not likely to be sensitive to the high stress gradients around a notch. (This usually requires the use of hundreds of tiny elements to represent the stress raiser, which is costly in terms of man hours and computer resources, or the use of specially formulated elements, which may not be available in a given code.) Therefore other methods must be used to estimate the degree of stress concentration associated with the geometry. These methods are the handbook approach, the theory of elasticity, numerical approximation or physical testing. Only the use of handbooks will be discussed here.

8.2.3.1 The Handbook Approach

A useful way to represent the effect of a stress raiser on nominal stresses is with theoretical stress concentration factors. The nominal alternating stress is then multiplied by this factor to approximate the local alternating stress.

A significant number of theoretical stress concentration factors have been generated for many common geometries. Some abridged data appear in most machine design textbooks. The most comprehensive and authoritative collection of theoretical stress concentration factors is Peterson (see Ref. 17 in the Topical Bibliography). These factors were generated using the theory of elasticity and physical testing, although this is transparent to the user. It is likely that reasonable approximations to most design-related stress concentrations can be found in this reference.

In the event that the local geometry cannot be approximated by a handbook case, the choices are less appealing. If the problem is readily solvable by application of the theory of elasticity, it has probably been done already and has appeared in the literature. Other possibilities are testing of physical models using photoelasticity, or advanced computer modeling. If the geometry includes a fatigue crack, a fracture mechanics approach should be used.

Real materials exhibit a wide range of sensitivity to the effects of a local stress raiser. Therefore, the theoretical stress concentration factor must be modified by a notch sensitivity factor for different materials, to obtain a somewhat lower value called the fatigue stress concentration factor. Direct application of the theoretical stress concentration factor to the alternating nominal stresses to estimate local stresses is nearly always conservative and, in some cases, is unduly conservative.

8.2.4 *Linear Elastic Fracture Mechanics*

It is an unavoidable fact that most structures contain defects or flaws. These flaws (that are conveniently approximated as cracks) may be design details, metallurgical defects or actual cracks that have grown in service. Fracture mechanics address the relationship between service loads, material properties, crack size and fatigue crack propagation. At the most basic level, fracture mechanics states that there is a stress intensity factor, K, that is a function of crack length, applied nominal stress and local geometry. As the service load is increased or cycled, the crack will grow incrementally. Eventually, a combination of crack length and stress is reached where the value of K equals some critical value K_C, that is a material property. At that point the crack propagates in an uncontrolled manner, and the structure fails.

Linear elastic fracture mechanics provides a means of describing the stress field around the crack tip, and hence the value of K. For a tensile, crack-opening mode (Mode I by convention), K_I is a function of the nominal stress, the square root of crack length, and geometry, e.g. $K_I = \sigma\sqrt{\pi a} \cdot f(geom)$. The geometry function is influenced by whether the crack is embedded within the structure, near an edge or at the surface, the size of the crack relative to the nearby structural geometry and the nominal stress distribution around the crack.

Stress intensity functions may be derived from elasticity or by numerical approximations. However, as with stress concentration factors, this is usually not necessary, because several references are available that present K_I values for numerous crack configurations. Furthermore, since the entire concept is based on linear elasticity, it is possible to superimpose the effects of individual crack configurations in order to approximate a more complex combination of crack shape, location and stress distribution. This is accomplished by adding K values directly or multiplying correction factors. The range of K values corresponding to the range of nominal stresses are then input directly to the fatigue crack growth and fatigue life assessment process.

8.2.5 *Compensation for Plasticity Effects*

The methods of conventional stress analysis are based on the assumptions that the material is isotropic and linear elastic, and that deflections are very small compared to the dimensions of the structure itself. Calculations may then be made based on the initial undeformed geometry. Superposition may be applied to obtain solutions to more complex loadings.

If the stresses in a material exceed the yield stress, the assumptions that are the basis for linear elastic stress analysis do not apply. This is because the relationship between stress and strain ceases to be either linear or elastic. This change in behavior may occur smoothly or abruptly in different materials. The behavior of the material after yielding also varies for different materials.

Various simple models have been devised to describe the idealized behavior of the elastic-plastic material, such as the elastic-perfectly plastic model and the elastic-strain hardening model. If plastic strains are much greater than the elastic strains, the elastic portion of the stress-strain curve may be treated as perfectly rigid. Empirical equations have been developed to approximate the plastic part of the stress-strain curve. The simplest is an exponential (power law) function, $\sigma = K\epsilon^n$, where σ and ϵ are the local stress and strain, respectively, and K and n are constants. Other more complex and accurate approximations exist, but they are more difficult to use.

Long fatigue lives are generally achieved by maintaining nominal stresses well below the yield stress (i.e., the *cyclic yield strength*). However, low-cycle fatigue is characterized by significant amounts of cyclic plastic deformation. This implies that if only a short life is required in terms of cyclic loading, a significant amount of plastic deformation may be tolerable; it may even be desirable if it can be used to advantage to save material.

In bending and torsion of solid sections, yielding occurs first at the outer fibers, progressing inward toward the neutral axis as the load is increased. The stiffness remains almost constant until the elastic core is greatly reduced in area. The load at which this occurs may be 10% to 100% of the load at which the outer fibers initially yielded, depending on the cross section. (If the load is increased beyond this level, the structure forms a plastic hinge and deforms in an uncontrolled manner until it collapses, or transfers load to some other part of the structure. At this point, stability is the issue, not fatigue.)

It is possible to estimate the distribution of stresses through a section from a static analysis, by assuming elastic-perfectly plastic or bilinear strain hardening behavior. In bending of unsymmetrical cross sections, it is necessary to account for the shift of the neutral axis. Guidance for such calculations are found in various references.

Elastic unloading of the partially yielded member results in a nonuniform profile of residual stresses through the cross section, that can also be estimated approximately. Subsequent reversal and release of the load results in a new residual stress pattern. Due to the strain hardening properties present in most engineering materials, it becomes nearly impossible to realistically predict the changing residual stress patterns. However, this is not usually necessary for two reasons. First, if the loads are maintained below the collapse loads, and the range of reversed stresses which are calculated elastically are less than approximately twice the yield stress, the structure will normally "shake down" to stable elastic behavior after a few repetitions of the load. Second, the fatigue performance is really controlled by the *cyclic* strains. In bending and torsion, the strains will vary approximately linearly through the section even though the stresses do not.

8.2.7 Topical Bibliography

Many readily available references contain discussions and information on "hand" stress analysis. The following references offer useful information on the following topics.

Topic	Reference Number
1) Elasticity	(1), (2), (8), (9)
2) Strength of materials	(3–8)
3) Energy methods	(2–5), (8)
4) Analysis of structural components	(3–13), (18)
5) Stress concentration factors	(5–8), (12), (17)
6) Fracture mechanics, stress intensity factors	(14–16)
7) Plasticity	(3–5), (8), (19), (20)

8.3 The Finite Element Method

8.3.1 Background

The finite element method was developed in the 1950s by structural engineers for the analysis of large-scale structural systems [21]. It was given a theoretical basis by mathematicians in the mid-1960s when various variational formulations of the method were presented [22,23]. In contrast to other numerical structural analysis methods, the finite element method has evolved to its present level of diversity due to the digital computer. The method possesses certain characteristics that take advantage of the special facilities offered by the modern digital computer. In particular, the method can be systematically programmed to accommodate such complex and difficult problems as nonhomogeneous materials, complicated time dependent boundary conditions, and nonlinear constitutive behavior. The variational formulations coupled with the amenability of the finite element method to digital computation allows application of the method in other fields, such as fluid mechanics, aerodynamics, heat transfer, quantum mechanics, and in general problems of mathematical physics. In this section, the finite element method is presented as a numerical technique for the analysis of complex structural systems. Discussion will be limited to the application of the method to small deformation linear elastic cases and to cases where elastic or elastic-plastic material response can be assumed [24-26].

The following notation will be employed. Forces and displacements are listed in column matrices $\{F\}$ and $\{\Delta\}$, respectively. Braces, $\{\}$ denote column vectors. Analogously, stresses and strains are listed in column matrixes $\{\sigma\}$ and $\{\epsilon\}$ where $\{\sigma\}^T = \lfloor \sigma_x, \sigma_y, \sigma_z, \sigma_{xz}, \sigma_{yz}, \sigma_{xy} \rfloor$ are components of stress, $\{\epsilon\}^T = \lfloor \epsilon_x, \epsilon_y, \epsilon_z, \epsilon_{xz}, \epsilon_{yz}, \epsilon_{xy} \rfloor$, are components of strain, the Roman superscript T denotes the transpose of a matrix, and the half brackets $\lfloor \ \rfloor$ denotes a row vector.

8.3.2 Theoretical Basis

The basic philosophy of the finite element method is piecewise approximation. That is, a solution to a complicated problem is approximated by sub-dividing the body of interest and representing the solution within each subdivision by a relatively simple function to approximate the distribution of the actual displacements and/or actual stresses, as shown in Fig. 8-1. These subdivisions are called "finite elements," and they are interconnected at points on and within the body at points which are called "nodes" or "nodal points." Finite element discretization techniques are usually categorized by one of three approaches: the "displacement method," the "equilibrium method," or the "mixed method." Displacements are assumed as primary unknown quantities in the displacement method; stresses are assumed as primary unknown quantities in the equilibrium method; and some displacements and some stresses are assumed as unknown quantities in the mixed method. The unknown magnitudes and amplitudes of the assumed functions (displacements, stresses, etc.) are the values of the functions evaluated at the nodal points. These unknowns are calculated from a system of algebraic equilibrium equations for the entire body. Discussion hereafter-will be limited to the displacement method.

8.3.2.1 The Element Stiffness Matrix

The finite element method of analysis is based on the mathematical formulation of basic structural elements and their use in the idealization of general structural problems. Commonly used elements include bars, beams, plates, shells, and solids. Each element type is represented numerically by an element stiffness matrix [k] that is used to relate the vector of element nodal forces, {f} to the vector of nodal displacements, {q} as follows

$$\{f\} = [k]\{q\} \quad (8\text{-}1)$$

These stiffness matrices are formulated using an assumed *displacement function* for the element, then applying any of the variational principles for mechanics to determine the force-displacement relationships.

In general the *displacement function* or *displacement field* for a finite element can be represented as

$$\{u\} = [N]\{q\} \quad (8\text{-}2)$$

where $[N]$ is a matrix of known functions and $\{q\}$ is a vector of *generalized nodal displacements* for the element. Dimensions of the matrix $[N]$ and the vector $\{q\}$ depend on the number of degrees of freedom of the element. Thus, for example, plane strain elements based on a quadratic *displacement field*, $[N]$ is a 2×12 matrix, $\{u\}$ is a 2×1 vector and $\{q\}$ is a

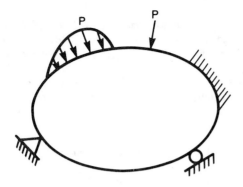

(a) Deformable Structure and Loading

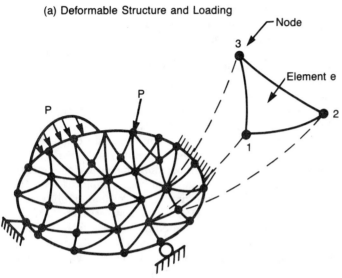

(b) Finite Element Idealization

Figure 8-1. A deformable structure under arbitrary loading and displacement boundary conditions and a finite element representation of the problem.

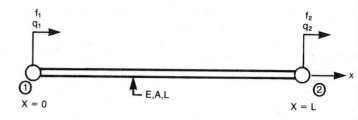

$$u(x) = (1 - \frac{x}{L})q_1 + \frac{x}{L}q_2$$

$$\epsilon_x = \frac{\partial u}{\partial x} = -\frac{1}{L}q_1 + \frac{1}{L}q_2$$

(a) Linear Displacement Field

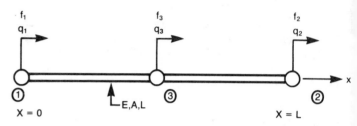

$$u(x) = \left[2\left(\frac{x}{L}\right)^2 - 3\left(\frac{x}{L}\right) + 1\right]q_1 - 4\left[\left(\frac{x}{L}\right)^2 - \left(\frac{x}{L}\right)\right]q_3 - \left[2\left(\frac{x}{L}\right)^2 - \left(\frac{x}{L}\right)\right]q_2$$

$$\epsilon_x = \partial u/\partial x = \frac{\partial u}{\partial x} = \left[4x-3\right]\frac{q_1}{L} - 4\left[2x-1\right]\frac{q_3}{L} - 2\left[4x-1\right]\frac{q_2}{L}$$

(b) Parabolic Displacement Field

Figure 8-2. One-dimensional truss elements with different displacement fields.

$$[N] = \begin{bmatrix} (1-\xi)(1-\eta) & 0 & \xi(1-\eta) & 0 & \xi\eta & 0 & (1-\xi)\eta & 0 \\ 0 & (1-\xi)(1-\eta) & 0 & \xi(1-\eta) & 0 & \xi\eta & 0 & (1-\xi)\eta \end{bmatrix} \quad (8\text{-}4)$$

where $\xi = x/a$ and $\eta = y/b$.

The displacement field of Eq. 2 can be used to obtain the strains

$$\{\epsilon\} = [B]\{q\} \quad (8\text{-}5)$$

where $\{\epsilon\}^T = \lfloor \epsilon_x, \epsilon_y, \epsilon_z, \epsilon_{xy}, \epsilon_{yz}, \epsilon_{zx} \rfloor$ is the *vector of strains*, and $[B]$ is a matrix that is obtained from differentiation of terms in the shape function $[N]$. Thus Eq. 5 expresses the strain field for the element in terms of the generalized nodal displacements $\{q\}$. For the bar element of Fig. 8-2(a), only the axial strain ($\epsilon_x = \partial u/\partial x$) is significant and the matrix $[B]$ of Eq. 5 is just

$$[B] = \frac{1}{L}\lfloor -1, 1 \rfloor \quad (8\text{-}6)$$

12×1 vector. The matrix $[N]$ is also known as the shape function.

As examples of shape functions, consider the bar and planar elements shown in Figs. 8-2 and 8-3, respectively. In these figures, displacements and forces are given in an element or local coordinate system. For the bar element of Fig. 8-2(a) displacement components inside the element are functions of only the axial coordinate, x. The displacement function of Eq. 2 in this simple case is given by

$$u = \lfloor (1-\xi), \xi \rfloor \begin{Bmatrix} q_1 \\ q_2 \end{Bmatrix} \quad (8\text{-}3)$$

where $\xi = x/L$. From Eq. 3 the shape function $[N]$ can be readily identified. Similarly, for the two dimensional element of Fig. 8-3, the displacement function is $\{u\}^T = \lfloor u(x,y), v(x,y) \rfloor$. The shape function $[N(x,y)]$ is obtained by solving the two-dimensional equations of linear elasticity. The shape function is given as

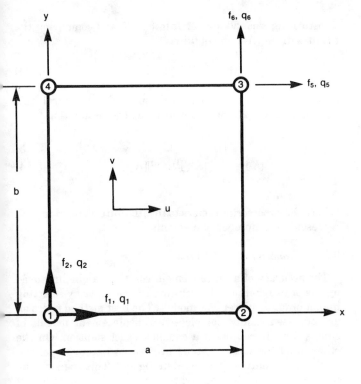

$u(x,y) = (1 - \frac{x}{a})(1 - \frac{y}{b})q_1 + (\frac{x}{a})(1 - \frac{y}{b})q_3 + (\frac{x}{a})(\frac{y}{b})q_5 + (1 - \frac{x}{a})(\frac{y}{b})q_7$

$v(x,y) = (1 - \frac{x}{a})(1 - \frac{y}{b})q_2 + (\frac{x}{a})(1 - \frac{y}{b})q_4 + (\frac{x}{a})(\frac{y}{b})q_6 + (1 - \frac{x}{a})(\frac{y}{b})q_8$

$\varepsilon_x = \partial u/\partial x = -(1 - \frac{y}{b})(\frac{q_1}{a}) + (1 - \frac{y}{b})(\frac{q_3}{a}) + (\frac{y}{b})\frac{q_5}{a} - (\frac{y}{b})\frac{q_7}{a}$

Figure 8-3. Two dimensional elements.

For the two dimensional element of Fig. 8-3, there are three strain components ($\varepsilon_x = \partial u/\partial x$, $\varepsilon_y = \partial v/\partial y$, $\varepsilon_{xy} = \frac{1}{2}(\partial v/\partial x + \partial u/\partial y)$) and the matrix $[B]$ is identified as

$[B] =$
$$\begin{bmatrix} -(1-\eta)\frac{1}{a} & 0 & (1-\eta)\frac{1}{a} & 0 & \eta\frac{1}{a} & 0 & -\eta\frac{1}{a} & 0 \\ 0 & -(1-\xi)\frac{1}{b} & 0 & -\xi\frac{1}{b} & 0 & \xi\frac{1}{b} & 0 & (1-\xi)\frac{1}{b} \\ -(1-\xi)\frac{1}{2b} & -(1-\eta)\frac{1}{2a} & -\xi\frac{1}{2b} & (1-\eta)\frac{1}{2a} & \xi\frac{1}{2a} & \eta\frac{1}{2a} & (1-\xi)\frac{1}{2b} & -\eta\frac{1}{2a} \end{bmatrix} \quad (8\text{-}7)$$

where $\xi = x/a$ and $\eta = y/b$.

The *equilibrium equations for an element* can now be obtained from the *principle of virtual work*, or *stationary potential energy*. Writing the virtual work of all the forces and equating it to zero, the following equation can be constructed

$$\int_V \{\delta\varepsilon\}^T\{\sigma\}dV - \{\delta q\}^T\{f\} = 0 \quad (8\text{-}8)$$

where $\{f\}$ is a vector of generalized nodal forces that correspond to the nodal displacements $\{q\}$. In order to calculate the element stiffness matrix $[k]$ using Eq. 8 a material constitutive equation is needed. If the material obeys the generalized form of Hooke's law, then

$$\{\sigma\} = [C]\{\varepsilon\} \quad (8\text{-}9)$$

where $\{\sigma\}^T = \lfloor \sigma_x, \sigma_y, \sigma_z, \sigma_{xy}, \sigma_{yz}, \sigma_{zx} \rfloor$ is a *vector of stresses*, and the *matrix of elastic constants*. Substituting Eqs. 5 and 9 into 8, the principle of virtual work is expressed as

$$\{\delta q\}^T[(\int_V [B]^T[C][B]dv)\{q\} - \{f\}] = \{0\} \quad (8\text{-}10)$$

Since all components of $\{\delta q\}^T$ are abitrary, Eq. 10 is satisfied only when the term in square brackets is zero, or

$$\{f\} = [k]\{q\} \quad (8\text{-}11)$$

where the matrix $[k]$ is given as

$$[k] = \int_V ([B]^T[C][B])dV. \quad (8\text{-}12)$$

The symmetric matrix $[k]$ is called the *element stiffness matrix* in a local coordinate system. The element k_{ij} in the matrix $[k]$ can be interpreted as the nodal force in the ith direction due to a unit displacement in the jth direction with all other displacements zero.

For the truss and two-dimensional elements of Figs. 8-2 and 8-3, respectively, the material is assumed to have the same properties in all directions (isotropic media). In the case of the truss element the matrix $[C]$ is replaced by the scalar, E representing Young's modulus. If it is assumed that the two-dimensional element is a *plane stress element* then $[C]$ is given by

$$[C] = \frac{E}{1-\nu^2}\begin{bmatrix} 1 & \nu & 0 \\ \nu & 1 & 0 \\ 0 & 0 & 1-\nu \end{bmatrix}$$

where E is Young's modulus and ν is Poisson's ratio. Substituting appropriate matrices $[B]$ and $[C]$ into Eq. 12 and carrying out the indicated volume integrations, it is possible to obtain the element stiffness matrix for the truss element as

$$[k] = \frac{AE}{L}\begin{bmatrix} 1 & -1 \\ -1 & 1 \end{bmatrix}. \quad (8\text{-}13)$$

Similarly the stiffness matrix for the plane stress element (Fig. 8-3) using

$$[k] = tab \int_0^1 \int_0^1 [B]^T [C] [B] \, d\xi dN \quad (8\text{-}14)$$

where t is the element thickness. In this case [k] turns out to be an 8 × 8 matrix, the details of which will not be presented here because of the complicated expressions involved in each element of the matrix.

Before a stiffness matrix for the entire structure can be constructed, element stiffness matrices must be transformed into a global coordinate system. The element force-displacement relationship in the global coordinate system is written as

$$\{f'\} = [k']\{q'\} \quad (8\text{-}15)$$

where $\{f'\}$ is a vector of element nodal forces, $\{q'\}$ is a vector of nodal displacements, and $[k']$ is an *element stiffness matrix in the global coordinate* system. Let the transformation from a local to the global system be defined as

$$\{q\} = [\Omega]\{q'\} \quad (8\text{-}16)$$

where $[\Omega]$ is a transformation matrix that is constructed from direction cosines of the local coordinate system relative to the global coordinate system. From equivalence of the virtual work of forces $\{f'\}$ and $\{f\}$, the following equation is obtained

$$\{f'\} = [\Omega]^T\{f\} \quad (8\text{-}17)$$

Substituting Eqs. 11 and 16 into 17 and comparing the result with Eq. 15 the element stiffness matrix $[k']$ is identified as

$$[k'] = [\Omega]^T[k][\Omega] \quad (8\text{-}18)$$

8.3.2.2 *Structural Stiffness Matrix*

Let $\{\Delta\}$ be a vector of independent generalized displacements and $\{F\}$ be a corresponding vector of equivalent nodal forces for the entire structure. Then the equilibrium equation in terms of displacements for the entire structure is given as:

$$\{F\} = [K]\{\Delta\} \quad (8\text{-}19)$$

where $[K]$ is a structural stiffness matrix that is to be synthesized from the element stiffness matrices. Let $\{\bar{f}\}$ and $\{\bar{q}\}$ represent composite vectors of element forces and displacements, respectively, and $[\bar{k}]$ be a composite stiffness matrix for all elements of the structure. Then the equilibrium equation for all elements can be compactly written in matrix form as

$$\{\bar{f}\} = [\bar{k}]\{\bar{q}\} \quad (8\text{-}20)$$

Now, a compatibility relationship between $\{\bar{q}\}$ and $\{\Delta\}$ can be expressed as

$$\{\bar{q}\} = [A]\{\Delta\} \quad (8\text{-}21)$$

where $[A]$ is a Boolean matrix, each row of which has a unit element in only one position, that relates a component of $\{\bar{q}\}$ to a component of $\{\Delta\}$. The equilibrium equation between $\{F\}$ and $\{\bar{f}\}$ is obtained using the principle of virtual work as

$$\{F\} = [A]^T\{\bar{f}\} \quad (8\text{-}22)$$

Substituting Eqs. 20 and 21 into Eq. 22 and comparing the result with Eq. 19, one obtains

$$[K] = [A]^T[\bar{k}][A] \quad (8\text{-}23)$$

that is a general formula for synthesizing the structural stiffness matrix. Eq. 23 may be written as the summation

$$[K] = \sum_{i=1}^{N_e} [A^{(i)}]^T[k'^{(i)}][A^{(i)}] \quad (8\text{-}24)$$

Where the superscript i refers to the ith finite element and N_e represents the number of elements.

8.3.2.3 *Isoparametric Elements*

The accuracy of a finite element solution can be improved by one of two methods: by refining the mesh, or by selecting higher order displacement models. The first method relies on an increased number of piecewise displacement models of simple form to represent a complex exact solution. On the other hand the latter method utilizes an improved polynomial approximation illustrated in Fig. 8-4. This suggests the number of elements required to achieve a desired degree of accuracy may be reduced with resulting savings in modeling time and computation time. It is soon recognized, however, that the real benefits of higher-order displacement fields over mesh refinement cannot be realized when the finite element retains simple straight-line geometric boundaries. This is because most structures have curved boundaries and contoured shapes and still require a large number of straight-sided elements to simply model complex structural shapes. The development of the element stiffness matrix presented in Section 8.3.2.1 was based on straight-sided elements.

In an effort to solve the problem of modeling the curvature of actual components, the family of so-called *isoparametric finite elements* was developed. The term isoparametric has been used to describe a conformal mapping technique whereby a simple element in the form of a rectangle or cube is warped or distorted in such a fashion as to conform to the curved shape of the actual structural component being modeled (Fig. 8-5).

Mathematically, the mapping process is accomplished by use of mapping equations such as those illustrated in Fig. 8-5(a). It should be noted that the x and y coordinates within the domain of the defined finite element are obtained by summing the products of a parameter N_i called the *nodal shape function*, and the corresponding x and y coordinates at each node:

$$x = \sum_{i=1}^{n} N_i(\xi, \eta) x_i$$

$$y = \sum_{i=1}^{n} N_i(\xi, \eta) y_i \quad (8\text{-}25)$$

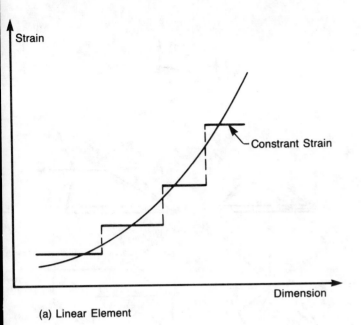

(a) Linear Element

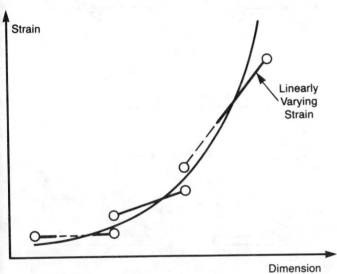

(b) Parabolic Element

Figure 8-4. Approximation of strain field with different displacement fields.

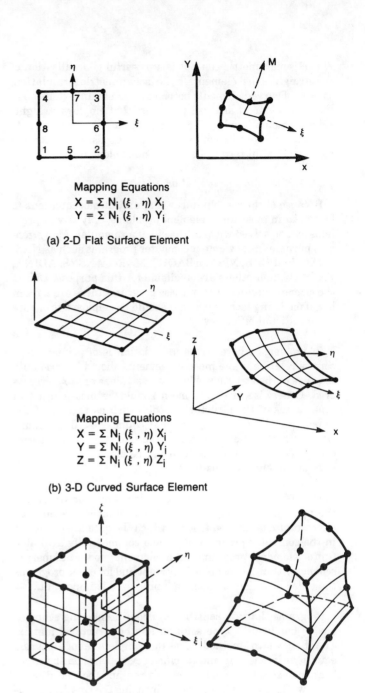

Figure 8-5. Conformal mapping of various finite elements.

where x_i and y_i are the coordinates of node i. Shape functions are defined in terms of dimensionless parameters ξ and η in the mapping equations. The order of the polynomial expressions is, of course, a function of the number of nodes that have been defined around the perimeter of the element. The range of the values for both ξ and η is defined to be from -1 to $+1$.

If Eqs. 2 and 25 are compared, it can be seen that these equations are of the same general form

$$\{u\} = [N]\{q\} \text{ and } \{x\} = [N]^T\{x_i\} .$$

An element is termed *isoparametric* if both the geometry and displacement fields of the element are described in terms of the same shape function $[N]$. If fewer nodes are used to interpolate geometry than are used to interpolate the displacement field, the element is termed a "subparametric element". A "superparametric" element results when the interpolating polynomials for geometry are of higher order than those used for interpolating the displacement field.

The isoparametric concept is a powerful generalization of the straight-sided element. In the isoparametric formulation, the use of curved elements becomes a systematic extension of the finite element analysis procedure that employs straight-sided elements (see Fig. 8-6). For this reason, it is possible to continue, without loss of generality, to discuss the finite element method in terms of straight-sided elements.

8.3.2.4 *Modeling Consideration*

Because of the large number of simultaneous equations to be solved in most finite element problems, only trivial problems can be solved without the aid of a computer. Many general purpose finite element computer programs such as NASTRAN, SUPERB, NISA, ABAQUS, MARC, ANSYS, ADINA, PATRAN, and others are available for this purpose. Due to the size and architecture of these programs there are a number of modeling techniques that should be employed to ensure computational efficiency.

In problems where stress information is of primary concern, a relatively fine mesh of elements should be used only in areas of the structure where stress values are to be recovered. Otherwise, a coarse mesh should be incorporated in other areas of the structural model, such as regions of the imposed loading and displacement boundaries. In problems where structural displacement response is of foremost concern, a relatively fine mesh should be employed in areas of high strain energy density.

The efficiency of a solution depends not only on the size of the problem (i.e., degrees of freedom, number of elements, etc.) but also on the sequence in which the data are processed in the computer program. For some commercially available finite element programs, the element node numbering scheme within the model is important and for other programs the order in which elements are processed is important.

Programs that are sensitive to nodal numbering schemes use a "bandwidth" type solution for the structural matrix. The "bandwidth" method solves the complete set of algebraic equations in one step and its efficiency depends on the number of non-zero terms in each matrix row and their position in the row. Bandwidth is minimized by using the smallest possible node number difference between nodes in a particular element.

Programs that depend on element entry order for solution efficiency use a "wave-front" type of solution. In a "wave-front" solution elements are brought into the solution process in the order in which they appear in the model data, intermediate results are obtained, and then more elements are brought in for solution. An element is then dropped from the solution "wave" after all elements attached to it have entered the problem. Wavefront solution times can be minimized by entering adjacent elements sequentially into the model file starting at a boundary.

For simple structures the bandwidth or wavefront can be minimized during the modeling process. For complex structures, ease of modeling should be considered first and

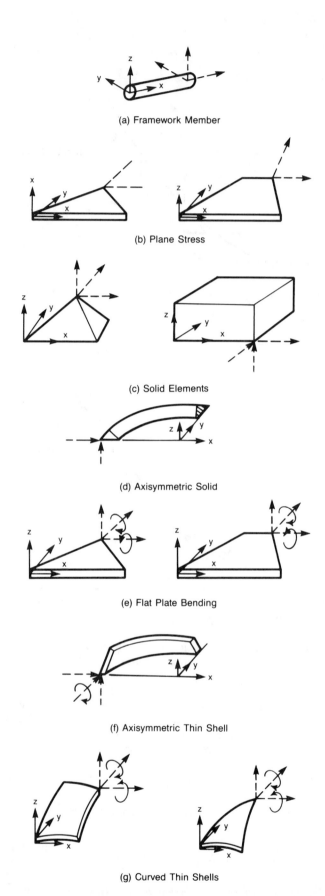

Figure 8-6. Types of finite elements.

minimization can be done with a preprocess wavefront or bandwidth computer program.

The basic steps in a linear elastic finite element analysis are summarized in Table 8-1

8.3.3 Elastic Plastic Analysis

8.3.3.1 Background

The stiffness formulation, Eq. 12, assumes linear elastic behavior for the complete range of loading. However, metals obey Hooke's law only in a certain small range of strain. When a metal is strained beyond the proportional limit (Fig. 8-7a), Hooke's law no longer applies. The behavior of metals beyond yielding is often quite complicated. Numerous experimental results show typical uniaxial stress-strain behavior, as indicated in Fig. 8-7a. The initial region appears as a straight line. This is the region in which Hooke's law is expected to apply. Mild steels demonstrate a "flat" stress-strain behavior after yield, which on the microstructure level of the material is caused by discontinuous slippage along slip planes of the cystrals. Most other metals demonstrate hardening behavior.

Table 8-1 Summary of Steps in a Finite Element Analysis Assuming Linear Elastic Material Behavior

Step	Activity
1.	Idealize the continuum by introducing the finite element mesh, numbering the nodes, and numbering the elements.
2.	Select the shape functions for the finite elements as in Eq. 2.
3.	Compute the element stiffness matrix using Eq. 12.
4.	Assemble the system stiffness matrix using Eq. 24.
5.	Impose boundary conditions and solve matrix Eq. 19.
6.	Calculate stresses, strains, and displacements, as required.

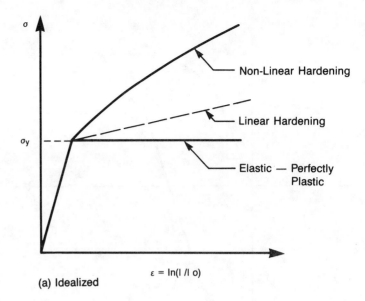

(a) Idealized

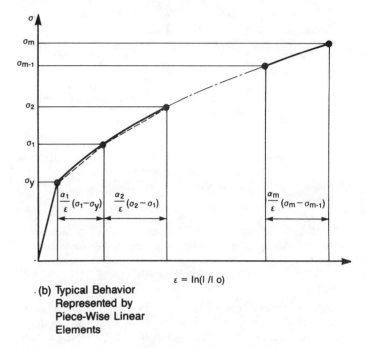

(b) Typical Behavior Represented by Piece-Wise Linear Elements

Figure 8-7. Monotonic yield behavior of metals.

The yield behavior of metals is further complicated by cyclic loading. While metals like aluminum often exhibit a similar strain hardening behavior for re- yielding after the initial yield, some metals exhibit different forms of hardening behavior for each cycle of loading even after several cycles of loading. Typical uniaxial experimental data are shown in Fig. 8-8.

The motivation to understand and predict cyclic plasticity in metals comes from basic investigations of low cycle fatigue [27,28]. The behavior of materials under cyclic plastic deformation is complex from the viewpoint of the material microstructure.

No universally applicable stress-strain laws exist that govern metallic plasticity on the macro-scale. Most existing plasticity theories are heuristic because they are based purely on experimental observations without direct reference to the material microstructure. Nevertheless these theories can often be applied successfully to solve elastic-plastic problems. In this section some aspects of how plasticity is considered in finite element formulations will be summarized.

Plasticity theories fall into two major categories, incremental or flow theories and deformation theories. The former category is more general than the latter. In a particular deformation theory, the plastic strains are uniquely defined by

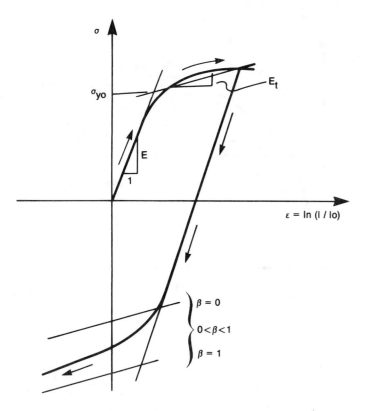

Figure 8-8. Typical cyclic stress-strain behavior of metals.

the state of stress, or vice versa; whereas in a given incremental theory, the plastic strains depend upon a combination of factors, including the plastic strain history and the state of stress.

In both theories the total strain vector $\{\varepsilon\}$ is decomposed into elastic, $\{\varepsilon^e\}$, and plastic, $\{\varepsilon^p\}$, components as follows:

$$\{\varepsilon\} = \{\varepsilon^e\} + \{\varepsilon^p\} \ . \tag{8-26}$$

Most theories of plasticity follow the experimental observation that there is no permanent change in volume due to plastic strains (plastic incompressibility). This can be stated mathematically in terms of the normal components of the strain vector $\{\varepsilon^p\}$ as

$$\dot{\varepsilon}^p_x + \dot{\varepsilon}^p_y + \dot{\varepsilon}^p_z = 0 \ , \tag{8-27}$$

where the superposed dots denote differentiation with respect to time, t.

8.3.3.2 Incremental or Flow Theory of Plasticity

Most flow theories have four major components:

1. A *constitutive* equation relating increments of stress to increments of the elastic strain.

2. A *yield criterion* or condition to specify the onset of plastic deformation.

3. A *hardening rule* that prescribes the strain-hardening of the material and the modification of the *yield criteria* as plastic flow progresses.

4. A *flow rule* that specifies the component of strain that is plastic or nonrecoverable.

Regardless of the theory used, linear incremental constitutive relations can be developed and incorporated into an equation relating generalized loads to generalized displacements of an arbitrary structure. These constitutive relations can be conveniently represented in matrix form.

First, a relation between stress increments and elastic strain increments must be assumed

$$\{\dot{\sigma}\} = [C^e]\{\dot{\varepsilon}^e\} = [C]\{\dot{\varepsilon} - \dot{\varepsilon}^p\} \tag{8-28}$$

where $[C^e]$ is the matrix of elastic constants. The superposed dot denotes time differentiation along the path of a material particle. It must be noted that the time derivative provides a desired ordered sequence for the incremental stress-strain relationship. Eq. 28 may be rewritten as

$$\{\dot{\sigma}\} = [C^{ep}]\{\dot{\varepsilon}\} \tag{8-29}$$

where $[C^{ep}]$ is called the *elastic-plastic matrix*, and is expressed as

$$[C^{ep}] = [C^e] - [C^p] \tag{8-30}$$

and $[C^p]$ is the *plastic matrix*. Calculation of $[C^p]$ requires computation of the plastic strain increments $\{\dot{\varepsilon}^p\}$ that will be discussed in the following paragraphs.

Secondly, a *yield criteria* or yield function describing the onset of plastic deformation must be used. A possible functional form supported by experiment is given by

$$F = f(\{S - \alpha\}) = k^2(\varepsilon^p_{eq}) \tag{8-31}$$

where $\{S\}$ represents the deviator components of $\{\sigma\}$; $\{\alpha\}$ and k represent the center and radius of the yield surface ($R = \sqrt{2}k$) respectively; and $\varepsilon^p_{eq} = \int \dot{\varepsilon}^p_{eq} dt$ is the history of the equivalent uniaxial plastic strain calculated from

$$\dot{\varepsilon}^p_{eq} = \frac{\sqrt{2}}{3} \{(\dot{\varepsilon}^p_x - \dot{\varepsilon}^p_y)^2 + (\dot{\varepsilon}^p_y - \dot{\varepsilon}^p_z)^2 + (\dot{\varepsilon}^p_z - \dot{\varepsilon}^p_x)^2 + \frac{3}{2}[(\dot{\varepsilon}^p_{xy})^2 + (\dot{\varepsilon}^p_{zx})^2 + (\dot{\varepsilon}^p_{yz})^2]\}^{1/2} \ . \tag{8-32}$$

A hardening rule prescribes the strain hardening of the material and the modifications of the yield surface during plastic flow. For simplicity the example will be limited to linear hardening and softening, that can be stated in a combined isotropic-kinematic form as

$$\{\dot{\alpha}\} = \frac{2}{3}(1-\beta)H'\{\dot{\varepsilon}^p\} \qquad (8\text{-}33)$$

for the *back stress* (the center of the yield surface), $\{\alpha\}$, and

$$\dot{R} = \frac{\sqrt{2}}{\sqrt{3}} \beta H' \dot{\varepsilon}^p_{eq} \qquad (8\text{-}34)$$

for the radius of the yield surface. In Eqs. 33 and 34 $H'(\varepsilon^p_{eq})$ is the *plastic modulus* (a superposed 'prime' denotes differentiation with respect to ε^p_{eq}) and β is the ratio of isotropic to kinematic hardening such that $\beta = 0$ corresponds to a kinematic hardening where the yield surface retains its initial size, shape, and orientation. Isotropic hardening, $\beta = 1$, indicates that during plastic flow the yield surface expands uniformly about the origin and never translates. A value for the hardening ratio within the range $0 \leq \beta \leq 1$ represents a combined hardening rule where the yield surface is allowed to both expand and translate. Figure 8-8 illustrates, on the basis of uniaxial experiments, the stress-strain behavior of materials described here.

If yielding does occur further information is required concerning the increment of deformation to complete the description of the material behavior. This information is provided by the *flow rule* that defines the component of strain that is plastic or nonrecoverable and can be written as

$$\{\dot{\varepsilon}^p\} = \begin{cases} \Lambda \dfrac{\partial F}{\partial \{\sigma\}^T} & \text{``plastic process''} \\ \{0\} & \text{``elastic process''} \end{cases} \qquad (8\text{-}35)$$

where Λ is a scalar determined from material data. Eq. 35 is referred to as an *associated flow rule* because it involves taking the partial derivative of the yield function. It is also known as the *normality condition* since it states that the plastic strain increment vector is normal to the yield surface in stress space.

The *von Mises yield criterion* is one form of yield surface that is based on the assumption that yielding is caused by the maximum distortional energy. The yield surface is

$$F = \sigma_{eq}^2 - 2k^2 = 0 = J_2 - k^2 \qquad (8\text{-}36)$$

where $\sqrt{2}k = \sigma_{yo}$ denotes an experimentally determined constant yield stress in simple tension, and σ_{eq} is the equivalent stress defined in principal stresses as

$$\sigma_{eq} = \frac{1}{\sqrt{2}}[(\sigma_1-\sigma_2)^2 + (\sigma_2-\sigma_3)^2 + (\sigma_3-\sigma_1)^2]^{\frac{1}{2}}$$

$$= \sqrt{\frac{3}{2}}[S_1^2 + S_2^2 + S_3^2]^{\frac{1}{2}} . \qquad (8\text{-}37)$$

In $\sigma_1, \sigma_2, \sigma_3$ space Eq. 37 describes the surface of a straight circular cylinder. The axis of this cylinder is inclined at aequal angles in the $\sigma_1, \sigma_2, \sigma_3$ system of coordinates as shown in Fig. 8-9. Using these definitions, Fig. 8-9 shows the *von Mises yield* surfaces in the plane $S_1 + S_2 + S_3 = 0$.

The *Tresca yield criterion* is another form of yield surface. The yield surface in the three-dimensional case is

$$|\sigma_2 - \sigma_3| \leq \sqrt{2}\sigma_{yo} = 2k$$
$$|\sigma_3 - \sigma_1| \leq \sqrt{2}\sigma_{yo} = 2k \qquad (8\text{-}38)$$
$$|\sigma_1 - \sigma_2| \leq \sqrt{2}\sigma_{yo} = 2k .$$

In the *elastic state* all the conditions in Eq. 38 are satisfied with the inequality signs. In the *yield condition* the equality signs must hold in one or two of these conditions, but not all three simultaneously. The equality conditions define a right hexahedral prism with axis $\sigma_1 = \sigma_2 = \sigma_3$ perpendicular to the deviatoric plane, $S_1 + S_2 + S_3 = 0$. The trace of the prism on the deviatoric plane is a right hexagon (Fig. 8-10). The impossibility of satisfying simultaneously all three equality signs in Eq. 38 is obvious from the geometry.

The above two criteria presume elastic-perfectly plastic material behavior, i.e., that there is no change in the yield conditions due to increases in the plastic strain or cyclic behavior. The von Mises and Tresca criteria may be modified to account for these effects. For example, strain hardening effects can be introduced by expanding the yield surface as a function of equivalent plastic strain $k = k(\varepsilon^p_{eq})$ and kinematically displacing it as a function of the history of plastic strain.

The modified *von Mises criteria* is

$$F = \frac{1}{2}(\sigma'_{eq})^2 - k^2(\varepsilon^p_{eq}) = 0 \qquad (8\text{-}39)$$

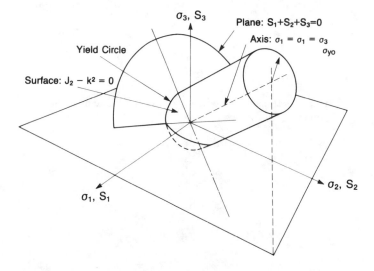

Figure 8-9. Geometrical representation of von Mises yield criterion in principal stress space.

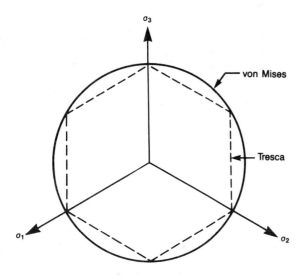

Figure 8-10. Intersection of yield surfaces with the plane: $S_1+S_2+S_3=0$.

where

$$(\sigma'_{eq}) = \frac{3}{2}\{(S_x-\alpha_x)^2 + (S_y-\alpha_y)^2 + (S_z-\alpha_z)^2$$
$$+ 2[(\sigma_{xy}-\alpha_{xy})^2 + (\sigma_{yz}-\alpha_{yz})^2 + (\sigma_{zx}-\alpha_{zx})^2]\} \;, \quad (8\text{-}40)$$

and $\alpha_x, \alpha_y, ..., \alpha_{xy}, ... \alpha_{zx}$ denote the components of the *back stress vector* $\{\alpha\}$.

Without going into the details of the derivation of Λ and of the elastic-plastic matrix $[C^{ep}]$, the final expression for the plasticity matrix $[C^p]$ is presented below for an isotropic elastic-plastic material with a von-Mises yield surface modified for combined isotropic and kinematic hardening:

$$[C^p] = \frac{2G}{\frac{2}{3}\sigma_y^2\left(1+\frac{H'}{3G}\right)} \begin{bmatrix} (S'_x)^2 & S'_xS'_y & S'_xS'_z & S'_x\sigma'_{xy} & S'_x\sigma'_{yz} & S'_x\sigma'_{zx} \\ & (S'_y)^2 & S'_yS'_z & S'_y\sigma'_{xy} & S'_y\sigma'_{yz} & S'_y\sigma'_{zx} \\ & & (S'_z)^2 & S'_z\sigma'_{xy} & S'_z\sigma'_{yz} & S'_z\sigma'_{zx} \\ & & & (\sigma'_{xy})^2 & \sigma'_{xy}\sigma_{yz} & \sigma'_{xy}\sigma'_{zx} \\ & & & & (\sigma'_{yz})^2 & \sigma'_{yz}\sigma'_{zx} \\ \text{Symmetric} & & & & & (\sigma'_{zx})^2 \end{bmatrix} \quad (8\text{-}41)$$

where G is the shear modulus, H' is the plastic modulus, and the primed quantities denote

$$S'_x = S_x - \alpha_x$$
$$S'_y = S_y - \alpha_y$$
$$S'_z = S_z - \alpha_z \quad (8\text{-}42)$$
$$\sigma'_{xy} = \sigma_{xy} - \alpha_{xy}$$
$$\sigma'_{yz} = \sigma_{yz} - \alpha_{yz}$$
$$\sigma'_{zx} = \sigma_{zx} - \alpha_{zx}$$

8.3.3.3 Deformation Theory of Plasticity

The deformation (or Hencky) theory is based on the following propositions:

(1) The body is isotropic

(2) The volumetric change is purely elastic and proportional to the mean pressure:

$$\varepsilon = 3Kp = \frac{(1-2\nu)}{E} p \quad (8\text{-}43)$$

(3) The stress and strain deviators are proportional:

$$\{e\} = \psi\{S\} \quad (8\text{-}44)$$

For an isotropic material, ψ is a function of the state of stress, in particular of the stress invariants J_2 and J_3. Eq. 44 is the flow rule that uniquely relates the plastic strain to the deviatoric stress that characterizes the deformation theory.

Eliminating the volumetric dilatation from Eq. 43, Hencky's relations can be derived:

$$\{\varepsilon\} = \frac{(1-2\nu)}{3E}p\{I\} + \psi\{S\} \quad (8\text{-}45)$$

where $\{I\}^T = \lfloor 1,1,1,0,0,0 \rfloor$ is the unit vector. The strain relations can be solved with respect to the stresses:

$$\{\sigma\} = \frac{E}{(1-2\nu)}\epsilon\{I\} + \frac{1}{\psi}\{e\} \quad (8\text{-}46)$$

Specific forms for ψ may be determined from the uniaxial stress-strain material response curve. For a uniaxial applied stress, $\sigma = \sigma_e$, the following important relation is obtained

$$\varepsilon_{eq} = \frac{2}{3}\psi\sigma_{eq} \;. \quad (8\text{-}47)$$

An experimentally determined stress-strain curve can be approximated using empirical models or a piecewise-linear model to describe the uniaxial response.

The strain hardening behavior of many metals can be simulated by the Ramberg-Osgood relation [29] given as

$$\varepsilon = \begin{cases} \sigma/E & \text{for } \sigma \leq \sigma_{y0} \\ \sigma/E + \alpha\left[\frac{\sigma}{E}\left(\frac{\sigma}{\sigma_{y0}}\right)^{n-1} - \frac{\sigma_{y0}}{E}\right] & \text{for } \sigma > \sigma_{y0} \end{cases} \quad (8\text{-}48)$$

where n, α, and σ_{y0} (the initial yield stress) are chosen to best model the observed behavior. Eq. 48 corresponds to

$$\psi = \frac{3}{2}\frac{\alpha}{E}\left(\frac{\sigma_{eq}}{\sigma_{y0}}\right)^{n-1} \quad (8\text{-}49)$$

The mathematical form for the piecewise linear representation of material behavior is

$$\varepsilon = \frac{\sigma_{y0}}{E} + \frac{\alpha_1}{E}(\sigma_1 - \sigma_{y0}) + \frac{\alpha_2}{E}(\sigma_2 - \sigma_1) + \ldots + \frac{\alpha_m}{E}(\sigma - \sigma_{m-1}) \quad (8\text{-}50)$$

where $\sigma_{m-1} < \sigma \leq \sigma_m$ and α_m is given by

$$\alpha_m = \frac{(E\Delta\varepsilon_m - \Delta\sigma_m)}{\Delta\sigma_m} \quad (8\text{-}51)$$

For this material model, the plastic strain rate is given by

$$\dot{\varepsilon}^p_{eq} = \alpha_m \dot{\sigma}_{eq}/E \quad (8\text{-}52)$$

and ψ is approximated by

$$\psi = \frac{3}{2} \frac{\alpha_m}{\sigma_{eq}} \quad (8\text{-}53)$$

8.3.3.4 Numerical Analysis Procedures

The elastic-plastic constitutive equations can be solved numerically and iteratively by allowing the stresses and strains to increase (decrease) in some increment, i.e., $\{\Delta\varepsilon\} = \{\dot{\varepsilon}\}\Delta t$, $\{\Delta\sigma\} = \{\dot{\sigma}\}\Delta t$. Considering this and the incremental or flow theory of plasticity presented in 8.3.3.2, or the deformation theory of plasticity presented in 8.3.3.3, it becomes apparent that upon the selection of an appropriate theory the constitutive equations can be represented in terms of a linear matrix relationship. There are several alternatives for incorporating this relationship into the final equations relating loads to deformations. Several approaches are discussed in Refs. [30-32]. These techniques have been divided into two general categories: a) the pseudo-force and b) the tangent modulus methods.

In the pseudo-force method, the well-known thermal analogy between temperature gradients and body forces in causing a strain field is extended to include plastic strains. Thus plastic effects are treated by interpreting plastic strains as "pseudo" body forces or pseudo forces. This analogy, first introduced by Mendelson and Manson [33] reduces the nonlinear analysis to the analysis of an elastic body of identical shape and boundary conditions, but with an additional set of loads:

$$[k]^k \{\Delta q\}^k = \{\Delta f\}^k + \{\Delta Q\}^{k-1} \quad (8\text{-}54)$$

where the superscript "k" denotes the "k^{th}" load increment, $k-1$ is the preceding step and the pseudo-force is calculated by using Eq. 55

$$\{\Delta Q\}^{k-1} = \iiint_V [B]^T [C] \{\Delta\varepsilon^p\}^{k-1} dV \quad (8\text{-}55)$$

The pseudo-force method is advantageous because of the ease with which it can be implemented in a finite element analysis and because the stiffness matrix developed for the elastic behavior of the structure remains unaffected by changes in material properties.

In the tangent-modulus method (outlined by Marcal [34]), the incremental linear constitutive relation, based on the plasticity theory, is introduced directly into the governing equilibrium equations. The elastic-plastic matrix $[C^{ep}]$ is updated for each increment of load by modifying its components. In other words, the tangent stiffness is computed at the end of each increment and used for the succeeding increment. The increments of plastic strain $\{\Delta\varepsilon^p\}$ are computed from Eq. 35, i.e., from

$$\{\Delta\varepsilon^p\} = \begin{cases} \Lambda \dfrac{\partial F}{\partial\{\sigma\}^T} \Delta t & \text{"plastic process"} \\[2mm] \{\sigma\} & \text{"elastic process"} \end{cases} \quad (8\text{-}56)$$

Once $[C^{ep}]$ is obtained, the element stiffness can be computed by writing Eq. 12 as

$$[k] = \iiint_V [B]^T [C^{ep}][B] dV \quad (8\text{-}57)$$

The principal drawback of the tangent modulus method is the necessity to compute new constitutive matrices and element stiffnesses and to assemble and invert an overall stiffness matrix for each step of the analysis.

Despite differences in application of the pseudo-force and the tangent-modulus concepts, there is, in some respects, a close relationship between the two methods. This relationship and the comparative merits of the two methods as applied to plane stress problems is discussed in Ref. [34].

Table 8-2 Summary of Steps in a Finite Element Analysis of an Elastic-Plastic Material with a von Mises Yield Surface, Associative Flow Rule, and a Linear Combination of Isotropic and Kinematic Hardening (Softening)

Step	Activity
A.	From the *converged* solution at time $t = t_n$, compute an *elastic trial stress* $\{\sigma^{tr}\}_{n+1}$. Check if this trial stress state causes yielding.
	1. Compute the element stiffness matrix using Eq. 12
	2. Assemble the system stiffness matrix using Eq. 24
	3. Apply a load increment and solve matrix Eq. 54
	4. Calculate the total strain increment $\{\Delta\varepsilon\}$ for this time step using matrix Eq. 5
	5. Calculate the trial stress: $$\{\sigma^{tr}\}_{n+1} = \{\sigma\}_n + [C]\{\Delta\varepsilon\} \quad (8\text{-}58)$$

($\{\sigma\}_n$ is the approximation of the stress at time t_n. The subscript refers to the time level. The strain

Table 8-2 Summary of Steps in a Finite Element Analysis of an Elastic-Plastic Material with a von Mises Yield Surface, Associative Flow Rule, and a Linear Combination of Isotropic and Kinematic Hardening (Softening) (Cont)

Step	Activity

increment $\{\Delta\varepsilon\}$ corresponds to the change in strain during the time interval $[t_n, t_{n+1}]$.)

6. Calculate the deviatoric trial stress: $\{S^{tr}\}_{n+1} = \{\sigma^{tr}\}_{n+1} - p\{I\}$

   where $(p = \frac{1}{3}(\sigma_x^{tr} + \sigma_y^{tr} + \sigma_z^{tr}))$

7. Check for yielding using

$$F = \frac{3}{2}\{S - \alpha\}\lfloor S - \alpha \rfloor - 2k^2(\varepsilon_{eq}^p) \qquad (8\text{-}59)$$

B. If $F > 0$, or if $F = 0$ and if the *trial rate-of-stress*, $\{\dot{\sigma}^{tr}\}_n = [C]\{\dot{\varepsilon}\}$ points towards the exterior side of the tangent plane to the yield surface, go to D.

C. If $F < 0$, or if $F = 0$ and if the *trial rate-of-stress* points towards the interior side of the yield surface to the tangent plane, the material behaves elastically for this time step and the update stress is given simply by the trial stress: $\{\sigma\}_{n+1} = \{\sigma^{tr}\}_{n+1}$. If $F = 0$, then the material yields at the end of this time step. Go to H.

D. If $\{\dot{\sigma}^{tr}\}_n$ points towards the exterior side of the yield surface tangent plane, then the material is yielding throughout this time step, go to G.

E. If $\{\dot{\sigma}^{tr}\}_n$ points towards the interior side of the yield surface tangent plane, then the material was behaving elastically at the beginning of this time step, but will yield during this strain increment. Compute *trial rate-of-stress* at end of increment, $\{\dot{\sigma}^{tr}\}_{n+1}$.

F. Update the stress state at the yield surface. See algorithms by Wilkins [31].

G. Using the total strain increment, less the amount required to cause yielding in (F)

   1. Calculate $[C^p]$ using Eq. 41

   2. Determine the hardening modulus, H', from the current value of ε_{eq}^p, the input uniaxial stress-strain curve

   3. Calculate the elastic-plastic modulus using Eq. 30.

   4. Calculate the stress increment and the new material stress state from Eq. 29

   5. Calculate the plastic strain increment from the stress increment and the known total strain

   6. Update ε_{eq}^p using Eq. 32 and $\varepsilon_{eq}^p = \int \dot{\varepsilon}_{eq}^p dt$.

   7. Determine the new size of the subsequent yield surface by using Eq. 34 and $R = \int \dot{R} dt$

   8. Calculate the new position vector of the yield surface center from Eq. 33

H. If the plastic strains are needed at the end of this time step (pseudo force method), calculate their material stress and total strain states.

I. Determine the elastic-plastic modulus $[C^{ep}]$ at the end of this step if necessary.

   1. If $\{\dot{\sigma}^{tr}\}_{n+1}$ points towards the interior side of the yield surface tangent plane, then: $[C^{ep}] = [C^e]$.

   2. If $\{\dot{\sigma}^{tr}\}_{n+1}$ points towards the exterior side of the yield surface tangent plane, then repeat steps in G.

8.3.4 General Applications and Examples

Now a few representative structural analysis applications of the finite element method will be examined in order to illustrate the way in which the formulations in the previous sections are employed.

8.3.4.1 *Engine Block Analysis*

Engine blocks usually consist of thick-walled materials reinforced with heavy section ribbing. Experience has shown that when determining overall stress and distortion fields in engine blocks, solid elements must be employed for the most part. The finite element analysis of a V-6 aluminum engine block required 5,700 eight-noded solid elements and 900 four-noded quadrilateral bending elements (Fig. 8-11). Only the right cylinder bank (cylinders 1, 2, and 3) was modeled as the block was assumed to be symmetrical. The elements were interconnected at a total of 12,000 nodal points. This model resulted in a total of over 70,000 unknowns. The size of this problem was prohibitive from the standpoint of computation time, data handling, and with respect to the identification of possible sources of error in the analysis. It, therefore, was convenient to divide the block structure into 15 regions, or "substructures or super elements," and each of these was analyzed by the finite element method in such a way as to produce a "super structure." The super elements were tied together via a conventional finite element procedure in the final phase of analysis. Modeling was designed so that each interior bulkhead was treated as a super-element.

In this analysis distortions and stresses induced by bolting the cylinder heads to the engine were calculated. The magnitude of the loads were obtained from laboratory tests. Fig. 8-12 depicts stress contours for maximum principal stresses on the surface of the interior bulkhead.

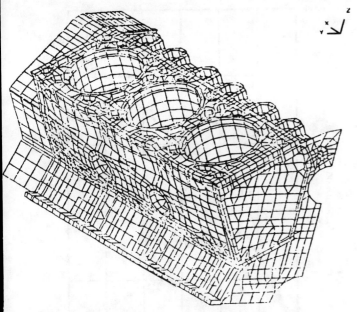

(a) Finite Element Idealization of One Half of Engine Block.

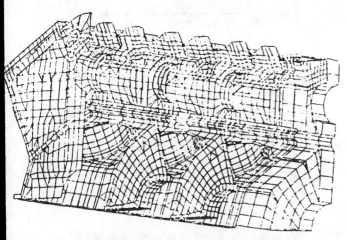

(b) Interior View of Finite Element Idealization.

Figure 8-11. Finite element analysis of a V-6 engine block.

8.3.4.2 SAE Keyhole Specimen

The Cumulative Damage Division of the SAE Fatigue Design and Evaluation Committee has employed a notched specimen as a test case for a number of experimental and analytical studies [37]. The configuration of this specimen, called the SAE Keyhole Specimen is shown in Fig. 8-13(a). As a result of the efforts of this committee, test data have been published on the relation between applied load and strain at the root of the notch for several kinds of steel [37,38]. For this reason, the specimen provided a good source for evaluating numerical analysis procedures.

Using symmetry arguments, a finite element model of the specimen was constructed, consisting of 170 nodes and 237

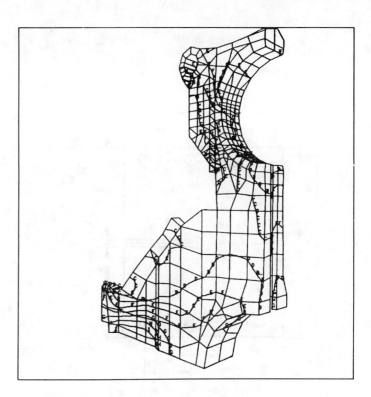

Figure 8-12. Maximum principal stress contours on interior bulkhead.

triangular and qualrilateral constant strain, plane stress elements (Fig. 8-13(b)). An elastic plastic analysis was performed using the deformation theory of plasticity. A comparison of predicted and actual load versus strain behavior is shown in Fig. 8-13c.

8.3.4.3 Uniformly Loaded Sheet with a Central Crack

The finite element idealization of a quadrant of the sheet, as shown in Fig. 8-14(a) involved 455 plane stress elements, 219 vertex nodes, and 250 mid- side nodes, resulting in 902 degrees-of-freedom when symmetry boundary conditions were applied [38]. A mixture of quadratic displacement elements (used in the region surrounding the crack tip), "transition" elements, and linear displacement elements were used to make up the mesh shown.

Results from the finite element analysis for the elastic behavior compared favorably with a continuum solution. Results for the plastic behavior are shown in Fig. 8-14(b) for five levels of loading corresponding to stress levels for which the gross section stress, σ_o was defined to be 37.3 percent, 49.5, 55.0, 70.0 and 75 percent of the material yield stress. The residual stresses in the vicinity of the crack tip were found to be significant and could, therefore, be expected to play an important role in determining the fatigue life of this structure. The propagation of the plastic zone for monotonically increasing and decreasing loads are depicted. As indicated in Fig. 8-14(b), the model predicted that [prior to the removal of all the load ($\sigma_o/\sigma_y = 0.345$)], a compressive plastic zone would

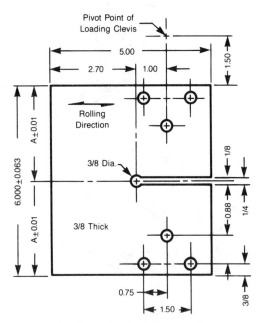

(a) Sketch of SAE Keyhole Specimen

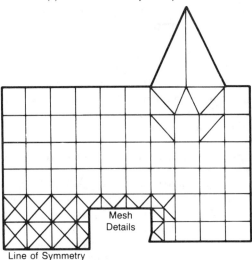

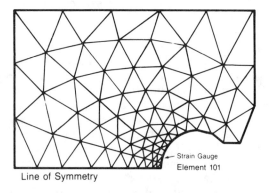

(b) Basic Finite Element Mesh for Half of the SAE Keyhole Specimen

Figure 8-13. Finite element grid for the SAE Keyhole Specimen. (A-B)

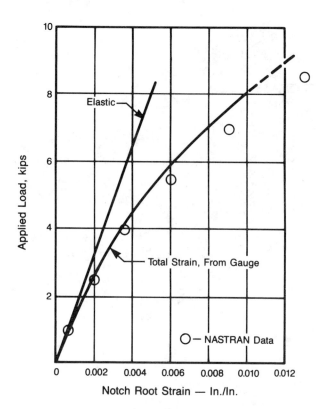

Keyhole Compressive Load-lb	Man-Ten Material Stress 10^{-3} –lb/in²			
	σ_x	σ_y	τ_{xy}	σ_e
1000	-0.734	-22.167	-0.566	21.832
7000	-3.417	-67.210	-2.635	65.727
8500	-3.722	-73.036	-2.870	71.421
Load-lb	Total Strain 10^3			
	ε_x	ε_y	γ_{xy}	ε_e
1000	0.160	-0.733	-0.047	0.749
7000	3.682	-9.196	-1.008	9.732
8500	5.477	-13.199	-1.488	13.890

(c) NASTRAN Stress and Strain Results for Element No. 101.

Figure 8-13. Finite element grid for the SAE Keyhole Specimen. (C)

develop in the region of the crack tip. The plastic zone that was predicted upon complete removal of the loads is also shown in this figure.

8.3.5 Application to Linear Elastic Fracture Mechanics

8.3.5.1 The Stress Intensity Factor

The concept of the stress intensity factor (SIF) proposed by Irwin [39] is now well established in linear elastic fraction mechanics (LEFM) theory as a means of assessing the strength of structures containing crack-life defects (see e.g. [40]). The calculation of the SIF involves solution of a boundary value problem in which the stress and displacement fields exhibit singular behavior at the crack tip.

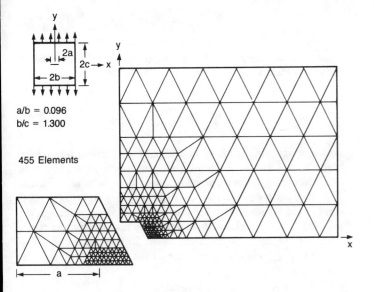

(a) Idealization of a Quadrant of a Sheet With a Central Crack.

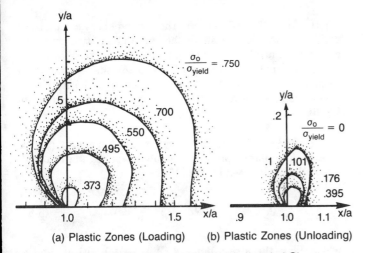

(a) Plastic Zones (Loading)   (b) Plastic Zones (Unloading)

(b) Propagation of Plastic Zone in a Uniformly Loaded Sheet

Figure 8-14. Elastic-plastic analysis of a sheet with a central crack.

Two-dimensional elasticity theory [See e.g. 41, 42] can be used to describe the stresses and displacements in the immediate vicinity of a crack tip for conditions of plane strain, generalized plane stress, antiplane shear and axisymmetry. For the in-plane situation, two modes of crack extension are possible, mode I which is a crack opening case and mode II, in which there is in-plane sliding of one crack face over the other. The antiplane shear case is referred to as mode III. In all these situations, which are illustrated in Fig. 8-15, the stress and displacement fields in the immediate neighborhood of the crack tip are described by expressions of the type [43].

$$_m O_{ij} = K_m r^{-\frac{1}{2}} \, _m f_{ij}(\theta) \quad i, j = 1, 2; \, m = I, II, III \quad (8\text{-}60)$$

$$_m U_i = (K_m r^{\frac{1}{2}}/G) \, _m g_i(\nu, \theta) \quad i = 1, 2; \, m = I, II, III \quad (8\text{-}61)$$

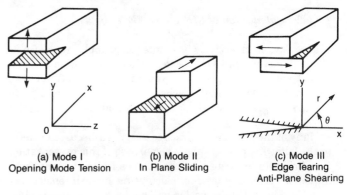

(a) Basic Modes of Crack Extension.

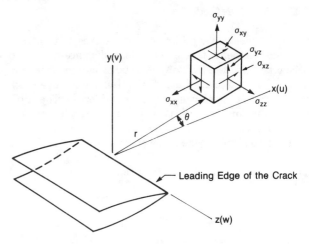

(b) Elastic Stresses Near the Crack Tip.

Figure 8-15. Linear elastic fracture mechanics (LEFM).

where G is the shear modulus, ν is Poisson's ratio, the $_m f_{ij}(\theta)$ and $_m g_i(\nu, \theta)$ are known functions and K_m is the stress intensity for the mth mode.

Thus, in the Mode I situation for example,

$$_I f_{11} = \frac{1}{\sqrt{2}} \cos\left[\frac{\theta}{2}\left(1 - \sin\frac{\theta}{2}\sin\frac{3\theta}{2}\right)\right] \quad (8\text{-}62)$$

and

$$_I g_1 = \frac{\sqrt{2}}{8}\left[(2x - 1)\cos\frac{\theta}{2} - \cos\frac{3\theta}{2}\right], \quad (8\text{-}63)$$

where $x = (3 - 4\nu)$ or $(3 - \nu)/(1 + \nu)$ for plane strain or generalized plane stress respectively. A mixed mode situation may be described by a superposition of expressions of the form of Eqs. 60 and 61. A number of solutions for different crack geometries may be found in handbooks of SIF by Tada, et al [44] and Shih [45]. Many of the handbook solutions, however, are restricted to idealized loadings and geometry.

8.3.5.2 *Calculation of SIF Using FE Methods*

Recently, much activity has centered on applying the finite element method [46,47] to calculating stress intensity factors. Several excellent survey papers and reports [48-52] have reviewed the various approaches. A survey of various finite element programs for calculating SIF's may be found in [53].

Many of the schemes in practice today are discussed briefly in the following paragraphs. The purpose of this presentation is to bring an awareness of the various finite element techniques which can be used for calculating stress intensity factors. Therefore, the discussion will be limited to a linear elastic analysis, with the majority of the discussion directed at mode I type crack opening.

8.3.5.2.1 *Direct Methods*

One of the most straightforward techniques is to utilize conventional elements to find a stress or displacement field near the crack tip. The SIF may then be calculated from Eq. 60 (if considering a load which only causes mode I opening), or from Eq. 61 (if displacements are considered). The difficulty with this method arises in having a mesh size fine enough to accurately calculate the rapidly increasing stress as the crack tip is approached. The engineering rule of thumb as discussed by A. Kobayashi [54] defines the region where sufficiently accurate results are required for the conventional elements as

$$\text{"region"} < \frac{a}{20} \quad (8\text{-}64)$$

where, in this case, a is the half crack length. For example, [54], a through center-cracked plate was modeled and subjected to a uniaxial stress. A constant strain element was used with a mesh size smaller than $a/10$ for all cracks considered (which represented a slight relaxation to the above rule of thumb for mesh size). Using the stress results, the SIF was under-estimated by 10%-15% due to the inability of the model to represent the stress gradient near the crack. However, examination of the displacements for the same computer run provided considerably more accurate results. The utilization of displacements rather than stresses appears to be a good choice if conventional elements are to be used.

8.3.5.2.2 *Energy Methods*

The strain energy of a body may also be used to calculate stress intensity factors. As in the conventional method, no special features of the finite element programs are required other than the calculation of the strain energy, which is a function of the element stiffness matrix, and the grid point displacements, which are generated during the solution procedures in finite element codes.

For the case of plane stress, the strain energy release rate, , for a crack opening in a mode I fashion is related to the corresponding stress intensity factor by

$$_I = (K_I^2/E) \quad (8\text{-}65)$$

where E is the elastic modulus. The corresponding expression for plane strain is

$$_I = \frac{1-\upsilon}{2G} K_I^2 \quad (8\text{-}66)$$

noting that is the shear modulus. A similar expression for mode II and mode III opening may also be derived [see e.g. 54]. The finite element procedure can then be used to calculate the strain energy release rate, , by computing a total component strain energy for a given crack shape with known area. The crack can then be extended by remodeling the crack tip and the strain energy recomputed for the component. The strain energy release rate may then be computed as

$$= \lim_{\Delta a \to 0} \frac{\Delta U_\sigma}{\Delta A} \quad (8\text{-}67)$$

where ΔU_σ is the difference in the strain energy calculations divided by the change in surface area of the two crack geometries. It has been shown that the stress intensity factors may be computed to within 1%-3% using the strain energy release rate, and at the same time relaxing the minimum mesh size relative to the conventional element as stated in Eq. 64.

A method which is comparable to the strain energy release rate for 2-D planar problems involves application of the J-integral as developed by Rice [55].

$$J = = \int_\Gamma \left(W dy - T \cdot \frac{\partial u}{\partial x} ds \right) \quad (8\text{-}68)$$

where W is the strain density ($W = \frac{1}{2} \sigma_{ij} \epsilon_{ij}$) with σ_{ij} and ϵ_{ij} being the linear elastic stress and strain tensors, and Γ is an arbitrary contour surrounding the crack tip. T is a traction vector with an outward normal relative to the contour Γ, u is the displacement vector and ds is a differential element along Γ. The integral is independent of the path of integration and, therefore, may be chosen far from the tip of the crack where the accuracy of the finite element procedure is acceptable. Once J is known, the SIF may be determined. Note that application of the J integral does not require a second solution as does the energy release rate calculation [67].

An area analogue of the J-integral, for plane problems was introduced by Parks [56]. The method is referred to as a stiffness derivative finite element technique and is based on the energy release rate. It has the advantage, as does the J-integral, that only a single crack length need be investigated. Grid points are moved rather than remodeling the crack surface.

Parks applied the stiffness derivative procedure to a double edge-notched, tensile specimen and obtained results within

5% of those of Bowie [57]. This procedure may also be applied to three-dimensional cracks.

8.3.5.2.3 Superposition Method

Yet another method is available for calculating SIF's that utilizes the conventional element and is referred to as the superposition technique. The method employs a previously derived analytical expression for the SIF as a function of the far field stress. One may, for example, pick an expression from a handbook that best fits a particular fracture problem. The finite element model can then be used to calculate the stress field in the region near the assumed crack, which is *not* modeled.

Grandt [58] used this basic procedure along with experimental data, in this case to determine SIF's for flaws in fasteners. The method has also been used to predict fatigue lives in butt welds, [59] as well as many other component evaluations.

8.3.5.2.4 Special Crack Tip Elements

Special finite elements that adequately represent the singularity appear to be very promising. Coarse mesh patterns are capable of giving very accurate results separating out the opening and sliding SIF. The conventional element stiffness calculations are modified to account for the 1/(crack length) singularity which exists at the crack tip. Several basic approaches have been applied to crack tip elements as reviewed in [60] for plane problems and for three-dimensional cracks in [61,62]. For example, Benzley [60] utilizes a 2D isoparametric quadrilateral element with a singular point at one of the corner grid points. The singularity is introduced by enriching a bilinear displacement function via the addition of K_I and K_{II} as unknowns. Writing the global governing equation in matrix form yields

$$\begin{bmatrix} K^{11} & K^{12} \\ K^{21} & K^{22} \end{bmatrix} \begin{Bmatrix} \bar{u} \\ K_I \\ K_{II} \end{Bmatrix} = \begin{Bmatrix} F \\ \bar{F} \end{Bmatrix} \qquad (8\text{-}69)$$

where $[K^{11}]$ represents the conventional element formulation with no singularity present, $[K^{22}]$ represents the singular terms and $[K^{12}]$ accounts for the coupled terms, noting that $[K^{12}]$ equals the transpose of $[K^{21}]$. If no forces exist on the boundaries of the special elements, then $\{\bar{F}\}$ is a null vector and $\{F\}$ represents the classical load vector.

In order to ensure displacement compatibility between the singular and conventional elements, a transition element can be used which drives the singular component of the enriched element to zero on the boundaries adjacent to the conventional elements.

A quarter model of a side cracked panel consisting of 110 elements is illustrated in Fig. 8-16. The finite element results with the special cracked elements are compared to the closed-form solution reported in [63] and shown in Figs. 8-16(b) and 8-16(c).

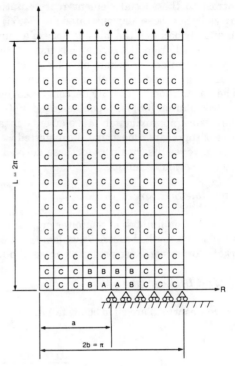

(a) Finite element definition of side cracked panel

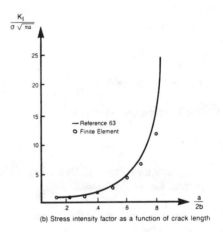

(b) Stress intensity factor as a function of crack length

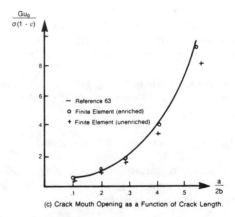

(c) Crack Mouth Opening as a Function of Crack Length.

Figure 8-16. Comparison of finite element with analytical results for side cracked panel (from Ref. 8.60).

In contrast to the special element representation of the singularity, it has been demonstrated in [64,65] that the same singularity occurs if the midside node of a conventional element is moved from its normal position at the center. Crack tip stress intensity factors may be generated in this manner using a "standard" eight-noded isoparametric element. The calculation of the SIF is completed by using a Westergaard solution [66] for a double edged, notched strip in tension using approximately 40 elements in a quarter model. Use of this coarse mesh and the distorted isoparametric element gave results within 9% of those derived using the boundary collocation technique.

8.3.6 Available Software

There are a large number of computer programs based on the finite element method that are available on the commercial market. Surveys of these programs are given in [67–71].

8.3.7 Topical Bibliography

The references which are cited below fall into the following topic areas:

Topic	Reference Numbers
1) Background	21–26
2) Elasto-Plastic Analysis	27–36
3) General Applications and Examples	37–38
4) Linear Elastic Fracture Mechanics	39–66
5) Available Software (Surveys of)	67–71

8.4 The Finite Difference Method

8.4.1 Background

8.4.1.1 Formulations and Types of Boundary Value and Initial Value Problems

The finite difference method (FDM) is an old numerical method for solution of a broad class of so called boundary-value or initial-value problems [72]. For any numerical method used, including the FDM, it is important to know the type of problem considered since each solution approach has some features that are problem-type dependent.

Roughly speaking, the following types of problems may be distinguished [73] the following types of problems – elliptic (see Fig. 8-17), e.g.

$$u_{xx} + u_{yy} = 0 \quad \text{in V, and} \quad u = g \text{ on } S_u \qquad (8\text{-}69)$$

where $u = u(x,y)$. A variety of engineering problems (e.g. static analysis of beams, flanges, plates, membranes, shells) fall in this group, e.g. hyperbolic.

$$u_{xx} - u_{tt} = 0 \text{ in V for the } t > 0, \text{ and } u(x,o) = f(x), u_t(x,o) = g(x), u(a,t) = h_1(t), u(b,t) = h_2(t) \qquad (8\text{-}70)$$

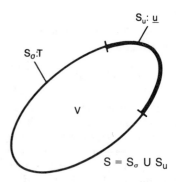

Figure 8-17. General elliptic surface.

Problems involving the dynamic analysis of structures, wave propagation, non-steady fluid flow, ... belong in this group, e.g. parabolic.

$$u_{xx} - u_t = 0 \text{ in V for } t > 0, \text{ and } u(x,o) = f(x), u(a,t) = h_1(t), u(b,t) = h_2(t) \qquad (8\text{-}71)$$

A variety of heat conduction problems, static analysis of structures involving viscosity serve as engineering examples of this type of problem.

The solution approach also depends on the way a given problem is formulated. It is helpful to distinguish between so-called local and global formulations.

All examples presented above are given in the local formulation i.e. in the form of a differential equation

$$Au = f \text{ in V}, u = u(P,t), P \epsilon V \subset \mathbb{R}^n, t \epsilon \mathbb{R} \qquad (8\text{-}72)$$

and appropriate boundary and/or initial conditions, e.g. for an elliptical operator A

$$Gu = g \text{ on S}, u = u(Q), Q \epsilon S \qquad (8\text{-}73)$$

where $f = f(P)$ and $g = g(Q)$ are given functions, while A and G are given differential operators.

Many of those problems may be also expressed in a global (sometimes called energy) formulation where a functional, e.g. total potential energy, total complementary energy, or a variational principle, e.g. the principle of virtual work, is given.

Each of the functionals known in mechanics may be presented in a form

$$I(\mu) = \frac{1}{2} b(\mu,u) - \iota(\mu) \qquad (8\text{-}74)$$

where the product $b(\mu,u)$ is the internal energy in a whole body, $\iota(\mu)$ presents the work of all external forces, μ denotes either displacements $\underline{\mu}$, stresses σ, or strains ϵ (or a combination of them) depending on the functional used. Though both

$b(\mu,u)$ and $\iota(\mu)$ are integrals, they usually involve differential operations as well.

In order to find the unknown function u in the global formuation we either have to minimize a functional $I(\mu)$

$$\text{find} \quad \min_{u} I(\mu), \quad (8\text{-}75)$$

or satisfy a variational principle

$$\delta I(\mu) \geq 0 \rightarrow b(\mu,\delta\mu) \geq \iota(\delta\mu) \quad (8\text{-}76)$$

for all admissible variations $\delta\mu$.

The finite difference method is a discretization technique that may be applied to both initial-value and boundary value problems given in the local formulation, Eqs. 72 and 73, as well as in the global one (Eq. 75 or 76).

8.4.1.2 The Basic Idea of the FDM

The basic idea of the FDM is simple: derivatives of functions are replaced by the corresponding finite difference of their values. For example, the value of the first derivative of a smooth function f(x) evaluated at a point $x = x_i$ may be replaced in the simplest FDM approach by the first difference Δf_i, divided by the interval Δx_i

$$\left.\frac{df}{dx}\right|_{x=xi} \approx \frac{f(x_i+\Delta x_i) - f(x_i)}{\Delta x_i} \equiv \frac{\Delta f_i}{\Delta x_i}. \quad (8\text{-}77)$$

Now this result may be used to replace a given differential equation with an unknown function y(x)

$$\frac{dy}{dx} = g(x,y), \quad y(o) = \bar{y}_0 \quad (8\text{-}78)$$

by so called difference equations (simultaneous algebraic equations)

$$[y(x_i + h_i) - y(x_i)]\frac{1}{h_i} =$$

$$g(x_i,y_i) \Rightarrow y_{i+1} - y_i = h_i g_i, \quad y_o = \bar{y}_0 \quad (8\text{-}79)$$

with the unknown values y_i at points (called nodal) $x_i = 1, 2, \ldots$ chosen in advance. In this way it is possible to discretize the initial value problem stated in Eq. 78.

Using the first difference and the same technique, Eq. 77 may be used to find second and higher order differences; e.g. for equal intervals Δx

$$\left.\frac{d^2 t}{dx^2}\right|_{x=x_i} \approx \frac{f_{i+1} - 2f_i + f_{i-1}}{\Delta x^2} \equiv \frac{\Delta^2 f_i}{\Delta x^2} \quad (8\text{-}80)$$

The same approach also holds for partial derivatives of a smooth function f(x,y), as shown in Fig. 8-18

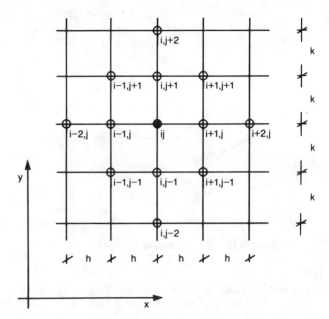

Figure 8-18. 2-D FD Mesh.

$$\left.\frac{\partial f}{\partial x}\right|_{ij} \approx \frac{f_{i+1,j} - f_{ij}}{h}, \quad \left.\frac{\delta f}{\delta y}\right|_{ij} \approx \frac{f_{i,j+1} - f_{ij}}{k}, \quad (8\text{-}81)$$

$$\left.\frac{\partial^2 f}{\partial x^2}\right|_{ij} \approx \frac{f_{i+1,j} - 2f_{i,j} + f_{i-1,j}}{h^2},$$

$$\left.\frac{\partial^2 f}{\partial x \partial y}\right|_{ij} \approx h^{-1}k^{-1}(f_{i+1,j+1} + f_{i-1,j-1} - f_{i-1,j+1} - f_{i+1,j-1}).$$

The above procedure, though very simple, is neither the only way, nor always the best one, to apply in order to get required finite difference operators.

Two examples of the FDM are included in Section 8.4.4.

8.4.1.3 Advantages and Limitations

So far the FDM has been considered with regularly distributed nodes, as in Fig. 8-18. Such a FDM version, commonly called classical, has been known for a long time, is well based, and is presented in many monographs and handbooks [72- 77]. Besides its simplicity, it has many essential advantages - mesh generation is easy, the same FD schema can be used in the whole domain, and its mathematical background (stability, convergence, ...) is well based.

The classical FDM is suitable in most cases and has been successfully applied for problems defined on regular domains like rectangles, circles, rings, etc. In the general case, however, the FDM approach is not particularly versatile, especially when compared with the finite element method (FEM). Intrinsic difficulties arise when dealing with:

- domains of arbitrary shape, especially with curvilinear boundaries [73,78] (see Fig. 8-19),

- local mesh refinement,

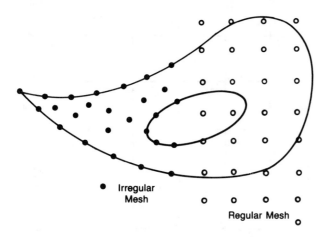

Figure 8-19. Regular and irregular grids.

- problems defined in domains composed of several subdomains of different dimensionality which are joined together,

- General, fully automatic computer approaches do not exist. Despite some recent attempts [78-80]. In most cases a manual approach is needed to generate new FD operators (e.g. in the case of an irregular domain boundary).

Because of these disadvantages once prevailing in common use, the classical FDM fell far behind the FEM as a general solver. On the other hand, a new FD approach called the generalized FDM (GFDM) has recently been developed and is now being developed by many authors [81-100]. It is based on arbitrarily irregular meshes, that remove all of the difficulties mentioned above, and which allow a general, fully-computerized approach.

8.4.2 Theoretical Basis

8.4.2.1 The Classical FDM

The most important information about both the classical and generalized FDM procedures are presented in the next two sections. The FDM applied to regular meshes requires the following steps:

1. Assume mesh type and nodes $P_{i, i=1,2,...,n}$

2. Derive, or assume if available, a FD operator appropriate and mesh type for the given differential operators (boundary operators included)

3. Generate the FD equations

4. Impose the boundary conditions and generate resulting FD equations (in the local formulation)

5. Solve the simultaneous algebraic FD equations which are obtained

6. Find the final quantities required.

The assumptions and constraints inherent in each of the steps used with the classical FDM are as follows:

Step 1. Regular meshes of several types are used (Fig. 8-20) in order to fix the best domain boundaries. Rectangular and triangular grids are the most frequently applied.

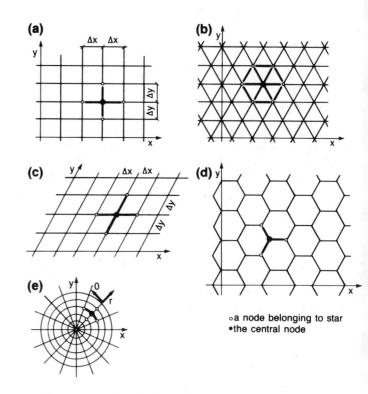

Figure 8-20. Examples of regular 2-D FD meshes: (a)-rectangular; (b)-triangular; (c)-paralellogram type, (d)-hexagonal, (e)-radial.

Step 2. First, a number m and a configuration of nodal points $P_{i+j}, j = 0, ..., m-1$ called "star" or "stencil" must be assumed and used to generate a FD schema for a central node P_i and a given differential operator L:

$$\left. Lu \right|_i \approx \sum_{j=0}^{m-1} B_{j(i)} u(P_{i+j}) \ . \qquad (8\text{-}82)$$

The operator L may represent an original differential equation (L ≡ A). Boundary conditions (L ≡ G), may appear in a functional I(u) or in evaluation of final results. Its FD equivalent has a general form, Eq. 82, as it may be concluded from Eq. 77, 80 and 81. Unknown FD coefficients $B_{j(i)}$ may be found, e.g. by using the simple approach shown earlier, or methods applied in the GFDM.

Both the number and configuration of nodes P_{i+j} in a star depend on the original operator L, and the mesh type assumed. As an example, some stencils for the Laplace's operator $L \equiv \Delta$ in 2-D space are shown in Fig. 8-20.

Some typical FD operators for the 2-D squared mesh are presented in Fig. 8-21.

Step 3. In the local formulation the collocation

$$(Au - f)\Big|_i = \sum_{j=0}^{m-1} B_{j(i)} u_{i+j} = 0, \quad i = 1, 2, \ldots, \eta \quad (8\text{-}83)$$

is used for direct generation of simultaneous algebraic difference equations.

In the global formulation the minimum of a functional

$$\min_{u_i} I(u_i, \ldots, u_n) \quad i = 1, 2, \ldots, \eta \quad (8\text{-}84)$$

is imposed, or the corresponding variational principle is used and satisfied for all δu_i. In both cases boundary conditions have to be included in advance. Unlike the local formulation, an assembly procedure of a FEM type is required.

Step 4. In the local formulation discrete equations corresponding to given boundary conditions $(Gu - g)_i = 0$, $i = n + 1, n + 2, \ldots, n + s$ are obtained following steps 2 and 3. However, in more complex cases satisfaction of boundary conditions may require individual treatment of each boundary node [73,75,77,78].

Step 5. Finally, simultaneous FD (algebraic) equations are obtained from steps 3 and 4. In the local formulation they may not be symmetric. They may be solved by routine procedures, e.g. elimination Gauss method.

Step 6. The results obtained from step 5 correspond to nodes and may not be of final interest. Therefore, further work is usually required, e.g. knowing nodal displacements to find stresses everywhere. An approximation by the minimization technique [90] discussed later for the GFDM is of great value here.

There are many possible ways to increase the accuracy of the results obtained by the above FDM procedure. One obvious way is to increase the total number of nodal points involved. The other option is to improve the quality of the FD approximation using higher order FD formulas. In the global formulation quality of the numerical integration also counts, and procedures of appropriate order should be used (see example 2, section 8.4.4.2).

8.4.1.2 *The Generalized FDM*

In the GFDM the mesh may be arbitrarily irregular (Fig. 8-19), though like the FEM, a relatively smooth transition zone between fine and coarse mesh regions is helpful in obtaining better quality results.(*)

Mesh irregularity leads to a number of new problems that must be addressed. They appear in the following steps of the GFDM procedure:

1. Mesh generation,

2. Domain partition into subdomains assigned to each individual node,

3. Selection and classification of stars,

4. GFDM formulas generation,

5. Generation of the FDM equations in local and global (including approximation, numerical integration and assembling) formulations,

6. Discretization of boundary conditions,

7. Solving for the required final results.

All these problems can be solved in a way which provides for a fully- automatic computer analysis, including processing of the initial data and final output. The most difficult problems are

- avoiding singular and ill-conditioned FD formulas

- formulating the best and most objective criteria for optimal mesh generation, star selection, FD formulas generation, and domain partitioning.

Though there are many papers on the GFDM [81-100] presenting different concepts of the method, several form the basis of this presentation [90-92]. The basic procedure is as follows:

Step 1. Any reasonable FE mesh generator may be applied, although there are meshes specifically designed for the GFDM.

Step 2. The domain may be partitioned into so-called Thiessen's polygons. Thiessen's polygon for a node P_i is the one obtained as the innermost figure formed by the lines of symmetry between a node P_i and all other nodes P_j in the domain (Fig. 8-22). Thiessen polygons are always convex. They cover the whole domain and do not overlap each other, having at most, a common boundary. Two nodes whose Thiessen polygons have at least one common point are called neighbors. Examples of domain partitioning into Thiessen's polygons are shown in the Fig. 8-23.

(*)Such a concept was introduced for the first time by McNeal [93], but significant new developments have occurred in recent years. A somewhat different concept is also used in fluid mechanics where irregular meshes are fully transformable onto a regular reference grid.

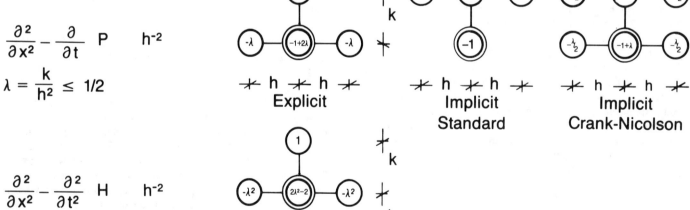

Figure 8-21. Some typical FD operators: E=elliptical, P=parabolic, H=hyperbolic.

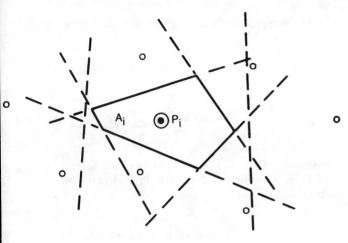

Figure 8-22. Thiessen polygon.

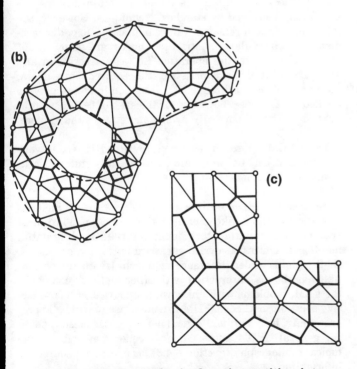

Figure 8-23. Examples in domain partition into Thiessen's polygons (a) smooth transition from a course to a finer mesh.

Step 3. Since the mesh is irregular both stars and FD formulas may differ for each node. The two currently most popular criteria for star selection, useful for the general difference operator of the second order defined in the 2-D space, are:

- the star should be formed by the Thiessen's neighbors, [Fig. 8-24(a)], and

- the cross criterion [Fig. 8-24(b)]; the domain around the central point should be divided into four quadrants, and two points nearest the central node should be included into the star from each quadrant.

Step 4. In the GFDM a different approach is adopted than in the classical FDM. Rather than for any specific differential operator L, the FD formulas are obtained at the same time for all the first and the second order derivatives of u [85]

$$Du \equiv \{u, u_x, u_y, u_{xx}, u_{yy}, u_{xy}\} \equiv \{^q u\} \ , \ q = 0, 1, ..., 5 \quad (8\text{-}85)$$

In order to evaluate the vector, Du_i, in a central node, P_i, using an m-nodes star, it is assumed that $u(x,y)$ is a sufficiently smooth function, and u_{i+j} may be expanded into a Taylor series as follows

$$u_{i+j} = u_i + h_{ij} \frac{\partial u_i}{\partial x} + k_{ij} \frac{\partial u_i}{\partial y} + \frac{h_{ij}}{2} \frac{\partial^2 u_i}{\partial x^2} + \frac{k_{ij}}{2} \frac{\partial^2 u_i}{\partial y^2} + h_{ij}k_{ij} \frac{\partial^2 u_i}{\partial x \partial y}$$
$$+ 0(\rho_{ij}^3) \equiv \widetilde{u}_{i+j} + 0(\rho_{ij}^3) \ , \quad (8\text{-}86)$$

where $\widetilde{u}_{i+j}$ is the second order approximation of u_{i+j}, and

$$h_{ij} = x_j - x_i, \ k_{ij} = y_j - y_i, \ \rho_{ij} = (h_{ij}^2 + k_{ij}^2)^{\frac{1}{2}}, \ u_{i+j} = u(x_{i+j}, y_{i+j}) \ .$$

In order to improve the quality of the approximation, the number of nodes in the star should usually exceed the minimum required ($m \geq 6$) to determine the full second order approximation of the vector Du. Then the solution is found [90] by minimization of the norm

$$B = \sum_{j=1}^{m-1} (u_{i+j} - \widetilde{u}_{i+j})^2 \, w_{j(i)}^2 \ , \quad (8\text{-}87)$$

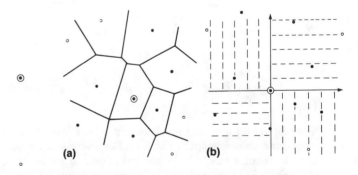

Figure 8-24. Star selection: (a)-by Thiessen neighbours; (b)-by the cross criterion; node belonging to the star central node other nodes.

where $w_{j(i)}$ is a weighting factor, mostly $w_{j(i)} = \rho_{ij}^{-3}$. Thus from

$$\frac{\partial B}{\partial \{^q u_i\}} = 0 \Rightarrow {^q u_i} = \sum_{j=0}^{m-1} {^q B_{j(i)}} u_{i+j}, \quad q = 0, 1, ..., 5 \ . \quad (8\text{-}87)$$

Once coefficients ${^q B_{j(i)}}$ are found, a second order FD operator may be generated. In the case of an operator of higher order than two, the relevant FD formula may be obtained by means of superimposition of Eq. 87, e.g.

$$\frac{\partial^3 u}{\partial x^3} = \frac{\partial}{\partial x}\left(\frac{\partial^2 u}{\partial x^2}\right) \ .$$

Step 5. Boundary conditions

$$\alpha(Q)u + \beta(Q)\frac{\partial u}{\partial n} = g(Q) \qquad Q \epsilon C \quad (8\text{-}88)$$

are considered, where $\frac{\partial u}{\partial n}$ denotes the derivative normal to a boundary C.

If $\beta = 0$ it is sufficient to situate nodes on the boundary C in order to satisfy exactly prescribed boundary conditions. Otherwise the following options may be considered:

- additional external (fictitious) nodes may be introduced

- only internal nodes may be used

- besides u_i new unknowns $u_i^1 = \frac{\partial u}{\partial n}\Big|_i$ may be introduced on the boundary

Usually, the last option is the most useful. Eq. 87 can be used for discretization of Eq. 88. Thus, handling boundary conditions can be made as easy as in the FEM.

Step 6. Having obtained Eq. 87 it is possible to compose any required FD operator. Thus, e.g. for the Laplace operator it is possible to construct the following relationship:

$$\Delta u|_i = \left(0 + 0\frac{\partial}{\partial x} + 0\frac{\partial}{\partial y} + 1\frac{\partial^2}{\partial x^2} + 1\frac{\partial^2}{\partial y^2} + 0\frac{\partial^2}{\partial x \partial y}\right) u|_i$$

$$\approx \sum_{j=0}^{m-1} \left({^3 B_{j(i)}} + {^4 B_{j(i)}}\right) u_{j(i)} = 0 \quad (8\text{-}89)$$

Using Eq. 89 for a square mesh an improved FD operator is constructed, as presented in Fig. 8-21. For an irregular mesh, such an operator has to be generated in each node unless a star is classified as one of the configurations that has already been considered before.

Thus, using Eq. 87 in the local formulation simultaneous FD equations are formed directly. In the global formulation integration is also needed. Then the best approach to solution of 2-D problems is:

- Perform triangulation of the domain using existing nodes, and

- Apply in each triangle the proper order of integration as in the FEM.

$$I_e(u) \approx \int_{A_e} F(u) dA \approx \sum_e \left(\sum_k H_k^e \overline{F}_k^e\right) \quad (8\text{-}90)$$

The integrand value in each Gaussian point, k, can be evaluated using the approximation technique already described in step 4, hence

$$\overline{F}_k^e \equiv F^e\left(\sum_j {^0 B_{j(k)}} u_{k+j}, ..., \sum_j {^5 B_{j(k)}} u_{k+j}\right)$$

$$= \overline{F}^e(u_k, ..., u_{k+m}) \ , \quad (8\text{-}91)$$

and finally it can be expressed in terms of unknown nodal values of an appropriate star.

- Assemble results from all triangles

$$I = \sum_e I_e(u) = I(u_i, ..., u_n) \quad (8\text{-}92)$$

and minimize that functional: $\min_{u_i} I, i = 1, 2, ..., n \rightarrow u_1, ..., u_n$.

Step 7. The FD simultaneous equations obtained in steps 5 and 6 may be solved by standard techniques. However, the following problem remains: values of $u_i, ..., u_n$ (e.g. displacements) in irregularly distributed nodes are known, but values of a differential operation Lu (e.g. stresses) in any arbitrary point of the domain are still required. Once again, the minimization technique [90] described in step 4 may be applied. In this case, however, for the sake of better smoothness, stars should normally involve more nodes than before.

The GFDM approach, as briefly outlined above, is very general and can be programmed to be fully automatic in computer implementation.

Besides the GFDM outline given here, its theory includes many other topics like: mathematical bases, non-elliptical operators, formulation techniques [89,100] on differential manifolds (arbitrary curvilinear coordinates, tensor objects, shells) higher order difference approach, treatment of nonlinear problems, interaction with other methods especially the FEM, problems with unknown boundaries, special effective solution methods GFDM oriented, use of symbolic programming, fully automatic version for regular meshes using the general GFDM approach. The reader may refer to the topical bibliography (Section 8.4.8) for additional sources of information.

8.4.3 The GFDM Approach to Nonlinear Problems

The FDM applied to nonlinear problems yields simultaneous nonlinear algebraic equations. When compared to other methods; e.g. the FEM, the GFDM is especially convenient in dealing with nonlinear problems. This is because the stiffness matrix K for each given operator F may always be found automatically, and each update of that matrix (usually required in an iterative solution approach), is fast, since it needs far fewer changes than in other methods.

Consider Eq. 72 as nonlinear and denote

$$F \equiv Au - f \equiv F(u, u_x, ..., u_{xy}) \equiv F(Du) . \qquad (8\text{-}93)$$

Using the FD approximation, Eqs. 87 and 91 the following can be constructed for each central node P_i

$$F(Du)|_i \approx \overline{F}(u_i), \ u_i \equiv \{u_i \ldots u_{i+m}\} , \qquad (8\text{-}94)$$

and for the whole system of n simultaneous FD equations

$$\overline{F} \equiv \overline{F}(u) = \{\overline{F}_i\}, \ u = \{u_1 \ldots u_n\} . \qquad (8\text{-}95)$$

Expanding Eq. 95 in a Taylor series with respect to a solution $u^{(\iota-1)}$ obtained in step L-1 of an iteration, the following is developed

$$\overline{F}^{\iota} = \overline{F}^{\iota-1} + \frac{\partial \overline{F}}{\partial u} (u^{\iota} - u^{\iota-1}) + \ldots . \qquad (8\text{-}96)$$

Thus, the stiffness matrix is

$$K = \frac{\partial \overline{F}}{\partial u} = \frac{\partial \overline{F}}{\partial (Du^t)} \frac{\partial (Du)}{\partial u} = [K_{ij}], \ K_{ij} = \sum_{q=0}^{5} \overline{F}_{i,q} \ ^qB_{j(i)} \qquad (8\text{-}97)$$

Here the matrix $^qB_{j(i)}$ presents the FD formulas (Eq. 87) already derived - they are constant and do not change through iterations. Thus only the vector

$$[F, q] \equiv \frac{\partial F}{\partial (Du^t)} = \left[\frac{\partial F}{\partial u} \ \frac{\partial F}{\partial u_x} \ \frac{\partial F}{\partial u_y} \ \frac{\partial F}{\partial u_{xx}} \ \frac{\partial F}{\partial u_{yy}} \ \frac{\partial F}{\partial u_{xy}} \right] \qquad (8\text{-}98)$$

is solution dependent, and has to be evaluated in each node, P_i, as $[\overline{F}_{i,q}]$ for each iteration ι. E.g. for $F(u) = u^3 + u_x^2 + u_x u_y + u_{yy}$ it follows that

$$[F, q] = [3u \ 2u_x + u_y \ u_x \ 0 \ 1 \ 0] .$$

Vector $[F, q]$ can be derived with a computer by means of symbolic programming.

Finally, using the Newton-Raphson procedure, the well known type-formulas are developed

$$K^{\iota-1} u^{\iota} = K^{\iota-1} u^{\iota-1} - F^{\iota-1} \qquad (8\text{-}99)$$

for iterative solution of the nonlinear FD simultaneous equations $F = 0$.

A similar approach may be used for the global formulation.

The approach shown above holds well for problems with smooth solutions e.g. for most geometrically nonlinear problems. In the case of certain discontinuities e.g. for elastic-perfectly plastic bodies, the problem of excessive smoothness is introduced by the FDM approximation, when compared with the exact solution. Some smearing effects must be accepted when the elastic- plastic boundary intersects the GFDM stars (Fig. 8-25(a)). Otherwise, it may be assumed that the stars are disconnected (Fig. 8-25(b)) by that boundary and move together with it while plastic effects develop. The first solution proved to be adequate in the elastic-plastic torsion of prismatic bars [91]. The second approach, though more precise, is much more troublesome. Another example of the application of the GFDM (in this case as applied to metal forming) is given [99].

8.4.4 General Applications and Examples

The classical FDM has been successfully applied to a variety of problems in the past. Nowadays, however, due to the strong competition coming from other methods, its use has been restricted mainly to

- problems defined in regular domains, or where there are no difficulties with boundaries when using a regular mesh, or

- problems where functionals are not available e.g. many initial value problems in fluid mechanics.

For elliptical type problems the FDM has been used in plate analyses, although even there, it has often been replaced in recent years by the GFDM [81,91,95,98]. However, the energy FDM has been successfully applied by D. Bushnell [101-103] to shell analyses.

In initial-value problems of the parabolic type, besides fluid mechanics, the FDM has been applied to various heat transfer problems [73]. The FD method of characteristics [73,74] for solution of hyperbolic problems is widely applied in fluid dynamics (gas dynamics included). The method of

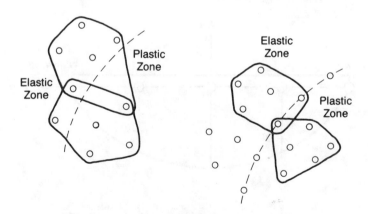

Figure 8-25. (a) Elastic-plastic boundary (intersect) stars); (b) Elastic-Plastic Boundary (does not intersect stars).

characteristics has also frequently been applied in nonlinear analyses of perfectly-plastic bodies.

Prospective applications of the GFDM are much more diverse, since the method is free of restrictions that hamper the classical FDM. Its versatility and potential applicability is comparable to that of the FEM. Some examples of current applications [91,94,95,97-99] are presented in Table 8-3. These examples range from linear to nonlinear and from elliptic to parabolic.

Several examples of solutions for highly sensitive, variable type problems involving large deformation of physically nonlinear membranes of arbitrary shape have been published [94,100].

Two simple examples are presented in the following paragraphs which illustrate the application of the FDM in the solution of boundary-value problems.

8.4.4.1 *Deflection of a Beam*

Find deflections $w = w(x)$ of a simply supported, uniformly loaded beam as shown in Fig. 8-26; q = load, EI = bending stiffness.

Original Formulation

Local

$$\frac{d^2w}{dx^2} = -m(x), \quad w\left(-\frac{L}{2}\right) = w\left(\frac{L}{2}\right) = 0,$$

$$m(x) = -\frac{qL^2}{8EJ}\left(1 - 4\frac{x^2}{L^2}\right).$$

Global

Find $w(x)$, to find the minimum of the total potential energy of the bent beam

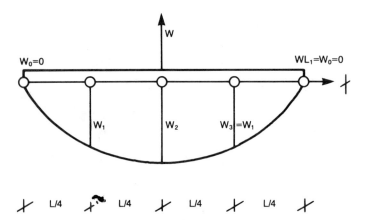

Figure 8-26. Deflections of a free supported, uniformly loaded beam.

$$I(w) = \int_{-\frac{L}{2}}^{\frac{L}{2}} \left[\frac{1}{2}\left(\frac{dw}{dx}\right)^2 - m(x)w\right]dx \quad w\left(-\frac{L}{2}\right) = w\left(\frac{1}{2}\right) = 0.$$

The Finite Difference Solution Approach

Local

Basic Finite Difference Equation

$$w_{i-1} - 2w_i + w_{i+1} = \left(\frac{L}{4}\right)^2 m_i \quad i = 1, 2, 3, \quad w_0 = w_4 = 0,$$

Using symmetry and denoting $Q = \dfrac{qL^4}{128EJ}$ the following equations with unknowns w_1 and w_2 are developed:

$$\left.\begin{array}{l} 0 - 2w_1 + w_2 = \dfrac{3}{4}Q \\[4pt] w_1 - 2w_2 + w_1 = Q \end{array}\right\} \Rightarrow \begin{cases} w_1 = -\dfrac{5}{4}Q = -\dfrac{5}{512}\dfrac{qL^4}{EI} = \dfrac{20}{19}w_{1\,exact} \\[6pt] w_2 = -\dfrac{7}{4}Q = -\dfrac{7}{512}\dfrac{qL^4}{EI} = \dfrac{21}{20}w_{2\,exact} \end{cases}$$

Global

Using symmetry and applying the simple trapezoidal rule for numerical integration, the following equation is constructed:

$$I(w) = 2\left\{\frac{1}{2}\left[\left(4\frac{w_1-w_0}{L}\right)^2\frac{L}{4} + \left(4\frac{w_2-w_1}{L}\right)^2\frac{L}{4}\right] - (m_0w_0 + m_1w_1)\frac{L}{8}\right.$$
$$\left. - (m_1w_1 + m_2w_2)\frac{L}{8}\right\} = \frac{4}{L}\left[w_1^2 + (w_2-w_1)^2\right]$$
$$+ \left(\frac{3}{2}w_1 + w_2\right)Q.$$

Calculating $\dfrac{\partial I}{\partial w_1} = 0$ and $\dfrac{\partial I}{\partial w_2} = 0$, the same results are found for w_1 and w_2 as those obtained by the local approach (though it need not always be so).

8.4.4.2 *Shear Stresses in a Prismatic Bar Subjected to Torsion*

Find the maximum shear stress max in a prismatic bar of rectangular cross-section subjected to torsion by a moment M_t.

Original Formulation

Local

$$\Delta F = -2G\theta \text{ in } A, \; F = 0 \text{ on } C \text{ (see Fig. 8-27)}$$

Global

$$I(\theta) = \int_{-\frac{a}{2}}^{\frac{a}{2}} \int_{-\frac{a}{2}}^{\frac{a}{2}} \left\{\frac{1}{2}\left[\left(\frac{\partial F}{\partial x}\right)^2 + \left(\frac{\partial F}{\partial y}\right)^2\right] - 2G\theta\right\} dxdy, \; F|_C = 0,$$

Mechanical problem	Domain	Type of boundary-value problem
torion of prismatic bars		elliptic type equation with constant coefficients
torsion of non-prismatic, axially-symmetric bars		elliptic type equation with variable coefficients
bending of plates of arbitrary shape		binarmonic equation
		variational approah
stationary heat flow in a mechanical device		elliptic type equation in a multiconnected domain
non-stationary heat flow		parabolic type equation
plane stress analysis, stress concentration		set of elliptic type equation
elasto-plastic torsion of prismatic bars		nonlinear elliptic type equation
large deflections of membranes		nonlinear elliptic type equation
large defomations of membrane shells		set of nonlinear equations of variable type ell-hyp.

Table 8-3. Some representative examples of the GFDM solutions.

where

$$\tau_{yz} = -\frac{\partial \theta}{\partial x}, \quad \tau_{xz} = \frac{\partial \theta}{\partial y}, \quad M_t = 2 \int_{-\frac{a}{2}}^{\frac{a}{2}} \int_{-\frac{a}{2}}^{\frac{a}{2}} F \, dxdy.$$

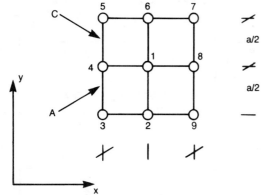

Figure 8-27. Prismatic bar subjected to torsion.

The FDM Solution Approach

A very simple, coarse mesh is used here with only one unknown value F_1, since $F_j = 0$, $j = 2, 3, ..., 9$, due to the boundary conditions. In order to illustrate the level of precision of the FDM solution two different FD schema are applied (Fig. 8-21) approximating the Laplace operator, ∇^2 = the standard schema is defined as follows:

$$F_2 + F_4 + F_6 + F_8 - 4F_1 = -2G\theta\left(\frac{a}{2}\right)^2 \to F_1$$

$$= \frac{1}{8} G\theta a^2 = 0.848 \, F_{1\text{exact}}$$

and the improved one is defined as:

$$\frac{1}{6}(F_3 + F_5 + F_7 + F_9) + \frac{2}{3}(F_2 + F_4 + F_6 + F_8) - \frac{10}{8}F_1 =$$

$$-2G\theta\left(\frac{a}{2}\right) \to$$

$$\to F_1 = \frac{3}{20} G\theta a^2 = 1.018 \, F_{1\text{exact}}.$$

Global Formulation

Using symmetry it follows that

$$I(F) = 4 \left\{ \frac{1}{2} \left[\frac{1}{2}\left(2\frac{F_2 - F_3}{a}\right)^2 + \frac{1}{2}\left(2\frac{F_1 - F_4}{a}\right)^2 \right. \right.$$

$$\left. \left. + \frac{1}{2}\left(2\frac{F_4 - F_3}{a}\right)^2 + \frac{1}{2}\left(2\frac{F_1 - F_2}{a}\right)^2 \right] \frac{a^4}{4}$$

$$- 2G\theta \left[\frac{1}{4}F_1 + \frac{1}{4}F_2 + \frac{1}{4}F_3 + \frac{1}{4}F_4 \right] \frac{a^2}{4} \right\} = 2F_1^2 - \frac{1}{2}G\theta a^2 F_1,$$

$$\frac{\partial I}{\partial F_1} = 4F_1 - \frac{1}{2}G\theta a^2 = 0 \to F_1 = \frac{1}{8} G\theta a^2.$$

Using the simplest linear approximation,

$$\tau_{\max} = \tau_{x2|2} \approx 2\frac{F_1 - F_2}{a} = \frac{2}{a}F_1, \quad M_t = \frac{2}{3}a^2 F_1,$$

while for the quadratic approximation applied for both differentiation and integration (Simpson rule)

$$\tau_{\max} \approx \frac{2}{a}(-3F_2 + 4F_1 - F_6)\frac{1}{2} = \frac{4}{a}F_1,$$

$$M_t = 2\frac{a^2}{4 \cdot 9}[F_3 + F_5 + F_7 + F_8 + 4(F_2 + F_4 + F_6 + F_8) + 16 F_1] = \frac{8}{9} a^2 F_1.$$

Final results for all possible combinations are presented in Table 8-4.

Table 8-4 Results of Example 2 Analysis

Approximation	$\tau_{\max}/\tau_{\max\,\text{exact}}$		$M_t/M_{t\,\text{exact}}$	
	standard	improved	standard	improved
Linear	.370	.444	.593	.711
Quadratic	.740	.888	.790	.948

8.4.5 Application to Linear Elastic Fracture Mechanics

There are two main objectives of the LEFM: evaluation of stress intensity factors and crack propagation rates.

Evaluation of stress intensity factors in a body results in the solution of a boundary value problem for that body. This might, of course, be done by using the FDM. However due to the stress concentration in the neighborhood of a crack tip, local mesh refinement would be desirable. Such effects may be more easily achieved by the GFDM than by the classical FDM. A test solution of that type is presented in Ref. [91].

Some solutions based on a 3-D fracture mechanics approach were obtained by the FDM for evaluation of stress concentrations in composite materials [104].

A similar situation holds for crack propagation, where only a few solutions (e.g. [105]) are available.

Thus, despite some attempts, the classical FDM does not seem to be an appropriate solution tool for LEFM analysis, while solutions obtained by the potentially powerful, but very new, GFDM approach are still to come.

8.4.6 *Available Software*

The availability of computer software for the FDM is far different from that of the FEM, where there are a large variety of general purpose programs.

Because of the way the classical FDM is used, it yields special purpose programs rather then general ones, though a few attempts have also been made in that direction.

One approach has been to develop programs which solve a particular class of linear partial differential equations [79,80]. The other approach worthy of mention, used by D. Bushnell [101-103] was to build computer programs called BOSOR oriented on stress, buckling and vibration analysis of complex shells of revolution. Otherwise, the many specialized programs which are available focus mainly on plate analyses.

The GFDM pretends to be a general solver like the FEM. However, only the FIDAMF package [91,100] (new professional version written in the standard FORTRAN code) enables an average user to solve a broad class of problems like those presented in Table 4-1. Other general packages of that type, which would enable mutual interaction of both FEM and GFDM, are still being developed.

In summary, despite similar potential powers of both methods, professional software currently available for the FDM is far behind that for the FEM.

8.4.7 *Summary*

Two versions of the FDM have been presented, namely

- the classical FDM that is an old, well based method, however suffering from serious restrictions coming from the requirement of a regular mesh,

- the GFDM that is a new, fully automatic, general purpose method adopting arbitrarily irregular meshes.

Mainly elliptical problems were discussed, although parabolic and hyperbolic problems were also mentioned.

Comparisons between the FDM and FEM were presented. Besides some previous attempts [102] there are very recent ones [86,87] to bring the GFDM and the FEM together, including a common computer program. This is being done in the following ways:

- use the GFDM to generate a "Finite Difference Element" and apply it in the FEM routine;

- use the FEM in order to derive the FE characteristics for the vector of derivatives Du, as in Eq. 85, rather than for any specific operator, and apply them in the GFDM procedure;

- use both methods at the same time, but in different parts of a common domain;

- use the GFDM approximation in order to get improved final results based on a rough FEM solution obtained in nodes.

In general, the FEM has clear advantages over the classical FDM. On the other hand, potential power, versatility as a general solver, and fully automatic computer approaches make the GFDM competitive with the FEM and BIM. Unfortunately, there are many more software packages available for the FEM than for the FDM or GFDM.

8.4.8 *Topical Bibliography*

The references which are cited below fall into the following topic areas:

Topic	Reference Numbers
1) Background and Theoretical Basis	72-78, 81-93, 95, 96, 100, 102, 107-111
2) Elasto-Plastic Analysis	91, 95, 96, 99
3) General Applications and Examples	74, 81, 86, 91, 92, 94, 95, 97, 103, 106, 108
4) Linear Elastic Fracture Mechanics	91, 104, 105
5) Available Software (Surveys of)	79, 80, 87, 91, 100-103

8.5 The Boundary Element Method

8.5.1 *Background*

The Boundary Element Method (BEM) is based on the numerical solution of boundary integral equations. Although boundary integral equation techniques have been used in engineering and the physical sciences for many years, the method is comparatively recent and its emergence is in great part due to the advances in computational mechanics, which were a consequence of the rapid development of the finite element method.

The BEM is based on the so-called direct formulation of boundary integral equations and the use of general curved elements, all of them comparatively recent advances taking place in the 1960's and 70's. For a theoretical and historical background the reader is referred to Ref. [112].

The current literature on BEM research is expanding rapidly. Recent conference proceedings, monographs, and research summary books illustrate the energy being devoted to this methodology, e.g., Refs. [113-126]. The subjects covered in these works, as well as the archival literature now appear to cover most significant problems in applied mechanics. The topical bibliography included in Section 8.5.6 provides additional details.

8.5.2 Theoretical Basis

The term "boundary elements" is used to indicate that the external surface of a domain is divided into a series of elements, over which the functions under consideration can vary in different ways (in much the same manner as in finite elements). This characteristic is important, since in the past integral equation techniques were generally restricted to constant sources, assumed to be concentrated at a series of points on the surface of the body, without due consideration to its proper geometry.

Finite differences and finite elements are called domain techniques because the whole volume of the body needs to be discretized. By contrast, boundary element methods only require discretization of the external surface of the body under consideration. This is because this technique is based on the use of influence functions which satisfy the governing equations of the problem. These influence functions are called fundamental solutions when they apply for infinite domains. The use of these functions has some important advantages.

1) The data required to run a problem are considerably reduced, making the technique ideally suited for Computer Aided Engineering applications. Furthermore inter-element continuity for displacements or tractions is not needed, which allows the construction of discontinuous elements. (This advantage incidentally, does not seem to have been fully appreciated by code developers, many of whom still seem tied down to classical finite element continuity requirements.)

2) The results obtained using boundary elements tend to be more accurate than those using finite elements, making the technique well suited for problems involving stress concentrations. Boundary element results are given in terms of surface displacements and surface stresses, both with a similar degree of accuracy (as occurs with so-called mixed finite elements). The higher accuracy of the method can be easily understood if one interprets the finite element matrices as the product of two rather inaccurate polynomial functions and the boundary elements as the product of a polynomial with a fundamental solution (which is the exact solution of the governing equations in an infinite domain, i.e. without boundary conditions).

3) Boundary Elements have no special problems in representing boundary conditions at infinity as they are implicit in the fundamental solution. Finite elements, by contrast (because of their "finiteness" cannot be properly used in these cases.

8.5.2.1 Governing Integral Equation for Elastostatics

To understand the development of the boundary element method in linear elastostatics it is convenient to start with the principle of virtual work in its so-called reciprocity form, for surface tractions

$$\int_\Omega b_k^* t_k \, d\Omega + \int_\Gamma u_k^* t_k \, d\Gamma = \int_\Gamma t_k u_k^* \, d\Gamma + \int_\Omega b_k u_k^* \, d\Omega . \quad (8\text{-}100)$$

Two different fields have been defined, one the approximate field represented by components of displacement, stress, strain or traction (u_k, σ_{ij}, ε_{ij} or t_k) and the virtual field (u_k^*, σ_{ij}^*, ε_{ij}^*, t_k^*). Ω represents the domain under consideration and Γ the boundary. b_k and b_k^* are the body forces corresponding to the two systems. Notice that the strain components can be expressed in terms of displacements, i.e.

$$\varepsilon_{ij} = \frac{1}{2}\left(\frac{\partial u_i}{\partial x_j} + \frac{\partial u_j}{\partial x_i}\right) \quad (8\text{-}101)$$

for the approximate strains, and similarly for the components of the virtual field. The relationship between stresses and strains is given by the generalized Hooke's law, i.e.

$$\sigma_{ij} = d_{ijk\ell}\, \varepsilon_{k\ell} \quad (8\text{-}102)$$

(similarly $\sigma_{ij}^* = d_{ijk\ell}\, \varepsilon_{k\ell}^*$).

The field identified with an asterisk "*" corresponds to the solution of the equilibrium equations under concentrated forces, i.e.

$$\frac{\partial \sigma_{jk}^*}{\partial x_j} + \Delta_\ell^i = 0 , \quad (8\text{-}103)$$

where Δ_ℓ^i is a Dirac delta function and represents a unit load at a point i in the ℓ direction ($b_\ell^* = \Delta_\ell^i$). This type of solution will produce, for each direction 'ℓ the following equation,

$$u_\ell^i + \int_\Gamma u_k\, t_k^*\, d\Gamma = \int_\Gamma u_k^*\, t_k\, d\Gamma + \int_\Omega u_k^*\, b_k\, d\Omega , \quad (8\text{-}104)$$

where u_ℓ^i represents the displacement at i in the 'ℓ direction. Notice that u_k^* and t_k^* are the fundamental solutions, i.e. the displacements and tractions due to a unit concentrated load acting at the point 'i' in the 'ℓ direction. If unit forces are assumed to act in the 3 directions and body forces, b_k, are neglected for simplicity, it follows that

$$u_\ell^i + \int_\Gamma u_k\, t_{\ell k}^*\, d\Gamma = \int_\Gamma t_k\, u_{\ell k}^*\, d\Gamma , \quad (8\text{-}105)$$

where $t_{\ell k}^*$ and $u_{\ell k}^*$ represent the tractions and displacements in the k direction due to unit forces acting in the ℓ direction. Eq. 105 is valid for the particular point 'i' where these forces are applied.

The fundamental solution for a 3-D isotropic body is

$$u_{\ell k}^* = \frac{1}{16\pi G(1-\nu)r}\left[(3-4\nu)\delta_{\ell k} + \frac{\partial r}{\partial x_\ell}\frac{\partial r}{\partial x_k}\right]$$

$$p_{\ell k}^* = -\frac{1}{8\pi(1-\nu)r^2}\left[\frac{\partial r}{\partial n}\left\{(1-2\nu)\delta_{\ell k} + 3\frac{\partial r}{\partial x_\ell}\frac{\partial r}{\partial x_k}\right\}\right. \quad (106)$$

$$\left. -(1-2\nu)\left\{\frac{\partial r}{\partial x_\ell}n_k - \frac{\partial r}{\partial x_k}n_\ell\right\}\right]$$

where n is the outward unit normal to the surface of the body. $\Delta_{\ell k}$ is the Kronecker delta, r is the distance from the point of application of the load to the point under consideration and the n_j's are the direction cosines.

The derivatives of r can be written as

$$\frac{\partial r}{\partial x} = \frac{r_\ell}{r} \quad (8\text{-}107)$$

8.5.2.2 Boundary Point

Eq. 105 can be specialized for a smooth boundary (i.e. for which the tangent plane to the surface is continuous at point i). This operation takes the singularity to the boundary, and the result is that half of it can be neglected (for a mathematical discussion the reader is referred to Refs. [112,127], giving

$$c^i u_\ell^i + \int_\Gamma u_k \cdot t_{\ell k}^* \, d\Gamma = \int_\Gamma t_k \, u_{\ell k}^* \, d\Gamma \quad (8\text{-}108)$$

where $c^i = \frac{1}{2}$. For non smooth boundaries the value of c^i is no alonger $\frac{1}{2}$ but its explicit calculation is not required, as it can be obtained using rigid body considerations [127].

8.5.2.3 Boundary Elements

Eq. 108 can now be applied on the boundary of the domain under consideration. For simplicity consider a two dimensional domain with its boundary divided into n straight linear elements or segments. The points where the unknown values are considered are called 'nodes' and taken to be in the middle of each segment for constant elements, or at the intersection (for constant continuous elements), or even inside the elements (constant discontinous elements). In addition, continuous or discontinuous curvilinear elements can be used. For these cases an extra mid-element node is required.

Only the case of constant size elements will be considered here for simplicity. The boundary can now be discretized into n elements, of which n_1 can have displacement boundary conditions and for the other n_2 elements ($n_2 = n - n_1$) the tractions are known. (Mixed boundary conditions combining displacements and tractions do not present any special difficulty but their discussion would complicate the notation in what follows.)

The values of the u and t components are assumed to be constant within each element and equal to the value at the mid-node of the element.

Eq. 108 for a given point 'i' becomes (in matrix form)

$$\underset{\sim}{c^i}\, \underset{\sim}{u^i} + \int_\Gamma \underset{\sim}{t^*}\, \underset{\sim}{u}\, d\Gamma = \int_\Gamma \underset{\sim}{u^*}\, \underset{\sim}{t}\, d\Gamma \quad (8\text{-}109)$$

where

$\underset{\sim}{u^i} = \begin{Bmatrix} u_1 \\ u_2 \\ u_3 \end{Bmatrix}$: displacement vector of point 'i' with components in $x_1\, x_2\, x_3$ directions.

u: displacement vectors at a point on the boundary Γ.

$\underset{\sim}{t} = \begin{Bmatrix} t_1 \\ t_2 \\ t_3 \end{Bmatrix}$: tractions at any point on the boundary Γ;

$$\underset{\sim}{c_i} = \begin{bmatrix} \tfrac{1}{2} & 0 & 0 \\ 0 & \tfrac{1}{2} & 0 \\ 0 & 0 & \tfrac{1}{2} \end{bmatrix}$$

$\underset{\sim}{t^*} = \begin{bmatrix} t_{11}^* & t_{12}^* & t_{13}^* \\ t_{21}^* & t_{22}^* & t_{23}^* \\ t_{31}^* & t_{32}^* & t_{33}^* \end{bmatrix}$ matrix whose coefficients $t_{\ell k}^*$ are the surface tractions in k direction due to a unit force acting in the 'ℓ' direction applied at point 'i'.

$\underset{\sim}{u^*} = \begin{bmatrix} u_{11}^* & u_{12}^* & u_{13}^* \\ u_{21}^* & u_{22}^* & u_{23}^* \\ u_{31}^* & u_{32}^* & u_{33}^* \end{bmatrix}$ matrix whose coefficients $u_{\ell k}^*$ are the displacements in the 'k' direction due to a unit force at 'i' in the 'ℓ' direction.

Eq. 109 can now be divided into 'n' elements, which gives

$$\underset{\sim}{c^i}\, \underset{\sim}{u^i} + \sum_{j=1}^{n} \left(\int_{\Gamma_j} \underset{\sim}{t^*}\, \underset{\sim}{u}\, d\Gamma \right) = \sum_{j=1}^{n} \left(\int_{\Gamma_j} \underset{\sim}{u^*}\, \underset{\sim}{t}\, d\Gamma \right) \quad (8\text{-}110)$$

For the case of constant elements the values of u and t can be taken outside the integrals which give

$$\underset{\sim}{c^i}\, \underset{\sim}{u^i} + \sum_{j=1}^{n} \left\{ \int_{\Gamma_j} \underset{\sim}{t^*}\, d\Gamma \right\} \underset{\sim}{u_j} = \sum_{j=1}^{n} \left\{ \int_{\Gamma_j} \underset{\sim}{u^*}\, d\Gamma \right\} \underset{\sim}{p_j} \quad (8\text{-}111)$$

Notice that this equation applies for a particular node 'i', where the fundamental solution is assumed to be acting. The integrals of t^* and u^* inside the brackets relate the 'i' node to the segment 'j' over which the singularity is carried out.

Influence matrices h and g can be defined by,

$$\hat{\underset{\sim}{h}}_{ij} = \int_{\Gamma_j} \underset{\sim}{t^*}\, d\Gamma \quad (8\text{-}112)$$

$$\hat{\underset{\sim}{g}}_{ij} = \int_{\Gamma_j} \underset{\sim}{u^*}\, d\Gamma$$

These integrations can be carried out numerically [127], as well as analytically. The equation for node 'i' can be written as,

$$\underset{\sim}{c^i}\, \underset{\sim}{u^i} + \sum_{j=1}^{n} \hat{\underset{\sim}{h}}_{ij}\, \underset{\sim}{u_j} = \sum_{j=1}^{n} \underset{\sim}{g}_{ij}\, \underset{\sim}{u_j} \quad (8\text{-}113)$$

With the following definitions

$$\underline{h}_{ij} = \hat{\underline{h}}_{ij} \quad \text{where } i \neq j$$
$$\underline{h}_{ij} = \hat{\underline{h}}_{ij} + \underline{c}^i \quad \text{where } i = j \tag{8-114}$$

Eq. 113 can be written as,

$$\sum_{j=1}^{n} \underline{h}_{ij}\, \underline{u}_j = \sum_{j=1}^{n} \underline{g}_{ij}\, \underline{t}_j \tag{8-115}$$

The whole set of equations for the n nodes can be found by displaying the singularity from one node to the next and the final system of n equations (Eq. 115) can be written in matrix form as,

$$\underline{H}\,\underline{U} = \underline{G}\,\underline{T} \tag{8-116}$$

Note that the values of displacements are assumed to be known at n_1 nodes and the values of tractions at the other n_2 points. Hence, by rearrangement

$$\underline{A}\,\underline{X} = \underline{F}, \tag{8-117}$$

where X is a vector of unknown displacement and traction components.

Solution of Eq. 117, using a standard Gauss elimination method, will result in the values of X and hence the full solution on the boundary (i.e. displacement and traction components) is known. Because tractions are obtained as well as displacements the formulation is sometimes called 'mixed'. It has the advantage that it produces very accurate results for stresses as well as displacements.

8.5.2.4 *Internal Results*

If the displacements are required at an internal point one can use Eq. 105 with the singularity applied at 'i' which is now an internal point. The procedure requires the calculation of a new set of h and g coefficients for each different internal point, but this is usually not too time consuming since it does not involve solution of a system of equations and only a limited number of internal points are required.

To obtain internal stresses it is necessary to apply the derivatives of Eq. 118 to first obtain the gradients of displacements, combine them to find the strain components and finally multiply these by the material characteristics using Hooke's law. For instance, for isotropic materials,

$$\sigma_{ij} = \frac{2G\nu}{1-2\nu}\delta_{ij}\frac{\partial u_\ell}{\partial u_\ell} + G\left(\frac{\partial u_i}{\partial x_j} + \frac{\partial u_j}{\partial x_i}\right) \tag{8-118}$$

where G and ν are the shear modulus and Poisson's ratio, respectively, for the material.

The final result is well known and gives the following integral expression,

$$\sigma_{ij} = \int_\Gamma D_{kij}\, t_k\, d\Gamma - \int_\Gamma S_{kij}\, u_k\, d\Gamma \tag{8-119}$$

The values of the third order tensor D_{kij} and S_{kij} are well known and can be found in Refs. [112,127].

8.5.2.5 *Higher Order Elements*

The introduction of higher order elements requires the use of transformation formulae for curvilinear surfaces similar to those used in the so-called 'isoparametric' finite elements. For these cases all integrations are usually carried out numerically.

Higher order elements combined with discontinuous elements converts the Boundary Element technique into a very efficient tool for engineering analysis. One advantage of the method over finite elements is that it does not require as many elements on the boundary as a finite element grid.

8.5.2.6 *Non-Smooth Boundaries*

It has already been mentioned that for the case of non-smooth boundaries, such as at edges and corners, the matrix $\underline{C}^i$ is no longer a diagonal matrix with coefficients 1/2. The correct formulation of the $\underline{c}^i$ terms in this case can be mathematically very cumbersome for elasticity problems and it is customary in this case to deal with the problem numerically.

Consider the system

$$\underline{H}\,\underline{U} = \underline{G}\,\underline{T} \tag{8-120}$$

before applying any boundary conditions. The coefficients of the diagonal submatrices in H can be obtained by applying rigid body conditions. If unit rigid body displacements are assumed in any direction, Eq. 120 becomes,

$$\underline{H}\,\underline{I}_\ell = \underline{O} \tag{8-121}$$

where $\underline{I}_\ell$ is a vector of unit displacements in the chosen 'ℓ' direction. All terms in the diagonal submatrices of H can be obtained by applying the 3 (in 3-dimensional analysis) or 2 (in two dimensional case) rigid body movements. These movements are a function of all the off diagonal terms which must be computed (usually using numerical integration).

8.5.3 *Other Considerations*

Although the above description of the method has neglected body forces, centrifugal forces and thermal effects, it is possible to introduce them without having to compute any domain integrals [127]. These integrals can be reduced to the boundary using some elegant theoretical devices (which are beyond the scope of this chapter). Unfortunately, not all boundary element codes allow for the inclusion of these effects without having to discretize the volume - which defeats the objective of the new method. The reader should be aware of these deficiencies if they exist in a boundary element code.

Checking the code's user manual carefully at the beginning and looking for these pitfalls will avoid problems later on.

8.5.4 Available Software

There are a limited number of sources of software for boundary integral analysis in comparison to the large number for finite element codes. Reviews of available programs are given in Refs. [69,71,150,153].

8.5.5 Summary

The boundary element method is a computationally more sophisticated method than the finite element method because it uses singular functions as influence solutions. This results in very accurate results when the method is well applied, but can give rise to substantial errors if a code is poorly written. Special care should be taken in using accurate numerical integration techniques which are also computationally efficient.

The solution of the boundary element equation system which can be fully populated and is generally non symmetric, also requires special care. Substantial savings can be achieved in many cases by using well written solvers. The emergence of relatively inexpensive parallel processing machines have also resulted in an increase of interest in the new method, as the matrices obtained are ideally suited to this type of hardware.

In general, the reader is well advised to check a boundary element code before using it. These checks can be very simple and should involve tests such as rigid body movements, rigid body rotations, constant strain states and others for which the exact solution is known. These tests are generally sufficient to determine the validity of the code.

8.5.6 Topical Bibliography

The references which are cited below fall into the following topic areas:

Topic	Reference Numbers
1) Background and Theoretical Basis	112–135, 161, 178, 179
2) Elasto-Plastic Analysis	136-138, 162, 175, 176
3) General Applications and Examples	139-153
4) Fracture Mechanics	154-159, 160, 163-174, 177
5) Available Software (Surveys of)	69, 71, 150, 153

8.6 Comparison of the Methods

8.6.1 Historical Background

The finite element method is a variant of several well known classical processes of approximation. The finite difference method is an older procedure that has been used in one form or another almost since the inception of differential calculus. It was first adapted for the purpose of stress analysis problem solving by Southwell and his collaborates [180]. Ever since the initial usage of finite elements, arguments about their relative merits versus the more established finite-difference methods have been made [181-182].

In the following paragraphs the advantages and limitations of the finite difference, finite element and boundary integral methods will be reviewed. It is important to remember that all three methods are approximate methods, and consequently, must be checked for accuracy or convergence of their results. All three methods normally require at least two analyses with different grids to establish whether the size of the grid mesh is yielding satisfactory solutions.

8.6.2 The Finite Element Method

Finite element analysis involves both mathematical approximations (those associated with the displacement function of a particular element) and geometric approximations to the component shape. As previously noted, an advantage of the finite element method is that it may be employed to analyze arbitrary structural problems in a routine manner - thus the case of non-homogeneous material behavior or geometry is not a problem because the direct assembly of elements with different properties presents no difficulties.

8.6.3 The Finite Difference Method

8.6.3.1 Derivative Approximation

The derivative approximation approach to finite-difference differs manifestly from the finite element method in two ways. First, in the final system of equations to be solved each equation appears *immediately* in its assembled form. Second, unsymmetric coefficient matrices often appear due to boundary condition approximations. While the first difference from the standard finite element formulation may be a point in favor of the finite difference approach the second is, from a computational point-of-view, a distinct disadvantage. Which process gives a better approximation for the same effort? This question cannot be answered in general. Croll and Walker [181] found that in the case of plane elasticity, the differences using a regular mesh of simplest triangular elements and corresponding simplest finite difference expressions appear to favor the finite element formulation.

A serious difficulty with the derivative approximation approach is the instability of the solution at fine mesh sizes, and the slow convergence of the results obtained from single precision computer programs. It should be noted that the solution may not always converge asymptotically as the number of finite difference stations increases (See e.g., [183]). In [184] it was indicated that the error associated with finite difference approximation of a given order is larger close to the boundary. Thus, the overall accuracy of the results may be increased if derivative approximations of higher order are used at the boundaries.

Discontinuous interfaces such as abrupt changes in material properties or structural geometry require special treatment, which by itself limits the generality of the derivative approximates approach [185-187].

8.6.3.2 *Variational Formulation*

This approach has many of the advantages of the finite element method with the serious disadvantage of producing a discontinuous displacement field over the elements involved.

8.6.4 *The Boundary Integral Equation Method*

Because this method allows the reduction of the problem dimension, the size of the algebraic problem is smaller than for the same problem formulated in terms of either the finite element method or the finite difference method. The method has the advantage that considerably better resolution of interior stresses is possible than by the other considered methods. The boundary-integral equation approach has the computational disadvantage that it leads to an unsymmetric set of algebraic equations involving an unsymmetric, densely packed coefficient matrix. This disadvantage may, in some applications, involve more computing effort than either the finite element method or the finite-difference method. The boundary integral equation is particularly suited for application to problems with high stress gradients such as the fracture mechanics problems discussed in Section 8.5.5 and geometries involving small ratios of surface area to volume. The advantages include reduced size of the computational problem, and increased solution accuracy [188-191].

8.7 References

[1] Timoshenko, S. P., and Goodier, J. N., Theory of Elasticity, 3rd Ed., McGraw-Hill Book Co., 1970.

[2] Whames, I. H., Mechanics of Deformable Solids, R. E. Krieger Publishing Co., 1979.

[3] Timoshenko, S., Strength of Materials, Parts I and II, 3rd Ed., R. E. Krieger Publishing Co., 1958 and 1976.

[4] Timoshenko, S., and Young, D. H., Elements of Strength of Materials, 5th Ed., Van Nostrand Reinhold, 1968.

[5] Juvinall, R. C., Engineering Considerations of Stress Strain and Strength, McGraw-Hill Book Co., 1967. See also Langhaar, H., *Energy Methods*, Wiley & Co.

[6] Shigley, J. E., Mechanical Engineering Design, 3rd Ed., McGraw-Hill Book Co., 1977.

[7] Spotts, M. F., Design of Machine Elements, 5th Ed., Prentice-Hall, (1978).

[8] Faupel, J. H., Engineering Design, John Wiley & Sons, 1964.

[9] Timoshenko, S., and Woinowsky-Krieger, S., Theory of Plates and Shells, 2nd Ed., McGraw-Hill Book Co., 1959 and 1968.

[10] Baker, E. H., et. al., Shell Analysis Manual, NASA CR-912, U.S. Dept. of Commerce, National Technical Information Center, 1968.

[11] Blodgett, O. W., Design of Welded Structures, The James F. Lincoln Arc Welding Foundation, 1966.

[12] Roark, R. J., and Young, W. C., Formulas for Stress and Strain, 5th Ed., McGraw-Hill Book Co., 1975.

[13] Kleinlogel, A., Rigid Frame Formulas, Frederich Ungar Publishing Co., 1952.

[14] Tada, H., Paris, P., and Irwin, G., The Stress Analysis of Cracks Handbook, Del Research Corp., 1985.

[15] Sih, G. C., Handbook of Stress Intensity Factors, Institute of Fracture and Solid Mechanics, Lehigh University, 1973.

[16] Rolfe, S. T., and Barsom, J. M., Fracture and Fatigue Control of Structures, Prentice-Hall, 1977.

[17] Peterson, R. E., Stress Concentration Factors, John Wiley & Sons, (1974).

[18] Jawad, M. H., and Farr, J. R., Structural Analysis and Design of Process Equipment, John Wiley & Sons, 1984.

[19] Johnson, W., and Mellor, P. R., Engineering Plasticity, Halsted Press, 1984.

[20] Mendelson, A., Plasticity: Theory and Application, Krieger Publishing Co., 1983.

[21] Turner, M. K., Clough R., Martin H., and Topp L., "Stiffness and Deflection Analysis of Complete Structures," *J. Aero Sci.*, Vol. 23, No. 9, September 1956, pp. 805-823.

[22] Melosh, R. J., and Bamford, R. M., "Efficient Solution of Load-Deflection Equations", ASCE Journal of the Structural Division, Vol. 95, No. ST4, ASCE April 1, 1969, pp. 661-676.

[23] Zienkiewicz, O.C., The Finite Element Method in Engineering Science, McGraw-Hill, Book Co., 1970.

[24] Desai, C. S., Introduction to the Finite Element Method, Van Nostrand Reinhold, 1972.

[25] Gallagher, R. H., Finite Element Analysis Fundamentals, Prentice-Hall, 1975.

[26] Zienkiewicz, O.C., The Finite Element Method, McGraw-Hill Book Co. (1977).

[27] Broom, T., and R. Ham, "The Hardening and Softening of Metals by Cyclic Stresses," Proceedings of the Royal Society, Vol. 242A, 1957, pp. 166-179.

[28] Morrow, J., Cyclic Plastic Strain Energy and Fatigue of Metals, ASTM STP No. 378, ASTM, 1965.

[29] Ramberg, W., and W. R. Osgood, "The Description of Stress/Strain Curves by Three Parameters," NACA TN902, July, 1943.

[30] Wilkins, M. L., "Modeling the Behavior of Materials."

[31] Wilkins, M. L., "Calculation of Elastic-Plastic Flow," in Methods of Computational Physics, 3, (B. Alder et al, eds.) Academic Press, 1964.

[32] Hughes, T. J. R., "Numerical Implementation of Constitutive Models: Rate-Independent Deviatoric Plasticity," in Theoretical Foundation for Large-Scale Computations for Nonlinear Material Behavior.

[33] Mendelson, A. and S. S. Manson, Practical Solution of Plastic Deformation Problems in the Elastic-Plastic Range, NASA TR-R-28, 1959.

[34] Marcal, P., "A Comparative Study of Numerical Methods of Elastic-Plastic Analysis," AIAA Journal, Vol. 6, No. 1, AIAA pp. 157, January 1968.

[35] Armen, H., Levine, H., Pifko, A., and Levy, A., Nonlinear Analysis of Structures, Grumman Research Department Report RE-454, May 1973.

[36] Kachanov, L. M., Foundations of the Theory of Plasticity, North-Holland, 1971.

[37] Fatigue Under Complex Loading: Analyses and Experiments AE-6, SAE (1977).

[38] Barron, G. E., "A Finite Element and Cumulative Damage Analysis of a Keyhole Test Specimen," SAE Technical Paper 750041, 1975.

[39] Irwin, G. R., "Fracture," Handbuch der Physik, Bd 4, Springer, pp. 551-590, 1958.

[40] Rolfe, S. T., and Barsom, J. M., "Fracture and Fatigue Control in Structures," Applications of Fracture Mechanics, Prentice-Hall, 1977.

[41] Sokolnikoff, I. S., Mathematical Theory of Elasticity, McGraw-Hill Book Co., 1956.

[42] Some Basic Problems of the Mathematical Theory of Elasticity, P. Noordhoff Ltd., 1954.

[43] Sih, G. C., and Liebowitz, H., "Mathematical Theories of Brittle Fracture," Fracture, Vol. II, Academic Press, 1968.

[44] Tada, H., Paris, P., Irwin, G., "The Stress Analysis of Cracks Handbook," Del Research Corp., 1973.

[45] Sih, G. C., "Handbook of Stress Intensity Factors," Institute of Fracture and Solid Mechanics," Lehigh University, 1974.

[46] Chan, S. K., Tuba, I. S., and Wilson, W. K., "On the Finite Element Method in Linear Elastic Fracture Mechanics," *Engineering Fracture Mechanics,* Vol. 2, pp. 1-17, 1970.

[47] Stanley, P., Fracture Mechanics in Engineering Practice, Applied Science Publishers Ltd., 1977.

[48] Gallagher, R. H., "Survey and Evaluation of the Finite Element Method in Fracture Mechanics Analysis," First International Conference in Reaction Technology, Berlin, 1971.

[49] Jeram, K., and Hellen, T. K., "Finite Element Techniques in Fracture Mechanics," International Conference on Welding Research Related to Power Plants, Southampton, 1972.

[50] Oglesby, J.J., and Lomacky, O., "An Evaluation of Finite Element Methods for the Computation of Elastic Stress Intensity Factors," *Journal of Engineering for Industry,* 95, pp. 177-185, 1973.

[51] Apostal, M. C., Jordan, S., and P. V. Marcal, "Finite Element Techniques for Postulated Flaws in Shell Structures, Special Report No. 22. Prepared for Electric Power Research Institute, 1975.

[52] Gallagher, R. H., Proceedings of the First International Conference in Numerical Methods in Fracture Mechanics, Swansea, 1-25, 1970.

[53] Benzley, S. E., Parks, D. M., "Numerical Methods in Fracture Mechanics," The National Symposium on Structural Mechanics Software, June 1974.

[54] Kobayashi, A. S., "Experimental Techniques in Fracture Mechanics," Iowa State University Press, SESA.

[55] Rice, J. R., "Plane Strain Deformations Near a Crack Tip in a Power-Law Hardening Material, *Journal of Mechanics of Physics and Solids,* 1967.

[56] Parks, D. M., "A Stiffness Derivative Finite Element Technique for Determination of Elastic and Tip Stress Intensity Factors," Technical Report NASA NGL 40-002-080/13, May 1973.

[57] Bowie, O. L., "Rectangular Tensile Sheet with Symmetric Edge Cracks," *Journal of Applied Mechanics*, 31, 1964, pp. 208-212.

[58] Grandt, Jr., A. F., Hinnerichs, T. D., "Stress Intensity Factor Measurement for Flanged Fastener Holes," Army Symposium on Solid Mechanics, September 1974.

[59] Lawrence, F. V., "Estimation of Fatigue-Crack Propagation Life in Butt Welds," Welding Research Supplement, pp. 212-220, May 1973.

[60] Benzley, S. E., "Representation of Singularities with Isoparametric Finite Elements," *International Journal for Numerical Methods in Engineering,"* Vol. 8, pp. 537-545, 1974.

[61] Blackburn, W. S., Hellen, T. K., "Calculation of Stress Intensity Factors in Three Dimensions by Finite Element Methods," *International Journal for Numerical Methods in Engineering*, Vol. 2, pp. 211-229, 1977.

[62] Ingrafea, A. R., and Manu, C., "Stress-Intensity Factor Computation in Three Dimensions with Quarter-Point Elements," *International Journal for Numerical Methods in Engineering*, Vol. 15, pp. 1427-1445, 1980.

[63] Keer, L. M. and Freedman, J. M., "Tensile Strip with Edge Cracks," *International Journal Engrg. Sci.*, Vol. 11, pp. 1265-1275, 1973.

[64] Henshell, R. D., Shaw, K. G., "Crack Tip Finite Elements are Unnecessary," *International Journal for Numerical Methods in Engineering*, Vol. 9, pp. 495-507, 1975.

[65] Barsoum, R. S., "On the Use of Isoparametric Finite Elements in Linear Fracture Mechanics," *International Journal for Numerical Methods in Engineering,"* Vol. 10, pp. 25-37, 1976.

[66] Westergaard, H. M., "Bearing Pressure and Cracks," Transactions, *ASME, Journal of Applied Mechanics*, 1939.

[67] Gallagher, R. H., "Large-Scale Computer Programs for Structural Analysis," in On General Purpose Finite Element Computer Programs, ASME Special Publication, pp. 3-3, ASME, 1970.

[68] Marcal, P. V., "Survey of General Purpose Programs for Finite Element Analysis," Advances in Computational Methods in Structural Mechanics and Design, University of Alabama Press, 1972.

[69] Structural Mechanics Computer Programs, The Uniaversity Press of Virginia, 1974.

[70] Fredriksson, B., and Mackerle, J., "Overview and Evaluation of Some Versatile General Purpose Finite Element Computer Programs," Finite Element Methods in the Commercial Environment, Vol. 2, 1979, pp. 390-419.

[71] Noor, A. K., "Survey of Computer Programs for Solution of Nonlinear Structural and Solid Mechanic Problems," *Computers and Structures,* Vol. 13, pp. 425-465.

[72] Richtmyer, R. D., and Morton, K. W., Difference Methods for Initial Value Problems, John Wiley & Sons, 1967.

[73] Forsythe, G. E, and Wasow, W. R., Finite-Difference Methods for Partial Differential Equations, John Wiley & Sons, 1967.

[74] Anderson, D. A., Tannehill, J. C., and Fletcher, R. H., Computational Fluid Mechanics and Heat Transfer, McGraw-Hill Book Co., 1984.

[75] Collatz, L., Nuinerische Behandlung von Differential Gleichungen, Springer-Verlag, Berlin-Gottingen-Heidelberg (1955). English edition: The Numerical Treatment of Differential Equations, New York (1966). Z. Csendes, A. FORTRAN Program to Generate Finite Difference Formulas. *International Journal for Numerical Methods in Engineering*, 9(3), pp. 581-599 (1975).

[76] Jain, M. K., Numerical Solution of Differential Equations, John Wiley & Sons, 1984.

[77] Samarski, A. A., Theory of Difference Schemas (in Russian), Nauka, Printing House, Moscow, 1977.

[78] Hunt, B., Finite Difference Approximation of Boundary Conditions Along Irregular Boundaries, *International Journal for Numerical Methods in Engineering*, Vol. 12, pp. 229-235, 1978.

[79] Chugh, A. K., Gesund, H., Automatic Generation of the Coefficient Matrix of Finite Difference Equations, *International Journal for Numerical Methods in Engineering*, Vol. 8(3), pp. 662-671, 1974.

[80] Petravic, M., Petravik, G. K., and Roberts, K. V., Automatic Production of Programmes for Solving Partical Differential Equations by Finite Difference Methods, Comp. Phys. Comm. Vol. 4, pp. 82-88, 1972.

[81] Barve, V. D., and Dey, S. S., Isoparametric Finite Difference Energy Method for Plate Bending Problems, *Computers and Structures,* Vol. 17(3), pp. 459-465, 1983.

[82] Dekker, K., Semi-Discretization Methods for Partial Differential Equations on Non-Rectangular Grids,

International Journal for Numerical Methods in Engineering, Vol. 15, pp. 405-419, 1980.

[83] Demkowicz, L., Karafiat, A., and Liszka, T., On Some Convergence Results for RDM with Irregular Mesh, *Comp. Meth. Appl. Mech. Engrg.*, Vol. 42(3), pp. 343-355, 1984.

[84] Frey, W. H., Flexible Finite-Difference Stencils from Isoparametric Finite Elements, *International Journal for Numerical Methods in Engineering*, Vol. 11, 1653-1665, 1977.

[85] Jensen, P. S., Finite Difference Techniques for Variable Grids, *Computers and Structures*, Vol. 2, pp. 17-29, 1972.

[86] Krok, J., and Orkisz, J., Application of the Generalized FD Approach to Stress Evaluation in the FE Solution, Proceeding of International Conference on Computational Mechanics, May 25-29, Springer Verlag, 1986.

[87] Krok, J., and Orkisz, J., On Unified Approach to the Finite Element and Finite Difference Methods, Proceedings of the First World Congress on Computational Mechanics, Sept. 22-26, Texas University Press, 1986.

[88] Godoy, L. A., Ill-Conditioned Stars in the Finite Difference Method for Arbitrary Meshes, *Computers and Structures*, Vol. 22(3), pp. 469-473, 1986.

[89] Kwok, S. K., An Improved Curvilinear Finite Difference (CFD) Method for Arbitrary Mesh Systems, *Computers and Structures*, Vol. 18(4), pp. 719-731, and Vol. (6), pp. 1087-1097, 1984.

[90] Liszka, T., An Interpolation Method for an Irregular Net of Nodes, *International Journal for Numerical Methods in Engineering*, Vol. 20, pp. 1599-1612, 1984.

[91] Liszka, T., and Orkisz, J., The Finite Difference Method at Arbitrary Irregular Grids and Its Application in Applied Mechanics, *Computers and Structures*, Vol. 11, pp. 83-95, 1980.

[92] Liszka, T., and Orkisz, J., The Finite Difference Method for Arbitrary Irregular Meshes - A Variational Approach to Applied Mechanics Problems, Proceedings of the second international congress on numerical methods for engineering, GAMNI, Dunod, Paris, pp. 277-285, 1980.

[93] MacNeal, R. H., An Asymmetrical Finite Difference Network, *Quarterly of Applied Mathematics*, Vol. 11, pp. 295-310, 1953.

[94] Orkisz, J., and Tworzydlo, W., Numerical Analysis of Arbitrary Membrane Spells by the Generalized Finite Difference Method, Proceedings of the First World Congress on Computational Mechanics, September 22-26, Texas University Press, 1986.

[95] Pavlin, V., and Perrone, N., Finite Difference Energy Techniques for Arbitrary Meshes Applied to Linear Plate Problems, *International Journal for Numerical Methods in Engineering*, Vol. 14, pp. 647-664, 1979.

[96] Perrone, N., and Kao, R., A General Finite Difference Method for Arbitrary Meshes, *Computers and Structures*, Vol. 5, pp. 45-58, 1975.

[97] Snell, C., Vesey, D. G., and Mullord, P., The Application of a General Finite Difference Method to Some Boundary Value Problems, *Computers and Structures*, Vol. 13, pp. 547-552, 1981.

[98] Syczewski, N., and Tribitto, R., Division of an Area in the Analysis of Surface Plates by the Mesh Method, *Computers and Structures*, Vol. 18(5), pp. 813-818, 1984.

[99] Tseng, A. A., A Generalized Finite Difference Scheme for Convention Dominated Metal Forming Problems, *International Journal for Numerical Methods in Engineering*, Vol. 20, pp. 1885-1900, 1984.

[100] Tworzydto, W., The General Finite Difference Method on a Differential Manifold, Proceedings of the First World Congress on Computational Mechanics, September 22-26, Texas University Press, 1986.

[101] Bushnell, D., BOSOR4 Program for Stress, Buckling and Vibration of Complex Shells of Revolution in Structural Mechanics Software Series, University Press of Virginia, Vol. 1, pp. 11-143, 1977.

[102] Bushnell, D., Finite Difference Energy Method Versus Finite Element Models: Two Variational Approaches in One Computer Program, Numerical and Computer Methods in Structural Mechanics, Academic Press, pp. 291- 336, 1973.

[103] Bushnell, D., Almroth, B. O., and Brogan, F., Finite Difference Energy Method for Nonlinear Shell Analysis, *Computers and Structures*, Vol. 3(1), pp. 361-388, 1971.

[104] Altus, E., and Rotem, A., A 3-D Fracture Mechanics Approach to the Strength of Composite Materials, *Engineering Mechanics*, Vol. 14, pp. 637-650, 1981.

[105] Shmueli, M., and Alterman, Z. C., Crack Propagation Analysis by Finite Difference, *Journal of Applied Mechanics*, Vol. 40, pp. 901-908, 1973.

[106] Davies, G., Ford, B., Mullord, P., and Snell, O., Application of an Irregular Mesh Finite Difference Approximation to the Plate Buckling Problem, in Variational Methods in Engineering, Southampton University Press, pp. 11.1-11.12, 1973.

[107] Kwok, S. K., Numerical Computations of Tensor Quantities on a Curved Surface by the Curvilinear Finite Difference (CFD) Method, *Computers and Structures*, Vol. 18(6), pp. 1087-1097, 1984.

[108] Lau, P. C. M., Numerical Solution of Poisson's Equations Using Curvilinear Finite Difference Method, *Applied Mathematical Modelling* Vol. 1, pp. 349-350, 1977.

[109] Mullord, P., A General Mesh Finite Difference Method Using Combined Nodal and Elemental Interpolation, *Applied Mathematical Modelling*, Vol. 3, pp. 433-440, 1979.

[110] Nay, R. A., and Utku, S., An Alternative for the Finite Element Method in Variational Methods in Engineering, Southampton Univ. Press, pp. 3.62-3.74, 1973.

[111] Wyatt, M. J., Davies, G., and Snell, C., A New Difference Based Finite Element Method, Proceedings of the Institute of Civil Engineers, Vol. 59(2), pp. 395-409, 1975.

[112] Brebbia, C. A., "The Boundary Element Method for Engineers", Pentech Press, 1978, Computational Mechanics Publications, 1984.

[113] Banerjee, P. K., and Butterfield, R., Boundary Element Methods in Engineering Science, McGraw-Hill Book Co., 1981.

[114] Banerjee, P. K., and Butterfield, R., Developments in Boundary Element Methods - 1, Applied Science Publishers, London (1982).

[115] Banerjee, P. K., and Shaw, R. P., Developments in Boundary Element Methods - 2, Applied Science Publishers, London (1982).

[116] Banerjee, P. K., and Mukherjee, S., Developments in Boundary Element Methods - 3, Applied Science Publishers, Ltd., 1984.

[117] Brebbia, C. A., Boundary Element Methods, Springer-Verlag, 1981.

[118] Brebbia, C. A., Boundary Element Methods in Engineering, Springer-Verlag, 1982.

[119] Brebbia, C. A., Futagami, T., and Tanaka, M., Boundary Elements, Springer-Verlag, 1983.

[120] Crouch, S. L., and Starfield, A. M., Boundary Element Methods in Solid Mechanics, George Allen and Unwin, 1983.

[121] Ingham, Derek B., and Kelmanson, Mark A., Boundary Integral Equation Analysis of Singular, Potential and Biharmonic Problems, Lecture Notes in Engineering, Vol. 7, Springer-Verlag, 1984.

[122] Kitahara, Michihiro, Boundary Integral Equation Methods in Eigenvalue Problems of Elastodynamics and Thin Plates, *Studies in Applied Mechanics*, Vol. 10, Elsevier, Science Publishing Co., 1985.

[123] Mukherjee, S., Boundary Element Methods in Creep and Fracture, Applied Science Publishers, 1982.

[124] Shaw, R., Pilkey, W., Pilkey, B., Wilson, R., Lakis, A., Chandouet, A., and Marino, C., Innovative Numerical Analysis for the Engineering Sciences, University Press of Virginia, 1980.

[125] Telles, J. C. F., The Boundary Element Method Applied to Inelastic Problems, Lecture Notes in Engineering, Vol. 1, Springer-Verlag, 1983.

[126] Venturini, W. S., Boundary Element Method in Geomechanics, Lecture Notes in Engineering, Springer-Verlag, 1983.

[127] Brebbia, C. A., Telles, J., and Wrobel, L., "Boundary Element Techniques - Theory and Applications in Engineering", Springer-Verlag, 1984.

[128] Kellogg, O. D., Foundations of Potential Theory, Dover Publications, Inc., 1929.

[129] Kupradze, V. D., Potential Methods in the Theory of Elasticity, Daniel Davey and Co., Inc. 1965.

[130] Mikhlin, S. G., Integral Equations, MacMillan Co., 1964.

[131] Cruse, T. A., An Improved Boundary-Integral Equation Method for Three Dimensional Elastic Stress Analysis, *Computers and Structures*, Vol. 4, pp. 741-754, 1974.

[132] Cruse, T. A., Numerical Solutions in Three-Dimensional Elastostatics, *International Journal of Solids and Structures*, Vol. 5, pp. 1259-1274, 1969.

[133] Lachat, J. C., and Watson, J. O., Effective Numerical Treatment of Boundary Integral Equations: A Formulation for Three-Dimensional Elastostatics, *International Journal for Numerical Methods in Engineering*, Vol. 10, pp. 991-1005, 1976.

[134] Rizzo, F. J., and Shippy, D. J., An Advanced Boundary Integral Equation Method for Three-Dimensional Thermoelasticity, *International Journal for Numerical Methods in Engineering,* Vol. 11, pp. 1753-1768, 1977.

[135] Cruse, T. A., and Wilson, R. B., Advanced Applications of Boundary Integral Equation Methods, *Nuclear Engineering and Design,* Vol. 46, pp. 223-234, 1978.

[136] Swedlow, J. L., and Cruse, T. A., Formulation of Boundary Integral Equations for Three-Dimensional Elasto-Plastic Flow, *International Journal of Solids and Structures,* Vol. 7, pp. 1673-1683, 1971.

[137] Mukherjee, S., Corrected Boundary-Integral Equations fin Planar Thermoelastoplasticity, *International Journal of Solids and Structures,* Vol. 13, pp. 331-335, 1977.

[138] Bui, H. D., Some Remarks About the Formulation of Three-Dimensional Thermoelastoplastic Problems by Integral Equations, *International Journal of Solids and Structures,* Vol. 14, pp. 935-939, 1978.

[139] Riccardella, P. C., An Implementation of the Bounday-Integral Technique for Planar Problems of Elasticity and Elasto-Plasticity, Ph.D. Thesis, Carnegie-Mellon University, 1973.

[140] Mendelson, A., Boundary-Integral Methods in Elasticity and Plasticity, NASA TN D-7418, NASA, 1973.

[141] Kamiya, N., and Sawaki, Y., An Integral Equation Approach to Finite Deflection of Elastic Plates, *International Journal of Non-Linear Mechanics,* Vol. 17, No. 3, pp. 187-194, 1982.

[142] Rizzo, F. J., Shippy, D. J., and Rezayat, M., Boundary Integral Equation Analysis for a Class of Earth-Structure Interaction Problems, Final Report for NSF Research Grant CEE-8013461, University of Kentucky, 1985.

[143] Cruse, T. A., and Rizzo, F. J., A Direct Formulation and Numerical Solution of the General Transient Elastodynamic Problem I, *Journal of Mathematical Analysis and Applications,* Vol. 22, No. 1, pp. 244-259, 1968.

[144] Cruse, T. A., A Direct Formulation and Numerical Solution of the General Transient Elastodynamic Problem II, *Journal of Mathematical Analysis and Applications,* Vol. 22, No. 2, pp. 341-355, 1968.

[145] Deruntz, J. A., and Geers, T. L., Added Mass Computation by the Boundary Integral Method, *International Journal for Numerical Methods in Engineering,* Vol. 12, pp. 531-549, 1978.

[146] Manolis, G. D., and Beskos, D. E., Dynamic Stress Concentration Studies by Boundary Integrals and Laplace Transform, *International Journal for Numerical Methods in Engineering,* Vol. 17, pp. 573-599, 1981.

[147] Geers, T. L., Boundary-Element Methods for Transient Response Analysis, Computational Methods for Transient Analysis Computational Methods in Mechanics North-Holland, pp. 221-243, 1983.

[148] Manolis, G. D., A Comparative Study on Three Boundary Element Method Approaches to Problems in Elastodynamics, *International Journal for Numerical Methods in Engineering,* Vol. 19, pp. 73-91, 1983.

[149] Rizzo, F. J., Shippy, D. J., and Rezayat, M., A Boundary Integral Equation Method for Radiation and Scattering of Elastic Waves in Three Dimensions, *International Journal for Numerical Methods in Engineering,* Vol. 21, pp. 115-129, 1985.

[150] Benitez, F. G., Some Considerations on Dynamic BIEM - I, II, A Dynamic Program for Two-Dimensional Fracture Mechanics Problems, *Engineering Fracture Mechanics,* Vol. 20, No. 4, pp. 599-611, pp. 613-622, 1984.

[151] Nardini, D., and Brebbia, C. A., A New Approach to Free Vibration Analysis Using Boundary Elements, *Applied Mathematical Modelling,* Vol. 7, pp. 157-162, 1983.

[152] Papoulis, A., A New Method of Inversion of the Laplace Transform, *Quarterly of Applied Mathematics,* Vol. 14, pp. 405-414, 1957.

[153] Durbin, F., Numerical Inversion of Laplace Transforms: An Efficient Improvement to Dubner and Abate's Method, *Computer Journal,* Vol. 17, pp. 371-376, 1974.

[154] Cruse, T. A., Fracture Mechanics, Boundary Element Methods in Mechanics, North-Holland.

[155] Blandford, G. E., Ingraffea, A. R., and Liggett, J. A., Two-Dimensional Stress Intensity Factor Computations Using the Boundary Element Method, *International Journal for Numerical Methods in Engineering,* Vol. 17, pp. 387-404, 1981.

[156] Heliot, J., Labbens, R. C., and Pellissier-Tanon, A., Results for Benchmark Problem 1, The Surface Flaw, *International Journal of Fracture,* Vol. 15, pp. R197-R202, 1979.

[157] Morjaria, M., and Mukherjee, S., Numerical Analysis of Planar, Time-Dependent Inelastic Deformation of Plates with Cracks by the Boundary Element Method, *International Journal of Solids and Structures,* Vol. 17, pp. 127-143, 1981.

[158] Cruse, T. A., and Polch, E. Z., Elastoplastic BIE Analysis of Cracked Plates and Related Problems, *International Journal for Numerical Analysis in Engineering*, submitted for publication.

[159] Snyder, M. D., and Cruse, T. A., Boundary-Integral Equation Analysis of Cracked Anisotropic Plates, A Contour Integral Computation of Mixed-Mode Stress Intensity Factors, *International Journal of Fracture*, Vol. 11, pp. 315-328, 1975.

[160] Abdul-Mihsein, M. J., and Fenner, R. T., Some Boundary Integral Equation Solutions for Three-Dimensional Stress Concentration Problems, *Journal of Strain Analysis*, Vol. 18, No. 4, pp. 207-215, 1983.

[161] Kuhn, G., and Mohrmann, W., Boundary Element Method in Elastostatics: Theory and Applications, *Applied Mathematical Modelling*, Vol. 7, pp. 97-105, 1983.

[162] Radaj, D., Mohrmann, W., and Schilberth, G., Economy and Convergence of Notch Stress Analysis Using Boundary and Finite Element Methods, *International Journal for Numerical Methods in Engineering*, Vol. 20, pp. 565-572, 1984.

[163] Bui, H. D., An Integral Equation Method for Solving the Problem of a Plane Crack of Arbitrary Shape, *Journal of the Mechanics and Physics of Solids*, Vol. 25, pp. 29-39, 1977.

[164] Cruse, T. A., and Meyers, G. J., Three-Dimensional Fracture Mechanics Analysis, *Journal of the Structural Division*, Vol. 103, ASCE, pp. 309-320, 1977.

[165] Rudolphi, T. J., and Ashbaugh, N. E., An Integral-Equation Solution for a Bounded Elastic Body Containing a Crack: Mode I Deformation, *International Journal of Fracture*, Vol. 14, No. 5, pp. 527-541, 1978.

[166] Heliot, J., Labbens, R. C., and Pellissier-Tanon, A., Semi-Elliptical Cracks in a Cylinder Subjected to Stress Gradients, Fracture Mechanics, STP 677, ASTM, pp. 341-364, 1979.

[167] Tan, C. L., and Fenner, R. T., Elastic Fracture Mechanics Analysis by the Boundary Integral Equation Method, Proceedings, Royal Society of London, pp. 243-260, 1979.

[168] Rudolphi, T., A Boundary Element Solution of the Edge Crack Problem, *International Journal of Fracture*, Vol. 18, No. 3, pp. 179-190, 1982.

[169] Watson, J. O., Hermitian Cubic Boundary Elements for Plane Problems of Fracture Mechanics, Research Mechanics, Vol. 4, pp. 23-42, 1982.

[170] van der Ween, F., Mixed Mode Fracture Analysis of Rectilinear Anisotropic Plates Using Singular Boundary Elements, *Computers and Structures*, Vol. 17, No. 4, pp. 469-474, 1983.

[171] Annigeri, B. S., and Cleary, M. P., Surface Integral Finite Element Hybrid (SIFEH) Method for Fracture Mechanics, *International Journal for Numerical Methods in Engineering*, Vol. 20, pp. 869-885, 1984.

[172] Martinez, J., and Dominguez, J., On the Use of Quarter-Point Boundary Elements for Stress Intensity Factor Computations, *International Journal for Numerical Methods in Engineering*, Vol. 20, pp. 1941-1950, 1984.

[173] Sato, Yoshihisa, Tanaka, Masataka, and Nakamura, Masayuki, Stress Intensity Factor Computation in 3-D Elastostatics by Boundary Element Method, *Engineering Analysis*, Vol. 1, No. 4, pp. 200-205, 1984.

[174] Fan, T. Y., and Hahn, H. G., An Application of the Boundary Integral Equation Method to Dynamic Fracture Mechanics, *Engineering Fracture Mechanics*, Vol. 21, No. 2, pp. 307-313, 1985.

[175] Banerjee, P. K., and Cathie, D. N., A Direct Formulation and Numerical Implementation of the Boundary Element Method for Two-Dimensional Problems of Elasto-Plasticity, *International Journal of Mechanical Sciences*, Vol. 22, pp. 233-245, 1980.

[176] Mukherjee, S., and Morjaria, M., A Boundary Element Formulation for Planar Time-Dependent Inelastic Deformation of Plates with Cutouts, *International Journal of Solids and Structures*, Vol. 17, pp. 115-126, 1981.

[177] Atkinson, C., Xanthis, L. S., and Bernal, M. J. M., Boundary Integral Equation Crack-Tip Analysis and Applications to Elastic Media with Spatially Varying Elastic Properties, *Computer Methods in Applied Mechanics and Engineering*, Vol. 29, pp. 35-49, 1981.

[178] Stern, M., A General Boundary Integral Formulation for the Numerical Solution of Plate Bending Problems, *International Journal of Solids and Structures*, Vol. 15, pp. 769-782, 1979.

[179] van der Ween, F., Application of the Boundary Integral Equation Method to Reissner's Plate Model, *International Journal for Numerical Methods in Engineering*, Vol. 18, pp. 1-10, 1982.

[180] Southwell, R. V., Relaxation Methods in Theoretical Physics, Clarendon Press, 1946.

[181] Croll, J. G., and Walker, A.C., "The Finite-Difference and Localized Ritz Methods," *International Journal for Numerical Methods in Engineering*, Vol. 3, pp. 155-160, 1971.

[182] Key, S. W., and Krieg, R. D., "Comparison of Finite Element and Finite Difference Methodology," ONR Symposium on Structural Mechanics, University of Illinois, 1971.

[183] Brogan, F., Fosberg, K., and Smith, S., "Experimental and Analytical Investigation of the Dynamic Behavior of a Cylinder with a Cutout," Paper No. 68-318, AIAA/ASME 9, Structural Dynamics and Materials Conference, April 1968.

[184] Croll, J. G., and Scrivener, J., "Convergence of Hypar Finite - Difference Solutions," ASCE Journal of the Structural Division, Vol. 95, pp. 809-830, 1969.

[185] de G. Allen, Relaxation Methods, McGraw-Hill Book Co., 1959.

[186] Ely, J. F., and Zienkiewicz, O. C., "Torsion of Compound Bars - A Relaxation Solution," *International Journal of Mechanical Science*, Vol. 1, pp. 356-365, 1960.

[187] Zienkiewicz, O. C., and Schimming, B., "Torsion of Nonhomogeneous Bars with Applied Symmetry," *International Journal of Mechanical Science*, Vol. 4, pp. 15-23, 1962.

[188] Cruse, T. A., "An Improved Boundary-Integral Equation Method for Three Dimensional Elastic Stress Analysis," *Computers and Structures,* Vol. 4, pp. 741-754, 1974.

[189] Bozek, D. G., Morman, K. N., and Kline, K. A., "Three-Dimensional Analysis Using Combined Finite Element-Boundary Integral Techniques," SAE Technical Paper 790978, SAE, 1979.

[190] Bozek, D. G., Katnik, R. B., Partika, J. E., Kowalskki, M. F., and Kline, K. A., "Application of the Boundary Integral Equation method (BIEM) to Complex Automotive Structural Components," SAE Technical Paper 811321, SAE, 1981.

[191] Cruse, T. A., "Numerical Evaluation of Elastic Stress Intensity factors by the Boundary-Integral Equation Method," The Surface Crack: Physical Problems and Computational Solutions, ASME, pp. 153-170, 1972.

[192] Malvern, L. E., *Introduction to the Mechanics of a Continuous Media*, Chapter 2, Prentice Hall, 1969.

Chapter 9 Structural Life Evaluation

9.1 Introduction
    9.1.1 Scope
    9.1.2 Relationship to Other Chapters
    9.1.3 Chapter Plan

9.2 The Role of Testing in Design
    9.2.1 Support of Analysis
        9.2.1.1 Material Properties
        9.2.1.2 Model Parameters
        9.2.1.3 Environmental Parameters
    9.2.2 Prototype Evaluation
        9.2.2.1 Functional Evaluation
        9.2.2.2 Parameter Evaluation
        9.2.2.3 Environmental Response Evaluation
        9.2.2.4 Durability Evaluation
    9.2.3 Validation of the Production Process
    9.2.4 In-Service Monitoring
        9.2.4.1 Service Use Monitoring
        9.2.4.2 Service Failure Monitoring
        9.2.4.3 Customer Satisfaction
        9.2.4.4 QC Testing

9.3 Analytical and Experimental Tools for Structural Life Evaluation
    9.3.1 Level 1: System Control
    9.3.2 Level 2: Test Site Control
    9.3.3 Level 3: Data Analysis
    9.3.4 Level 4: Laboratory Management

9.4 Vehicle Durability Testing
    9.4.1 Test Objective
    9.4.2 Test Specimen
    9.4.3 Environmental Input Exciters
    9.4.4 Data Recording Locations
    9.4.5 Data Collection Techniques
    9.4.6 Test System Programming and Control
    9.4.7 Durability Test Monitoring and Analysis

9.5 Modal Testing and Analysis

9.6 Summary

9.7 References

9.1 Introduction

9.1.1 Scope

A test engineer faces a series of considerations (decisions) when designing a program to conduct service history simulation testing of full scale vehicles (structures) in the laboratory. This chapter proposes logical decision paths with pertinent discussion of tradeoffs that are intended to serve as design guides in this process.

The chapter begins by presenting a model of how a test important in establishing a proper division of responsibility for the validity and usefulness of test results. This is followed by specific discussions of service history simulation testing for fatigue durability evaluation.

9.1.2 *Relationship to Other Chapters*

This chapter on structural life evaluation is closely linked to several other chapters. Fatigue testing of structural components is discussed for the purpose of either supporting an analytical fatigue life prediction, as discussed in Chapter 10, or to evaluate prototype or production components. Prototype or production component assessments depend heavily on estimates of field service histories determined by techniques described in Chapter 5. The techniques for strain measurement and crack detection described in Chapter 7 are commonly used during structural life predictions. Abnormal or unexpected component failures that occur during laboratory testing can be examined by the techniques described in Chapter 11.

9.1.3 *Chapter Plan*

The intention of this chapter is to give general guidelines for the proper use of some of today's testing technology. In spite of the sophistication available in testing systems, their use as part of the total product engineering process requires use of engineering judgement based upon past experience. The decisions involved in making a tradeoff between analytical and experimental techniques are good examples [1].

The communication between design and testing groups needs to be very close. This link is the subject of the first part of the chapter. Fig. 9-1 provides an illustration of the information flow between the significant engineering activities throughout a product life cycle. Note that the design department is at the hub of the communications process and as such, communication to the test department is done through the design department. For example, the test laboratory might perform tests in response to a service failure, but the service department should report the failure to the design department initially.

The most rapid advances in testing technology in recent years have occurred in the area of service history simulation [2-4]. This is especially significant in full scale vehicle durability testing using systems similar to the one shown in Fig. 9-2. A discussion of pertinent concepts in the use of such systems provides the principal emphasis of the third technical section of this chapter. The last section addresses the related technical disciplines of modal testing and analysis.

9.2 The Role of Testing in Design

Successful testing starts with a clear definition of the test objective. This allows the test engineer to "keep his eye on the ball," so to speak. It is possible to categorize test objectives by taking an overview of the total design process. As shown in Fig. 9-3, test laboratory support of the vehicle design process

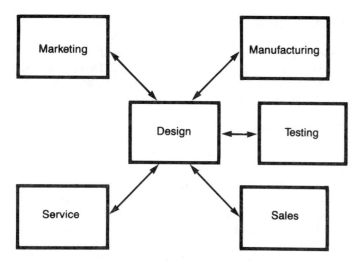

Fig. 9-1 Information flow between engineering activities during a product life cycle.

Fig. 9-2 A modern service duplication test of a passenger car. Ref [25]

can be separated into four categories: 1) support of analysis tools and techniques, 2) evaluation of prototypes, 3) validation of production processes (assemblies) and 4) in-service monitoring. It is proper to consider the test laboratory as providing the "window to the outside world" from the design process.

9.2.1 *Support of Analysis*

The heart of the design process is the set of analysis tools that the designer employs. An analysis tool could be a mental comparison of the present requirements to past experience, or it could be a sophisticated finite-element analysis system with color graphics or scale model testing using simulated service loads. The use of finite element analysis and/or scale model testing (static or dynamic) using simulated service loads should normally be considered in the early stages of the design process. Many of the design deficiencies can thus be eliminated by using either one or both of these techniques before the first prototype is built.

In most cases, the test laboratory is called on to provide data to be used in the analysis procedures. The process is illustrated graphically in Fig. 9-4. The role of the test laboratory in supporting the design process at this point could be thought of as a "front-end" activity in that the necessary information to initiate the process is being generated. This area is not the principal subject of this paper, but will be discussed briefly for completeness.

9.2.1.1 *Material Properties*

The supplying of material properties is a traditional test laboratory activity. There are, however, new requirements being imposed in this area as reported for example by Landgraf and Conle [5]. Emphasis is being placed on insuring that

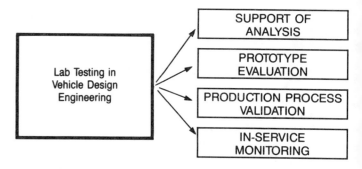

Fig. 9-3 Test laboratory support of the vehicle design process.

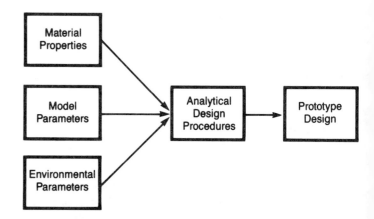

Fig. 9-4 Test laboratory design analysis support functions.

useful information is being generated and stored in a manner that makes it readily available when and where it is needed. (The usefulness of the data is mainly established by correlating analytically predicted results with measured results.)

9.2.1.2 Model Parameters

The generation of empirical model parameters at the front end of the design process applies principally to subsystems. For example, rubber (pneumatic) tire data is in the form of parameters for empirical models of tire behavior. A specific component can be entered into an analytical model in empirical form if it remains unchanged from a previous design. This can include very complicated subsystems such as the use of a modal model obtained from modal analysis techniques for a vehicle body [6].

A computer system model can be built from the empirical and/or experimental modal data on components and subsystems. This model can then be used to study the total system behavior using the simulated service loads. The product design can thus be optimized before a prototype is built. If the test results from the prototype do not agree, then the computer model can be modified to closely duplicate the real life situation. The computer simulation is repeated until satisfactory agreement is obtained. This model can then be used in the future on a similar design. Another application of the computer simulation is in the analysis of interference and packaging. An example is the interference that may occur when a mechanism travels through its motions or defined orbit.

9.2.1.3 Environmental Parameters

The lack of proper environmental input parameters can often produce the weak link in an analytical analysis. Providing such information becomes especially difficult in the case of multiple inputs such as a vehicle traversing a roadway. Environmental input descriptions must be reducible to a form that can be processed rapidly by the analysis routines in order to allow iteration of the analysis process. Need for environmental input is also becoming increasingly important as composite materials find their place in structural components. These components, in many cases, have to meet very stringent performance requirements at very high and/or cold temperatures. Added to these requirements are the effects of humidity and salt spray.

Environmental input descriptions have a double valued importance in that they are needed in the testing process itself. By definition, service history simulation testing requires this information.

9.2.2 Prototype Evaluation

Once the analytical design procedures reach an initial conclusion, a prototype can be designed and built. It is a test laboratory function to evaluate the suitability of the prototype for the defined purpose. This should be done, however, with close coordination between all areas of the design process in order to achieve maximum benefit.

Prototype evaluation should be viewed as an audit of the analytical design process. The different types of evaluation tests are shown in Fig. 9-5. It should be understood which of the elements in Fig. 9-4 are being checked in each test. This check can be made using a computer program during design stage.

9.2.2.1 Functional Evaluation

One of the first tests normally conducted is a basic functional test. Some of the errors which are most obvious in a prototype are difficult to discern analytically. An example is interference of parts as a mechanism. This may occur when a suspension travels through its motions. This is a check on the analysis procedures.

9.2.2.2 Parameter Evaluation

Another possible test is experimental measurement of model parameters. A simple example would be measurement of parameters needed to define the lateral control characteristics of a vehicle, such as suspension deflection steer, roll steer, etc. A modal survey is an empirical means of checking dynamic or vibration analysis of a structure, that in some cases may reveal points of high dynamic strains that may not be detected by static analysis. This function might, thus, provide a check on model parameters that were entered into the process initially, a check on the analysis procedures, and in some cases provide an extension of the computer model of a structure or substructure.

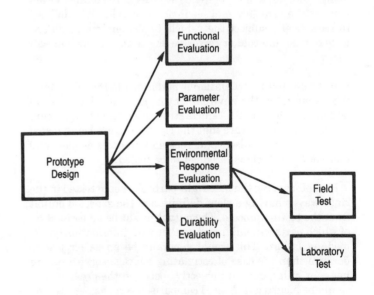

Fig. 9-5 Test laboratory prototype evaluation functions.

9.2.2.3 *Environmental Response Evaluation*

A third type of test can be an environmental response evaluation. This involves instrumenting the specimen to some extent and then subjecting it to environmental inputs. In the case of a prototype vehicle the environmental inputs can come from a test track or from the public roadways.

In the simplest form the test could be a subjective evaluation of dynamic characteristics (ride and handling) by the test driver. A more complex effort could be the recording of strain at a suspected critical location and subsequent fatigue damage analysis. Referring to Fig. 9-4, this test category would provide a check on the combination of the environmental parameters and the analytical design procedure.

Noise, vibration and harshness (NVH) analysis is another type of environmental evaluation. NVH characteristics are being added as performance criteria in the design of many new vehicles. With the availability of the microcomputer-based portable real time analyzer it is becoming easier to measure this index. In the past this type of index was measured by "seat of the pants" type analyses.

9.2.2.4 *Durability Evaluation*

A durability evaluation test is the most difficult, time consuming, and expensive of all types. It can be performed by subjecting the specimen to actual environmental inputs (test track or public highways) or by performing a simulation test in the laboratory.

If the test is performed by driving a vehicle, then the complete vehicle is functionally involved in the test. The environment, however, will normally be a subset of the total environment that a vehicle would see in actual service. It is difficult to realistically subject the vehicle to the environmental effects of heat and cold or corrosive environments such as salt spray.

If the durability evaluation is performed in the laboratory, then a subset of the environment is again involved, and in addition, a subset of the vehicle structure is tested. Great care must be taken to insure that the right parts are being tested in the right way. This gets increasingly complex as more and more of the vehicle is intended to be tested.

Part of insuring that the right parts are being tested in the right way is having a clear definition of the expected output from the testing process. The output might be an estimation of which parts will fail first, together with information on the mode of failure. This would generally be an easier test to perform than the case of estimating average service life (in hours or miles, etc.). An objective between the previous two would be incurred in A vs. B comparison tests for service life, as commonly are performed on consummable parts such as brake lining materials.

A durability evaluation test provides a check on all four of the process blocks that lead to prototype design as shown in Fig. 9-4, with the possible exception of environmental parameters. A durability evaluation test usually involves recording the effects of an actual environment on the test specimen as a means of establishing the test inputs. In this case, the check would include environmental parameters.

It is a common desire to be able to test a prototype without gathering field data with the prototype, i.e., testing with the same environmental parameter information as was available to the analytical design process. The check may not include realistic environmental parameters.

9.2.3 *Validation of the Production Process*

Types of tests performed are similar to those used for prototype evaluation. The main difference is that these parts, components, or assemblies are selected randomly as initial production begins. Results of this testing are used to determine whether the parts from the production line are as good as or better than the prototype parts. The successful completion of this test phase is necessary before actual production begins. This testing should be repeated whenever there is a significant change in component design or any part of the production process.

9.2.4 *In-Service Monitoring*

The role of the test laboratory should not end when a product reaches production. Product in-service monitoring requires facilities and skills that are similar to those needed in other test laboratory activities. This is the principal motivation for including in-service monitoring in this discussion. As shown in Fig. 9-6, it is convenient to divide in-service monitoring into four categories.

9.2.4.1 *Service Use Monitoring*

Service use monitoring is a very important and also very different task to perform. This ties back to Fig. 9-4 in the area of generation of environmental parameters, particularly customer correlation information such as percent of time on freeways, rough roads, etc. As mentioned earlier, lack of information in this area can often produce the "weak link" in a design effort. Today's electronic technology allows powerful data reduction and recording systems to be made rugged, yet highly portable [7-10].

9.2.4.2 *Service Failure Monitoring*

Service failure monitoring is largely a data base management task. The test laboratory provides a convenient location for inspection and examination of failed parts. This failure analysis effort often leads to further test laboratory work in other areas. This is discussed in more detail in Chapter 11.

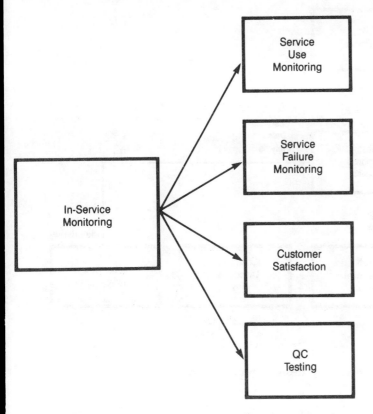

Fig. 9-6 Test laboratory in-service monitoring functions.

9.2.4.3 *Customer Satisfaction*

Customer satisfaction measurement is on the fringe of the main stream of test laboratory activity. It should be coordinated with other test laboratory activities, however, and must be considered in total laboratory operation.

9.2.4.4 *QC Testing*

Quality control testing appears in two different areas. It is common to perform QC tests on vendor supplied materials and parts that are used in the manufacture of final products. A strong emphasis is currently being placed on the quality control of ground vehicles. This emphasis is leading to sophisticated QC testing on the production line.

This completes the overview of the tie between design and testing. A proper balance and intercommunication between analytical design and experimentation is a key to optimizing return on investment. A further step discussing the actual flow of information and areas where digital technology tools may be applied is warranted.

9.3 Analytical and Experimental Tools for Structural Life Evaluation

The Engineering Design and Analysis Process flow chart given in Fig. 9-7 illustrates design activities and information flow used in product design.

Computer based automation tools are used throughout this process. Through data networks and compatible software, information can be shared among any of the functional areas.

Five tools are typically used:

- Data acquisition and classification tools determine load data for a fatigue test or for analysis.

- Data editing and modification tools accelerate a test and improve results by removing insignificant, non-fatigue contributing data.

- Laboratory testing tools simulate or evaluate a product's response to a load environment.

- Test data monitoring tools determine how the specimen reacts to the load environment over a period of time.

- Laboratory data analysis tools analyze the results of field or laboratory testing to determine a product's life.

Three paths are evident in the engineering design and analysis process. Two of these paths require use of analysis tools developed to predict the fatigue life a product sees in service; editing tools to take a recorded service environment and edit it into meaningful data; and a materials database to determine fatigue life. All of these elements are closely related to the traditional fatigue laboratory. The third path, in addition to the tools for the first two, requires test excitation tools to take the field recorded data and program laboratory fatigue tests; analysis tools to take field recorded load data and edit these data to accelerate laboratory testing; monitoring tools to assure that the field recorded load data are being reproduced accurately and to determine specimen failure; and analysis of laboratory acquired data to predict the life of the product.

Path 1 illustrated in Fig. 9-8 shows how to analyze a product design before a prototype is available by using an analytical model. This model can be used to determine a product's strain sensitive regions and resulting product life. Analyzing a model requires a large database of loads if Path 1 is to be used for product design. This path is also used for A-B comparison testing since changes in a material or design are easily evaluated for comparative design validation.

Path 2 illustrated in Fig. 9-9 shows how to field test a prototype and determine its life by measuring its actual response. Actual field testing duplicates an environment that generates the loads a product will experience.

Prototypes of the design are often subjected to field testing to both record a service history and determine actual life of the product. Although loads are accurately reproduced, lack of repeatability and ability to do accelerated testing are drawbacks to field testing. Data gathered during field testing can be used to help evaluate the product's life, especially where a load database is insufficient.

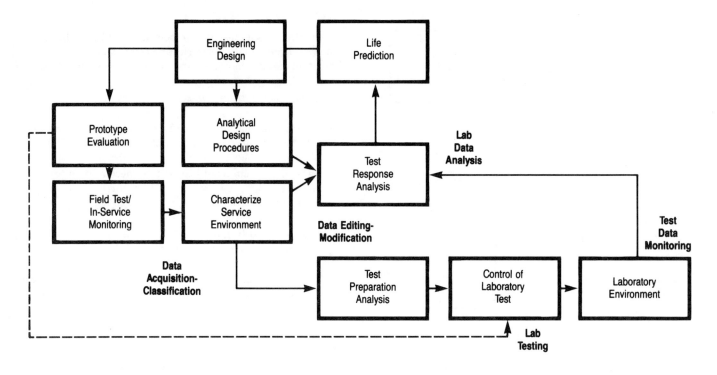

Fig. 9-7 Functional diagram—engineering design and analysis.

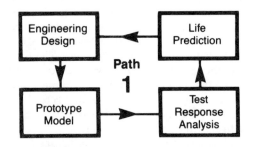

Fig. 9-8 Analysis of a product design before a prototype is available by using an analytical model.

Path 3 illustrated in Fig. 9-10 shows how to measure a typical field service history and reproduce the loads in the laboratory to determine life. This path can be used whenever a load history and a prototype are available. Laboratory testing is repeatable and can be accelerated by simulating only the high fatigue loads. Laboratory testing can be used to monitor a specimen response load history for product fatigue life prediction.

The engineering design and analysis diagram illustrates where laboratory automation tools can be used in the design process. In analyzing requirements for fatigue laboratory automation, it serves as an overview to determine the necessary elements required for successful communication of data and integration of analysis tools to the actual test.

The engineering design process requires laboratory automation tools with varying functional requirements, yet intimately linked with each other. Four levels of automation can be identified, as shown in Fig. 9-11. Software and hardware tools developed for these four levels must be designed to consider both the functions to be performed and the response time necessary to perform the functions for an integrated, well-designed system of laboratory automation.

9.3.1 Level 1: System Control

Direct control of the electro-hydraulic closed-loop requires that corrections between a command to the servoloop and a response from the controlled variable occur almost instantly. For typical fatigue tests, the response of the closed-loop must be within microseconds. Analog servocontrollers are almost universally used today because their response time meets all the requirements of closed-loop control and they are relatively inexpensive to design and manufacture.

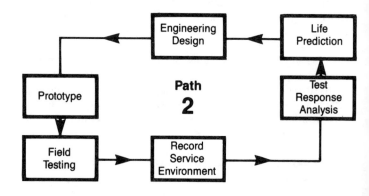

Fig. 9-9 Field testing of a prototype to determine its life and/or structural response.

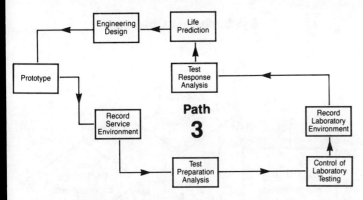

Fig. 9-10 Measurement of a field service history and reproduction of the loads in the laboratory to determine life.

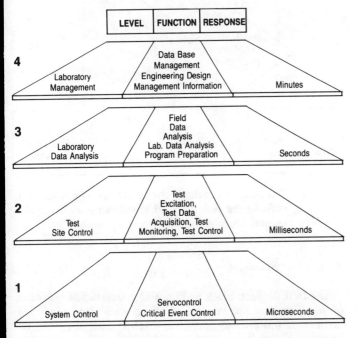

Fig. 9-11 Four levels of automation in the engineering design process.

In the very near future, microprocessors will offer greater speed and control, software algorithms will be developed to reduce the burden on the processor, and direct digital control will become feasible.

9.3.2 Level 2: Test Site Control

Controlling a fatigue test, test setup, programming, data acquisition and test monitoring, are functions traditionally performed by a digital computer. Test site control functions require computer response in milliseconds or "real time" for fatigue test system programming and data acquisition needs. Typical functions performed at this level are:

- Program simple or complex waveforms into single test stations or multiple independent test stations.

- Program field recorded service time histories into single test stations or multiple independent test stations.

- Acquire data from monitored test station transducers.

- Monitor single test stations to compare critical time history responses or compare spectral density plots between responses.

- Monitor multiple test stations to take action on critical events.

9.3.3 Level 3: Data Analysis

Data analysis capability is vital to successfully, repeatably and quickly perform several laboratory test functions including:

- Preparing data for fatigue test programs.

- Classifying field or laboratory data for use in life prediction programs.

- Monitoring response from fatigue tests.

- Measuring system responses.

These analysis functions share a common characteristic that is important to laboratory automation. That is, analysis is not dependent on real time. It can be accomplished when computer time is available.

Many versatile software programs are commercially available for complete laboratory multi-task/multi-channel data analysis, including:

- Time domain analysis

- Frequency domain analysis

- Life prediction

- Data classification

- Data editing

These types of analysis programs are often run in an engineering workstation environment.

9.3.4 Level 4: Laboratory Management

Computer-based automated systems frequently are required to provide an interface with more extensive, in-house fatigue design data management systems.

Data acquired by a test control system and/or analyzed by a laboratory analysis program, are often transmitted to a shared resource computer system for engineering design from a common data base.

9.4 Vehicle Durability Testing

Up to this point ideas have been expressed concerning the interrelationships of product design and testing. The remainder of this chapter focuses on durability testing by service history simulation in the laboratory. The general benefits of laboratory tests over field tests are covered in available literature [2,11] and will not be enhanced in this discussion.

The figures in the following sections are offered as possible decision path guidelines for use in organizing a service history simulation test program. The decision flow paths are not meant to be independent, but rather to be used in a highly interactive and iterative fashion as the design of the test program progresses.

9.4.1 Test Objective

The importance of a clear test objective in terms of expected results from the test cannot be over-emphasized. The process of conducting a durability evaluation test consists of subjecting the critical parts of a structure to a realistic fatigue environment. The extent to which the environment is known influences the level of result that can be obtained from the test. This is illustrated by the decision path shown in Fig. 9-12.

If the expected test result is an estimate of which parts are likely to fail, and what their failure mode is likely to be, then less environmental information is necessary. The most extensive test result would include a quantitative measure of service life. This is a much more stringent requirement in that a quantitative description of the total expected (average) in-service environment is necessary. The in-service environment is likely to be a highly random process. Significant effort has been applied to this problem [12].

A practical solution to this dilemma that has evolved in the ground vehicle industry has been the establishment of "standard" test roads. Correlation to acceptable service life can be established through extensive experience (several model years) in many cases. Often such "standard" roads are first established from sections of public highways, and then reproduced on a proving ground [13].

In the case where "standard" roads are established for field durability evaluation purposes, the fatigue environment for a laboratory durability evaluation can be established from this "road." The laboratory test will then have the same tie to service life as the corresponding field test.

When a considerable experience base of previous durability test results is available, A vs. B comparison testing can be pursued. This allows qualitative judgements to be made on the in-service suitability of the article under test.

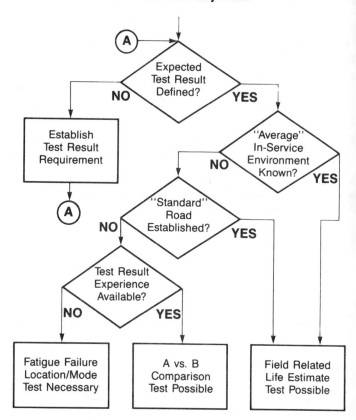

Fig. 9-12 The extent to which the environment is known influences the level of result that can be obtained from a test.

9.4.2 Test Specimen

The first decision block in Fig. 9-13 concerns the question of whether it is known apriori which structural parts are critical. If the critical parts are known, then they can be tested individually in a component or substructure test. If a life estimate is possible, and the critical part in the vehicle has been determined, then a sample from this area can be a candidate for specimen testing [14,16]. This would be the case when the fatigue damaging conditions at the critical location are well understood. If this is the case, material coupon samples subjected to service histories could satisfy test objectives.

If the critical parts are not known, as many of the parts as possible should be included in a "complete" structural test. An example would be a test system for durability evaluation of a vehicle structure while subjected to roadway inputs similar to the one shown in Fig. 9-2.

If the validity of the fatigue environment is doubtful for any part, it should become a candidate to be instrumented and monitored during the test. A separate test would be in order for any parts that are not being subjected to a proper fatigue environment.

9.4.3 Environmental Input Exciters

Once the specimen is isolated, the next task is to define the environmetal exciter input points. If the specimen is a substructure, some of its inputs would be internal forces with respect to the total structure. It is paramount that the exciters be able to load the specimen in a manner that is as close as possible to the loading it receives in the actual environment.

Shown in Fig. 9-14 is a decision path guide for specimen restraint considerations, which are closely aligned with exciter input definition, in that restraints apply loads to the specimen as well. In a vertical road simulator the restraint system should keep the vehicle body located while applying a minimum of lateral force. The body is usually considered to be a critical part in such a test.

If it is known that the vehicle does not receive significant inputs through the suspension in the lateral direction relative to the vertical direction, then suspension parts should be seeing a realistic fatigue environment also. This would tend to be true for large trucks which do not see high lateral accelerations, especially when fully loaded. It would not be true for automobiles that might see high lateral acceleration on rough roads.

In the case of automobiles, horizontal inputs are included when fatigue testing the suspension. This leads to a restraint tradeoff in that restraining the body will reduce the validity of the body test, while improving the validity of a suspension test.

A possible alternative is to perform a separate test for each restraint condition. With the body restrained, it becomes part of the fixture for testing the suspension and its mounting points. In this case, very low frequency or mean loads during cornering can be imposed on the suspension simultaneously with higher frequency loads. The body (new specimen) can be tested using the suspension as part of the fixture. Since fatigue of the body is mainly due to bending and twisting or racking, it is largely insensitive to lateral low frequency or mean loads. It is generally true that as the number of critical parts included in a test increases, the difficulty of achieving the proper fatigue environment on any given part also increases.

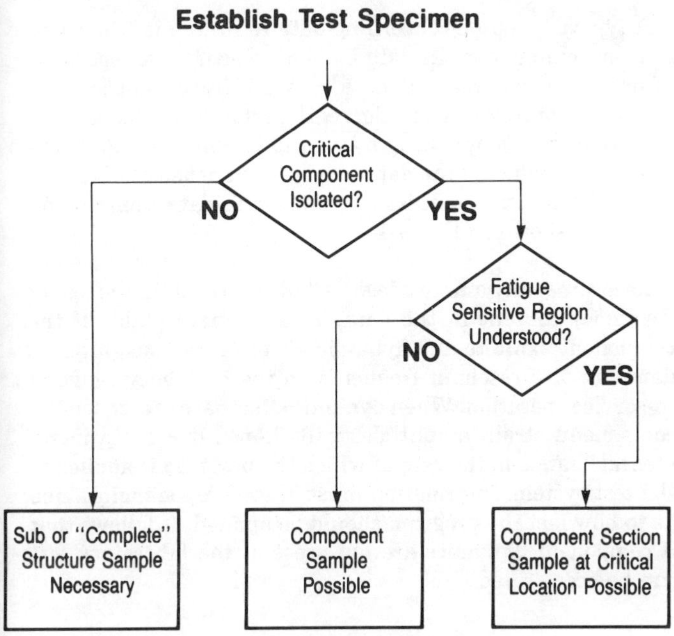

The mode of servo-control of the exciters should be chosen to be consistent with the specimen characteristics and the restraint system. The servo-control system should maintain the fatigue environment during the durability test with minimum sensitivity to such things as calibration drift or specimen wear. In general, internal loads should be controlled with load or strain control and specimen motions (inertial loads) should be controlled with stroke control.

9.4.4 Data Recording Locations

Once environmental input points are determined, a series of tradeoffs must be made concerning data recording (including transducer mounting), test system programming, and

Fig. 9-13 Establish test specimen.

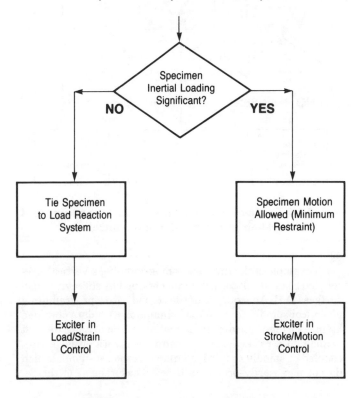

Fig. 9-14 Decision path guide for specimen restraint considerations.

test system control as shown in Fig. 9-15. Recorded data must contain sufficient information to define the test inputs to simulate the fatigue environment on all critical parts included in the test. Data which will do this can be obtained in several ways:

1. Knowledge of the environment.

2. Direct measurement of the required inputs while the specimen is being subjected to actual environments.

3. Measurement of specimen response signals while the specimen is being subjected to actual environments.

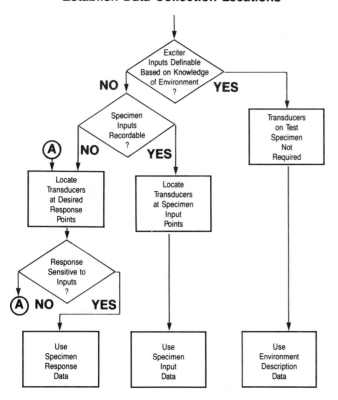

Fig. 9-15 Trade-offs concerning data recording, text system programming and control.

An example of the first case—programming a vertical tire-coupled road simulator with data obtained by surveying road surfaces [17]. An example of the second case—programming a vertical spindle-coupled road simulator with data obtained by measuring vertical spindle motion, while driving the test vehicle on a roadway. In the third case, with either a tire-coupled or spindle-coupled simulator, response signals which do not necessarily provide a *direct* measurement of the required program inputs can be recorded. For example, strains in the shock towers are suitable.

The first case is the most desirable, in that the data will be specimen independent. However, this approach is often very difficult or impractical. For example, it has not been accomplished for vehicle horizontal inputs to the authors' knowledge. The third case is easiest from the data recording point of view, but necessitates a greater test control capability. Analog servocontrol could suffice in case 2, while digital compensation control is normally required in case 3 in order to establish the inputs from implicit information provided by response signals [2-4].

It is important to remember that it is the *specimen inputs* that are to be simulated. It is not sufficient to reproduce the recorded response signals to ensure a good test. The response signals must implicitly define the required inputs in an accurate fashion. This may be evaluated by qualifying the response signal sensitivity to exciter input perturbations. Erroneous results will be obtained if a recorded response signal contains significant content due to engine induced vibration. In this case, the spindle exciters would generate inputs to reproduce this content in the response signal, resulting in incorrect spindle loads.

It is clear that the response measurements should be sensitive only to the simulated inputs. Error should be avoided by not attempting to simulate events that include non-simulated inputs. For example, avoid high lateral acceleration and high fore-aft acceleration (braking and traction) events on a vertical tire-coupled simulator. More complex conditions such as proper independence of response measurements are presented in reference [4].

9.4.5 *Data Collection Techniques*

A typical question concerning data recording is "How, and in what manner can the data be compressed?" It is especially desirable to compress the data recorded by an on-board in-service recording device. Currently established techniques involve either compressing by recording amplitude distributions or by reducing the data to a spectral density characterization. A decision process for evaluating data compression choices is shown in Fig. 9-16.

Recording only amplitude distributions result in the loss of frequency content in the data. This is unacceptable if the component exhibits any dynamic effects such as significant damping or a resonant frequency in the frequency range of in-service operation. When dynamic effects are present in the component, strain amplitudes will change due to an incremental change in the rate at which the program is applied to the test system. Information must, therefore, be maintained as to how fast the program should be applied. It follows that if component dynamics are not present, the laboratory test can be accelerated.

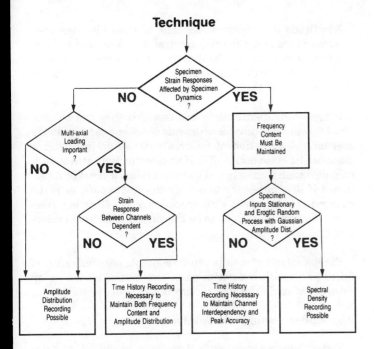

Fig. 9-16 Decision process for evaluating data compression choices.

Amplitude distributions must always be known in fatigue work since they directly affect fatigue damage. These data can be characterized by spectral densities for test control purposes when the amplitude distribution of the specimen inputs is *known* to be Gaussian. This would not be the case for spindle motions on a rough road. Work is in progress to relieve the Gaussian requirement [18], but the techniques are not yet fully established, especially in the multiple-channel case.

A further requirement for spectral density characterization is that the recorded event must be obtained from a stationary and ergodic random process [19]. Such a situation occurs when driving at a constant velocity on a fixed type of roadway. A change in vehicle velocity or a change in the nature of the roadway necessitates a new spectral density description, i.e., several descriptions for different roads at different velocities can be formed. About ten minutes of data is required for each condition.

All events which do not meet the requirements must be recorded as time histories. All transient events such as railroad crossings, sporadic chuckholes, etc., must be recorded as time histories.

When the specimen does not exhibit dynamic effects, as shown at the top of Fig. 9-16, it is possible to characterize the environment from amplitude distributions for a single channel system. If it is necessary to maintain an interrelationship between channels in a multiple channel system, the complete time history must be recorded.

9.4.6 *Test System Programming and Control*

The test system must be programmed with time histories. Thus, any data compression technique used to characterize data recorded for test control purposes must allow reconstruction of time histories. Time history reconstruction from spectral densities with a Gaussian amplitude distribution has been well established for a long period of time. Recording and subsequent reconstruction from amplitude information for a single channel signal, such that quality is maintained is a contemporary topic with some significant recent advances [20,21]. Multiple channel signals with unknown interrelationships have not yet been handled.

It is desirable that the reconstruction technique be implemented in an on-line manner during the durability test. This eliminates the need for repetitively programming a stored time history.

Referring to Fig. 9-17, the first question is whether or not the specimen exhibits significant dynamics. If not, the test can be accelerated compared to real time. If the exciter inputs are to be established from recorded response signals that are remote from the inputs, then compensation control of the test setup is required. Handling response signals that are remote from the inputs implies significant linear or non-linear mechanical impedance or cross-coupling. These conditions make direct servo-control of measured responses difficult or impossible. This is a common situation when testing a complex specimen such as a vehicle.

Either time history or spectral density representations of responses can be controlled directly [2,4]. The case of responses measured in terms of amplitude distributions presents a difficulty. Techniques have not been established for controlling response amplitude distributions by adjusting input amplitude distributions. A time history response can be generated from the recorded amplitude response distribution and then time history control can be applied.

If the data are recorded at the specimen inputs, then direct playback is possible if the excitation system dynamics are insignificant. Otherwise, compensation for exciter dynamics is in order. When the data are characterized from amplitude distributions, a technique such as null pacing can be used to compensate for exciter dynamics. Null pacing requires that the servoloop be controlled on the recorded signal.

9.4.7 *Durability Test Monitoring and Analysis*

Monitoring and analysis tie directly to the desired test result. Monitoring implies on-line observation of specimen response to test conditions. It can be as simple as maintaining elapsed time or the number of passes through a sequence.

If the objective is to subject certain parts to a realistic fatigue environment, then this should be checked by monitoring. As mentioned before, this is especially true if there is a question concerning a particular critical part. A typical analysis output of this type of monitoring is shown in Fig. 9-18.

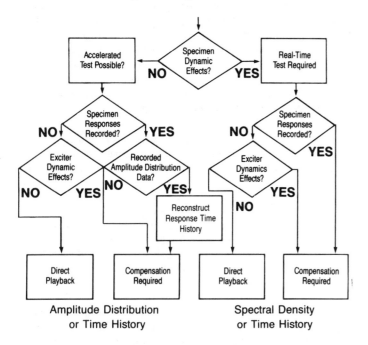

Fig. 9-17 Decision path for establishing programming and control method.

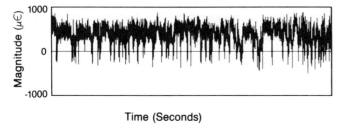

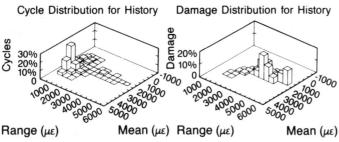

For this strain-time history, note the difference between the cycles that constitute the majority of the signal and the cycles that constitute the majority of the damage.

Fig. 9-18 Typical analysis output from on-line observation of a component.

Amplitude distributions, sections of time histories from significant transient events, spectral densities, and so forth can be monitored. It can be convenient to compare responses during the test to the first or early responses as a measure of time variance of test conditions.

If amplitude distributions are recorded they can be analyzed to contain estimates of the rate of fatigue damage on the specimen. Correlation of failures with an on-line fatigue clock can be quite useful [22]. If monitoring indicates significant deviation from original test conditions, it must be determined if it is necessary to modify the test inputs, or if the changes are consistent with specimen aging. This is a complex area that is destined to be the subject of future development.

Final analysis of results can be a simple pass-fail, a list of failed parts with a description of the failure modes, or an estimate of useful fatigue service life in hours or miles.

9.5 Modal Testing and Analysis

Today there is a wide range of products for which structural vibration problems present a major design limitation. For structures such as turbine blades and suspension bridges the structural integrity is of great concern. Optimum designs of these structures require a thorough knowledge of their dynamic characteristics. There is also another class of structures for which vibration is directly related to performance, either by reason of causing temporary malfunction during excessive motion, or by creating disturbance or discomfort, including that due to noise. Airplanes, automobiles, trucks and even computers fall in the latter class. It is essential that the vibration levels encountered in service or during operation be anticipated and controlled. Therefore, experimental vibration measurements are normally done to:

1. determine the nature and extent of vibration response levels, or to

2. verify theoretical models and predictions.

The two vibration measurements indicated above require two different types of tests. In the first case, excitation forces are simply measured under actual operating conditions. In the second case, the test must also be conducted under closely controlled conditions to yield additional detailed information. This second type of test, that includes both time domain data acquisition and its transformation into the frequency domain, is called "modal testing". It involves testing of components or structures with the objective of obtaining a mathematical description of their vibration or dynamic behavior. This description is obtained in the form of a frequency response function (FRF). This description is then used to obtain the component's or structure's natural frequencies, the mode shapes, and the modal mass, damping and stiffness properties. This process of modal testing and analysis is shown in Fig. 9-19.

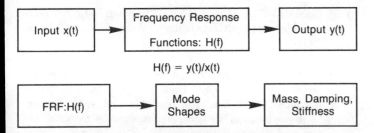

Fig. 9-19 Modal testing and analysis.

It can be seen from this figure that the mathematical description of FRF is output/input, or it is a measure of output for unit input. For example, if the output is strain and the input is force in pounds, the frequency response function will be a measure of strain per pound. FRF, in general, provides information on both the absolute magnitude of the response quantity and the phase lag between the response and the excitation. Knowledge of the FRF is a prerequisite to perform a system response simulation for any defined or arbitrary excitation, whether it is performed theoretically on a computer or experimentally in a laboratory. Modal testing and analysis has also contributed to the success of remote parameter control (RPC) testing [23].

A few of the many applications of modal testing are listed below.

1. Natural frequencies and mode shapes obtained from a modal test can be compared with the corresponding data produced by a finite element model. This is usually done to validate the theoretical model before it is used for predicting a structure's response magnitudes to complex excitations. The damping calculated for each mode shape from the modal test data can be incorporated into the theoretical model before it is used to predict response to any specific excitation.

2. A component mathematical model can be created using the modal properties obtained from the modal test and then this model can be incorporated into a total structure model. This is similar to "substructuring" as used in assembly of a finite element model of a complex structure. In this situation it is important that all the component vibration modes in the vicinity of the frequency range of interest be included, since it is quite possible that out-of-range frequency modes may affect the structure's dynamic response in the frequency range of interest.

3. Modal models developed from experimental data may be used to predict the effects of modifications to the original tested structure. These modifications can be relatively minor such as adding a spring, damper or a stiffener for fine-tuning the structure's dynamic response, or they can include moderate design changes made to improve performance.

4. In certain situations it may be required to have knowledge of the dynamic forces responsible for a given type of structure's behavior but it may be impractical or even impossible to directly measure these forces. For such situations, the mathematical description of the FRF can be combined with the measured response in order to deduce the dynamic forces.

In summary, modal testing can be used to

1. help formulate a test program based on RPC, or to

2. validate a finite element model which can be used later for studying a structure's response to any type of excitation, or to

3. evaluate the effects of any design modifications on that structure's dynamic response.

9.6 Summary

A model of how a test facility should interface to the engineering process has been developed. This is important in establishing proper test objectives at the beginning of a program with a clear definition of expected test results.

Full scale service simulation testing to establish vehicle (structural) durability has become very sophisticated, with a number of complex tradeoffs required to establish a test program. Tradeoffs involve such questions as:

"Should the complete structure be tested at one time or should it be separated into components or substructures?"

"Can the test be accelerated?"

"Can the recorded data from the environment be compressed?"

Questions of this type do not have simple answers. Rather, a logical decision path [24] has been proposed as a guide in the design of service history simulation test programs.

9.7 References

[1] Wetzel, R. M. and Donaldson, K. H., Jr., "Experimental and Analytical Considerations in Service Life Estimation." Society of Environmental Engineers, February 1977.

[2] Cryer, B. W., Nawrocki, P. E., and Lund, R. A., "A Road Simulation System for Heavy Duty Vehicles," SAE Technical Paper 760361, SAE, 1976.

[3] Jacoby, G., "Load Simulation of Wheel Axle and Complete Vehicle Load," Symposium on Experimental Simulation Tests of Motor Vehicles and Their Elements," October 1979.

[4] Lund, R. A., "Multiple Channel Environmental Simulation Techniques," Symposium on Experimental Simulation Tests of Motor Vehicles and Their Elements," October 1979.

[5] Landgraf, R. W. and Conle, A., "The Development and Use of Mechanical Properties in Industry," MPC-14, ASME, November 1980.

[6] Peterson, E. L., "Integrating Mechanical Testing into the Design Development Process," SAE Technical Paper 791077, SAE, 1979.

[7] Baker, M., "Unattached Field Measurement Instrumentation," SAE Technical Paper 740940, SAE, 1974.

[8] Downing, S., Galliart, D., and Berenyi, T., "A Neuber's Rule Fatigue Analysis Procedure for Use with a Mobile Computer," SAE Technical Paper 760317, SAE, 1976.

[9] Socie, D., Shifflet, G., and Berns, H., "A Field Recording System with Applications to Fatigue Analysis," Symposium on Service Fatigue Loads Monitoring, Simulation and Analysis," ASTM, 1977.

[10] Flis, T. J. and Wallace, B., "A Micro-computer Based Data Acquisition System for Versatile Mobile Data Processing," SAE Technical Paper 780151, SAE, 1978.

[11] Manduzzi, D. J. and Reid, K. E., "Ford's New Programmable Front Wheel Drive Half Shaft Test System," *Closed Loop*, MTS Systems Corp., June 1980.

[12] Buxbaum, O. and Zaschel, J. M., "Separation of Stress-Time Histories According to Their Origin," *Konstruktion*, Vol. 31, No. 9, 1979, pp. 345-351.

[13] Smith, K. V. and Stornant, R. F., "Cumulative Damage Approach to Durability Route Design," SAE Technical Paper 791033, SAE, 1979.

[14] Wetzel, R. M., "Smooth Specimen Simulation of Fatigue Behavior of Notches," *Journal of Materials*, JMLSA, Vol. 3, No. 3, September 1968, pp. 646-647.

[15] Dowling, N. E. and Wilson, W. K., "Geometry and Size Requirements for Fatigue Life Similitude Among Notched Members," Westinghouse R&D Center Scientific Paper 80-1D3-PALFA-P2, 1980.

[16] Seeger, T. and Heuler, P., "Generalized Application of Neuber's Rule," *Journal of Testing and Evaluation*, JTEVA, Vol. 8, No. 4, July 1980, pp. 199-204.

[17] Dodds, C. J., "The Laboratory Simulation of Vehicle Service Stress," ASME Paper 73-DET-24, ASME, 1973.

[18] Bily, M. and Bukoveczky, J., "Digital Simulation of Environmental Processes with Respect to Fatigue," *Journal of Sound and Vibration*, Vol. 49, No. 4, 1976, pp. 551-578.

[19] Bendat, J. S. and Piersol, A. G., "Random Data: Analysis and Measurement Procedures," John Wiley and Sons, 1971.

[20] Conle, A., "An Examination of Variable Amplitude Histories in Fatigue," Ph.D Thesis, University of Waterloo (Canada), 1979. See also: A. Conle and T. H. Topper, "Fatigue Service Histories: Techniques for Data Collection and History Reconstruction," SAE Technical Paper 820093, SAE, 1982.

[21] Thakkar, R. B. and McConnell, K. G., "Sinusoidal Simulation of Fatigue Under Random Loading," SAE Technical Paper 740217, SAE, 1974.

[22] Donaldson, K. H., Jr., "A Method for Real-Time Damage Assessment," M.S. Thesis, University of Illinois, 1978.

[23] Grote, P., and Greiner, G., "Taking the Test Track to the Lab," *Automotive Engineering*, Volume 95, No. 6, SAE, June 1987, pp. 61-61.

[24] Lund, R. A. and Donaldson, K. H., Jr., "Approaches to Vehicle Dynamics and Durability Testing," SAE Technical Paper 820092, SAE, 1982.

[25] Jaeckel, H. R., "Design Validation Testing," SAE Paper No. 820690, see also Proceedings of the SAE Fatigue Conference, P-109 (April 1982), p. 153.

Addresses of Publishers:

ASME, American Society of Mechanical Engineers, 345 E. 47th Street, New York, NY 10017

ASTM, American Society for Testing & Materials, 1916 Race Street, Philadelphia, PA 19103

MTS Systems Corp., Box 24012, Minneapolis, MN 55424

SAE, Society of Automotive Engineers, 400 Commonwealth Drive, Warrendale, PA 15096

Society of Environmental Engineers (Note: Merged with the Institute of Environmental Engineers to form the Institute of Environmental Sciences), 940 E. Northwest Highway, Mt. Prospect, IL 60056

University of Illinois, 249 Armory Boulevard, 505 E. Armory Street, Champaign, IL 61820

University of Waterloo, University Avenue W., Waterloo, Ontario N2L 3B8, Canada

Westinghouse R&D Center, 1310 Beulah Road, Pittsburgh, PA 15235

John Wiley & Sons, 605 Third Avenue, New York, NY 10158

Chapter 10 Fatigue Life Prediction

10.1 Introduction
    10.1.1 Scope
    10.1.2 Relationship to Other Chapters
    10.1.3 Chapter Plan

10.2 Background Considerations in Life Prediction
    10.2.1 Background
    10.2.2 Crack Initiation and Crack Propagation Approaches to Life Prediction

10.3 Crack Initiation Approach
    10.3.1 Overview
    10.3.2 Notch Stress Analysis
    10.3.3 Damage Analysis
    10.3.4 Methods of Crack Initiation Life Prediction
        10.3.4.1 Load-Life Method, Constant Amplitude Loading
        10.3.4.2 Load-Life Method, Variable Amplitude Loading
        10.3.4.3 Stress-Life Method, Variable Amplitude Cycling
        10.3.4.4 Strain-Life Method, Variable Amplitude Cycling
    10.3.5 Selection of a Crack Initiation Life Prediction Method

10.4 Crack Propagation Approach
    10.4.1 Stress Intensity Factors
    10.4.2 Fracture Toughness
    10.4.3 Critical Crack Size
    10.4.4 Fatigue Crack Growth
    10.4.5 Examples of Crack Propagation Life Prediction for Constant Amplitude Loading
    10.4.6 Crack Propagation Life Prediction for Variable Amplitude Loading
    10.4.7 Summary

10.5 Practical Aspects of Life Prediction
    10.5.1 Crack Initiation vs. Crack Propagation Analysis/Small Cracks
    10.5.2 Multiaxial Effects
        10.5.2.1 Multiaxial Loading - Initiation Effects
        10.5.2.2 Multiaxial Conditions - Initiation Effects
        10.5.2.3 Multiaxial Conditions - Propagation Effects
    10.5.3 Environmental Effects
    10.5.4 Fabrication Effects
    10.5.5 Processing Effects
    10.5.6 Load Sequence and Crack Closure Effects
    10.5.7 Residual Stress Effects
    10.5.8 Cracks Growing from Notches
    10.5.9 Cracks at Welds

10.6 Applications of Life Prediction in Design Analysis
    10.6.1 Fatigue Assessment of Cast Axle Housings
    10.6.2 Fatigue Assessment of a Piston Rod in a Forg-
    10.6.3 Analysis of Crack Growth in a 1025 HR Welded Frame Structure
        10.6.3.1 Background
        10.6.3.2 Test Description
        10.6.3.3 Test Results
        10.6.3.4 Crack Growth Prediction Methodology
        10.6.3.5 Prediction
        10.6.3.6 Discussion

10.7 Conclusion

10.8 References

Appendix 10A SAE Cumulative Fatigue Damage Test Program
    10A.1 Load Histories
    10A.2 Specimens and Materials
    10A.3 References

Appendix 10B Stress-Strain Simulation at a Notch
    10B.1 Introduction
    10B.2 Notch Analysis
    10B.3 References

10.1 Introduction

Fatigue life prediction techniques play an ever-expanding role in the design of components in the ground vehicle industry [1,2]. Many companies employ such predictive techniques for applications ranging from initial sizing through prototype development and product verification. Fig. 10-1 illustrates the role of life prediction in both preliminary design and in subsequent evaluation-redesign cycles—first on paper, then in component laboratory tests, and finally in field proving ground tests of assemblies or complete vehicles.

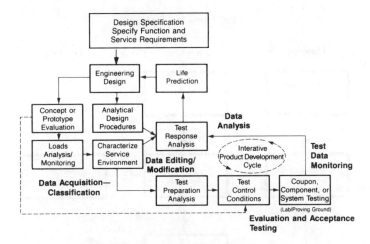

Fig. 10-1 Functional diagram—engineering design and analysis [3].

Life prediction is useful in either the analytical or experimental routes to design for durability as illustrated in Fig. 10-2.

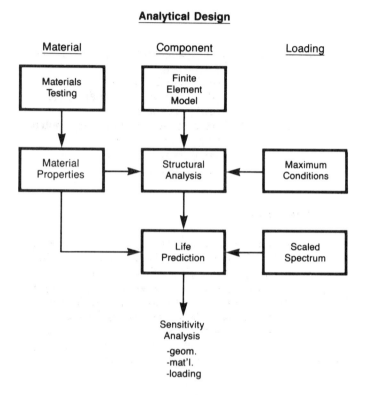

Fig. 10-2 Analytical and experimental routes to life prediction [4].

Experimental design deals with hardware and results in geometry and application-specific answers. In contrast, analytical design deals with paper studies and so can generate solutions for a wide range of geometries and applications to provide results for sensitivity studies. Combining sensitivity studies on geometry, material, and loading with the confidence gained through experience and iterative evaluation as shown in Fig. 10-3, leads to a sound foundation for successful design. Note, as illustrated in Fig. 10-3, that life prediction is the focal point for many of the activities involved in design for durability.

10.1.1 *Scope*

The purpose of this chapter is to outline the fatigue life prediction process and provide sufficient background to allow an engineer to develop an appreciation for the important considerations in fatigue life prediction. This chapter also cites references that provide additional details, should they be required. The chapter describes the commonly used life prediction techniques and assesses their potential for different applications.

This chapter begins with consideration of methods for predicting crack initiation life, i.e., the life to the formation of cracks approximately two or three millimeters in size. However, the remaining life spent in growing these cracks to a critical size is also considered. In the latter case, a critical flaw is taken as one of a size that poses a threat to structural integrity or causes a component to malfunction. Some components contain initial crack-like defects or discontinuities from the time of fabrication (e.g., welded or cast components). In such cases, a fatigue life estimate based solely on crack initiation considerations may be non-conservative. A crack propagation evaluation may be a worthwhile supplement to, or replacement for, the crack initiation prediction.

10.1.2 *Relationship to Other Chapters*

This chapter on life prediction integrates many of the issues discussed in other chapters, as shown in Fig. 10-3. This does not imply that considerations such as strain measurement are not important by themselves. Rather, it is to emphasize that activities like strain measurement and material properties development have a common end purpose—fatigue life prediction directed at ensuring safety and durability. In this respect, this chapter is the focal point for other chapters that provide inputs to the life prediction process. Chapter 5, Service History Determination, Chapter 8, Numerical Analysis Methods, and Chapter 7, Strain Measurement and Flaw Detection, provide input to the life prediction process, that must be application-oriented and define a specific component, its service environment, boundary conditions, and notch or crack features. Chapter 3, Material Properties, and Chapter 4, Effects of Processing on Fatigue Performance, also provide application-oriented input that define a material's behavior, subject to specific fabrication or other processing aspects.

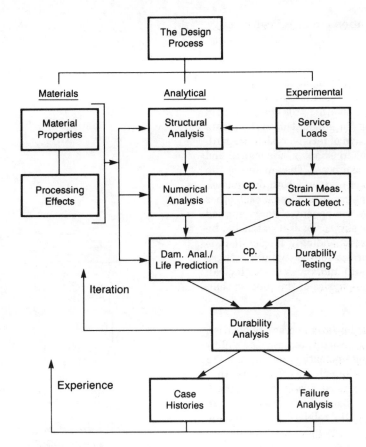

Fig. 10-3 Interplay between elements of the design process, the chapters of this handbook and the life prediction process.

In the same way that preceding chapters provide input to this chapter, this chapter compliments Chapter 11, Failure Analysis, by providing an analytical means to verify the results of a given analysis. More importantly, life prediction provides the means to assess which of several "fixes" to the cause of a failure offer the best potential prior to redesign and retesting.

10.1.3 *Chapter Plan*

The organization for this chapter is outlined in Fig. 10-4. The chapter begins with an overview of the fatigue failure process and, from this, develops approaches to life prediction. Essential features of methods to implement approaches to life prediction are presented first for crack initiation and then for crack growth (or crack propagation). Advantages and disadvantages of the various methods are discussed. Complicating factors such as mean stress, stress-strain state, processing effects, and environment are also discussed. References are made to other chapters that deal with such effects as processing in greater detail. Finally, several examples are given to illustrate the use of life prediction methods in product development.

Overview of Life Prediction
- Fatigue Process
- Approaches to Model Fatigue Process
  —Initiation
  —Propagation
  —Combinations
- Choosing the Right Approach (es)
- Inputs and Operations
- Data Requirements and Availability

Approaches to Life Prediction

Initiation Methods
- Load-Life
- Nominal Stress Life
- Local Stress-Strain
- Processing
- Multiaxiality
- Environment

Propagation Methods
- LEFM
- Sequence and Threshold
- Short Cracks
- Processing
- Environment

Combined Method
- Failure Criteria

Applications of Life Prediction in Design/Analysis
- Cast Axle Housing
- Forging Hammer Piston
- Frame Structure

Summary

Fig. 10-4 Overview of Chapter 10.

10.2 Background Considerations in Life Prediction

10.2.1 Background

In metals of interest to the ground vehicle industry, fatigue damage and crack growth typically occurs by the process of reversed-slip (or reversed-plastic strain) [5,6]. The amount of slip on a given cycle relates to the amount of cyclic strain imposed. Therefore, the correlation of fatigue resistance with strain (or plastic strain), as done in Chapter 3, is logical and consistent with the physics of the process. The amount of slip that develops under the action of a given strain depends on how easy it is for two planes of atoms to move past each other along the slip plane. Thus, the component of force applied normal to the slip plane makes slip easier if it is tensile, by separating the planes, and correspondingly harder if it is compression. For this reason, correlation of fatigue resistance with both strain (or plastic strain) and stress is also logical, and is more consistent with the physics of the process than just strain.

The slip process is responsible for fatigue crack nucleation and the micro- and macrocrack growth process. Key differences in the slip behavior during nucleation and growth are in the magnitude of slip and in the volume of material that undergoes slip. In crack nucleation, the strain level is small, but may be rather widespread compared to crack growth, which concentrates high strain in the relatively small plastic zone at the crack tip. Since slip is correlated with strain, slip during nucleation is widespread, but low level, when compared to the behavior during crack growth. Fig. 10-5 illustrates the role of slip in nucleation and macrocrack growth. In nucleation, the slip process occurs in the highly stressed locations. Once cracking begins, the deformation formerly accommodated by the slip distributed in the critical location, is taken up by deformation at, and ahead of, the crack tip and along the wake of the crack.

Life prediction involves calculating stresses and strains in highly stressed areas where slip concentrates from the input loads for a given material and geometry. The life prediction process is illustrated in Fig. 10-6. The calculated stresses and strains are transformed into fatigue damage or crack growth. Integrating damage or crack growth subject to an empirical failure criterion, through some service history, leads to a predicted life.

There are two different paths for life prediction shown in Fig. 10-6. One of these deals with prediction of crack initiation, whereas the second deals with crack growth. Each of these paths is elaborated on later in more detail. These paths may be followed as independent means to life prediction, or the results may be combined. Criteria to aid in choosing an independent path, or their combination, will also be discussed later.

Because slip occurs easiest at the surface, and is concentrated by material or geometric stress raisers, care must be taken in life prediction to account for processing and other factors that alter the surface and create stress raisers. Because slip can be effected by environment, and environment

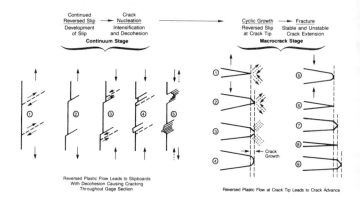

Fig. 10-5 Role of slip in fatigue initiation and crack propagation.

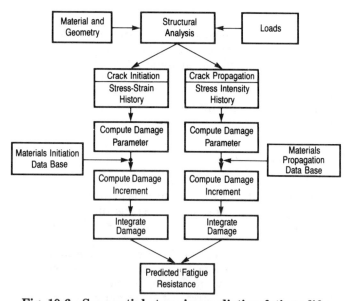

Fig. 10-6 Sequential steps in predicting fatigue life.

can cause pitting (stress raisers), etc., care also must be taken to account for environmental factors in life prediction. Finally, because slip involves plastic strain analyses, and fatigue involves inelastic action. Stress may not be linearly related to strain and the current stress and strain may depend on the prior deformation history. Consequently, tracking stress and strain for purposes of life prediction require cycle counting procedures like rainflow counting, as detailed in Chapter 5, and nonlinear analyses for stress raisers as detailed in Chapter 8.

10.2.2 Crack Initiation and Crack Propagation Approaches to Life Prediction

The fatigue process occurs everywhere in a body where the stress and strains are large enough to cause continued reversed-slip. The fatigue process is problematic at areas where stress and strain are the largest, since the rate of the process is proportional to the magnitude of slip. Thus, while fatigue may occur throughout a component, continued service is threatened by the areas that have the highest stresses and strains. These areas are localized at stress raisers (notches) and are critical to component durability and serviceability. The fatigue process involves a period of damage accumulation leading to crack initiation followed by a period of crack growth, until the critical flaw size is reached (or the residual strength is exceeded). *Total life* is thus the sum of an *initiation* and *propagation* phase.

The fact that crack initiation and propagation occur is important in life prediction, because the presence of a crack alters the stress field in a component. Once this stress field is disturbed significantly by the crack, the slip process concentrates at the crack tip. Stresses and strains located even a small distance from the crack tip no longer characterize the magnitude of slip that is now concentrated at the tip of the crack. Consequently, nominal stress and strains, or even those at a notch where the crack started, cannot be used to characterize the amount of slip at the crack tip. It follows, therefore, that once the crack is long enough to significantly disturb the stress field, the approach used to predict the life must include the crack's effect on the stress field and relate this to the rate of damage. Approaches that account for the effect of cracks on the stress field involve the use of fracture mechanics [7,8]. These approaches are used to characterize a materials resistance to fatigue in terms of crack growth behavior, as measured with precracked specimens. Examples of precracked specimens are shown in Fig. 10-7.

Until a crack long enough to disturb the stress field is formed, an approach based on stresses and strains at the site where the crack will develop can be used to simulate the fatigue process. Such an approach for life prediction is based on fatigue data developed from unnotched and uncracked specimens, such as those shown in Fig. 10-7.

Life prediction approaches that are based on the presence of a crack are termed crack growth methods, while those based on damage assessment during "initiation" (formation of a crack a few mm in size) are termed crack initiation methods.

As stated earlier, the *total life* of a component is the sum of the *initiation* and *propagation* phases. This is illustrated in Fig. 10-8, which also shows the results for the smooth specimen and cracked specimens discussed in Fig. 10-7. Life pre-

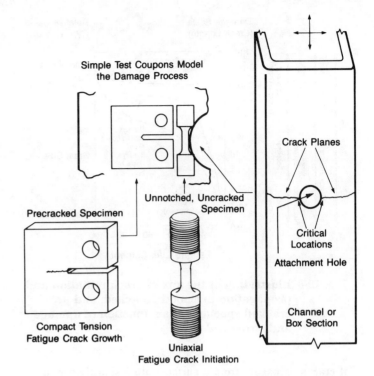

Fig. 10-7 Representation of structural behavior using uncracked and precracked specimens.

initiation life and a propagation life. Some components do exist, in which the fabrication process produces crack-like flaws. In these components, the crack-like flaw begins to grow almost immediately, so that the life is dominated by crack propagation. Life prediction, in such cases, logically involves only the crack growth approach. Many other components, that are made from high quality materials and are produced by carefully controlled fabrication processes, do not contain significant initial defects. These parts may spend most of their life initiating cracks. Life prediction, in these cases, logically involves only the crack initiation approach.

Estimates of crack propagation life can provide other useful design information. Suppose, for instance, that predicted crack initiation life at a given location meets a goal. Confidence in the structural adequacy of the design will increase if the life to propagate a crack from initiation size to a critical size is substantial. This provides an additional "design margin." As another example, suppose that the predicted initiation life just meets its goal at a location where the consequences of failure are serious. If the corresponding predicted crack growth life is very short, then further steps to minimize the chances of crack initiation should be considered. On the other hand, suppose that predicted initiation life at a less critical location nearly meets its goal but the corresponding crack propagation life puts the total life (initiation plus propagation) well past the design goal. In this case, the time and cost of re-design and/or re-testing could conceivably be saved.

It should also be noted that cracks will not always grow to critical sizes. There are some situations in which cracks form, grow a small distance and stop. This can happen, for instance,

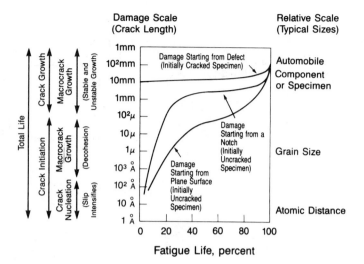

Fig. 10-8 Illustration of the mix of crack initiation and propagation in smooth, notched, and pre-cracked specimens as a function of damage and structural size.

if cracks propagate from a surface into a sharply declining gradient of stress such as that below the root of a sharp notch or if cracks grow into a compressive residual stress field of sufficient magnitude and depth below the surface. Knowledge of when this is likely to occur is not revealed by typical crack initiation evaluations.

The following sections elaborate on commonly used crack initiation and crack growth approaches. The format is the same for each approach. General features of each approach are presented first. Then, the method or methods commonly used to implement the approach are presented and discussed.

10.3 Crack Initiation Approach

10.3.1 *Overview*

The crack initiation approach involves two operations in converting the load history, component geometry, and material input into the predicted life. These operations must be performed sequentially, as shown in Fig. 10-9. First, stresses and strains at the critical site are estimated. Then, the critical location or local stresses and strains are used to compute damage, that is algebraically added up throughout the history until a critical damage sum (failure criteria) is reached. The point in the history at which the failure criteria is met is the predicted life.

Some methods implement the crack initiation approach by computing damage indirectly from remote loads or stresses, that are simply scaled to account for the notch. These methods that indirectly calculate damage using remote or "nominal" stresses and strains are sometimes called *nominal* methods. Other methods implement the crack initiation approach by calculating stresses and strains at the notch, so that damage is assessed directly in terms of local strains and

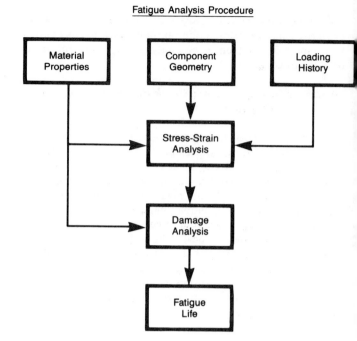

Fig. 10-9 Information path in a crack initiation life prediction.

stresses. These methods, that utilize local stresses and strains, are often called *local* methods or *critical location* methods. Smith first recognized in 1961 [9] the need to account for local inelastic action and related residual stresses. Subsequently, Morrow, et. al. [10] and Crews and Hardrath [11] further elaborated on this local stress-strain concept.

10.3.2 *Notch Stress Analysis*

All crack initiation methods use some form of notch stress analysis to complete the first operation in life prediction. The nominal or "indirect" methods scale nominal input by a notch factor, that may come from elastic stress analysis, using numerical or experimental methods. In either case, use is made of the theoretical or elastic stress concentration factor, K_t. As illustrated in Fig. 10-10, K_t is defined as:

$$K_t = \frac{\sigma_1}{S_1} \qquad (10\text{-}1)$$

In this equation, σ_1 is the maximum principal notch root (or local) stress, S_1 is the corresponding nominal stress. Values of σ_1 and S_1 also can be estimated from strain measurements. However, care must be taken to account for multiaxial effects in this calculation, and nominal strains must be defined consistent with nominal stress.

In some nominal approaches, the stress analysis operation is implicit. As shown in Fig. 10-10, load is sometimes related empirically to life by tests and actual hardware. This produces a load-life relationship for purposes of life prediction.) Both operations in life prediction—stress analysis and damage analysis—are implicit in the components behavior.

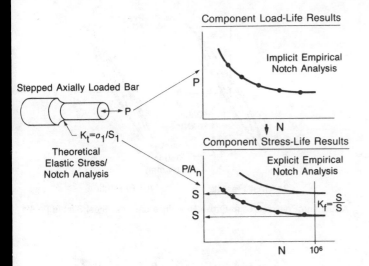

Fig. 10-10 Stress analysis in "nominal" methods of life prediction.

Alternatively, a fatigue notch factor, K_f, defined as

$$K_f = \frac{\text{stress in smooth specimen}}{\text{stress in notched specimen}} \qquad (10\text{-}2)$$

may be estimated from the load-life data transformed onto stress-life coordinates, for a particular fatigue life (usually at least 10^6 cycles). The factor K_f can then be used to scale smooth specimen results to predict life, as illustrated in Fig. 10-10.

The local method also employs notch stress analysis. However, there are some significant differences, most notably that inelastic action at the notch is directly accounted for. Inelastic notch effects are accounted for in several ways.

One way is to use Neuber's rule [12] as a "universal" notch analysis. Neuber's rule is

$$K_t = (K_\varepsilon \cdot K_\sigma)^{1/2} \qquad (10\text{-}3)$$

where $K_\varepsilon = \varepsilon_1/e_N$ and $K_\sigma = \sigma_1/S_1$, with ε_1 = maximum local principal strain* and e_N = the corresponding nominal strain, defined consistent with S*. Neuber's rule is often rewritten for fatigue, following Morrow, et. al. [10], as

$$K_f = (K_\varepsilon \cdot K_\sigma)^{1/2} \qquad (10\text{-}4)$$

where K_f is as defined in Eq. 2. Eq. 4 can be rewritten as

$$K_f^2 = \frac{\Delta\sigma}{\Delta S} \cdot \frac{\Delta\varepsilon}{\Delta e_N} \qquad (10\text{-}4\text{a})$$

or, *assuming* nominally elastic conditions,

$$\frac{\Delta S^2 K_f^2}{4E} = \frac{\Delta\sigma}{2} \cdot \frac{\Delta\varepsilon}{2} \qquad (10\text{-}5)$$

where E is the modulus of elasticity. Eq. 5 relates the local stress and strain to the imposed stress range, ΔS, a material constant, E, and an empirical factor, K_f, or more rigorously, the theoretical value, K_t. Eq. 5 is one equation with two unknowns, $\Delta\sigma$ and $\Delta\varepsilon$. The second equation is the relationship between stress and strain, given in Chapter 3, with the solution obtained as illustrated in Fig. 10-11a.

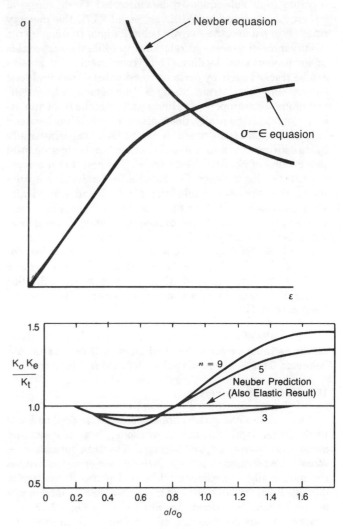

Fig. 10-11 Neuber's rule applied in a "universal" notch analysis.

The "universal notch analysis" of Eq. 5 is an approximation, and it may not work well for materials and geometries that are different from those used in its original derivation. For example, for a plate with a central hole, the analyses of Huang [13], reproduced in Fig. 10-11b, shows that Eq. 5 is systematically in error, depending on the nominal stress and the material. For some materials, such as aluminum alloys, the error is quite small, while for materials like mild steel, the error can be quite large [14]. Fortunately, the convenience of Neuber can be retained, and this error accounted for by using results of correct solutions, such as shown in Fig. 10-11, and developing a correction factor to account for this plasticity-related error [15].

Another way to account for inelastic notch behavior makes use of a nominal load or nominal stress-local strain response. As postulated by Lind and Mroz [16] following Ilyushin's theorum, and subsequently demonstrated for a range of structural elements by Williams, et. al. [17], the memory rules observed for stress-strain behavior hold in components for any force-displacement relationship while the material is under proportional loading. Thus, component local strains can be tracked cycle by cycle on coordinates of nominal load (stress) and local strain, using a "component calibration" technique. The component calibration technique is an analytically or experimentally determined relationship between load (stress) and local strain. It can be found experimentally by measuring strain as a function of load under incremental step cycling of nominal load (stress). The result of a component calibration is shown, for example, in Fig. 10-12. Alternatively, the component calibration can be found analytically via FEM results (Chapter 8), using the stable cyclic stress-strain curve. It is significant to note that nominal stress-local strain behavior seems to follow a "master curve" for a given material and notch type, over a range of notch severities. For example, Fig. 10-12 presents results for aluminum, for a range of circular notch geometries for $2 \leq K_f \leq 6$ [18]. Similar results have been found for steels, as shown by the work of Seeger, et al. [19].

The results of a nominal load (stress)-local strain analysis serves as input to calculate local stresses from local strain behavior. This is accomplished as detailed in Chapter 3.

10.3.3 Damage Analysis

Once stresses and strains are estimated from applied loads using either local or nominal methods, these stresses and strains are converted to "damage." The transformation of stresses and strains to damage is the second operation that is common to all life prediction schemes. The transformation to damage involves two steps—damage assessment and damage accumulation. These steps are illustrated in Fig. 10-13. The process of damage assessment involves determining the amount of damage done on a given reversal or cycle of loading. This is accomplished using some reference damage (fatigue resistance) curve. As illustrated, the reference curve plots a damage parameter on the ordinate and life (cycles or reversals or some normalized life) on the abscissa. The value of the damage parameter is calculated using the stress and strain for that event obtained from the notch analysis. As shown in the figure, the damage is found by entering the reference curve at the value of the damage parameter for that reversal or cycle. The damage done is computed as the inverse of the corresponding number of cycles (reversals) to failure. Some methods use rather complicated measures of damage, while others use history dependent resistance curves [20,21].

Damage is assessed in terms of some parameter that provides an equivalence condition between the test data, that define the resistance curve, and the practical problem. In

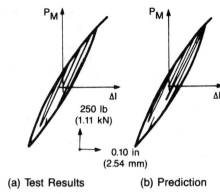

(a) Test Results (b) Prediction

Actual and Predicted Load-Stroke Response of Boron Steel Simple Frame

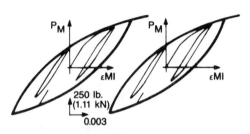

(a) Test Results (b) Prediction

Actual and Predicted Load-Surface Strain Response of Boron Steel Frame (Strain Measured Beside Central Loading Points).

a) **Cycle by Cycle Analysis of Force-Displacement Response and Force-Local Strain Response.**

Fig. 10-12 Methods of notch analysis. (A)

"nominal" methods, this parameter may be the load (stress) used in the test. In local methods, this parameter usually includes mean stress effects and may include multiaxial effects. Common mean stress parameters include [22-24]:

$$\frac{\Delta S}{2} \cdot \frac{S_f}{\sigma'_f - S_m} \qquad (10\text{-}6)$$

$$S_{mx} \Delta e/2 \qquad (10\text{-}7)$$

$$\frac{\Delta S}{2} + \frac{\Delta S + S_m}{2} \qquad (10\text{-}8)$$

where S_f is the fully reversed fatigue limit at some life (usually at least 10^6 cycles)
S_m is the mean stress
σ'_f is the fatigue strength coefficient
S_{mx} = max stress.

A recent review by Laflen [25] of mean stress parameters concluded the Smith, Watson, Topper parameter does not deal with compression effects well, and noted that an energy-based form [26]

$$S_m \Delta \varepsilon + \Delta S \Delta e \qquad (10\text{-}9)$$

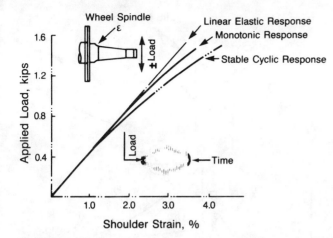

b) Example Of Experimental Component Calibration For Use In (a).

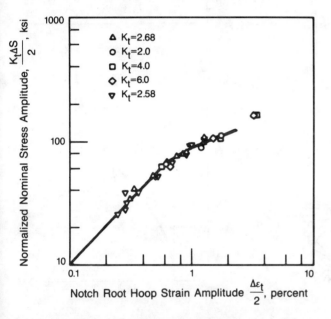

c) Master Curve For Circular Notch Geometries For $2 \leq k_t \leq b$: 2024-T351 And 24ST3 Aluminum Alloy.

Fig. 10-12 Methods of notch analysis. (B, C)

where $\Delta\epsilon$ is the local strain range and Δe is the nominal strain range, was the best predictive parameter of those examined (for the cases considered). Fig. 10-14 shows predicted trends for several mean stress parameters.

Damage accumulation is the second step in damage analysis. Criteria for damage accumulation vary, and a number have been proposed [17]. Of these, the linear criteria, often referred to as Miner's rule, remains the most popular. It has been shown to be reasonably effective when coupled with history dependent resistance curves (Fig. 10-13), and the failure criterion, $\Sigma D = 1$ (i.e. total accumulated damage equals unity). It has also been shown to be effective when the usual $\Sigma D = 1$ criterion is replaced by an empirically determined criteria for the material, geometry, and history of interest. This latter process—termed relative Miner—is popular in

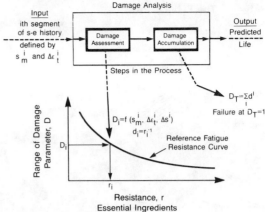

(a) Schematic of the Damage Analysis Procedure

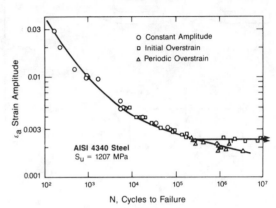

(b) Effect of Periodic Overstrain on Strain-Life Behavior of Aircraft Quality AISI 4340 Steel: Different Histories Give Rise to Different Resistance Curves.

Fig. 10-13 Stresses and strains at critical locations are transformed into damage increments and integrated subject to some failure criterion in making life predictions.

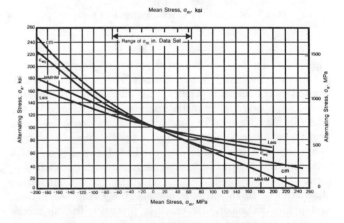

Fig. 10-14 Comparison of predicted trends for several damage parameters. Ref. [25]

Germany and some other European countries [28].

Various combinations of notch stress analysis and damage analysis exist. Each combination represents "a method" for

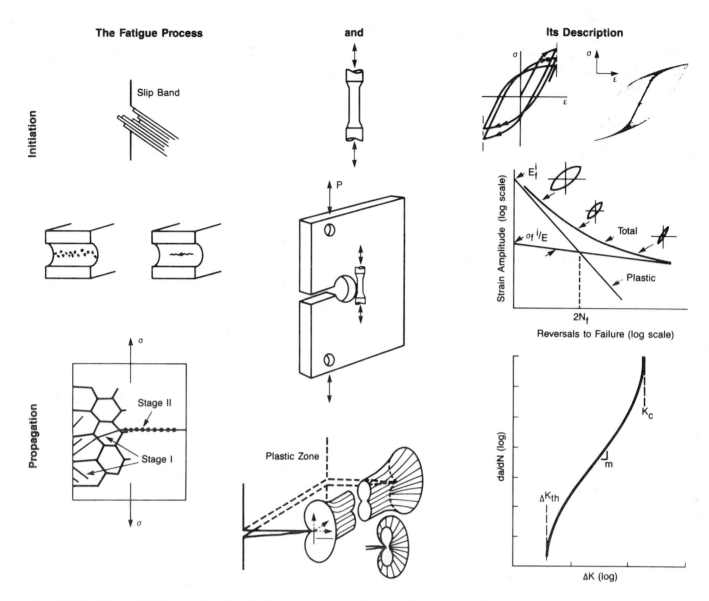

Fig. 10-15 Schematic illustrating the fatigue process and how it is characterized for purposes of life prediction.

life prediction based on the crack initiation concept. In components where the initiation of cracks less than about 2 mm dominates the life, this predicted life corresponds reasonably with component life. In other applications, predicted initiation life may have to be supplemented by predicted crack growth behavior to obtain realistic predictions of a component's total life. In the following, methods for predicting crack initiation life popular in ground vehicle applications are presented and discussed. Details on crack growth modeling, shown in Fig. 10-15, follow in Section 10.4. Combined models, and factors that modify the cracking process to include mechanisms other than fatigue are dealt with in Section 10.5.

10.3.4 *Methods of Crack Initiation Life Prediction*

The commonly used methods of crack initiation life prediction are presented in terms of examples. The first example deals with constant amplitude loading. Subsequent examples deal with variable amplitude loading. Advantages and limitations are noted with each example.

10.3.4.1 *Load-Life Method, Constant Amplitude Loading*

This model makes use of an empirical estimate of K_f [29] based on

$$K_f = 1 + \frac{(K_t - 1)}{(1 + a/r)} \qquad (10\text{-}10)$$

where "a" is a material constant and r is the notch root radius in the notch analysis. The constant "a" is an empirical best fit to relate smooth and notched specimen data in terms of K_f and K_t, such as shown in Fig. 10-10. In calibrating this relationship between K_f and K_t, as given in Eq. 10, a is found to depend on the ultimate strength S_u. The empirical relationship between S_u and "a," given by

$$a = 0.0254\,(2079\ (\text{MPa})/S_u\ (\text{MPa}))^{1.8} \qquad (10\text{-}11)$$

may be used to estimate "a" or Eqs. 1, 2 and 10 may be used when actual data are available for the material and geometry of interest. Smooth specimen and notch root crack initiation stresses (at a given life) are related by Eq. 2. The material considered in this example is ASTM A36 (see Table 10-1).

The desire is to estimate the fatigue strength for crack initiation in one-million cycles for a notched link made of hot-rolled, low carbon steel. It is assumed that the link will be subjected to completely reversed, (R = −1) axial loads. The notched link is shown in Fig. 10-16a. Axial fatigue data for ASTM A36, with an as-rolled surface, tested at R = −1, are presented in Table 10-1 for both unnotched and notched rotating bending conditions.

The fatigue notch factor, K_f, and the material constant, "a," *for the rotating bending tests* were determined using Eq. 2, and the data in Table 10-1a:

$$K_f = \frac{243 \text{ Ma}}{122 \text{ Ma}} = 1.99 \quad (10\text{-}12)$$

and from Eq. 10 and the data in Table 10-1a,

$$a = 0.254 \text{ mm} \left(\frac{2.55\text{-}1}{1.99\text{-}1} - 1\right) = 0.144 \text{ mm} \quad (10\text{-}13)$$

The ratio of the hole diameter to link-width, D/W, for the link considered is

$$\frac{D}{W} = \frac{20.6 \text{ m}}{76.2 \text{ m}} = 0.270 \quad (10\text{-}14)$$

and, from Fig. 10-16, the stress concentration factor of the notched link is,

$$K_t = 2.39 \quad (10\text{-}15)$$

The fatigue notch factor K_f *for the notched link* was computed from Eq. 11, where the notch root radius, r, is 1/2D (the hole radius):

$$K_f = \frac{2.39\text{-}1}{1 + \frac{0.144 \text{ m}}{10.3 \text{ m}}} = 2.37 \quad (10\text{-}16)$$

The S-N curve for the notched link was constructed by dividing the nominal net-section stress levels, $\Delta S/S_u$, for A36 (Table 10-1b and Fig. 10-16b) by $K_f = 2.37$, and plotting (in Fig. 10-16b) the quotients versus the crack initiation fatigue life, N_i. The damage analysis is implicit in constructing the S-N curve.

Table 10.1 Fatigue Data for Hot-Rolled Carbon Steel (A36)

10-1a. Rotating bending, A36

Fatigue limits, S_f, MPa

unnotched: 243
notched*: 122

10-1b. Axially loaded 19.1 mm A36 plate as-rolled surface, R = −1

$\Delta S/S_u$	N_f, cycles
	1.33×10^5
0.799	$2.02 \times$ ''
	$2.50 \times$ ''
0.769	$2.89 \times$ ''
0.719	$3.35 \times$ ''
0.703	$4.10 \times$ ''
	$9.45 \times$ ''
0.671	1.09×10^6
	$2.47 \times$ ''
	$2.01 \times$ ''
0.639	$3.76 \times$ ''**
	$5.87 \times$ ''**
0.607	$6.00 \times$ ''**

*circumferential V-notch, notch root radius = 0.254 mm.
 $K_f = 2.55$.
**Did not fail.

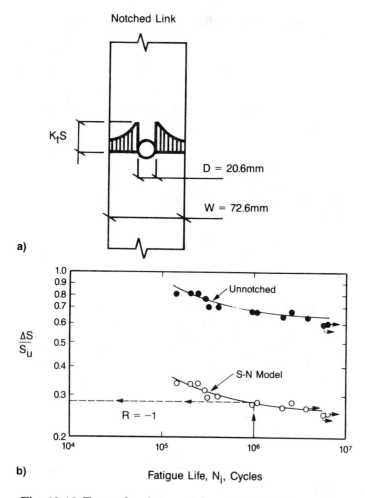

Fig. 10-16 Example of stress-life method of life prediction for a constant amplitude history.

The estimate of fatigue strength of the notched link for crack initiation at one-million cycles is calculated from Fig. 10-16b to be

$$\left.\frac{\Delta S}{S_u}\right|_{10^6 \text{ cycles}} = 0.28 \quad (10\text{-}17)$$

Advantages of this approach are:

(1) It is based on test data that may be found in the literature or may be based on actual test results for the material and geometry of interest.

(2) Stress analysis is simple and available from handbooks.

Limitations are:

(1) Accuracy is tied to data used to estimate K_f and analysis used to estimate K_t.

(2) The method applies only to constant amplitude cycling.

10.3.4.2 Load-Life Method, Variable Amplitude Loading

This method requires that the constant amplitude, load vs. life fatigue properties of the component and nominal loading history are known. Specifically, a load-life curve obtained from component testing and a rainflow counted histogram of the applied loads must be available to the designer. Reliable life estimates cannot be achieved from level crossing and other types of histograms. (Cycle counting is discussed in more detail in Chapter 5, Section 5.7.2.)

A simple example using the data provided in Appendix 10A will best illustrate this method. Field data must be converted into engineering units through use of appropriate scale factors. For this data, the conversion is given by

$$\Delta P = P_{max}/999 \times \text{Range}$$
ΔP = actual load range
P_{max} = maximum load in history (selected by designer)
Range = values of load from histogram

Constant amplitude load-life data for the component (in this case the SAE keyhole specimens) [30-32] may be fit to a simple power function.

$$\Delta P = P' (2N_f)^m$$
ΔP = load range
P' = intercept at 1 reversal
m = slope
$2N_f$ = reversals to failure (in this case failure means reversals to initiation of a crack 2.5 mm long)

Fatigue damage for the variable amplitude loading history can be calculated from Miner's linear damage rule.

$$D_i = \frac{2N}{2N_f}$$

D_i = fatigue damage for each load range
$2N$ = number of reversals at each load range
$2N_f$ = reversals to failure for each load range

The total damage for the history is obtained by summing the damage for each range in the history. At failure the total damage should equal 1, or

$$D = \sum_{i=1}^{N} D_i = 1 \text{ at failure} \quad (10\text{-}17)$$

Calculations are summarized in Table 10-2 for the keyhole specimen and a maximum load of 35.6 kN using Man-Ten material data. The calculated fatigue life represents the predicted number of times the loads in the histogram can be applied to the component before failure. Failure is defined, and is arbitrarily chosen to be the same in both variable amplitude and constant amplitude loading. For example, failure may be defined as the first detectable crack, or complete collapse of the structure. The definition of failure must be consistent in both types of loading. In this case, failure has been defined as a crack 2.5 mm long.

Table 10.2 Load Life Example

Range	Reversals	P (kN)	$2N_f$	Damage
200	346	7.1	5.48×10^7	6.30×10^6
250	486	8.9	2.02×10^7	2.40×10^5
300	268	10.7	8.91×10^6	3.01×10^5
350	182	12.5	4.46×10^6	4.08×10^5
400	136	14.3	2.45×10^6	5.54×10^5
450	68	16.0	1.45×10^6	4.70×10^5
500	42	17.8	9.01×10^5	4.66×10^5
550	36	19.6	5.88×10^5	6.12×10^5
600	20	21.4	3.98×10^5	5.02×10^5
650	6	23.2	2.78×10^5	2.15×10^5
700	6	24.9	1.99×10^5	3.01×10^5
750	8	26.7	1.46×10^5	5.46×10^5
800	0	28.5	1.09×10^5	0
850	16	30.3	8.35×10^4	1.91×10^4
900	8	32.0	6.46×10^4	1.23×10^4
950	14	33.9	5.07×10^4	2.76×10^4
1000	18	35.6	4.03×10^4	4.46×10^4
1050	18	37.4	3.24×10^4	5.56×10^4
1100	10	39.2	2.63×10^4	3.81×10^4
1150	6	41.0	2.15×10^4	2.79×10^4
1200	6	42.8	1.78×10^4	3.37×10^4
1250	4	44.5	1.48×10^4	2.70×10^4
1300	0	46.3	1.24×10^4	0
1350	2	48.1	1.05×10^4	1.91×10^4
1400	0	49.9	8.91×10^3	0
1450	0	51.7	7.61×10^3	0
1500	2	53.5	6.53×10^3	3.06×10^4
			TOTAL	3.83×10^3

LIFE $\dfrac{1}{3.83 \times 10^3}$ 261 BLOCKS

Experimental test data (solid symbols) and estimated fatigue lives (solid lines) for the SAE test program are shown in Fig. 10-17. One loading block represents one repetition of the loading history.

Advantages of this method are:

(1) It is an actual test of the component or structure of interest.

(2) Manufacturing effects and local stress concentration effects are automatically included.

(3) Stress analysis is not required.

Limitations are:

(1) The method cannot be used in the early stages of design before the first prototype is built.

(2) A new set of tests must be conducted whenever material or geometry changes are made.

(3) Mean stress effects cannot be included.

10.3.4.3 Stress-Life Method, Variable Amplitude Cycling

Stress-life analysis requires a stress-life curve for the material, stress concentration factor for the most highly stressed region, and a rainflow counted histogram of the nominal stress ranges.

Stress-life data from Appendix 10A can be related to the expected fatigue lives by the following relationship:

$$\frac{\Delta S}{2} = \sigma'_f (2N_f)^b \qquad (10\text{-}18)$$

$\dfrac{\Delta S}{2}$ = stress amplitude

σ'_f = intercept at 1 reversal (if the fatigue data were obtained from a strain controlled test, this would be the fatigue strength coefficient)

$2N_f$ = reversals to failure

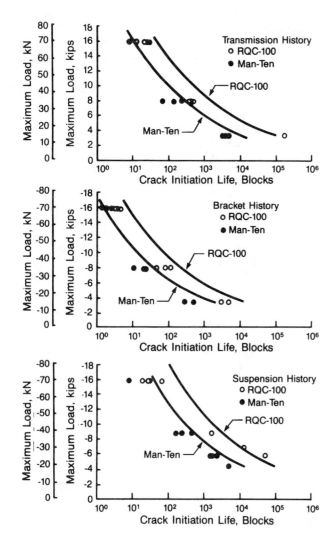

Fig. 10-17 **Experimental data and estimated fatigue lives for load-life analysis.**

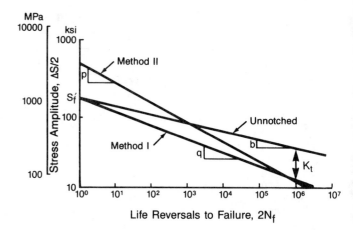

Fig. 10-18 **Methods for estimating notched stress-life curve.**

b = Slope of $S = 2N_f$ curve (if the stress life data was obtained from a strain controlled test, this would be the fatigue strength exponent)

Fatigue lives for the component are calculated from a notched stress-life curve. These curves may be estimated from unnotched smooth specimen data and the stress concentration factor. Two methods are often employed.

Method I: The long life fatigue strength at 10^6 reversals is divided by the stress concentration factor. The notched stress-life curve may then be approximated by a line drawn between this point and the unnotched fatigue strength at 1 reversal.

Method II: The long life fatigue strength at 10^6 reversals is divided by the stress concentration factor. The notched stress-life curve may then be approximated by a line drawn between this point and the unnotched fatigue strength at 10^3 reversals.

Both methods for estimating notched stress-life curves are shown in Fig. 10-18. In both methods, the stress concentration factor (K_t) can be replaced by the fatigue notch factor (K_f).

Stress-life curves for the notched component may be estimated from the following relationships.

Method I

$$2N_f = \frac{\Delta S}{2\sigma'_f}^{1/q}$$

$$q = b - (\log_{10} K_t)/6.0$$

(10-19)

Method II

$$2N_f = \left(\frac{\Delta S}{2 K_t \sigma'_f}\right)^{1/P}$$

(10-20)

$$P = b - (\log_{10} K_t)/3.0$$

A stress-life analysis of the same data used in the previous example is shown in Table 10-3 for a maximum load of 35.6 kN using the Man-Ten materials data shown in Appendix 10A.

First, load ranges in the histogram must be converted into nominal stress ranges. If the reduced section of the keyhole specimen (see Fig. 10A-2 in Appendix 10A) is considered a simple beam subjected to both tension and bending, the following relationship may be developed:

$\Delta S = 7.26\ \Delta P$
ΔS = nominal stress range
ΔP = nominal load range

When this method is used to obtain nominal stress ranges, the stress concentration factor, K_t, is equal to 3.0.

Fatigue damage for each nominal stress range is then calculated from Eqs. 19 or 20, and fatigue lives are computed

Table 10.3 Stress Life Example

Range	Reversals	S (MPa)	METHOD I $2N_f$	METHOD I Damage	METHOD II $2N_f$	METHOD II Damage
200	346	80	6.15×10^7	5.62×10^{-6}	1.69×10^7	2.04×10^{-5}
250	486	10	1.71×10^7	2.83×10^{-5}	7.04×10^6	6.90×10^{-5}
300	268	120	6.02×10^6	4.44×10^{-5}	3.43×10^6	7.80×10^{-5}
350	182	140	2.49×10^6	7.30×10^{-5}	1.87×10^6	9.71×10^{-5}
400	136	160	1.15×10^6	1.17×10^{-4}	1.10×10^6	1.23×10^{-4}
450	68	180	5.91×10^5	1.15×10^{-4}	6.96×10^5	9.76×10^{-5}
500	42	200	3.23×10^5	1.30×10^{-4}	4.60×10^5	9.13×10^{-5}
550	36	220	1.87×10^5	1.92×10^{-4}	3.16×10^5	1.13×10^{-4}
600	20	240	1.13×10^5	1.76×10^{-4}	2.24×10^5	8.91×10^{-5}
650	6	260	7.18×10^4	8.35×10^{-5}	1.64×10^5	3.66×10^{-5}
700	6	280	4.70×10^4	1.28×10^{-4}	1.22×10^5	4.90×10^{-5}
750	8	300	3.16×10^4	2.53×10^{-4}	9.32×10^4	8.58×10^{-5}
800	0	320	2.18×10^4	0	7.23×10^4	0
850	16	340	1.54×10^4	1.04×10^{-3}	5.70×10^4	2.80×10^{-4}
900	8	360	1.11×10^4	7.19×10^{-4}	4.55×10^4	1.75×10^{-4}
950	14	380	8.16×10^3	1.72×10^{-3}	3.68×10^4	3.81×10^{-4}
1000	18	400	6.08×10^3	2.96×10^{-3}	3.00×10^4	5.99×10^{-4}
1050	18	420	4.60×10^3	3.91×10^{-3}	2.48×10^4	7.26×10^{-4}
1100	10	440	3.52×10^3	2.84×10^{-3}	2.06×10^4	4.84×10^{-4}
1150	6	460	2.73×10^3	2.20×10^{-3}	1.73×10^4	3.46×10^{-4}
1200	6	480	2.14×10^3	2.80×10^{-3}	1.47×10^4	4.09×10^{-4}
1250	4	500	1.69×10^3	2.36×10^{-3}	1.25×10^4	3.20×10^{-4}
1300	0	520	1.35×10^3	0	1.07×10^4	0
1350	2	540	1.08×10^3	1.83×10^{-3}	9.22×10^3	2.16×10^{-4}
1400	0	560	8.85×10^2	0	7.99×10^3	0
1450	0	580	7.23×10^2	0	6.96×10^3	0
1500	2	600	5.96×10^2	3.36×10^{-3}	6.09×10^3	3.28×10^{-4}
			TOTAL	2.71×10^{-2}	TOTAL	5.21×10^{-3}

METHOD I LIFE = $1/2.71 \times 10^{-2}$ = 37 BLOCKS

METHOD II LIFE = $1/5.21 \times 10^{-3}$ = 191 BLOCKS

---

from Eq. 17. Experimental data and estimated fatigue lives employing the two methods for estimating notched stress-life curves are shown in Figs. 10-19 and 10-20.

Advantages of this approach are:

(1) It can be used for initial design.

(2) Changes in material and geometry can be evaluated.

Limitations are:

(1) It does not account for notch root plasticity, that is the cause of fatigue.

(2) Mean stress effects cannot be handled well.

(3) Requires empirical K_t factors for good results.

10.3.4.4 Strain-Life Method, Variable Amplitude Cycling

Strain-life analysis accounts for notch root plasticity and tory, and stress concentration factors for the component of interest. In addition, it requires an analysis relating nominal stresses and strains to the critical stresses and strains at the notch root [33].

Method I: If the loading data is in terms of rainflow counted histogram of nominal stress (load) ranges, a simple analysis similar to load-life and stress-life can be carried out. The procedures is illustrated in Table 10-4 for the same example used previously.

Neuber's rule may be employed to estimate notch stresses and strains from elastic nominal stresses and strains. Neuber's rule may be written in the following form:

$$\frac{(K_f \Delta S)^2}{4E} = \frac{\Delta \sigma}{2} \frac{\Delta \varepsilon}{2} \quad (10\text{-}22)$$

The left hand side of this equation is a constant for each nominal elastic stress range in Table 10-3. A fatigue notch factor of 3.0 was employed in Table 10-4.

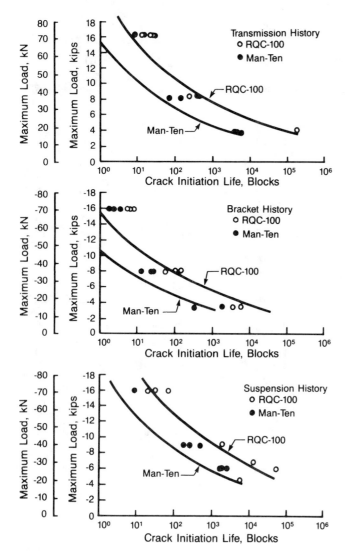

Fig. 10-19 Experimental data and estimated fatigue lives for stress-life I analysis.

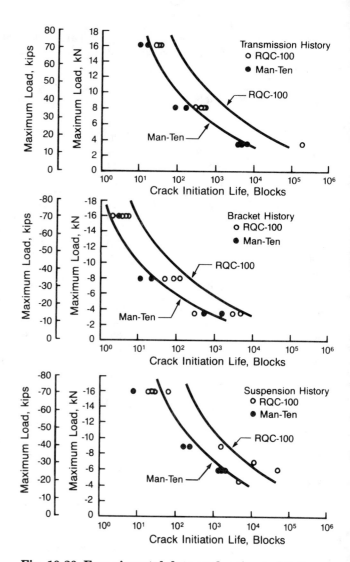

Fig. 10-20 Experimental data and estimated fatigue lives for stress-life II analysis.

An equation for the cyclic stress-strain curve may be substituted into Eq. 22

$$\frac{\Delta\varepsilon}{2} = \frac{\Delta\sigma}{2E} + \left(\frac{\Delta\sigma}{2K'}\right)^{1/n'} \quad (10\text{-}23)$$

with the following result:

$$\frac{(K_f \, \Delta S)^2}{4E} = \frac{\Delta\sigma^2}{4E} + \frac{\Delta\sigma}{2}\left(\frac{\Delta\sigma}{2K'}\right)^{1/n'} \quad (10\text{-}24)$$

Material properties for Eq. 24 may be found in Appendix 10A. Because of the exponents, Eq. 24 cannot be solved directly for the unknown stress range, $\Delta\sigma$. Trial and error or iteration techniques can be employed to solve this equation. Once the stresses have been determined, the unknown strain amplitudes can be obtained from Eq. 23.

After the notch strain amplitudes, $\Delta\varepsilon/2$, have been determined, the fatigue life, $2N_f$, may be calculated from the familiar strain-life equation:

$$\frac{\Delta\varepsilon}{2} = \left(\frac{\sigma'_f}{E}\right)(2N_f)^b + \varepsilon'_f(2N_f)^c \quad (10\text{-}25)$$

Again, trial and error or iterative techniques must be employed (for this reason, computers are often used to solve these equations). Fatigue damage and life is then calculated employing Miner's rule as before.

Experimental data and estimated fatigue lives from a strain-life analysis are shown in Fig. 10-21.

Although precise mean stresses cannot be calculated from a histogram, mean strains and strain ranges and the maximum/minimum effects of mean stress can be determined. The stress corresponding to the largest strain on the loading history will lie on the cyclic stress-strain curve. A hysteresis loop is formed from the largest to the smallest strain in the history. All other stresses and strains must lie inside this loop. If there are large plastic strains, the mean stress for a zero mean strain could be either tensile or compressive, as shown in Fig. 10-22. As the mean strains approach the maximum or

Table 10.4 Strain Life Example

Range	Reversals	$\frac{(K_f \Delta S)^2}{4E}$	$\frac{\Delta \sigma}{2}$ (MPa)	$\frac{\Delta \epsilon}{2}$	$2N_f$	Damage
200	346	.071	119	5.95×10^{-4}	2.17×10^9	1.59×10^{-7}
250	486	.111	148	7.51×10^{-4}	2.36×10^8	2.06×10^{-6}
300	268	.160	174	9.18×10^{-4}	4.23×10^7	6.33×10^{-6}
350	182	.217	198	1.10×10^{-3}	1.07×10^7	1.69×10^{-5}
400	136	.284	220	1.29×10^{-3}	3.58×10^6	3.80×10^{-5}
450	68	.359	239	1.50×10^{-3}	1.45×10^6	4.70×10^{-5}
500	42	.443	256	1.73×10^{-3}	6.81×10^5	6.17×10^{-5}
550	36	.536	271	1.97×10^{-3}	3.59×10^5	1.00×10^{-4}
600	20	.638	286	2.23×10^{-3}	2.07×10^5	9.66×10^{-5}
650	6	.749	299	2.51×10^{-3}	1.28×10^5	4.70×10^{-5}
700	6	.870	310	2.80×10^{-3}	8.33×10^4	7.20×10^{-5}
750	8	.998	321	3.11×10^{-3}	5.68×10^4	1.41×10^{-4}
800	0	1.13	331	3.43×10^{-3}	4.02×10^4	0
850	16	1.28	340	3.76×10^{-3}	2.94×10^4	5.45×10^{-4}
900	8	1.44	350	4.11×10^{-3}	2.20×10^4	3.64×10^{-4}
950	14	1.60	358	4.47×10^{-3}	1.68×10^4	8.31×10^{-4}
1000	18	1.77	366	4.84×10^{-3}	1.31×10^4	1.37×10^{-3}
1050	18	1.96	373	5.23×10^{-3}	1.04×10^4	1.73×10^{-3}
1100	10	2.15	381	5.63×10^{-3}	8.39×10^3	1.19×10^{-3}
1150	6	2.35	388	6.05×10^{-3}	6.83×10^3	8.77×10^{-4}
1200	6	2.55	395	6.47×10^{-3}	5.64×10^3	1.06×10^{-3}
1250	4	2.77	401	6.91×10^{-3}	4.70×10^3	8.51×10^{-4}
1300	0	3.00	407	7.36×10^{-3}	3.95×10^3	0
1350	2	3.23	413	7.81×10^{-3}	3.34×10^3	5.97×10^{-4}
1400	0	3.48	419	8.29×10^{-3}	2.85×10^3	0
1450	0	3.73	425	8.77×10^{-3}	2.46×10^3	0
1500	2	3.99	430	9.27×10^{-3}	2.13×10^3	9.40×10^{-4}
					TOTAL	1.10×10^{-2}

LIFE = $1/1.10 \times 10^{-2}$ = 91 BLOCKS

minimum strain, or if plastic strains are small, the calculated difference between maximum and minimum mean stress effects will be reduced.

Once mean stress and possible mean stress relaxation effects have been bounded, the test engineer can decide if detailed analysis, based on the actual sequence of peaks and valleys, is required.

Method II: The need to consider the *sequence* of actual peaks and valleys is best illustrated by referring to the load histories shown in Fig. 10-23.

Both load histories, A and B, have the same rainflow counted histogram, i.e, one large cycle and several thousand smaller cycles, all with zero mean. Actual test results for the two load histories are shown in Fig. 10-24.

The difference between the load histories is that history A

compressive residual stress at the notch root. Note that the nominal residual and mean stresses are zero, but the mean and residual stresses at the notch root may be tensile or compressive, depending on the exact sequence of events. If an accurate assessment of residual and mean stress effects is required, a sequential analysis is necessary.

Stresses and strains at the notch root must be determined on a reversal by reversal basis. First, a model for simulating the stress-strain response of the material is required. These models estimate the stresses that correspond to a given strain history. Second, a method for determining the stresses and strains at the critical location from the nominal stresses and strains is required. When the two models are combined, the notch root stress-strain response for any nominal loading history can be obtained. For example, the two stress-strain histories in the bottom of Fig. 10-23 were calculated from the nominal stress history. Details of these procedures can be found in Appendix 10B.

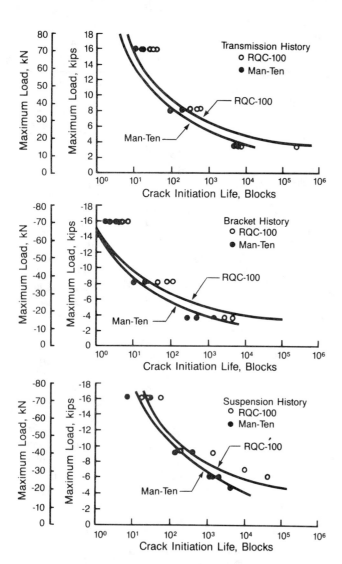

Fig. 10-21 Experimental data and estimated fatigue lives for strain-life analysis [30-32].

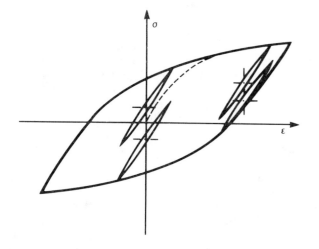

Fig. 10-22 Mean stress example.

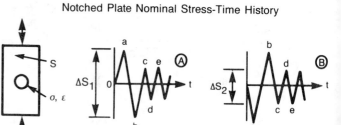

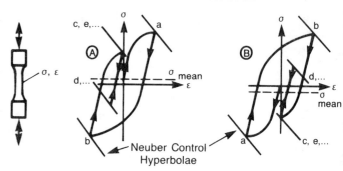

Fig. 10-23 Sequence effects example [34].

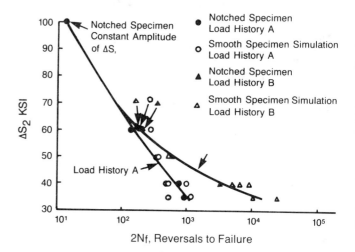

Life as a Function of ΔS_2 for Notched Specimens and Smooth Specimen Simulations (7075-T6 Aluminum, $K_f = 1.95$)

Fig. 10-24 Test results for sequence example [34].

Mean stress effects for each cycle can now be included. The strain range and mean stress of each cycle defined by rainflow counting is obtained from the notch root stress and strain histories. Damage is then computed from the strain-life equation modified for mean stress (after Morrow [22])

$$\frac{\Delta \varepsilon}{2} = \left(\frac{\sigma'_f - \sigma_0}{E}\right)(2N_f)^b + \varepsilon'_f (2N_f)^c \qquad (10\text{-}26)$$

where σ_0 = mean stress.

248

Fatigue damage is then calculated for each cycle in the loading history and summed using Miner's rule.

Experimental data for the SAE test program and estimated fatigue lives for the sequential analysis are shown in Fig. 10-25.

Advantages of strain-based analysis:

(1) Accounts for notch root plasticity.

(2) Correctly accounts for mean stress effects if a sequential analysis is performed.

(3) Results in more accurate life estimates for ductile metals.

Limitations are:

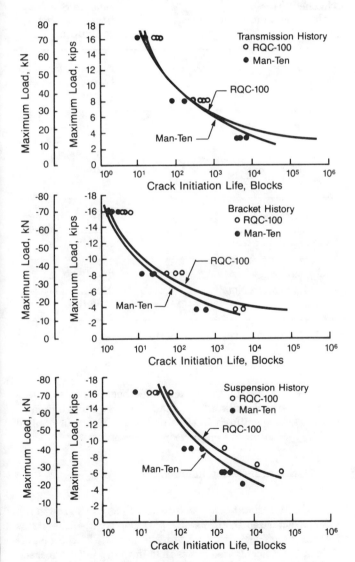

Fig. 10-25 Experimental data and estimated fatigue lives for sequential strain-life example

(1) Requires empirical K_t or K_f factors for best results.

(2) Not necessarily applicable to long life situations where surface finish and other processing variables have a large effect.

10.3.5 Selection of a Crack Initiation Life Prediction Method

Several methods have been presented to illustrate methods of crack initiation prediction that are popular in the ground vehicle industry. Other methods exist—differ in concept—but most differ in the specific way the notch analysis is performed and damage is calculated. As such, the methods illustrated represent a cross section of methods to implement crack initiation approaches. These methods, and a nominal strain variant of the nominal stress method are summarized with reference to the input history, the notch analysis, and the fatigue data used in the damage analysis in Table 10-5. The table shows that some methods are simpler than others to the extent that they require less data. Differences in input requirements have been noted in the examples discussed in the preceding section as illustrations of these prediction methods. Likewise, advantages and limitations of the methods have been outlined.

The obvious question facing the designer is, "Which of these methods is best suited to my current problem?" This question can best be answered in light of the situation being analyzed. Some important considerations are summarized in Table 10-6. During preliminary development where much uncertainty exists about material, geometry and loading it may be appropriate to use simpler, more approximate methods for sensitivity analysis. Later in product development, when loads are better understood and designs are formalized, more detailed methods may be fruitful if the needed input data are available. At this stage it may be appropriate to develop the missing input data including more detailed stress analyses, although priorities for such activities will always be subject to time and cost considerations and the skill in using and interpreting results from the more complex predictive schemes.

10.4 Crack Propagation Approach

In response to the need to assess the resistance to failure of components that contain initial defects or discontinuities that develop fatigue cracks, the field of fracture mechanics has been developed over the past several decades. In many practical situations, fracture mechanics evaluations permit the following quantities to be predicted: (a) the critical size of a crack, (b) the life expended in growing a crack from a specified initial size to a critical size, and (c) the likelihood that cracks will be non-propagating or arrest. To date, structural evaluations based on fracture mechanics have been used primarily in the aerospace and nuclear power industries, but increasing interest and applications are developing in the ground vehicle industry.

The following approaches for evaluating structural integrity in the presence of flaws and for predicting crack prop-

Table 10.5 Cumulative Damage Analysis Combinations

Analysis Type	Service History Input, n Source	Stress or Strain Notch Analysis	Fatigue Properties Input, N_f Source
Load-Life Analysis	Load Amplitude vs Time History for Component	None	Load Amplitude vs Cycles to Failure for the Component
Nominal Stress-Life Analysis	Nominal Stress Amplitude vs Time History	None	Notched Specimen Nominal Stress vs Cycles to Failure
		Notched Stresses Calculated from Nominal Stress	Smooth Specimen Stress vs Cycles to Failure
Notch Root Stress-Life Analysis	Notch Root Stress Amplitude vs Time History	None	Smooth Specimen Stress vs Cycles to Failure
Nominal Strain-Life Analysis	Nominal Strain Amplitude vs Time History	None	Notched Specimen Nominal Strain vs Cycles to Failure
		Notch Strains and Stresses* Calculated from Nominal Strain	Smooth Specimen Strain vs Cycles to Failure
Notch Root Strain-Life Analysis	Notch Root Strain Amplitude vs Time History	Notch Stresses* Calculated from Notch Strain	Smooth Specimen Strain vs Cycles to Failure

*Stresses are calculated to allow the effect of mean stress at the root to be included in the life estimate.

mechanics (LEFM). As with most analytical methods used in engineering, there are certain limitations on the use of LEFM. Broadly speaking, LEFM can be applied when flaws or cracks are in regions where the nominal stresses are elastic. If regions experience significant plastic straining such as may occur near a notch, or if the crack itself causes extensive yielding in lower strength metals, then different approaches are generally needed, based on elastic-plastic fracture mechanics. Also, LEFM can be applied only when cracks are large enough with respect to microstructural features, such as grain size, that concepts of continuum mechanics are applicable. Currently, there are no widely accepted guidelines for specifying minimum crack sizes for applicability. However, the sizes predicted by the initiation methods discussed in this chapter (several millimeters) should be large enough in most cases. The above limitations still permit LEFM to be applied in many situations of engineering interest.

10.4.1 Stress Intensity Factors

In LEFM, resistance to fracture as well as crack growth life are characterized in terms of a parameter called the stress intensity factor, K_I, that provides a measure of the magnitude of the concentrated stress field in the vicinity of a crack tip.* Suppose that a structural region containing a crack is

*Do not confuse the stress intensity factor, K_I, with K_t or K_f discussed previously with regard to stress concentration factors.

Table 10.6 Capabilities and Characteristics of Various Methods for Estimating Fatigue Life

Life Estimates Based On:	Level of Abstraction	Stage of Design	Expense and Delay	Effects Included in Life Estimate			
				Mean and Sequence	Fabrication	Dyn. and Vib. Loads	Environment
Materials Data							
(a) Local Strain	High	Paper	Small	(a) Yes	No	Maybe	No
(b) Nominal Stress				(b) No			
(c) Crack Growth				(c) Maybe			
Constant Amplitude Component Data	Medium	Paper to Middle	Medium	No	Yes	Maybe	No
Service Simulation							
(a) Loads Applied	Low	(a) Middle to Late	High	(a) Probably	Yes	(a) Probably	No
(b) Operational		(b) Late		(b) Yes		(b) Yes	
Actual Service	None	Complete	Very High	Yes	Yes	Yes	Yes

subjected to tensile stress normal to the crack plane. Referring to Fig. 10-26, the elastically computed crack tip stresses are of the form:

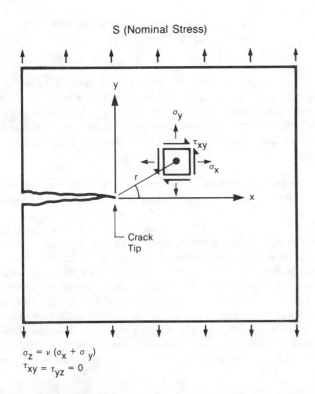

$\sigma_z = \nu (\sigma_x + \sigma_y)$
$\tau_{xy} = \tau_{yz} = 0$

Fig. 10-26 Mode I stress components in the vicinity of a crack tip.

$$\sigma_x = \frac{K_I}{\sqrt{2\pi r}} \cos \frac{\theta}{2} \left[1 - \sin \frac{\theta}{2} \sin \frac{3\theta}{2} \right]$$

$$\sigma_y = \frac{K_I}{\sqrt{2\pi r}} \cos \frac{\theta}{2} \left[1 + \sin \frac{\theta}{2} \sin \frac{3\theta}{2} \right] \quad (10\text{-}27)$$

$$\tau_{xy} = \frac{K_I}{\sqrt{2\pi r}} \sin \frac{\theta}{2} \cos \frac{\theta}{2} \cos \frac{3\theta}{2}$$

The stress intensity factor is the K_I term in Eq. 10-27. This factor depends primarily on crack size and the stress applied to a region, but also varies with different types of cracks and loadings. All K_I factors have the general form

$$K_I = FS\sqrt{a} \quad (10\text{-}28)$$

where: S = applied nominal stress
a = crack length (or depth)*
F = a dimensionless quantity used to account for the type of crack loading, and the ratio of crack size to specimen or component dimensions.

The value of S is that computed as if no cracks were present. The K_I factor has units of MPa - m$^{1/2}$ or ksi - in$^{1/2}$.

Fig. 10-27 shows K_I factors for edge cracks in a plate subjected to uniform tension, or to bending, while Fig. 10-28 shows the K_I factor for a surface crack subjected to uniform tension. Surface cracks are usually semi-elliptical in shape, as if someone had pressed a thumbnail into a surface. Both the crack depth and surface length are needed to describe K_I.

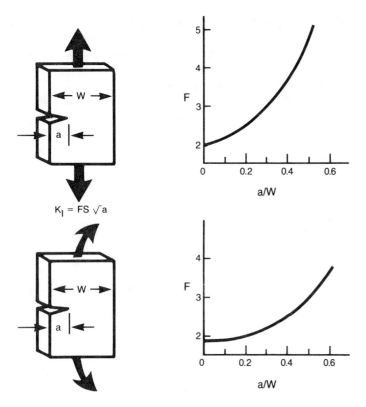

Fig. 10-27 Stress intensity factors for an edge crack subjected to tension or bending.

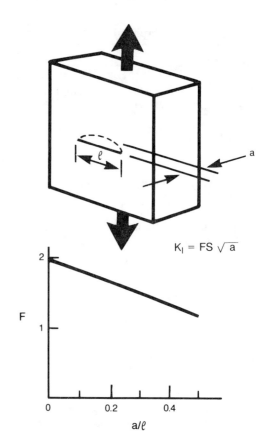

Fig. 10-28 Stress intensity factors for a surface crack subjected to tension (for cracks shallow relative to thickness).

Initial flaws found in components are frequently surface cracks. Closed form solutions for K_I factors for many other loading situations and crack types have been developed and are tabulated in handbooks [35-37]. They may also be computed by various numerical methods or determined experimentally. For instance, a number of widely used, commercially available finite element codes have such options.

The K_I factors depicted in Figs. 10-27 and 10-28 are for the so-called tensile opening mode and designated by the subscript $_I$. This mode is also illustrated in Fig. 10-29. Two other modes of crack surface displacement are modes II and III, the shear and tearing modes, respectively. Corresponding stress intensity factors K_{II} and K_{III} can be computed for these modes as well.

Although various proposals have been made, there are currently no widely accepted methods for taking into account (in a fracture analysis) the combined effect of K_I with either or both of the other two modes. In most situations, however, fracture and crack growth tend to occur in the tensile opening mode normal to the direction of maximum principal stress so that K_I tends to be the governing parameter. In cases where stresses from different types of loading (e.g., bending and uniform tension) are acting in the same direction, the different K_I factors may be added to obtain a total K_I value.

10.4.2 Fracture Toughness

When a conventional component stress analysis is conducted, the stresses are compared to material properties such as yield strength or ultimate tensile strength to see if the component is safe under static loading. If flaws or cracks are present, it is necessary to compare the appropriate computed K_I factor to some other material property to assess resistance to fracture. This property is called fracture toughness, denoted by K_c.

Fracture toughness is determined by preparing several specimens similar to those shown in Fig. 10-27 from the material of interest. A sharp notch is cut into each specimen, then the specimens are subjected to fatigue cycling until cracks form at the notches and grow some distance. With a crack of a known size present in each specimen, increasing tensile stress is applied until fracture occurs. With both the stress at fracture and crack size known, the corresponding value of K_I at fracture can be computed. This critical value gives the K_c for the material. It will differ slightly from specimen to specimen due to inherent variability in material behavior (as with any other mechanical property). Multiple specimens are normally tested to ascertain the variability in K_c.

An important feature of K_c is that, while determined with one type of specimen, it can be used to evaluate the resistance to fracture for other crack sizes and different component geometries and types of cracks by calculating the appropriate applied K_I and comparing it to the value of K_c.

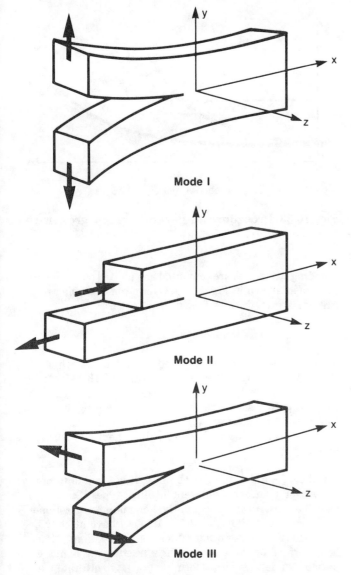

Fig. 10-29 Different modes of crack surface displacement.

Upon tensile load application, the concentrated stresses at a crack tip cause material to yield in a zone around it. If the extent of plastic deformation becomes too large during a fracture toughness test, then determination of plane strain K_{Ic} looses its validity. This is often the case for lower strength steels. Test methods and criteria for valid determination of K_{Ic} are given in Ref. [42]. Tabulated values of K_{Ic} are presented in many sources (e.g. [38-40]). Various methods of elastic-plastic fracture mechanics [43] have also been developed for assessing resistance to fracture when use of LEFM becomes questionable.

10.4.3 Critical Crack Size

The critical crack size for unstable or catastrophic crack growth can be computed by substituting the value of K_C for K_I into Eq. 28 and the greatest value of tensile stress in a loading history for S, and then solving for a. If handbook K_I relations are used, this requires the choice of a relation that approximates the loading situation and type of crack expected in the component at the region of interest. An example of this kind of fracture evaluation will be given later. Approximate knowledge of the critical crack size for fracture provides a useful supplement to traditional strength evaluations that assume a component is free of flaws. By pinpointing regions where critical crack sizes may be small, modifications can be considered to increase the critical size and/or minimize chances for crack initiation, such as by reducing local stress levels or using materials with higher K_c values. On the other hand, existence of locations where critical crack sizes are large provides some reassurance that if cracks do initiate, in spite of efforts to preclude them, they are not likely to cause imminent failure. Also, if a component contains initial defects as a result of fabrication processes or material imperfections, knowledge of critical crack sizes helps in assessing how serious the defects are and if steps are needed to reduce the frequency of their occurrence and/or their size.

10.4.4 Fatigue Crack Growth

The approach just described utilizes the greatest value of loading expected in service to predict critical crack sizes for

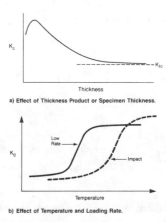

Fig. 10-30 Typical effects on fracture toughness in steels.

K_c has been tabulated [38-40] for a variety of structural alloys. If it is not available, it can be determined by preparing fracture toughness specimens from the alloy of interest. It is also feasible to prepare small specimens from material removed from a manufactured component. It should be noted that K_c is sensitive to temperature and loading rate, as is the case with other material properties such as yield strength. Fig. 10-30 portrays those effects on K_c for typical steels. There is a temperature, that varies from alloy to alloy, where a drastic reduction in K_c occurs. This so-called "nil ductility" temperature is raised by higher rates of loading, such as due to impact. As a result, K_c should be determined at a temperature and loading rate close to that expected in service, whenever possible. In addition, K_c tends to decrease with thickness, reaching a minimum value termed the plane strain fracture toughness, K_{Ic} (see Fig. 10-30).

fracture. Fracture mechanics can also be utilized to estimate the life spent in growing a crack from an initial size to a critical size, as depicted schematically in Fig. 10-31. Initial sizes can be taken as the largest defect size that could exist from the time a component is put in service, or (if significant initial defects are not likely to be present) as the size that would typically exist after the crack initiation life has been expended (e.g. 2 mm).

In order to predict crack growth for service loadings it is necessary to have basic crack growth rate data for the material of interest. Such data are similar in purpose to S-N (e-N) used to assess crack initiation life. As shown in Fig. 10-32, crack growth data are developed by subjecting a specimen with an initial crack to uniformly repeated cyclic stresses.

Fully-reversed cycling, as is typically done to generate S-N data, is generally not used since crack growth is primarily caused by cyclic tensile stresses. For a given applied stress range, ΔS, the crack size is monitored as a function of applied cycles and plotted as in Fig. 10-33. The rate of growth, i.e, the increase in crack size per cycle, is the slope of the curve shown in Fig. 10-33, termed da/dN.

As the crack size increases, the rate also increases (for constant amplitude loading). For a number of different crack sizes, the values of da/dN defined in Fig. 10-33 and corre-

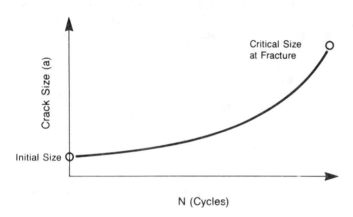

Fig. 10-31 Schematic of fatigue crack growth.

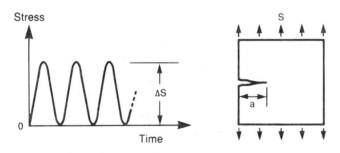

Fig. 10-32 Loading applied to generate crack growth data.

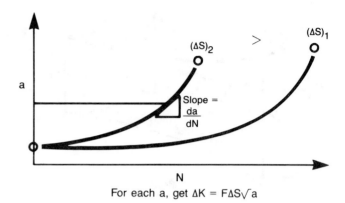

Fig. 10-33 Procedure to generate crack growth rate data.

sponding ΔK values are computed and plotted on log-log coordinates as shown in Fig. 10-34. Over a considerable range of ΔK, crack growth rates follow the relation

$$da/dN = C(\Delta K)^n \qquad (10\text{-}29)$$

where C and n are constants fitted to the data and which vary from alloy to alloy.* There is also a value of ΔK below which no growth typically occurs, called the threshold stress intensity, ΔK_{th}. It is somewhat analogous to the fatigue limit for S-N data. As cracks grow in size and ΔK increases, da/dN becomes very high and fracture is imminent as K_c or K_{Ic} is approached.

If da/dN – ΔK data are available, it is feasible to predict growth for a variety of different loadings, types of cracks, etc., as will be described subsequently. For a number of commonly used structural alloys, da/dN – ΔK data have already been compiled [38-40]. For other materials, such data may have to be generated by those conducting a fracture mechanics evaluation. Guidelines for performing crack growth rate experiments are given in Ref. [44].

For a wide variety of structural steels, it has been found [39] that upper bounds on crack growth rate can be described by the following relations over the linear region in Fig. 10-34:

$$\frac{da}{dN} = 6.9 \times 10^{-12} (\Delta K)^3 \qquad (10\text{-}30)$$

for ferritic-pearlitic steels, and

$$\frac{da}{dN} = 1.4 \times 10^{-10} (\Delta K)^{2.3} \qquad (10\text{-}31)$$

for martensitic steels. In these relations, da/dN has units of m/cycle and ΔK is in MPa – $m^{1/2}$. These equations might be useful in preliminary fracture mechanics evaluations when data for the particular alloy of interest are not readily available.

---

*The "n" used in Eq. 29 is not the same as used to define the monotonic strain hardening exponent.

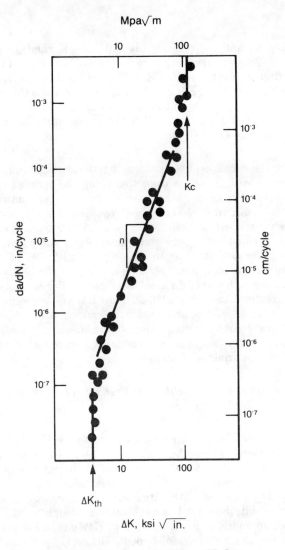

Fig. 10-34 Typical da/dN – ΔK data.

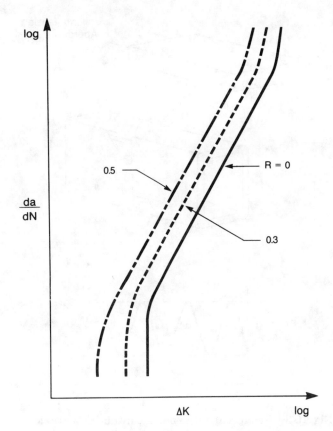

Fig. 10-35 Schematic of influence of R-ratio on crack growth rate.

To this point, correlations for crack growth data generated with R = 0 loading have been considered, where R is the ratio of the minimum to maximum stress applied in testing. Suppose that tests are performed with stress ratios greater than zero, but with the range of applied stress kept the same as in the R = 0 tests. In most cases, increased R values (and thus tensile mean stress levels) cause greater growth rates for the same ΔK, as depicted in Fig. 10-35.

Knowledge of R-ratio effects is often important in making crack growth life predictions, particularly for the variable amplitude or irregular type of loadings generally encountered in service. If data on R-ratio effects are not available for an alloy of interest, they can be generated by performing tests like those used to develop R = 0 data. If that is not feasible, relations such as the following can be used (with caution) to estimate R-ratio effects based only on R = 0 input data:

$$da/dN = \frac{A (\Delta K)^m}{(1 - R) K_c - \Delta K} \qquad (10\text{-}32)$$

$$da/dN = \frac{B (\Delta K - \Delta K_{th})^p}{(1 - R) K_c - \Delta K} \qquad (10\text{-}33)$$

where m, p, A and B are empirical constants fitted to R = 0 data. Eq. 33 represents an attempt to account for the effect of positive R-ratios on the threshold stress intensity range, ΔK_{th}. The extent to which these or similar relations provide a good approximation for R-ratio effects will vary from alloy to alloy.

The effects of negative R-ratios (i.e., compressive minimum load in the load cycle) on crack growth rates are either negligible, or cause an increase in growth rate relative to R = 0 behavior [47,48]. Increases are generally less than a factor of two in growth rate. Thus, crack growth life predictions are typically made using ΔK calculated only for the tensile portions of loading cycles, especially since data on the effects of compression frequently do not exist for a particular alloy. Given all of the other uncertainties involved in making fatigue life and crack growth predictions, neglecting the effects of compressive loads can often be justified, especially in "first cut" analyses.

10.4.5 *Examples of Crack Propagation Life Prediction for Constant Amplitude Loading*

Consider a section of a box-beam structure subjected to constant amplitude bending, as illustrated in Fig. 10-36. Suppose that a surface crack either exists from the time of fabrication or forms by fatigue at the location depicted in that figure.

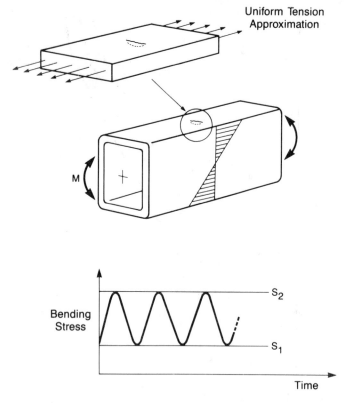

Fig. 10-36 Section of a box-beam containing a crack and subjected to constant amplitude bending.

First of all, an initial crack size must be specified. The size could be that corresponding to an existing defect or could be that typical of the size predicted by a crack initiation life evaluation (e.g., several millimeters) depending on which circumstance is of interest. Let the initial crack depth be denoted by a_1. Since a surface crack is being considered, it is also necessary to specify a ratio of crack depth to surface length, (a/ℓ). In practice, values of 0.3 to 0.5 are frequently used. It will be assumed that the ratio of (a/ℓ) selected remains the same during crack growth.

Next, a critical crack size must be determined. Referring to Fig. 10-36, the surface crack experiences stress due to bending. However, that stress can be considered as nearly uniform over the thickness of the wall in which the crack resides. As a result, the crack should behave as if it were in a plate subjected to uniform tension stress, and the K relation from Fig. 10-28 can be used as a reasonable approximation. That relation is set equal to the fracture toughness, K_c, for the specified ratio of (a/ℓ) and solved for the critical crack depth, denoted a_c, by substituting the highest level of tensile stress in the loading history, in this case S_2.

Having specified the initial and critical (final) crack sizes, the crack growth life between those two limits can now be estimated. Minimum and maximum values of K, designated K_1 and K_2, for the first cycle of loading are obtained by substi-

tuting S_1 and S_2, respectively, into the K relation from Fig. 10-28 using the initial crack size, a_1. The range of K for the first cycle is then $\Delta K = K_2 - K_1$, and the R-ratio is K_1/K_2. The increment of crack growth, Δa, for the cycle is the value of da/dN computed by substituting ΔK, R and K_c into a relation such as Eq. 32. The predicted crack size at the end of the cycle is thus $a_2 = a_1 + \Delta a$.

For the subsequent cycle of loading, the above procedure is repeated with K values computed using the updated crack size a_2. The procedure is continued cycle-after-cycle until the critical crack size, a_c, is reached, resulting in a relation between crack size and number of cycles. These types of crack growth calculations are well-suited for a computer and straightforward to program. Crack growth life for different types of flaws at different structural locations can be investigated through use of different K relations. For instance, in the example shown in Fig. 10-36, growth of a quarter-circular corner crack could be predicted using existing K relations for that crack configuration. Also, the influence on predicted crack growth life of various initial crack sizes or ratios of (a/1) could be determined if desired.

10.4.6 Crack Propagation Life Prediction for Variable Amplitude Loadings

Many structural components experience service loadings which resemble those depicted in Fig. 10-37. In order to predict crack growth for these types of loadings, several approaches are possible. None of the approaches can be said to be clearly superior at this time since the amount of testing done to date has not been substantial enough to indicate which approach might be preferable. However, all of the approaches share a common element. They rely upon constant amplitude da/dN − ΔK data as an input, just as crack initiation life prediction methods utilize constant amplitude ε-N or S-N input data.

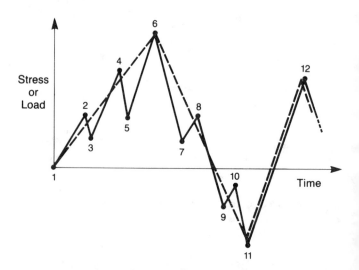

Fig. 10-37 Segment of an irregular load history.

Another factor that can potentially complicate crack growth prediction for irregular loadings is that occasional high tensile loads, interspersed with lower level loading, can cause crack growth to be much slower than it would have been without the high loads (assuming the high loads do not cause gross crack extension or failure). This is often termed "crack retardation." On the other hand, occasional large compressive loads exert an opposite effect, but not to the same extent as tensile loads of the same magnitude. Such phenomena are frequently referred to as "sequence effects" and will be described more fully in Section 10.5 (Practical Aspects of Life Prediction). Although methods have been proposed for taking such effects into account in crack growth predictions, they are either semi-empirical or require elastic-plastic analyses. Also, the sequence of loading experienced in service can vary considerably from vehicle to vehicle and thus knowledge of loading sequence is generally uncertain or unknown, making it difficult to try to take sequence effects into account. In view of those considerations, the approaches to be described will neglect sequence effects. This can be justified since neglecting the beneficial effect of intermittent high tensile loads is conservative, and the tendency of large compressive loads to accelerate growth is generally a second-order effect.

The *first* approach for predicting crack growth is to select an initial crack size and to determine a critical crack size based on the highest tensile stress in the history or block of loading, using methods previously described for constant amplitude loadings. Increments of crack growth are then computed successively for each rising tensile stress range, such as 1 to 2, 3 to 4, 5 to 6, and zero stress to 12 in Fig. 10-37. Since compressive stresses are neglected, the range 11 to 12 becomes zero to 12. For each range, the corresponding R-ratio is determined and used in computing the given crack growth increment from a relation such as Eq. 32. Crack size is updated at the end of each range and a running sum is kept for comparison with the initial crack length.

If the loading history consists of the same block repeated again and again, then it is also possible to determine crack growth vs. number of blocks by a procedure that reduces computations considerably. Referring to Fig. 10-38, crack growth for a block is computed for the initial crack size and a number of other sizes between the initial and critical sizes. This yields a set of values of crack growth increment per block, $\Delta a/\Delta B$, for the different crack sizes. A curve of crack growth vs. blocks can then be constructed as illustrated in Fig. 10-38, extending from initial to critical sizes. This can be done readily with a computer program.

A *second* approach to predicting crack growth life is to perform cycle counting on a load (stress) history using either the ordered overall range or rainflow methods. For example, in the case of the stress segment in Fig. 10-37, overall ranges are shown by the dashed lines. Crack growth increments would be calculated for the rising tensile overall ranges from 1 to 6 and zero stress to 12, using the same kind of procedure as described previously. Rainflow counting would also produce overall ranges as well as numerous smaller intervening

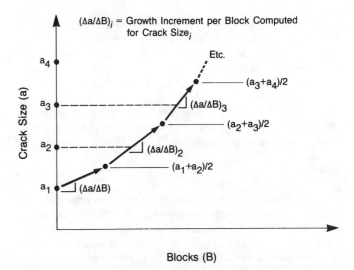

Fig. 10-38 A procedure for estimating crack life in terms of blocks.

cally on the order of three, neglecting the smaller ranges as in the ordered overall range methods leads to results that are generally close to those of rainflow counted load histories. Details of implementing crack growth predictions by the ordered overall range methods leads to results that are generally close to those of rainflow counted load histories. Details of implementing crack growth predictions by the ordered overall range method and by rainflow counting are given in Refs. [49] and [50], respectively. Use of either of these two cycle counting methods offers the advantage of compatibility with crack initiation life prediction procedures made with either method. In other words, a given load block or history can be cycle-counted once and the results used for both initiation and propagation analyses. The ordered overall range method retains the sequence of the original loading, while the rainflow method loses it. This difference would be important only if one wished to try to account for sequence effects. An evaluation of test data [49] on fatigue crack growth under irregular loading shows that the range-by-range approach described earlier, or the cycle counting approach just described, both produced predictions that were in reasonably good agreement with observed growth. It is thus not yet clear which approach is more accurate, as mentioned earlier.

The *third* and final approach to be considered here is intended for use primarily with random loadings (that do not contain infrequent high tensile loads). A loading spectrum can be characterized in terms of statistical parameters such as root-mean-square stress range. For example, it has been found in one study [51] that the average crack growth rate for random loadings correlated well with constant amplitude data through the relation

$$da/dN = C(\Delta K_{rms})^n \qquad (10\text{-}34)$$

where: ΔK_{rms} = the root-mean-square stress intensity range for the loading. Others [52,53] have also obtained good correlations with similar spectrum characterizations for random loadings. In these studies of crack growth under random loading, stress spectra were represented by a continuous, unimodal distribution, in particular, by a Rayleigh distribution function. The usefulness of this type of approach for load histories described by other types of distributions is uncertain. Also, such statistically-based approaches lose the loading sequence, but again that is only of concern if sequence effects cannot be neglected and an attempt to take them into account is thereby necessitated.

10.4.7 Summary

An outline of the general procedure recommended here for estimating crack growth life is shown in Fig. 10-39. There are a number of important details involved in implementing this procedure for cracks emanating from notches, or growing through residual stress fields, etc. Those details are discussed in Section 10.5 (Practical Aspects of Life Prediction). The selection of an initial crack size is in some ways the most uncertain aspect of a crack growth life evaluation. If initial defects are likely to be present from the time of fabrication, then the analyst must decide on an appropriate size based on knowledge from inspections or experience with the particular component and manufacturing processing being considered. If the initial size is taken as that predicted to exist after crack initiation life has been expended, then a size of about two to three millimeters could be used. That size corresponds approximately to the size of cracks that exist at fracture in the small lab specimens used to generate the input data (ε-N or S-N) for crack initiation life prediction methods. More sophisticated approaches [54,55] have been proposed for defining the transitional crack size between initiation and propagation, and the analyst may wish to investigate their use.

10.5 Practical Aspects of Life Prediction

10.5.1 *Crack Initiation vs. Crack Growth Analysis/Small Cracks*

The preceding sections have outlined methods of crack initiation and crack growth life prediction. The question of when to use a crack initiation analysis or a crack growth analysis or a combination of the two remains an issue that is best resolved case by case. However, some guidance can be given.

In applications where inspection dictates the existence of a material or geometric notch, crack growth analysis is appropriate in estimating remaining life. Likewise, in situations where significant defects exist and are considered acceptable or are undetectable as the part enters service, a crack growth analysis is appropriate. Crack growth analysis may also be relevant during design/analyses for materials selection and component sizing. Such is the case where significant defects are considered acceptable or may be undetectable. "Safe" life in these cases requires either no growth, or slow stable crack growth to crack sizes that are stable—that is well below the critical size at limit load conditions. In all crack growth analyses, care should be taken to ensure that the analysis procedure is valid. For example, the use of LEFM for small* flaws in ductile materials may underestimate growth rates and so overestimate the actual life. In such cases care should be taken to "benchmark" the analysis procedure using tests that simulate the component and the service.

In other applications where defects are avoided by careful selection of materials and quality fabrication, coupled with quality control, a crack growth analysis may underestimate actual service lives significantly. Crack initiation analysis is best suited for these cases. Care must be taken to account for geometric/fabrication related stress raisers.

The question of whether initiation predictions should be followed by growth predictions can be answered a number of ways, depending on how "initiation" is defined. If initiation is defined as a microscopic crack, periods of initiation and propagation tend to be balanced, or tipped toward propagation. However, recent literature suggests that there is a significant problem in adapting LEFM to small* cracks [56-60], particularly for soft steels. The reason for this problem is evident in Fig. 10-40, and is supported by data shown in the same figure for steels. The figure indicates the incompatibility of the LEFM crack growth threshold concept and the crack initiation threshold concept (or endurance limit).

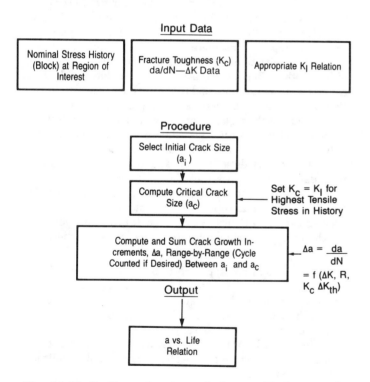

Fig. 10-39 Outline of approach for predicting crack growth life.

---

*A "small" crack, as used in this context, means one that is significantly smaller than the radius of the geometric notch that it may be growing from, or one that is approximately the same size (or smaller than) the plastic zone it creates. In this regard, the absolute size of a "small" crack cannot be defined in general, but must be defined for each case.

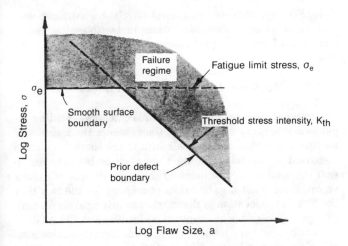

a) Failure boundary defined in the framework of stress intensity and fatigue limit for the two cases of smooth and defected surfaces.

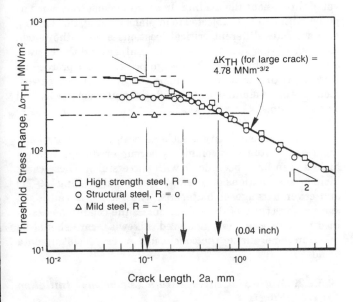

b) Observed on coordinates of log stress - log surface crack length (after [61])

Fig. 10-40 Fatigue crack growth rate behavior of short cracks near threshold.

On coordinates of stress and crack size it is apparent from the schematic that the LEFM threshold concept overestimates safe operating conditions (compared to the fatigue limit) when the flaw size is small. The constant amplitude data shown for several steels support the use of the fatigue limit or endurance concept, and so indicate problems in using LEFM at smaller crack sizes due to overestimating allowable stress or life. Problems evident in this format for near-fatigue limit conditions develop for similar reasons at higher stresses and finite lives. The finite life situation can be further complicated for LEFM analysis if significant notch root plasticity develops.

Data plotted in Fig. 10-40 also show that for "large" flaws the use of a crack initiation analysis overestimates the near-fatigue limit behavior. In this case, the data shown for the steels indicate that the fracture mechanics based prediction correctly portrays the cracking behavior. It follows from Fig. 10-40 that the choice of an initiation or propagation analysis for life prediction should depend on the flaw size. Moreover, as is evident in Fig. 10-40 for the three steels, the flaw size separating correct use of initiation and propagation analyses depends on the material—in this case the grade of steel. Examination of the figure shows that this transition flaw size ranges from about 0.1 mm (0.004 inch) to 1.0 mm (0.04 inch).

Guidance as to whether to use a crack initiation or crack growth analysis also may be drawn by analogy to Fig. 10-7 and also Fig. 10-15. These figures characterized notch root response in terms of notch stresses and strains imposed on a smooth specimen, with cracking beyond the notch represented by precracked specimen data. It follows that the upper bound crack length for initiation analysis is the length over which the cracking processes are similar at the notch, and in the smooth specimen (that is subjected to stresses and strains of the same magnitude as predicted in the notch). Assuming that this is the case, the behavior in the notch field can be represented using smooth specimen behavior. Beyond the notch field precracked specimen data can be used to characterize the crack. In this region the crack is no longer affected by the notch.

Fracture mechanics stress intensity factor solutions for notched geometries can be used to assess the distance from the notch (transition crack length a_t) where the crack no longer recognizes the effect of the notch. In 1976, Novak and Barsom [62] suggested a definition for the lower bound transition crack length:

$$a + \sqrt{dp} \qquad (10\text{-}35)$$

or $a_t = 0.25r$, for circular notches

where d is the semiaxis length of the notch and is the root radius (equal to r for circular notches).

In 1977, Smith and Miller [63] proposed

$$a_t = 0.13\sqrt{dp} \text{ or}$$
$$a_t = 0.13\ r, \text{ for circular notches}, \qquad (10\text{-}36)$$

following an analysis similar to that of Novak and Barsom. Also, following similar logic, Dowling [64] in 1980 suggested a transition length of

$$A_t \simeq 0.10r . \qquad (10\text{-}37)$$

It follows from fracture mechanics considerations that the transition from the notch field also depends on the local geometry. For radii on the order of 1 cm, these criteria for a_t

lead to values on the order of 1 mm (0.04 inch). This value is similar to the upper bound found in Fig. 10-40 based on material considerations.

Finally, guidance as to the value of a_t may be drawn from the sensitivity of errors in life predictions to different values of a_t. Such results are consistent with the findings of the fracture mechanics analysis in that, for a given material and constant net section, they show that a_t depends on the geometry (both root radius and depth of notch). Such an analysis shows that for deep, sharp notches significant nonconservative errors are made at values of $a_t \simeq r$ or greater. Test results for such specimens show almost immediate initiation, so that the nonconservative error arises in part from the contribution of the predicted (and over-estimated) crack initiation life. These results suggest that crack growth predictions offer the best approach to predict the life of components with sharp notches. In cases involving sharp notches, the best approach may be to benchmark the prediction process by lab testing on actual hardware, then empirically shift the predicted trends to match the benchmarks. Problems may also be encountered in applying LEFM to blunt notches in lower strength materials. The benchmarking procedure is also suggested in such applications.

In summary, both the material and the geometry must be considered in deciding to use an initiation or propagation analysis or a combination of the two as the basis for a life prediction. In some situations, the choice may be clearcut—but in others, when the choice is uncertain such as when dealing with lower strength materials, consideration should be given to benchmarking the analysis by testing of actual hardware. In critical applications this will probably be necessary regardless of the material used.

10.5.2 Multiaxial Effects

Some structures or components may be loaded by a single load or by a group of loads acting in phase that can be represented by a single load vector. Other components may be loaded by a group of loads that act out of phase. In these cases the loading is multiaxial.

At the site where fatigue cracks eventually form in a component the state-of-stress may be uniaxial or multiaxial and the state-of-strain may be biaxial or multiaxial, even if the remote loading is uniaxial. Thus, multiaxial effects may develop through the local state-of-stress or strain. The analysis and discussion to this point has tacitly assumed that the loading is uniaxial (or in phase multiaxial) and that the local state-of-stress and strain in the component of interest is similar to that in simple smooth or precracked laboratory specimens.

This section does not focus on the effect of multiaxiality as much as it does on the significance of the effects of multiaxiality on life prediction. The significance of the effect of multiaxiality differs for crack initiation and for crack propagation. Therefore, these topics are considered in separate sub sections. Multiaxial effects of loading and stress/strain state also differ, so these topics are dealt with independently when possible. Finally, because multiaxial effects are still not well understood in general, readers concerned with multiaxial effects are referred to recent literature and literature reviews [65-68].

10.5.2.1 Multiaxial Loading—Initiation Effects

Two separate cases exist within this category—loading in-phase and loading out-of-phase. Cases where the multiaxial loading is in-phase can be dealt with in the same manner as discussed for uniaxial loading. Care must be taken to represent the combination of loads by the equivalent (static) load vector. Care must also be taken to account for the fact that the K_t and the location of the maximum principal stress may change as a result of the multiaxial loading and the relative proportions between the loads [29]. Aside from these two complications that influence how the input load is computed, the analysis can proceed as developed earlier using the appropriate notch analysis. Out-of-phase loading, however, is not handled as easily. In this case different principal planes are activated throughout the loading. In certain geometries, such as plates with notches subjected to in-plane loading, this loading also activates different critical locations around the notch. Analysis of this situation is, in general, beyond the current state of the technology. The design process must proceed by very approximate analyses supported by extensive experimental studies simulating the material and geometry as well as the loading.

Despite the complexity associated with out-of-phase loading, certain situations commonly encountered in ground vehicle design have been dealt with successfully. These cases include combinations of tension, torsion and bending [69-71] and cover a range of materials. The reader is cautioned against direct use of these data unless the conditions examined match those being analyzed. Likewise, care should be taken in generalizing the conclusions or extrapolating data trends.

10.5.2.2 Multiaxial Stress-Strain Conditions—Initiation Effects

A number of studies have been completed on the effects of a multiaxial stress-strain state on unnotched and notched specimen fatigue behavior. Studies on unnotched specimens subjected to proportional stressing are numerous (i.e., components of stress increasing proportionally or in-phase). These results suggest that equivalence criteria such as Von Mises (or octahedral shear stress) and Tresca provide a viable basis to relate the multiaxial state of stress being analyzed to that which exists in the reference data. Recent data indicate that the quality of the correlation achieved between different states of stress or strain may be enhanced by including the effect of the hydrostatic component. Finally, correlation of data at long lives seems best in terms of stresses whereas at shorter lives correlations are best presented on a strain basis. Data of interest to the ground vehicle industry also represent combinations of axial loading, torsion, and bending [65,68].

A few studies have been made of multiaxial effects at notch roots. In this case multiaxiality may involve stress states that are uniaxial or biaxial, since the local strain field is always biaxial. Differences in stress state at notches arise through a Poisson effect or local constraint. In sheets that are thin compared to the diameter of the notch the absence of constraint through the thickness means that a significant stress does not develop in the thickness direction at the notch root. As the thickness, t, to diameter, D, ratio increases, constraint develops and the notch field tends from plane stress (uniaxial at the notch root) to plane strain. Accurate predictions of notched specimen fatigue performance have been achieved for proportional loading at notches [69,72] using equivalence criteria like Von Mises (octahedral shear stress) over a range of local stress states. The approach that was used is the same as detailed earlier except that damage was assessed in terms of equivalent notch root stress and strain.

10.5.2.3 Multiaxial Conditions—Propagation Effects

Multiaxial loading effects on crack growth rate and fracture toughness have been studied extensively for macroscopic cracks. In general, these studies suggest that as long as the inelastic zone at the crack is confined, i.e., small compared to the crack length, stress components parallel to the crack do not contribute significantly to the growth rate. Therefore, prediction of crack growth under proportional loading can follow the approach detailed earlier, with superposition of the loading resolved in the direction normal to the crack plane. Local stress state effects are not expected to be significant in that thickness (constraint) effects in the range of practical interest do not appear to be significant, or are, in any case, second order effects.

Situations for which the loading produces significant crack tip plasticity do not seem to fit the pattern for small scale yielding. In such cases, there is an effect of the component of load parallel to the crack plane. In particular, compression applied parallel to the crack plane accelerates the growth rate compared to the uniaxial case for Mode I cracking. Analysis [73,74] and experiment [75] show this acceleration may be significant. Further simulated service testing should be considered if this situation is encountered in design. Testing and empirical evaluation should also be considered as the basis for decisions involving the effect of nonproportional loading on crack growth if significant crack tip plasticity develops due to the loading component parallel to the crack plane.

10.5.3 Environmental Effects

The discussion up to this point regarding fatigue damage analysis has been based on the premise that reversed slip is the primary damage mechanism. Factors that alter the reversed slip process from that which typically occurs in the laboratory may change the rate of damage. The effect of factors like humidity, salt water, temperature, etc., can be assessed by comparing life or growth rate for these conditions with that for the ambient laboratory conditions assumed so far in this chapter.

Fig. 10-41 illustrates the effect of saltwater on fatigue life [76]. While characteristic of the effect of environment, these results represent specific conditions, i.e., specific salt concentrations and temperatures. These and other variables like pH and oxygen concentration strongly influence the observed trends. It is clear from Fig. 10-41 that a saltwater environment significantly reduced the life to initiation and increased the growth rate for both materials. Order of magnitude decreases in smooth and blunt notched specimen fatigue life imply significant surface effects of the environment. Apparently, the environment enhanced the decohesion process or concentrated slip. For example, some aggresive environments cause surface pitting that concentrates slip and increases the damage rate significantly. Alternatively, the environment may cause the initiation process, which in ambient conditions leads to transgranular cracking, to change to grain boundary attack, providing a ready-made crack-like surface defect very early in life. Once cracking develops, the environment may, for example, change the mechanism of growth from fatigue by reversed slip to brittle mechanisms that increase the growth rate.

Regardless of the specific mechanism, some environments produce significant changes in the damage rate and fatigue life, as compared to data developed for ambient conditions. When environments other than ambient are present at critical areas, care should be taken to develop application-specific data for use in life prediction. When this is not possible, guidance as to the effect of environment may be taken from the environments for aggressive literature for comparable materials and conditions. Care should also be taken to ensure that ambient laboratory conditions are representative of ambient service conditions, since, for some materials, ambient laboratory conditions may be more aggressive than service.

The life prediction procedure can be the same as detailed earlier, except that the in-environment data should be used in place of ambient results. Certain additional complexities may be encountered when the damage process is strongly rate-dependent. Life prediction in such cases stretches the technology to its limit. Such problems are best dealt with by combining current understanding with experiments, in support of the analysis, during the design phase. References useful in dealing with aqueous and gaseous environments include [75-85], while those dealing with temperature effects include [86-92].

10.5.4 Fabrication Effects

Two specific fabrication effects on fatigue performance are considered—welding and fretting fatigue. Welding is a means of fabrication whereas fretting is a consequence of the method of fabrication, the design, and the application.

Life prediction for weldments has been approached in three ways: (1) total life based on nominal stress, empirically calibrated by testing of a specific weld configuration and loading, represented by ower bound tread curves in structural and welding design codes [92]; (2) propagation from some assumed flaw size using LEFM calibrated by testing flawed specimens and weldments [94,95]; and (3) total life based on

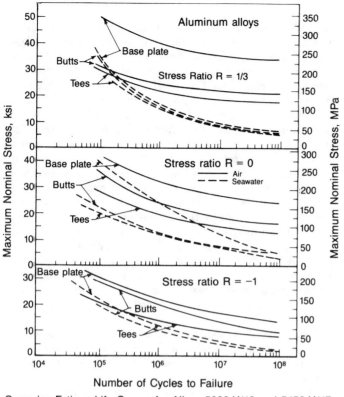

Corrosion-Fatigue-Life Curves for Alloys 5086-H116 and 5456-H117

aluminum

Fig. 10-41 (1) Effect of saltwater on fatigue performance.

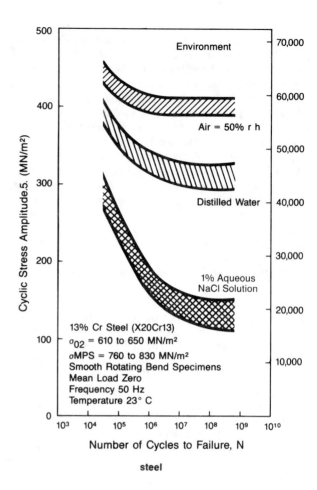

steel

a) on fatigue resistance

Fig. 10-41 (2) Effect of saltwater on fatigue performance.

using the critical location approach, calibrated by testing smooth and flawed specimens [96-98]. All three approaches are detailed in the indicated references. The purpose here is to indicate that these approaches exist, and to comment on which may be relevant in ground vehicle design.

Recall from §10.5.1 that the *total* fatigue life has been identified as the sum of *initiation* and *propagation* periods. In cases where sharp defects exist, or the cross section remaining after initiation is very large, the life may be dominated by crack propagation. In this case, the first two approaches to weld fatigue life estimation are reasonable, and indeed, for certain applications, these approaches share a common data base and are related through. When sharp, crack-like defects are absent from the weldment and secondary bending is absent, the first and third approaches are most appropriate. However, certain configurations used in the ground vehicle industry (e.g. thin section weldments) are not represented by the structural (bridge and building) configurations that form the basis for calibration of the first approach. Thus, the first method may be of limited practical utility unless weldments unique to the ground vehicle industry are tested to develop the necessary calibration data. Likewise, the third approach may be of limited value—although the recent work on spotwelds [99] and a recent generalization of the related notch analysis [97] will broaden the utility of this method significantly.

Fretting is a mechanism related to a small amplitude, differential displacement at the point of contact between adjacent components. Fretting fatigue refers to the combination of fretting and fatigue mechanisms and is an important consideration whenever there is load transfer between mating surfaces [100]. Because fretting involves the surface, its influence is on crack initiation with little or no effect on the subsequent propagation behavior. It is a common cause of failure because it results in a significant reduction in life, as is illustrated in Fig. 10-42. Life prediction, accounting for the presence of fretting-fatigue, should be considered in situations involving shrink-fits on rotating components and in low load-transfer situations between components attached with bolts or rivets and possibly spotwelds. The approach can be similar to that detailed earlier for reversed slip fatigue, except that damage should be summed for the effects of both damage mechanisms (e.g. see [101]).

10.5.5 Processing Effects

Processing effects refers to the affect on material properties of changes that alter the microstructure, the internal

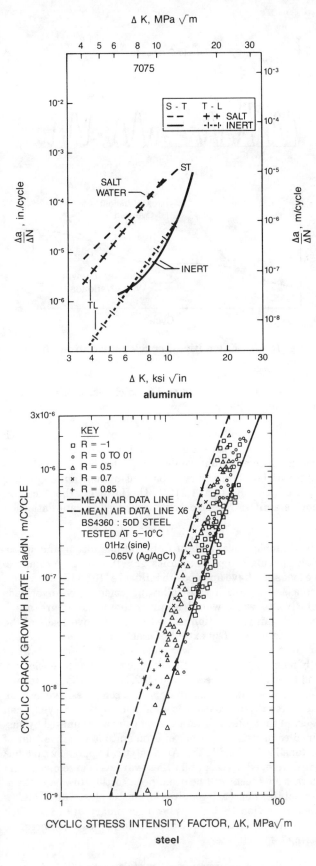

aluminum

b) on crack growth resistance

Fig. 10-41B Effect of saltwater on fatigue

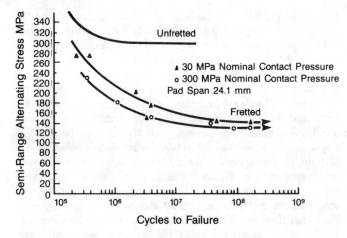

Fretting Fatigue 3-1/2 NiCrMoV Steel
Zero Mean Stress
Single Pair of Contact Pads

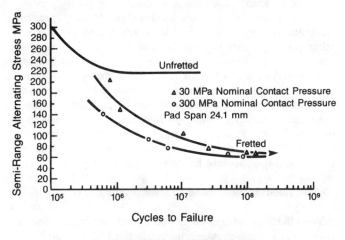

Fretting Fatigue 3-1/2 NiCrMoV Steel
300 MPa Mean Stress
Single Pair of Contact Pads

(Similar Results Develop at Shallow Notches ~ $K_t \leq 3.5$)

Fig. 10-42 Effect of fretting on fatigue behavior.

stresses, and the character of the surface. Processing effects should also be considered when components are tested, if the microstructure, internal stresses, and surface conditions differ throughout the body and are different from the conditions used to develop simple specimen reference data. A key factor to evaluate is whether gradients in microstructural size or changes in grain orientation exist in going from the critical location into the bulk. If such gradients exist and are significant, they generally preclude the development of simple specimen data representative of the overall cracking process for the component. If this situation exists, great care must be exercised in choosing or developing reference data. This is true for initiation and crack growth data.

For crack initiation analyses, particular care should be

by cooling gradients in casting, and which are subsequently cut in machining. Care should also be taken when dealing with surfaces that are torn during stamping or otherwise finished differently than the surface of the specimen used to develop baseline fatigue data. Handbooks exist that indicate how to degrade S-N data to account for selected processing operations. Chapter 4 presents a discussion of processing effects and should be consulted if appropriate data are not available. Processing effects, once transferred to an S-N basis, can be accounted for in life prediction by replacing the fatigue life curve discussed earlier by a curve which includes processing effects. Where residual stresses are important, they can be accounted for by changing the origin for the stress-strain analysis, assuming elastic material response. Note that this approach may be nonconservative for compressive residual surface stresses if the history causes significant local plasticity. In such cases, the prediction should be made without regard to the surface residual stress.

Crack propagation trends are also influenced by processing. Intermediate crack growth rates depend strongly on microstructural orientation, but they are not so sensitive to other variables such as grain size and the specific microstructure (e.g., pearlite vs. bainite vs. martensite). Very low and low crack growth rates (high and low K regimes) are affected by all of these factors. A factor of 10 increase in crack growth rates may occur when cracks are grown parallel to, rather than across, highly oriented microstructures. It follows then that, as for initiation, care must be taken in selecting data from the literature or in designing specimens and developing application-specific data for crack growth analysis.

10.5.6 Load Sequence and Crack Closure Effects

The effects of loading sequence on fatigue crack propagation behavior were first emphasized in the 1960's when it was found [102-104] that application of intermittent tensile overloads could cause crack growth rates following those overloads to be much less than it would have been in their absence, as illustrated in Fig. 10-43. Experiments with high-to-low load sequences also showed a longer crack growth life than would have been predicted on the basis of the summation of crack growth for each cycle using constant amplitude da/dN-ΔK input data.

This phenomenon has come to be known as crack retardation and received considerable study during the 1970's for a number of steels [105-107] and aluminum [108-112] alloys. Several empirical trends emerged from these studies. The amount of retardation increases as the ratio of the tensile overload to the level of subsequent loading increases. In fact, if the overload ratio is large enough, crack arrest can occur, with the ratio depending on material, specimen geometry, etc. Crack retardation tends to increase if several consecutive overloads are applied periodically instead of single overloads periodically. However, application of too many consecutive overloads will cause growth itself that counteracts any gains in life due to retardation. If the loading following tensile overloads has an R-ratio greater than zero (positive mean stress levels), retardation will be diminished with increases in the R-ratio. Finally, retardation is most pronounced in

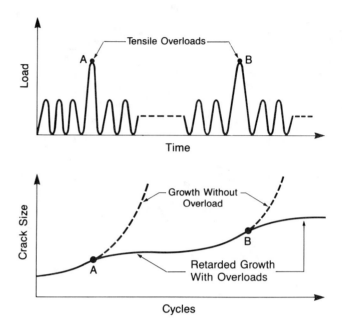

Fig. 10-43 Schematic of crack retardation effect.

thinner sections and is reduced in thicker sections or higher yield strength alloys.

Just as high-to-low loading sequences can produce crack retardation, a low-to-high sequence tends to accelerate growth relative to what would be expected based on calculating growth on a cycle-by-cycle basis using constant amplitude input data [108,110,111,113,114]. However, the "acceleration" stabilizes rather quickly compared to retardation effects.

While tensile overloads of sufficient magnitude and appropriate spacing can prolong crack growth life, large compressive loads can have an opposite influence [106,115]. However, increases in growth rate following compressive overloads (relative to what would occur without them) are much smaller than retardation caused by tensile overloads of the same magnitude. The extent of crack growth "acceleration" appears to be greater in lower yield strength alloys and varies with geometry and type of loading (e.g., uniform tension vs. bending). It has also been found [102,115,116] that if a tensile overload is followed immediately by a compressive overload, crack growth retardation is diminished. It has even been observed that effects of retardation can be eliminated by sustained compressive loading of a magnitude equal to that of the tensile overload [117]. An analytical approach that has been developed to account for these various load sequence and interaction effects will be described shortly. However, to implement this approach requires an elastic-plastic analysis. As noted in Section 10.2, it has been customary in LEFM crack growth life predictions to neglect the influence of compressive loadings.

A number of semi-empirical models [118-120] have been developed in attempts to account for crack retardation due to tensile overloads (but not compressive loading effects). A summary of the main features of Wheeler's model [118] will illustrate the semi-empirical nature of these approaches. Wheeler proposed that the growth rate for a given cycle, i, following a tensile overload could be represented by

$$(da/dN)_i = (C_p)_i [C(\Delta K)_i^n] \quad (10\text{-}38)$$

where: $(\Delta K)_i$ = applied stress intensity range
$(C_p)_i$ = a crack retardation parameter, with possible values ranging from 0 to 1, indicating crack arrest or no retardation at the extremes.

Referring to Fig. 10-44, the $(C_p)_i$ parameter is expressed in terms of the current crack tip yield zone size for cycle i relative to the yield zone size caused by the tensile overload, according to

$$(C_p)_i = [(r_y)_i/(a_p - a)]^m \quad (10\text{-}39)$$

where: $(r_y)_i$ = current yield zone size
$(a_p - a)$ = distance from the crack tip to the boundary of the yield zone caused by the tensile overload
m = an empirically-determined "shaping exponent."

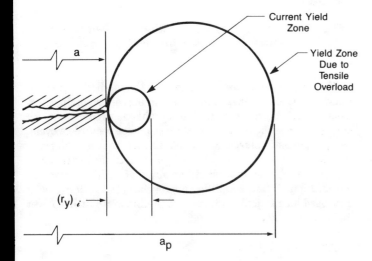

Fig. 10-44 Crack tip yield zones in Wheeler's crack growth retardation model.

Retardation is assumed to be at its maximum immediately after application of the overload and to diminish thereafter as the crack penetrates the yield zone induced by the overload. Reasonably good predictions have been achieved with this model through proper choice of the shaping exponent m, but testing usually must be done to guide selection of m, thus limiting general applicability of the model and other models similar to it.

The sequence effects described previously, as well as R-ratio (mean stress) effects, can be reconciled to a large extent by a physical mechanism known as crack closure. Traditionally it has been assumed that a crack tip would open and close in fatigue cycling at zero load. Elber [120] found that during constant amplitude loading, a crack actually closes when a tensile load is still being applied and it does not open again until a sufficiently large tensile load is applied in the next cycle, as illustrated in Fig. 10-45.

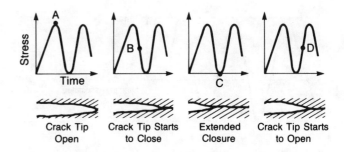

Fig. 10-45 Illustration of crack closure.

The mechanism causing crack closure can be explained as follows. A yield zone is always present surrounding a crack tip, as depicted in Fig. 10-46.

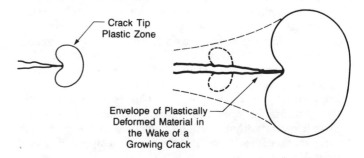

Fig. 10-46 Schematic of crack tip yield zones and the "plastic envelope" left in the wake of a growing crack.

As a crack grows through this zone (that increases in size with increased stress intensities), an envelope of plastically deformed material is left in the wake of the crack. Residual tensile deformations are present in this envelope. Upon unloading, the residual deformations cause crack surfaces to contact and close while the component or specimen is still experiencing tensile loading. The contact, in turn, produces compressive residual stresses behind the crack tip with no load applied. Upon subsequent tensile loading, the compressive residual stresses and corresponding residual displacements must be overcome before the crack will open again. Since a crack will grow only when open, the effective stress range driving crack growth is that portion above the crack opening stress level. Based on those considerations, Elber proposed that crack growth rates be correlated in terms of an effective stress intensity range illustrated in Fig. 10-47 and given by

$$da/dN = C(\Delta K_{eff})^n \quad (10\text{-}40)$$

where: $\Delta K_{eff} = K_{max} - K_{open}$ for $K_{open} > K_{min}$
$\Delta K_{eff} = K_{max} - K_{min}$ for $K_{min} > K_{open}$
C, n = empirical constants fitted to R = 0 input data

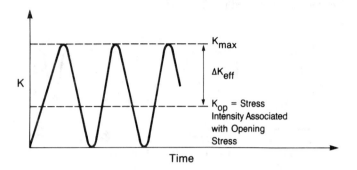

Fig. 10-47 Effective stress intensity range due to crack closure.

The value of K_{open} varies from alloy to alloy, with section thickness, and with the nature of the loading.

The crack closure-effective stress intensity range concept is useful in explaining R-ratio (mean stress) effects on crack growth. As shown in Fig. 10-48, K_{open} increases with increased R-ratios but ΔK_{eff}, as defined in Eq. 40, increases even more. For the same applied ΔK, the increase in ΔK_{eff} with R-ratio produces higher growth rate values than those at R = 0. The effects of R-ratio, as predicted by crack closure, are generally in good agreement with experimentally observed trends. Thus, once the constants C and n in Eq. 40 are fit to R = 0 data, that relation can be used to predict growth rates at other R-ratios without the need for empirically based formulations such as Eqs. 32 or 33.

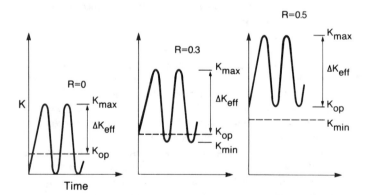

Fig. 10-48 Schematic of differences in effective stress intensity range with R-ratio of applied loading.

The crack closure concept can also be used to account for load sequence effects such as retardation due to tensile overloads or acceleration due to compressive overloads. For example, consider the low-high-low sequence shown in Fig. 10-49. When the loading is increased, a period of transient adjustment occurs in which the opening stress increases from its stabilized value at the lower load level to the higher stabilized value at the higher level. During the transition, ΔK_{eff} will temporarily be larger than the stabilized value at the higher load level, causing transient acceleration in growth rates relative to what would be expected at the higher load level. Later, when the load level is decreased, the opening stress will gradually return to the stabilized level at the lower load level, but during this period, ΔK_{eff} is much smaller than its stabilized value at the lower load level, causing crack growth retardation.

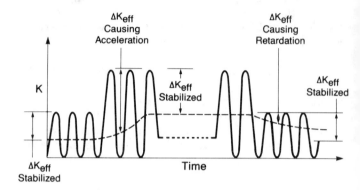

Fig. 10-49 Illustration of variation of effective stress intensity range with loading sequence.

The ΔK_{eff} approach is thus useful in the prediction of crack growth rates for arbitrary loading sequences. The main drawback to the approach is that to be able to predict how the opening stress varies with loading sequence requires a cycle-by-cycle, elastic-plastic analysis [121,122] that can be time-consuming and expensive for irregular loadings characteristic of those encountered in ground vehicle service. In recognition of that, simplified approaches [48,123] have been proposed for random loadings, but their general utility is yet to be determined.

10.5.7 Residual Stress Effects

Many, if not most, components contain residual stresses (self-stresses) from manufacturing processes and assembly operations. Residual stresses may also develop in service due to localized plastic straining at stress concentrations (notches) where cracks tend to form. The purpose of this section is to describe an analytical approach that can be used to account for the effects on crack growth of residual stresses present from the time of fabrication. This can be done using an elastic stress analysis and principles of LEFM. Residual stress effects on fatigue crack initiation resistance of materials are covered in Chapter 4.

To quantify residual stress effects on crack growth, the following items are needed in addition to the usual input data for a crack growth analysis: (1) an accurate estimate of the residual stress field into which a crack is expected to grow, including both the magnitude and distribution of stresses; and (2) a method for determining the influence of residual stresses on the stress intensity factors assumed to govern crack growth behavior. Item 1 can be accomplished by analytical methods in some cases, or by the experimental techniques described in Chapter 7 in most cases. For item 2, the approach employed most frequently involves superposition of the respective stress intensity factors for the applied stresses and for the residual stresses. The K factor for the residual stresses can be obtained analytically by loading the crack faces with the residual stresses that exist normal to the plane of potential crack growth in the uncracked components. Examples of methods for determining such residual stress intensities, denoted K_{res} here, as well as use of superposition are given in Refs. [124-133] Fig. 10-50 shows one such K_{res} factor for a crack growing into compressive residual stresses that decrease linearly from a maximum value at the surface. Note that K_{res} varies with crack length.

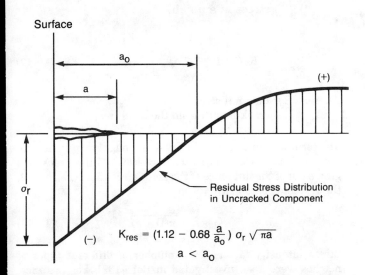

Fig. 10-50 Residual stress intensity factor for a straight-fronted edge crack growing into compressive residual stresses [126].

To utilize K_{res} factors in crack growth predictions, effective stress intensities are defined as:

$$K_{eff, max} = K_{max} + K_{res}$$
$$K_{eff, min} = K_{min} + K_{res}$$ (10-41)

where K_{min} and K_{max} are the stress intensity factors for the minimum and maximum values of applied stress. The term effective stress intensity is often used in describing the superposition approach and should not be confused with the same term used in descriptions of the crack closure approach, as in Section 10.5.1. If $K_{eff, min}$ is greater than zero,

$$\Delta K_{eff} = K_{eff, max} - K_{eff, min} = \Delta K_{applied}$$ (10-42)

and the effective R-ratio is

$$R_{eff} = (K_{min, eff}/K_{max, eff})$$ (10-43)

If $K_{eff, min}$ is less than or equal to zero,

$$\Delta K_{eff} = K_{eff, max}$$ (10-44)

and the effective R-ratio is taken as

$$R_{eff} = 0$$ (10-45)

The effect of residual tension is taken into account through the use of R_{eff} (that is greater than the applied R) in da/dN vs. ΔK relations such as Eqs. 32 and 33. Fig. 10-51 shows an example of how R_{eff} varies with crack growth and how ΔK_{eff} stays the same as $\Delta K_{applied}$ for crack growth through a region of residual tension.

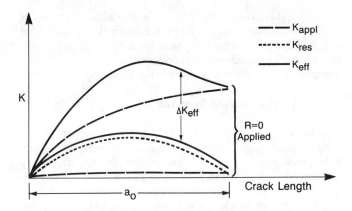

Fig. 10-51 Variation of effective stress intensities for crack growth through residual tension.

On the other hand, the effect of residual compression is to reduce ΔK_{eff} relative to $\Delta K_{applied}$, as illustrated in Fig. 10-52. In this case, R_{eff} is taken as zero, the same as the R-ratio for the applied loading.

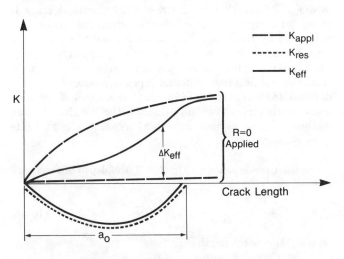

Fig. 10-52 Variation of effective stress intensities for crack growth through residual compression.

The superposition approach described above has been used successfully not only to account for the effect of residual stress fields on crack growth rates but also to predict whether or not cracks may stop growing as they grow into a region of residual compression [134,135]. This possibility can be assessed by comparing ΔK_{eff} to the threshold stress intensity range ΔK_{th}. If ΔK_{eff} drops below ΔK_{th} as a crack grows, it would be expected to arrest. This assumes that the crack is large enough (roughly 0.5 mm or so) so that ΔK_{th} as determined from tests of standard specimens containing large cracks is applicable. The superposition approach has also been able to predict situations in which tensile residual stresses are large enough to cause cracks to grow under applied cyclic compression and then eventually arrest, when they have grown out of the influence of the residual tension [136,137].

Finally, it should be noted that it is also possible to use the crack closure approach described in Section 10.5.1 to account for effects of pre-existing residual stresses on crack growth. However, this requires an elastic-plastic analysis and is thus much more involved than the superposition approach.

10.5.8 *Cracks Growing from Notches*

Fatigue cracks most often form at notches and either grow a short distance and arrest or continue growing to failure. Non-propagating cracks occasionally develop at sharp notches where the high stress concentration and steeply declining notch stress gradient combine to make cracks start quickly but then decelerate in growth and arrest as they grow into the much lower stresses in the bulk of a component. On the other hand, cracks emanating from blunt notches, where the stress concentration and stress gradient are milder, generally grow to failure. This section provides guidelines on predicting the crack growth behavior of fatigue cracks originating at notches, and assessing whether or not they are likely to be non-propagating.

Fig. 10-53 shows an edge notch in a plate under uniaxial loading. Suppose that notch stresses are elastic. A crack is growing through the stress field over which the stresses are magnified by the notch. The stress intensity factor at the crack tip is influenced by the notch stress gradient. Available literature is helpful in analyzing this problem. K_I factors for cracks emanating from different types of notches have been developed [35-37]. These factors can be computed for arbitrary notch stress gradients through the weight function method [130] or various numerical techniques (e.g., finite elements).

For the case depicted in Fig. 10-53, the depth of the notch stress field [138] is:

$$d \approx 0.13 \sqrt{Dr} \qquad (10\text{-}46)$$

where: D = notch depth
r = notch radius.

Within the field, the K_I factor [138] is given by

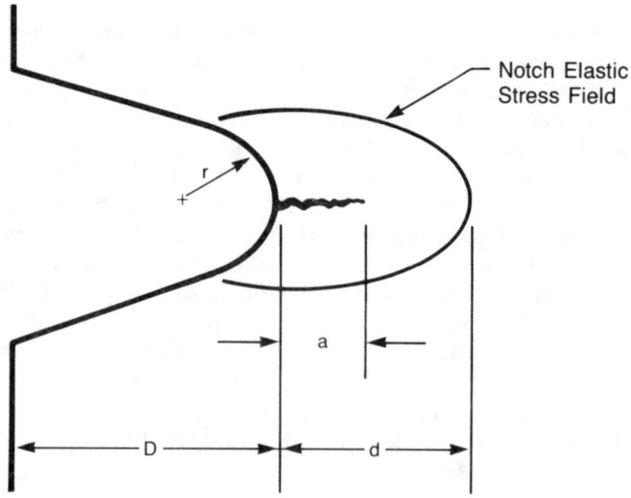

Fig. 10-53 Schematic of zone of concentrated elastic stress at a notch.

$$K_I = \left[1 + 7.7 \sqrt{\frac{D}{r}}\right]^{0.5} S \sqrt{\pi a} \qquad (10\text{-}47)$$

where: S = nominal stress
a = crack length from the notch root.

The term within brackets in Eq. 47 is roughly equivalent to the notch stress concentration factor K_t. If the crack has grown out of the influence of the notch stress field (i.e., a > d), the K_I factor becomes

$$K_I = S \sqrt{\pi(D + a)} \qquad (10\text{-}48)$$

Stress intensity factors for a number of different types of notches were also investigated in Ref. [139]. For instance, Fig. 10-54 shows a numerical solution for the K_I factor for cracks growing from both sides of a circular hole in a plate under uniaxial tension. The K_I factor for a crack that neglects the influence of the notch stress field agrees reasonably well with the numerical solution for cracks that have grown away from the notch. The K_I factor that is taken as the notch stress concentration factor, K_t, times the K_I factor for an edge crack agrees reasonably well with the numerical solution for smaller crack sizes. Thus, if "exact" solutions are not available for cracks growing from notches, the two bounding cases given below could be used in an LEFM analysis as long as notch stresses are elastic:

$$K_I = K_t \times (K_I \text{ without notch influence}) \text{ for } a < \ell \qquad (10\text{-}49)$$

$$K_I = K_I \text{ for crack of length } (a + D) \text{ for } a > \ell \qquad (10\text{-}50)$$

where: D = relevant notch dimension such as notch depth or hole radius
a = crack length from the notch surface
ℓ = transition size between solutions based on the limiting cases in Eqs. 49 and 50

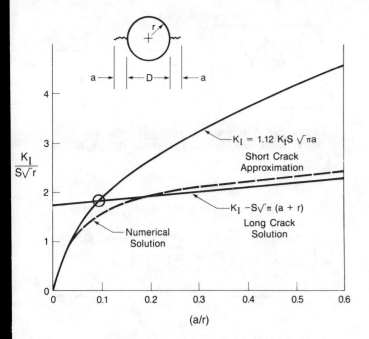

Fig. 10-54 Comparison of stress intensity factors for cracks emanating from a circular hole in a plate under uniaxial tension [139].

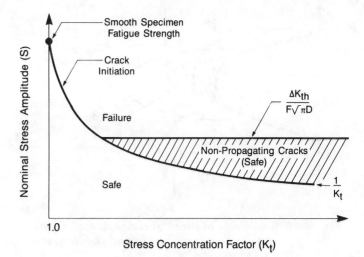

Fig. 10-55 Notch fatigue behavior in the long-life regime.

For various notch configurations, the value of ℓ is found to be 0.1 to 0.2 of the notch root radius [139]. This guideline produces values which are generaly in the same range as those of Eq. 46. For example, in the case of a circular hole of 5 mm radius in a wide plate, the guideline estimates ℓ to be on the order of 0.75 to 1.5 mm, while Eq. 46 gives a value of d of 1.38 mm. Knowledge of the extent of influence of the notch stress field can be important. If an initial crack size for use in an LEFM analysis is taken to be outside of this zone of influence, complications in trying to account for the influence are circumvented. In most cases, transitional crack sizes between crack initiation and crack growth (see Section 10.5.1) are outside of that influence.

In order to evaluate the potential for non-propagating cracks in elastic notch stress fields, the following criteria [138-140] has been proposed. Referring to Fig. 10-55, cracks are predicted to initiate in the long-life fatigue regime at values of nominal stress amplitude equal to the smooth specimen fatigue strength divided by the elastic stress concentration factor. At lower values of K_t (milder notches), cracks, once initiated, will grow to failure. However for higher stress concentrations, cracks should arrest if

$$s \leq \frac{\Delta K_{th}}{F\sqrt{\pi D}} \qquad (10\text{-}51)$$

where: S = nominal stress amplitude
  D = notch depth
  ΔK_{th} = threshold stress intensity range
  F = a dimensionless quantity used to account for the type of crack and loading, and the ratio of crack size to specimen or component dimensions.

Test data cited in Refs. [138-140] support the criteria depicted in Fig. 10-55, but tests conducted to date have been for constant amplitude loading so that application of relations such as Eq. 51 to irregular loading cases is uncertain, as of this writing.

The previous discussion has treated situations where notch stresses are elastic, but in many cases there is also a zone of elastic-plastic straining at a notch root as illustrated in Fig. 10-56. According to Ref. [139], the size of the zone can be estimated from

$$\frac{e}{r} = \left[\left(\frac{K_t S}{\sigma_y}\right)^{2/3} - 1\right] \qquad (10\text{-}52)$$

where: S = nominal stress
  σ_y = yield strength (cyclic)
  e = depth of elastic-plastic zone
  r = notch root radius.

According to Ref. [141], the size is

$$e = 1/4\left[\left(\frac{K_t S}{\sigma_y}\right)^2 - 1\right] \qquad (10\text{-}53)$$

Both of these relations give comparable results for typical notches and are applicable for plane stress (e.g., thin plates). For plane strain, values of e should be considered to be roughly one-third of the values computed from Eq. 52 or 53. As pointed out in Ref. [139], if the value of $(K_t S/\sigma_y)$ is between about 1.1 and 1.3, the elastic-plastic zone size engulfs the elastic size.

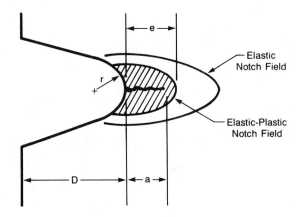

Fig. 10-56 Schematic of zones of concentrated elastic and elastic-plastic stress and strain at a notch.

When a crack is growing within the elastic-plastic zone, growth rates can no longer be predicted with an LEFM approach. Various methods [141-143] based on elastic-plastic analyses have been proposed for estimating crack growth but none has become widely accepted for use in fatigue design evaluations, at least as of this writing. In Ref. [139], it is suggested that an initial crack size be selected so that it is out of the elastic-plastic zone, or the elastic zone for that matter, so that crack growth life can be predicted using LEFM without accounting for notch effects. Again it should be noted that the crack sizes allowed by crack initiation methods (on the order of one to two mm) will generally satisfy this recommendation.

Finally, to account for the effect of notch elastic-plastic straining on the criteria for non-propagating cracks, the following modification of Eq. 51 is suggested in Ref. [4]:

$$S \leq \frac{\Delta K_{th}}{F \sqrt{\pi (D + e)}} \qquad (10\text{-}54)$$

10.5.9 Cracks at Welds

The use of fracture mechanics to predict the growth life of cracks starting at welds has been gaining increased use in recent years. Welds often have stress concentrations caused by toes and roots, as well as possible defects such as those depicted in Fig. 10-57 that act as sites for crack formation. Frequently, cracks start relatively quickly at such locations and most of the life of the welded component is spent in crack growth. Also, regions in which cracks form and grow often contain tensile residual stresses that promote early crack initiation and accelerate growth.

The approach used to estimate the crack growth life of welded structure is basically the same as that described in Section 10.4. The following items are needed to perform a crack growth life assessment for weldments:

(a) knowledge of the stress history applied at locations where cracks are likely to develop, obtained analytically or experimentally (e.g., by strain gages);

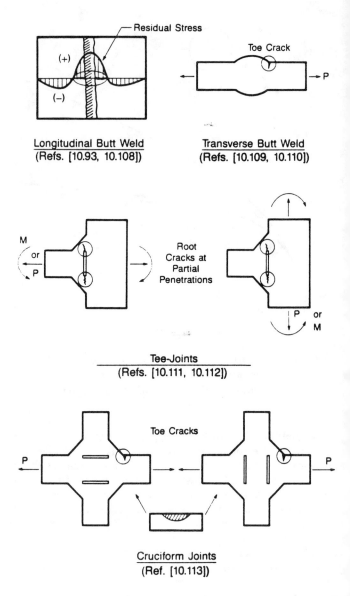

Fig. 10-57 Various weld defects and stress concentrations.

(b) a stress intensity factor that accounts for the particular type of weld geometry being considered, including the effects of stress concentration caused by toes, roots, etc., as appropriate;

(c) da/dN − ΔK input data for the material through which a crack is expected to grow, i.e., weld metal, heat affected zone (HAZ) or base metal, or a combination thereof; and

(d) a residual stress intensity factor for the plane of potential crack growth, if residual stresses normal to that plane are thought to be significant and steps are not taken to relieve them (if tensile).

Stress intensity factors have been developed for a variety of weld configurations, such as a crack growing perpendicular to a longitudinal butt weld while in the influence of the weld residual stress field [129,144], toe cracks in transverse butt welds [145,146], root cracks [147,148] and toe cracks [149] in fillet welds, and cracks in lap joints [147]. These cases are illustrated in Fig. 10-58.

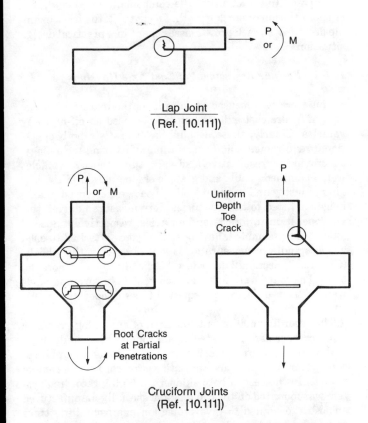

Fig. 10-58 Several weld geometries for which stress intensity factors have been developed.

Stress intensity factors have even been developed for cracks at spot welds [150]. An example of one of the above factors is shown in Fig. 10-59.

For other weld configurations, stress intensity factors can be developed by various numerical methods (e.g., finite elements) or by an approximate approach [149], that consists of multiplying the elastic stress distribution at a weld stress concentration by the stress intensity factor, without the presence of the stress concentration, as illustrated in Fig. 10-60. This is not a rigorous approach for determining the influence of stress concentrations but it may be of use if more accurate stress intensity factors cannot be obtained. It is similar in spirit to the simplified approach proposed in Eqs. 49 and 50 of Section 10.5.8.

For weldments that are geometrically complex, it is often feasible to use stress intensity factors for simpler geometries, as in Fig. 10-58, to approximate the actual situations encountered. For instance, Fig. 10-61 shows how available K_I factors could be applied to the evaluation of a welded structural detail in a construction vehicle.

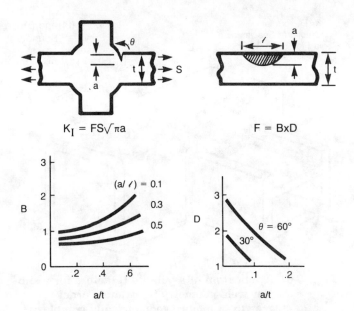

Fig. 10-59 Stress intensity relation for a cruciform weldment with a surface flaw at a weld toe.

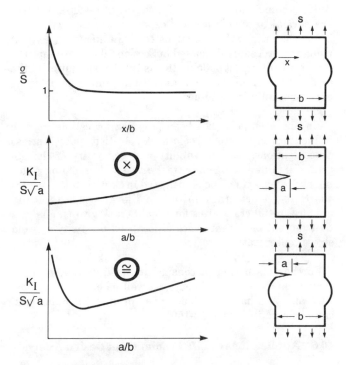

Fig. 10-60 Approximation to stress intensity factor for a toe crack in a butt weld.

The da/dN – ΔK input data for application in weld evaluations can be generated by standard methods as described in Section 10.4. If crack growth is expected to occur through the weld, base metal or heat affected zone, then appropriate input data should be used. In quite a few cases, the da/dN – ΔK behavior in those different materials is similar; however, this cannot be assumed in general.

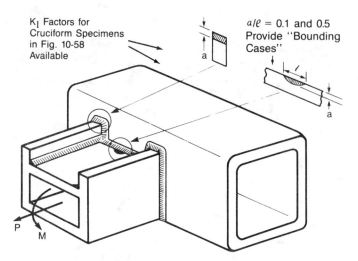

Fig. 10-61 Application of available I_I factors for a simpler weld geometry to estimate crack growth in a more geometrically complex weldment.

The presence of residual stresses in regions surrounding welds is almost unavoidable. If significant residual stresses are normal to the direction of expected crack growth, their influence on growth predictions can be taken into account using the procedure discussed in Section 10.5.7. Perhaps the most difficult aspect of doing this is the determination of the residual stress distribution, that usually must be obtained experimentally for weldments.

An approach [151] has been proposed for estimating the influence on crack growth of localized plasticity at weld stress concentrations, but the validity and generality of the approach are unknown. Similar to the case with cracks originating at the type of notches considered in Section 10.5.8, methods that predict crack initiation life at welds typically produce initial crack sizes for a weld crack growth analysis that are out of the influence of localized plasticity at weld stress concentrations.

Further discussion of considerations involved in crack growth analyses of weldments, as well as examples of how such analyses have been put to use in different industries, can be found in Refs. [152-156].

10.6 Applications of Life Prediction in Design Analysis

This chapter was introduced as being important in the ground vehicle industry because of the utility of life prediction in guiding design/analysis decisions. This section briefly illustrates three cases that support this assertion. These cases involve materials selection/substitution, local geometry, and production considerations. The Case Studies chapter provides still further support for the utility of life prediction in design analysis. The first two examples focus on crack initiation prediction based on a critical location method. It is noteworthy that materials selection/optimization using initiation methods emphasizes different properties than does crack growth predictions [157]. Since microstructural changes that enhance crack initiation resistance often run counter to changes that enhance crack growth resistance, different conclusions may have been drawn if these assessments were based on macrocrack growth. This also suggests that microstructures may be *tailored* to optimize the largest anticipated life fraction for the component, whether that is crack initiation or crack growth. In contrast, the third example describes a crack growth assessment in a practical design situation.

10.6.1 *Fatigue Assessment of Cast Axle Housings*

This work was focused at optimizing the material and design of future cast axle housings to be used on off-highway vehicles. Clearly, these designs had to satisfy both operational requirements and current manufacturing technology. To provide representative load data, three types of vehicles were examined: cable and grapple skidders used in the logging industry and a shovel loader. Loads experienced in axle housings fitted to these vehicles were measured during service operation in forests and quarries, respectively. Special axles were assembled, having strain gages attached to each wheel spindle to measure loads and torques at the wheels. (Chapter 7, Section 2 describes these operations in more detail.) The axles were fitted to each vehicle in turn and loads were measured during all expected types of operation.

The recording equipment consisted of an analog magnetic tape recorder and associated signal conditioning equipment, battery power sources, etc. For the shovel loader an umbilical cord system was used and the data were recorded in a remote vehicle. In the case of both skidders, a high-g recording system was mounted on the cab roof in a specially manufactured enclosure designed to give protection against falling trees and overhead branches. During all tests, a strip-chart recorder was used to monitor each data channel and to ensure the signal integrity would be maintained.

After digitization of the analog recordings, the data files were combined to represent daily duty cycles prior to analysis. Preliminary analyses established torque and loading spectra for each application. These spectra provided the designer with an indication of their relative severity in terms of axle loading. To optimize the designs and assess the validity of using the same axle units for each application, a more comprehensive analytical durability assessment was carried out. The complete axle casting was assessed using the local stress/strain approach. The major aim of this design exercise was to assess the wisdom of manufacturing these units from nodular cast iron as a replacement for the cast steel currently being used. This change would offer significant benefits from a production viewpoint. The effects of gross section changes, local geometry adjustments and material substitutions were evaluated to determine the most appropriate design to satisfy operational requirements. Table 10-7 lists fatigue life estimates for a range of alternative designs.

This brief assessment indicated that a change of material, to nodular iron, while retaining the same detail geometry, reduced the anticipated durability. However, analysis results suggested that improvements in local detail (i.e., reducing K_t

Table 10.7 Fatigue Life Estimates for Cast Axle Housing

Rear Axle Housing Fitted to Grapple Skidder Operating for One Representative Working Day

Critical Area	Material	Stress Concentration Factor	Life Estimate (Days)
A	Cast Steel	2.5	4755
	Nodular Cast Iron	2.5	450
	Nodular Cast Iron	1.8	4750
B	Cast Steel	2.5	15996
	Nodular Cast Iron	2.5	1600
	Nodular Cast Iron	1.9	16010

values) would produce a significant life improvement. Overall, by considering both design and material changes it was possible to improve the design of the axle castings and to manufacture them from nodular cast iron.

10.6.2 Fatigue Assessment of a Piston Rod in a Forging Hammer

This particular forging hammer is used for the manufacture of crankshafts and between 12 and 17 blows are required to produce one forging. The piston rod and tup attachment are shown in Fig. 10-62. Fatigue failures repeatedly occurred in the area indicated. Strain gages were attached to the rod of this and a similar forging hammer to measure the nominal stresses experienced during normal operation. Having ascertained that the signal frequency was within the response of the instrumentation system, the data were recorded on to an analog tape recorder. Loads in both forges were monitored over a period of about two hours to obtain representative samples. The strains experienced during each complete forging operation were digitized and stored in the service load history databank. A rainflow cycle count of the history was performed to establish a histogram of the number of cycles occurring at a given nominal stress range.

Subsequently, life predictions, based on the local strain-life approach, were carried out on a block of the load history representing one complete crankshaft forging operation. Fatigue assessments based on various potential combinations of material, gross section design and local design (different K_t values at the change of the section where the piston rod interfaces with the tup) were carried out.

On the basis of the results obtained, material changes were recommended that increased the durability threefold. Design and operational changes reduced the stresses by an additional 10 percent—increasing the durability by the modified parts over a factor of ten, compared to the original design.

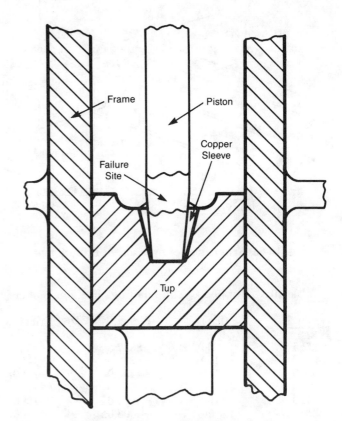

Fig. 10-62 Piston rod and tup attachment in a forging hammer, with typical crack initiation sites indicated.

10.6.3 Analysis of Crack Growth in a SAE 1025 HR Welded Frame Structure [158]

10.6.3.1 Background

The purpose of this analysis was to compare laboratory test data on fatigue crack propagation in a structure subjected to a loading representative of field service with a prediction based on a linear elastic fracture mechanics approach. In order to illustrate considerations involved in predicting crack growth, the approach is described and discussed in some detail.

10.6.3.2 Test Description

Ref. [159] describes laboratory fatigue tests of a frame structure consisting of a SAE 1025 HR steel I-beam welded to a square tube, as shown in Fig. 10-63. Loading was applied to the free end of the I-beam, producing cantilever bending. The load history consisted of repetition of a block intended to simulate the field loading on a C-11 field cultivator frame during a left turn. Each block contained the load sequence shown in Table 10-8.

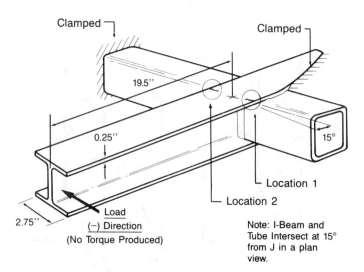

Fig. 10-63 Frame structure used in fatigue test.

Table 10.8 Applied Cyclic Loadings in Each Block

P min, N	(lbs.)	P max, N	(lbs.)
222.5	(50)	−8677.5	(−1950)
−445.0	(−100)	−6230.0	(−1400)
−445.0	(−100)	−4450.0	(−1000)
890.0	(200)	−7565.0	(−1700)
−445.0	(−100)	−6675.0	(−1500)
222.5	(50)	−4005.0	(− 900)
445.0	(100)	−5117.5	(−1150)
1112.5	(250)	−4450.0	(−1000)
1557.5	(350)	−4227.5	(− 950)
667.5	(150)	−5117.5	(−1150)
445.0	(100)	−4227.5	(− 950)
222.5	(50)		

10.6.3.3 Test Results

Cracks formed in the upper flange of the I-beam at locations 1 and 2, as shown in Fig. 10-63. The applied bending stresses at 1 were predominantly tension, and at 2 were primarily compression. The crack growth data reported in Ref. [159] are shown in Table 10-9. Cracking started first at location 2. After a crack had grown to approximately 7 mm, another crack initiated at location 1 and thereafter grew at a faster rate. The crack growth rate at location 2 eventually decreased dramatically, as shown in Table 10-9. Possible reasons for this behavior will be discussed later.

Table 10.9 Crack Growth Data

Crack Length, mm (in.)

Block No.	Location 1		Location 2	
1350	No data reported		First visual crack	
5202	No data reported		6.4	(0.25)
5708	6.4	(0.25)	7.9	(0.31)
6220	12.7	(0.50)	11.2	(0.44)
7070	16.0	(0.63)	12.0	(0.47)
7404	19.1	(0.75)	12.0	(0.47)
7500	25.4	(1.00)	12.1	(0.48)

10.6.3.4 Crack Growth Prediction Methodology

To predict crack growth, the following items were needed:

(1) A nominal, elastic stress history, either determined from strain gage readings, or from a loading forecast. (The stresses were calculated as if no crack was present.)

(2) A stress intensity factor that best approximated the cracked component geometry and loading situation found in the actual structure.

(3) Constant amplitude crack growth rate input data (da/dN vs. ΔK) for the structural material.

(4) A means of using the data from item 3 to compute and sum increments of growth for the stress history. For highly irregular histories, this involved application of a "cycle counting" method to convert the histories into "equivalent" cycles.

10.6.3.5 Prediction

A prediction of crack growth at location 1 was made, starting from the first recorded crack length of 6.4 mm (0.25 in.).

The elastically-calculated, outer fiber stress sequence per block at location 1 is shown in Fig. 10-64.

Stress was computed from the strength of materials beam formula $\sigma = Mc/I$, where I is for both flanges and the web. Note that σ is the gross-section, nominal stress, as if no cracks were present.

The top flange was modelled as a plate with a through-thickness, edge crack, subjected to pure bending. The stress intensity factor for this case [160] is shown in Fig. 10-65. Although a crack was present on the opposite side of the

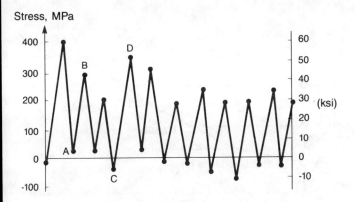

Fig. 10-64 Calculated outer fiber stresses per block at location 1 of the flange.

flange (location 2), was assumed to be closed under compressive loading when the crack at 1 was growing under applied tensile stress ranges. Since the crack at 2 could carry compressive load, it was assumed that the width, b, for the stress intensity factor was that of the entire flange.

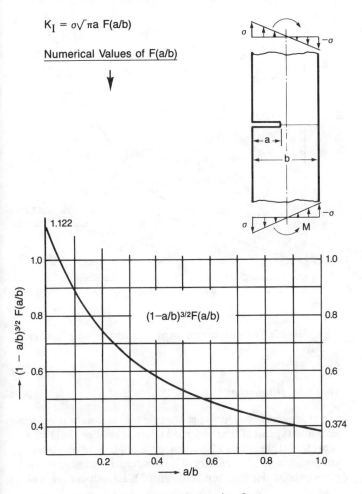

Fig. 10-65 Stress intensity factor.

Crack growth rate data for 1025 HR were not available for use in making the prediction. Because of expected similar fatigue cracking behavior, data for 1020 HR were used. These data were obtained as da/dN (mm/cycle) plotted vs. stress intensity range, ΔK (MPa $\sqrt{mm}$) on log-log paper. The data were then fit to the relation: $da/dN = C (\Delta K)_n$, where $C = 1.7 \times 10^{-14}$ and $n = 3.36$. The tests that generated these input data were conducted with zero to maximum tensile load cycling, that is very similar to the sequence in the load block being considered here (when the compressive portion of the cycles is clipped off). See Fig. 10-64. Thus, no crack growth rate correction for other mean stress levels was considered necessary. If the load sequence contained ranges applied at mean stresses other than that associated with zero to maximum tension, then the input data would have been fit to a relation such as:

$$da/dN = \frac{C(\Delta K)^m}{(1-R)K_c - \Delta K}$$

where: $R = K_{min}/K_{max}$
K_c = fracture toughness

When defined based on $R = 0$ data (0 to maximum tension load), this relation predicts the increase in growth rate for ranges applied at other R values.

While "cycle-counting" methods are generally needed to convert irregular service loadings into "equivalent" cycles, the relative uniformity of ranges in the repeated load block considered here obviated the need for cycle counting. Crack propagation rates were then computed and summed by the following procedure.

First, crack growth was calculated only for rising tensile ranges, such as from points A to B in Fig. 10-64. Second, compressive loading was assumed not to cause crack growth. Thus, a range such as from C to D in Fig. 10-64 was treated as if it were from zero to D.

Using the initial crack length of $a_0 = 6.4$ mm (0.25 in.), the rising tensile stress ranges of Fig. 10-64 were converted to stress intensity ranges, as shown in Table 10-10, using the relation for K from Fig. 10-65.

Table 10.10 Sample Crack Growth Calculation Per Block

$a_0 = 6.4$ mm (0.25 in.), $b = 70$ mm (2.75 in.)

$a_0/b = .09, \left(1 - \frac{a_0}{b}\right)^{1.5} = .87 \quad F(a/b) = 1.03$

$CF = (K /K = 90) = 0.92$ for $= 75$

$K = F(a/b) \times CF \times S \times \quad a$

S (MPa)	K (MPa $\sqrt{mm}$)	$\Delta a/\Delta N$ (mm/range)
416	1770	.00139
277	1180	.00035
192	815	.00010
362	1540	.00087
300	1275	.00046
192	815	.00010
245	1040	.00023
213	905	.00015
203	865	.00012
245	1040	.00023
203	865	.00012
		$\Delta a/\Delta Block$ = .00412 mm/Block

Since the crack grew along the intersection of the I-beam and square tube, that were at 75° to each other, an additional correction was needed to account for the crack not being perpendicular to the maximum principal stress direction in the flange. This correction is shown in Fig. 10-66. (In most other situations, such "additional corrections" are not required.)

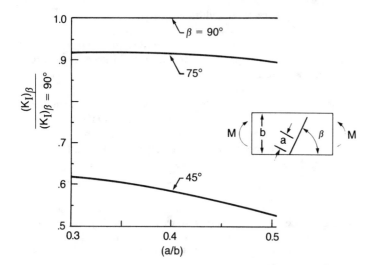

Fig. 10-66 Slanted crack correction factor [160].

For each K range in Table 10-10, an increment of growth, $\Delta a/\Delta N$, was determined from the input da/dN vs. ΔK data. These increments were summed to give the total increment of growth due to the block of loading, i.e., $\Delta a/\Delta Block$. At this point, the crack size was updated to $a_0 + (\Delta a/\Delta Block)$, and the above procedure was repeated to calculate the next increment of growth, and so forth, block after block. A quicker way of obtaining a curve of crack length vs. number of blocks, with little loss of accuracy, would have been to calculate the slopes ($\Delta a/\Delta Block$) at several pre-selected crack lengths, a_0, a_1, a_2, ..., a_f and then to perform a numerical integration of:

$$\text{Blocks} = \int_{a_0}^{a_f} \left(\frac{\Delta a}{\Delta Block}\right)^{-1} \Delta a$$

The curve can also be generated by hand, using the method illustrated in Fig. 10-67.

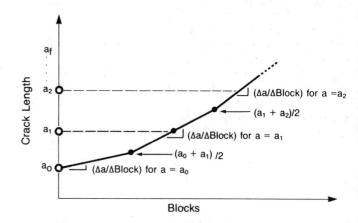

Fig. 10-67 Quick method for generating a crack propagation curve.

A prediction of crack growth based on the above procedure is shown in Fig. 10-68 for comparison with the test data from Table 10-9.

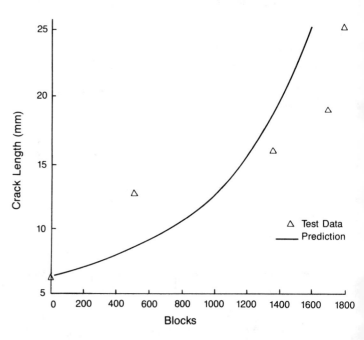

Fig. 10-68 Fatigue crack propagation prediction vs. test data.

10.6.3.6 Discussion

Agreement between the prediction and test data is reasonable and typical of that observed in other studies [161] where representative service loadings were applied to a relatively simple structural geometry. The irregularity in actual crack growth behavior (i.e., not "following" a smooth curve) is also typical.

It is significant that in all of the tests reported in Ref. [159], the crack initiation life for 1025 HR (life to the first visual crack) averaged about 1700 blocks, while the average crack propagation life was about 12,000 blocks. For tests with A13E, it was 1250 and 19,750, respectively. This is not an uncommon result for structural components subjected to actual service loadings. Many cycles are spent in growing a crack from a small size to one that would be considered objectionable (either by posing a threat to structural integrity or by causing user complaints).

In the tests considered here, several of the maximum elastic stresses far exceeded the monotonic yield strength of SAE 1025 HR ($\sigma_y \approx 276$ MPa (40 ksi)). Neglecting the effect of residual stresses induced by plastic bending, it was still considered permissible to predict crack growth based on elastic stress calculations, because the crack growth rate input data which were used were also correlated in terms of elastically-calculated stress intensity ranges, even when significant plastic deformation was occurring in the test specimen used to generate the data. In most structures, large scale yielding is avoided, of course, so that this consideration is rarely of practical concern in a design analysis.

Limit load calculations show that for elastic-perfectly plastic material, a rectangular cross-section will fully yield in bending if the elastically-calculated outer fiber stress reaches 1.5 σ_y. This was the case for the test considered here. The SAE 1025 HR steel had a relatively flat monotonic stress-strain curve, approximating elastic-perfectly plastic behavior. Referring to Fig. 10-64, the first cycle of loading had a maximum elastic stress of 416 MPa (60.3 ksi), almost exactly 1.5 times σ_y (276 MPa (40 ksi)). Upon unloading from this initial high loading, it is likely that residual stresses were present in the flange, as shown in Fig. 10-69. The assumed residual stress distribution helps provide an explanation for the crack growth behavior described earlier. Location 2 of the flange experienced *applied* cyclic stressing that was almost entirely compressive, with small portions of some of the cycles being tensile. (The stress sequence at 2 was the same as in Fig. 10-69, except with all signs reversed.) On the basis of applied stresses, crack growth at location 2 should have been exceedingly slow, if it occurred at all. Yet, a crack initiated first at this location and grew much faster than the crack at location 1, at least during the earlier stages of the test. An explanation for this probably lies in the observation that the sum of residual tensile stress and the small applied cyclic tensile stress at location 2 gave a local cyclic stress sequence that was actually more tensile than the sum of compressive residual stress and the much larger applied cyclic tensile stress at location 1. It is also interesting to note that the crack at location 2 grew only about 12 mm, then stopped. This distance corresponds to the initial size of the assumed tensile residual stress zone in Fig. 10-69.

The effect of residual stress on crack propagation can be very significant, as shown by the example above. The prediction of crack growth for location 1 neglected the effect of compressive residual stresses at that location because the starting point for the prediction was a crack depth of about 7 mm, where the residual stresses were undoubtedly at a relatively low value. (See Fig. 10-69.) Had test data been

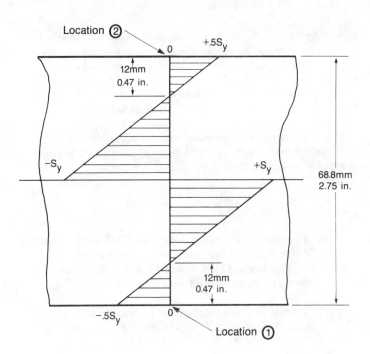

Fig. 10-69 Residual stresses in flange due to highest load in block (for elastic-perfectly plastic material).

location 1, then a prediction that neglected compressive residual stresses would probably have been in considerable error (i.e., too conservative). The residual stresses in these tests were due to gross yielding of the structure, that is a situation hopefully avoided in most service applications. On the other hand, residual stresses due to welding, metal forming, etc., cannot be easily avoided. Methods should be used, as discussed earlier, to account for the effect of such residual stresses on crack propagation rates.

10.7 Conclusion

A good conclusion to this chapter is a quote taken from a recent paper [162] examining the integration of design and analysis with life prediction as the focal point.

"The methods outlined are a powerful design tool which provides the means to avoid simple pitfalls at the design stage and eliminate costly failures in prototype and/or production structures and components. These techniques are now being used extensively to provide successful solutions to a wide variety of design problems and develop future generations of components and structures.

An increase in confidence that fatigue assessments are based on representative service loading data has accompanied the successful implementation of unattended on-board systems to analyze signals using on-line rainflow cycle counting. Currently, these systems only mathematically reduce the variable amplitude loading sequences to establish histograms. There is no attempt to model actual material behavior to assess the effects of both stresses and strains. A natural progression is to develop an analysis system capable of providing life assessments based on local stress-strain behavior at critical areas. In addition to providing nominal loading spectra for components, the continuously up-dated accumulated damage at critical areas would also be available. A permanent installation of this type of system on vehicles or production equipment would be extremely attractive.

The success of the technique and software developed to date, in terms of both applicability to design and ease of operation, suggests that their use will become more widespread."

10.8 References

[1] "Proceedings of the SAE Fatigue Conference," SAE P-109, Society of Automotive Engineers, Inc., 1982. See also Leese, G., Multiaxial Fatigue: *Analysis and Experiments* to be published late 1988 by SAE.

[2] "Digital Techniques in Fatigue," Society of Environmental Engineers, Fatigue Group, B. J. Dabell, Ed., 1983.

[3] Donaldson, K. and Dabell, B., "Decision Path Guidelines for Applying Digital Testing Techniques in Fatigue Service Simulation," in Digital Techniques in Fatigue, Society of Environmental Engineers, Fatigue Group, B. J. Dabell, Ed., 1983.

[4] Kocanda, S., "Fatigue Failure of Metals," Sijthoff & Noordhoff International Publishers, 1978.

[5] Klesnil, M. and Lukas, P., "Fatigue of Metallic Materials," Elsevier Scientific Publishing Company, 1980.

[6] Broek, D., "Elementary Engineering Fracture Mechanics," Noordhoff International Publishing, 1974.

[7] Ewalds, H. L. and Wanhill, R. J. H., "Fracture Mechanics," Edward Arnold and Delftse Uitgevers Maatschappik, 1984.

[8] Knott, J. F., "Fundamentals of Fracture Mechanics," John Wiley & Sons, 1973.

[9] Smith, C. R., "Small Specimen Data for Predicting Life of Full-Scale Structures," Fatigue Testing of Aircraft Structures, ASTM STP 338, pp. 241-240, 1962.

[10] Morrow, J. D., Wetzel, R. M., and Topper, T. H., "Laboratory Simulation of Structural Fatigue Behavior," Effects of Environment and Complex Load History on Fatigue Life, ASTM STP 462, pp. 74-91, 1970, Also NADC ST 6818. See also Topper, T. H., Wetzel, R. M., and Morrow, J-D., "Neuber's Rule Applied to Fatigue of Notched Specimens," JMLSA, ASTM, 4 (1), 200-209, March 1969.

[11] Crews, J. H., Jr. and Hardrath, H. F., "A Study of Cyclic Plastic Stresses at a Notch Root," Experimental Mechanics, 6 (6), pp. 313-320, June 1966.

[12] Neuber, H., "Theory of Stress Concentration for Shear Strained Prismatic Bodies with Arbitrary Nonlinear Stress-Strain Law," J. App. Mech., 28 (4), pp. 544-560, December 1961.

[13] Huang, W. C., "Theoretical Study of Stress Concentrations of Circular Holes and Inclusions in Strain Hardening Materials," Int. J. Solids Structures, 8 (2), 149-192, February 1972.

[14] Leis, B. N., Gowda, C. V. B., and Topper, T. H., "Some Studies of the Influence of Localized and Gross Plasticity on the Monotonic and Cyclic Concentration Factors," Testing and Evaluation, 1 (4), 135-136, July 1973.

[15] Truchon, M., "Application of Low-Cycle Fatigue Test Results to Crack Initiation from Notches," Low-Cycle Fatigue and Life Prediction, ASTM STP 770, C. Amzallag, B. N. Leis, and P. Rabbe, Eds., pp. 254-268, 1982.

[16] Lind, N. C. and Mroz, Z., "A Study of Cyclic Plasticity," Proc. IABSE Symp. on Resistance and Ultimate Deformability of Structures Acted on by Well- Defined Repeated Loads, Lisboa, pp. 9-14, 1973.

[17] Williams, D. P., Lind, N. C., Conle, A. Topper, T. H., and Leis, B. N., "Structural Cyclic Deformation Response Modelling," in Mechanics in Engineering, University of Waterloo Press, SM-11, p. 291, 1977.

[18] Leis, B. N. and Sampath, G. S., "Development of an Improved Method of Consolidating Fatigue Life Data," NASA CR-145312, February 1978.

[19] Beste, A. and Seeger, T., "Elasto-plastic Stress/Strain at Notches, Comparison of Test and Approximate Computations," in Sitzung des Arbeitskreises Betriebsfestigkeit, Deutscher Verband fur Materialprufung e.V., Stuttgart, p. 383, 1979.

[20] Kaechele, K., "Review and Analysis of Cumulative Fatigue-Damage Theories," The RAND Corporation, RM-3650-PR, August 1963.

[21] Leve, H. L., "Cumulative Damage Theories," in Metal Fatigue: Theory and Design, A. F. Madayag, Ed., John Wiley & Sons, Inc., p. 170.

[22] Morrow, JoDean, "Fatigue Properties of Metals," Manual Society of Automotive Engineers, ISTC Div. 4, April 1964.

[23] Smith, K. N., Watson, P., and Topper, T. H., "A STress-Strain Function for the Fatigue of Metals," SMD Report 21, University of Waterloo, October 1969; see also, Smith, K. N., Watson, P., and Topper, T. H., "A Stress-Strain Function for the Fatigue of Metals," J. of Matls., JMLSA, 5 (4), pp. 767-778, December 1970.

[24] Fuchs, H. O., "A Set of Fatigue Failure Criteria," J. Basic Eng Trans ASME, June 1965.

[25] Laflen, J. H. and Cook, T. S., "Equivalent Damage—A Critical Assessment," NASA CR-167874, November 1982.

[26] Leis, B. N., "An Energy-Based Fatigue and Creep-Fatigue Damage Parameter," J. of Pressure Vessel Tech., Trans. ASME, 99 (4), pp. 524-533, November 1977.

[27] Watson, P. and Plumtree, A., "Fatigue and Mean Stress—A Perspective," in Computer-Aided Testing and Model Analysis, Society for Experimental Mechanics, pp. 102-107, 1984.

[28] Schutz, D. and Lowak, H., "The Application of the Standardized Test Program for the Fatigue Life Estimation of Fighter Wing Components Falstaff (Part IV)," from Proceedings of the 8th ICAF Symposium on Problems with Fatigue in Aircraft, Lausanne, June 2-5, 1975, pp. 3.64/1-3.64/22, ICAF Doc. No. 801, compiled and edited by J. Branger and F. Berger.

[29] Peterson, R. E., Published discussion of paper by H. J. Grover, Proc. ASTM, 45, p. 522, see also Peterson, R. E., "Notch Sensitivity," Metal Fatigue, Chapter 13, Sines and Waisman, Eds., McGraw-Hill, 1959.

[30] Tucker, L. and Bussa, S., "The SAE Cumulative Fatigue Damage Test Program," Paper 750038 presented at the SAE Automotive Engineering Congress, Detroit, Michigan, February 1975.

[31] "Fatigue Under Complex Loading: Analysis and Experiments," Advances in Engineering, 6, Society of Automotive Engineers, Warrendale, PA, 1977.

[32] Berns, H. D., "Field Service History Analysis for Ground Vehicles," SAE Paper No. 750553, 1975.

[33] Mitchell, M. R., "Fundamentals of Modern Fatigue Analysis for Design," Fatigue and Microstructure, ASM, 1979.

[34] Stadnick, S. J. and Morrow, J. D., "Techniques for Smooth Specimen Simulation of the Fatigue Behavior of Notched Members," in Testing for Prediction of Material Performance in Components and Structures, ASTM STP 515, pp. 229-252, 1973.

[35] Sih, G. C., *Handbook of Stress Intensity Factors*, Institute of Fracture and Solid Mechanics, 1973.

[36] Rooke, D. P. and Cartwright, D. J., *Compendium of Stress Intensity Factors*, 1976.

[37] Tada, H., Paris, R. C. and Irwin, G. R., *The Stress Analysis of Cracks Handbook*, Paris Productions, Inc., 1986.

[38] *Metallic Materials and Elements for Aerospace Vehicle Structures*, MIL-HDBK-5, Naval Publications and Forms Center, 1981.

[39] Rolfe, S. T., and Barsom, J. M., *Fracture and Fatigue Control in Structures*, Prentice-Hall, Inc., 1977.

[40] Damage Tolerant Design Handbook, MCIC-HB-01R, Compiled by University of Dayton, Published and distributed by Battelle Columbus Division, Columbus, OH, 1983. See also Computer Data Base on Fatigue Thresholds and Crack Growth Rates, Engineering Materials Advisory Services, Ltd., West Midlands, U.K., 1986.

[41] Fuchs, H. O. and Stephens, R. I., *Metal Fatigue in Engineering*, John Wiley and Sons, 1980.

[42] "Standard Test Method for Plane-Strain Fracture Toughness of Metallic Materials," E 399-88, ASTM, 1988.

[43] *Elastic-Plastic Fracture Mechanics Technology*, STP 896, ASTM, 1986.

[44] "Standard Test Method for Constant-Load-Amplitude Fatigue Crack Growth Rates Above 10^{-8} m/cycle," E 647-83, ASTM, 1983.

[45] Forman, R. G., Kearney, V. E. and Engle, R. M., "Numerical Analysis of Crack Propagation in Cyclic Loaded Structures," *Journal of Basic Engineering*, Transactions, ASME, Vol. 89, No. 3, ASME, 1967, p. 459.

[46] Hoeppner, D. W. and Krupp, W. E., "Prediction of Component Life by Application of Fatigue Crack Growth Knowledge," *Engineering Fracture Mechanics*, Vol. 6, 1974, p. 47.

[47] Stephens, R. I., Benner, P. H., Mauritzen, G. and Tindall, G. W., "Constant and Variable Amplitude Fatigue Behavior of Eight Steels," *Journal of Testing and Evaluation*, Vol. 7, No. 2, 1979, p. 68.

[48] Nelson, D. V., "Review of Fatigue-Crack-Growth Prediction Methods," *Experimental Mechanics*, Vol. 17, No. 2, 1977, p. 41.

[49] Nelson, D. V. and Fuchs, H. O., "Prediction of Fatigue Crack Growth Under Irregular Loading" in *Fatigue Crack Growth Under Spectrum Loads*, STP 595, ASTM, 1976, p. 267.

[50] Socie, D. F., "Variable Amplitude Fatigue Life Estimation Models," SAE Technical Paper 820689, SAE, 1982.

[51] Barsom, J. M., "Fatigue Crack Growth Under Variable Amplitude Loading in ASTM A514-B Steel," in *Progress in Flaw Growth and Fracture Toughness Testing*, STP 536, ASTM, 1973, p. 147.

[52] Smith, S. H., "Random-Loading Fatigue Crack Growth Behavior of Some Aluminum and Titanium Alloys," in *Structural Fatigue in Aircraft*, ASTM STP 404, ASTM, 1966, p. 74.

[53] Swanson, S. R., Cicci, F. and Hoppe, W., "Crack Propagation in Clad 7079-T6 Aluminum Alloy Sheet Under Constant and Random Amplitude Fatigue Loading," in *Fatigue Crack Propagation*, ASTM STP 415, ASTM, 1967, p. 312.

[54] Dowling, N. E., "Notched Member Fatigue Life Predictions Combining Crack Imitation and Propagation," *Fatigue of Engineering Materials and Structures*, Vol. 2, 1979, p. 129.

[55] Socie, D. F., Morrow, J. and Chen, W., "A Procedure for Estimating the Total Fatigue Life of Notched and Cracked Members," *Engineering Fracture Mechanics*, Vol. 11, 1979, p. 851.

[56] Hudak, S. J., "Small Crack Behavior and the Prediction of Fatigue Life," 103 (1), 1981.

[57] Cameron, A. D. and Smith, R. A., "Fatigue Life Prediction for Notched Members, Int. J. Pres Ves. & Piping, 10, pp. 205-217, 1982.

[58] Dowling, N. E., "Growth of Short Fatigue Cracks in an Alloy Steel," Paper 82-1D7-STINE-P1, Westinghouse R&D Center, 1982.

[59] Leis, B. N., Hopper, A. T., Ahmad, J., Broek, D., and Kanninen, M. F., "Critical Review of the Fatigue Growth of Short Cracks," Engineering Fracture Mechanics, 23 (5), pp. 883-898, 1986. See also Leis, B. N., Kanninen, M. F., Hopper, A. T., Ahmad, J., and Broek, D., "A Critical Review of the Short Crack Problem in Fatigue," AFWAL-TR-83-4019, 1983.

[60] Behavior of Short Cracks in Airframe Components, AGARD Conference Proceedings No. 328, 1982.

[61] Kikukawa, M., Jons, M., and Adachi, M., "Direct Observation and Mechanism of Fatigue Crack Propagation," in Fatigue Mechanisms, ASTM STP 675, 1979.

[62] Novak, S. R. and Barsom, J. M., "Brittle Fracture (K_{Ic}) Behavior of Cracks Emanating from Notches," ASTM STP 601, pp. 409-447, 1976.

[63] Smith, R. A. and Miller, K. J., "Fatigue Cracks at Notches," International J. of Mech. Sci., 19, pp. 11-22, 1977.

[64] Dowling, N. E., "Fatigue at Notches and the Local Strain and Fracture Mechanics Approaches," in Fracture Mechanics, ASTM STP 677, pp. 243-273, 1979.

[65] Krempl, E., "The Influence of State of Stress on Low-Cycle Fatigue of Structural Materials: A Literature Survey and Interpretive Report," ASTM STP 549, 1971.

[66] Brown, M. W. and Miller, K. J., "Two Decades of Progress in the Assessment of Multiaxial Low-Cycle Fatigue Life," Low-Cycle Fatigue and Life Prediction, ASTM STP 770, C. Amzallag, B. N. Leis, and P. Rabbe, Eds., pp. 482-499, 1982.

[67] Downing, S., "Compilation of SAE Activity on Multiaxial Fatigue," to be published.

[68] Multiaxial Fatigue, ASTM STP 853, K. J. Miller and M. W. Brown, Eds., 1985.

[69] Tipton, S. M. and Nelson, D. V., "Fatigue Life Predictions for the Notched Shaft in Combined Bendingnand Torsion," Multiaxial Fatigue, ASTM STP 853, K. J. Miller and M. W. Brown, Eds., pp. 514-550, 1985.

[70] Lee, S. B., "A Criterion for Fully Reversed Out-of-Phase Torsion and Bending," Multiaxial Fatigue, ASTM STP 853, K. J. Miller and M. W. Brown, Eds., pp. 553-568, 1985.

[71] McDiarmid, D. L., "Fatigue Under Out-of-Phase Biaxial Stresses of Different Frequencies," Multiaxial Fatigue, ASTM STP 853, K. J. Miller and M. W. Brown, Eds., pp. 606-621, 1985.

[72] Leis, B. N. and Topper, T. H., "Long-Life Notch Strength Reduction in the Presence of Local Biaxial Stress," J. of Engineering Materials and Technology, Trans. ASME, 99 (3), pp. 215-221, July 1977.

[73] Kfouri, A. P. and Miller, K. J., "Crack Separation Energy Rates for Inclined Cracks in a Biaxial Stress Field of an Elastic-Plastic Material," Multiaxial Fatigue, ASTM STP 853, K. J. Miller and M. W. Brown, Eds., pp. 88-107, 1985.

[74] Ahmad, J., Leis, B. N., and Kanninen, M. F., "Analysis of Fatigue Crack Propagation Under Biaxial Loading Using an Inclined Strip Yield Zone Model of Crack Tip Plasticity," Fatigue Fract. Engng. Mater. Struct., 00 (0), 1986.

[75] Brown, M. W. and Miller, K. J., "Mode I Fatigue Crack Growth Under Biaxial Stress at Room and Elevated Temperature," Multiaxial Fatigue, ASTM STP 853, K. J. Miller and M. W. Brown, Eds., pp. 135-152, 1985.

[76] Jaske, C. E., Payer, J. H., and Balint, V. S., "Corrosion Fatigue of Metals in Marine Environments," MCIC Report MCIC-81-42, Metals and Ceramics Information Center, Battelle Columbus Division, July 1981. See also Mitchell, M. R., "Fatigue Considerations in Use of Aluminum Alloys," SAE Paper No. 820689 or P-109, pp. 249-272, 1982.

[77] Fundamental Aspects of Stress Corrosion Cracking, R. W. Staehle and D. van Rooyen, Eds., NACE, 1969.

[78] Corrosion-Fatigue Technology, ASTM STP 642, H. L. Craig, T. W. Crooker, and D. W. Hoeppner, Eds., 1978.

[79] Corrosion Fatigue: Mechanics, Metallurgy, Electrochemistry, and Engineering, ASTM STP 801, T. W. Crooker and B. N. Leis, 1883.

[80] Corrosion Fatigue, R. N. Parkins and Ya. M. Kolotyrkin, The Metals Society, London, 1980.

[81] Embrittlement by the Localized Crack Environment, R. P. Gangloff, Ed., AIME, 1985.

[82] Modeling Environmental Effects on Crack Growth Processes, R. H. Jones and W. W. Gerberich, Eds., AIME, 1986.

[83] The Theory of Stress Corrosion Cracking in Alloys, J. C. Scully, Ed., NATO, 1971.

[84] "Stress Corrosion Cracking and Hydrogen Embrittlement of Iron-Base Alloys," R. W. Staehle, et. al., Eds., NACE, 1977.

[85] See Ref. 1, pp. 249-272, Mitchell, M. R., et. al., "Fatigue Considerations in Use of Aluminum Alloys," SAE Paper No. 820699, 1982.

[86] Fatigue at Elevated Temperatures, ASTM STP 520, A. E. Carden, A. J. McEvily, and C. H. Wells, Eds., 1974.

[87] Thermal Fatigue of Materials and Components, ASTM STP 612, D. A. Spera and D. F. Mowbray, Eds., 1977.

[88] Thermal Stresses in Severe Environments, D. P. H. Hasselman and R. A. Heller, Eds., Plunum Press, 1980.

[89] Coffin, L. F., "Fatigue at High Temperature," Fatigue at Elevated Temperatures, ASTM STP 520, pp. Unify Treatment of High Temperature Fatigue - A Partisan Proposal Based on Strain Range Partitioning," Fatigue at Elevated Temperatures, ASTM STP 520, pp. 744-782, 1973.

[90] Structural Materials for Service at Elevated Temperatures in Nuclear Power Generation, ASME MPC-1, A. O. Schaefer, Ed., 1974.

[91] 1976 ASME - MPC Symposium on Creep-Fatigue Interaction, ASME MPC-3, R. M. Curran, Ed., 1976.

[92] Ductility and Toughness Considerations in Elevated Temperature Service, ASME MPC-8, G. V. Smith, Ed., 1978.

[93] AWS D1.1, Structural Welding Code-Steel, American Welding Society, 550 N.W. Legune Road, Miami, Florida, 1987.

[94] Maddox, S. J., "Fracture Mechanics Applied to Fatigue in Welded Structures," Proceedings of the Conference on Fatigue of Welded Structures, The Welding Institute, 1971.

[95] Gurney, T. R., "Fatigue of Welded Structures," 2nd Edition, Cambridge University Press.

[96] van der Zanden, A. M., Robins, D. B., and Topper, T. H., "Fatigue Life Prediction of Weldments with Internal Cavities," Testing for Prediction of Material Performance in Structures and Components, ASTM STP 515, pp. 268-284, 1972.

[97] Lawrence, F. V., Mattos, R. J. Higashida, Y., and Burk, J. D., "Estimating the Fatigue Crack Initiation Life of Welds," Fatigue Testing of Weldments, ASTM STP 648, D. W. Hoeppner, Ed., pp. 134-158, 1978.

[98] Yung, J.-Y. and Lawrence, F. V., "Analytical and Graphical Aids for the Fatigue Design of Weldments," SAE Technical Paper Series No. 850803, 1985.

[99] Lawrence, F. V., Wang, P. C., and Corten, H. T., "Improvement of Spot Weld Fatigue Resistance," SAE Technical Paper Series No. 840112, 1984.

[100] Waterhouse, R. B., "Fretting Corrosion," Pregamon Press, 1972.

[101] Leis, B. N., "A Concept for Fatigue Analyses of Complex Components," J. of Testing and Evaluation, 5 (4), pp. 309-319, July 1977.

[102] Schijve, J., "Fatigue Crack Propagation in Light Alloy Sheet Material and Structures," Report MP-195, National Aerospace Lab, The Netherlands, August 1960.

[103] Hudson, C. M. and Hardrath, H. F., "Investigation of the Effects of Variable Amplitude Loadings on Fatigue Crack Propagation Patterns," NASA TN-D-1803, NASA, 1963.

[104] Hudson, C. M. and Hardrath, H. F., "Effects of Changing Stress Amplitude on the Rate of Fatigue Crack Propagation in Two Aluminum Alloys," NASA TN-D-960, NASA, 1961.

[105] Vargas, L. G. and Stephens, R. I., "Subcritical Crack Growth Under Over-Loading in Cold-Rolled Steel," *Proceedings of the Third International Conference on Fractur*, Munich, April 1973.

[106] Rice, R. C. and Stephens, R. I., "Overload Effects on Subcritical Crack Growth in Austenitic Manganese Steel," in *Progress in Flaw Growth and Fracture Toughness Testing*, ASTM STP 536, ASTM, 1973, p. 95.

[107] Gallagher, J. P. and Hughes, T. F., "Influence of Yield Strength on Overload Affected Fatigue Crack Growth Behavior in 4340 Steel," AFFDL TR-74-27, Air Force Flight Dynamics Lab, 1974.

[108] Hudson, C. M. and Raju, K. N., "Investigation of Fatigue Crack Growth Under Simple Variable Amplitude Loading," NASA TN-D-5702, NASA, 1970.

[109] Corbly, D. M. and Packman, P. F., "On the Influence of Single and Multiple Peak Overloads on Fatigue Crack Propagation in 7075-T6511 Aluminum," *Engineering Fracture Mechanics*, Vol. 5, 1973, p. 479.

[110] Von Euw, E. F. J., Hertzberg, R. W. and Roberts, R., "Delay Effects in Fatigue Crack Propagation," in *Stress Analysis and Growth of Cracks*, ASTM STP 513, ASTM, 1972, p. 230.

[111] Trebules, V. W. Roberts, R. and Hertzberg, R. W., "Effects of Multiple Overloads on Fatigue Crack Propagation in 2024-T3 Aluminum Alloy," in *Progress in Flaw Growth and Fracture Toughness Testing*, ASTM STP 536, ASTM, 1973, p. 113.

[112] Probst, E. P. and Hillberry, B. M., "Fatigue Crack Delay and Arrest Due to Single Peak Overloads," *AIAA Journal*, Vol. 12, No. 3, AIAA, 1974, p. 330.

[113] McMillan, J. C. and Pelloux, R. M. N., "Fatigue Crack Propagation Under Program and Random Loads," in *Fatigue Crack Propagation*, ASTM STP 415, ASTM, 1967, p. 505.

[114] Matthews, W. T., Barratta, F. I. and Driscoll, G. W., "Experimental Observations of a Stress Intensity History Effect Upon Fatigue Crack Growth Rate," *International Journal of Fracture Mechanics*, Vol. 7, No. 2, 1971, p. 224.

[115] Hsu, T. M. and Lassiter, L. W., "Effects of Compressive Overloads on Fatigue Crack Growth," AIAA/ASME/SAE 15th Structures, Structural Dynamics and Materials Conference, AIAA, April 1974.

[116] Wei, R. P., Shih, T. T. and Fitzgerald, J. H., "Load Interaction Effects on Fatigue Crack Growth in Ti-6Al-4V Alloy," NASA CR-2239, NASA, April 1973.

[117] Stephens, R. I., McBurney, G. W. and Oliphant, L. J., "Fatigue Crack Growth with Negative R-Ratio Following Tensile Overloads," submitted to *International Journal of Fracture Mechanics*, 1975.

[118] Wheeler, O. E., "Spectrum Loading and Crack Growth," *Journal of Basic Engineering, Transactions*, ASME, Vol. 94, No. 1, ASME, 1972, p. 181.

[119] Willenborg, J., Engle, R. M. and Wood, H. A., "A Crack Growth Model Using an Effective Stress Intensity Concept," AFFDL-TM-71-1-FBR, Air Force Flight Dynamics Lab, January 1971.

[120] Elber, W., "The Significance of Fatigue Crack Closure," in *Damage Tolerance in Aircraft Structures*, ASTM STP 486, ASTM, 1971, p. 230.

[121] Newman, J. C., "Finite Element Analysis of Fatigue Crack Propagation—Including the Effects of Crack Closure," Ph.D. Dissertation, Virginia Polytechnic Institute and State University, 1974.

[122] Newman, J. C., "Prediction of Fatigue Crack Growth Under Variable Amplitude and Spectrum Loading Using a Closure Model," in *Design of Fatigue and Fracture Resistant Structures*, ASTM STP 761, ASTM, 1982, p. 255.

[123] Elber, W., "Equivalent Constant Amplitude Concept for Crack Growth Under Spectrum Loading," in *Fatigue Crack Growth Under Spectrum Loads*, ASTM STP 595, ASTM, 1976, p. 236.

[124] Cathey, W. H. and Grandt, A. F., "Fracture Mechanics Consideration of Residual Stresses Introduced by Coldworking Fastener Holes," *Journal of Engineering Materials and Technology*, Transactions, ASME, Vol. 102, ASME, 1980, p. 85.

[125] Chandawanich, N. and Sharpe, W. N., Jr., "An Experimental Study of Fatigue Crack Initiation and Growth from Coldworked Holes," *Engineering Fracture Mechanics*, Vol. 11, 1979, p. 609.

[126] Underwood, J. H. and Throop, J. F., "Surface Crack K-Estimates and Fatigue Life Calculations in Cannon Tubes," in *Part-Through Crack Fatigue Life Predictions*, ASTM STP 687, ASTM, 1979, p. 195.

[127] Rybicki, E. F., Stonesifer, R. B. and Olson, R. J., "Stress Intensity Factors Due to Residual Stresses in Thin-Walled Girth-Welded Pipes," *Journal of Pressure Vessel Technology*, Transactions, ASME, Vol. 103, No. 1, ASME, 1981, p. 66.

[128] Wu, X. R. and Carlsson, J., "Welding Residual Stress Intensity Factors for Half-Elliptical Surface Cracks in Thin and Thick Plates," *Engineering Fracture Mechanics*, Vol. 19, No. 3, 1984, p. 407.

[129] Glinka, G., "Effects of Residual Stresses on Fatigue Crack Growth in Steel Weldments Under Constant and Variable Loads," in *Fracture Mechanics*, ASTM STP 677, ASTM, 1979, p. 198.

[130] Paris, P. C., McMeeking, R. M. and Tada, H., "The Weight Function Method for Determining Stress Intensity Factors," in *Cracks and Fracture*, ASTM STP 601, ASTM, 1976, p. 471.

[131] Petroski, H. J. and Achenbach, J. D., "Computation of the Weight Function from a Stress Intensity Factor," *Engineering Fracture Mechanics*, Vol. 10, 1978, p. 257.

[132] Aamodt, B. and Bergan, P. G., "On the Principle of Superposition for Stress Intensity Factors," *Engineering Fracture Mechanics*, Vol. 8, No. 2, 1976, p. 437.

[133] Parker, A. P. and Bowie, O. L., "The Weight Function for Various Boundary Conditions," *Engineering Fracture Mechanics*, Vol. 18, No. 2, 1983, p. 473.

[134] Renzhi, W., "Effect of Residual Stresses of Shot Peening on the Fatigue Behavior of a High Strength Steel," *Fatigue of Engineering Materials and Structures*, Vol. 2, 1980, p. 413.

[135] Elber, W., "Effects of Shot Peening Residual Stresses on the Fracture and Crack Growth Properties of D6AC Steel," in *Fracture Toughness and Slow-Stable Cracking*, ASTM STP 559, ASTM, 1974, p. 45.

[136] Hubbard, R. P., "Crack Growth Under Cyclic Compression," *Journal of Basic Engineering*, Transactions, ASME, December 1969, p. 625.

[137] Saal, H., "Fatigue Crack Growth in Notched Parts with Compression Mean Load," *Journal of Basic Engineering*, Transactions, ASME, Vol. 94, ASME, 1972, p. 243.

[138] Smith, R. A. and Miller, K. J., "Prediction of Fatigue Regimes in Notched Components," *International Journal of Mechanical Science*, Vol. 20, 1978, p. 201.

[139] Dowling, N. E., "Fatigue at Notches and the Local Strain and Fracture Mechanics Approaches," in *Fracture Mechanics*, ASTM STP 677, ASTM, 1979, p. 247.

[140] Nelson, D. V. and Socie, D. F., "Crack Initiation and Propagation Approaches to Fatigue Analysis," in *Design of Fatigue and Fracture Resistant Structures*, ASTM STP 761, ASTM, 1982, p. 110.

[141] Hammouda, M. M. and Miller, K. J., "Prediction of Fatigue Lifetime of Notched Members," *Fatigue of Engineering Materials and Structures*, Vol. 2, 1980, p. 377.

[142] El Haddad, M. H., Smith, K. N. and Topper, T. H., "A Strain Based Intensity Factor Solution for Short Fatigue Cracks Initiating from Notches," in *Fracture Mechanics*, ASTM STP 677, ASTM, 1979, p. 274.

[143] El Haddad, M. H., Dowling, N. E., Topper, T. H. and Smith, K. N., "J-Integral Applications for Short Fatigue Cracks at Notches," *International Journal of of Fracture*, Vol. 16, No. 1, 1980, p. 15.

[144] Terada, H., "An Analysis of the Stress Intensity Factor of a Crack Perpendicular to the Welding Bead," *Engineering Fracture Mechanics*, Vol. 8, 1976, p. 441.

[145] Lawrence, F. V., "Estimation of Fatigue Crack Propagation Life in Butt Welds," *Welding Journal*, Vol. 52, 1973, pp. 212-5.

[146] Usani, S., Kusumoto, S., Kimoto, H., Kawakami, M., "Effects of Crack Length and Flank Angle Size on Fatigue Strength at Toes of Mild Steel Welded Joint," *Transactions of the Japan Welding Society*, Vol. 9, No. 1, 1978, p. 11.

[147] Usani, S. and Kasumoto, S., "Fatigue Strength at Roots of Cruciform, Tee and Lap Joints," *Transactions of the Japan Welding Society*, Vol. 9, No. 1, 1978, p. 3.

[148] Frank, K. H., "The Fatigue Strength of Fillet Welded Connections," Ph.D. Dissertation, Lehigh University, 1971.

[149] Maddox, S. J., "An Analysis of Fatigue Cracks in Fillet Welded Joints," *International Journal of Fracture*, Vol. 11, 1975, p. 221.

[150] Pook, L. P., "Fracture Mechanics Analysis of the Fatigue Behavior of Spot Welds," *International Journal of Fracture*, Vol. 11, 1975, p. 173 and discussion, p. 531.

[151] El Haddad, M. H., Topper, T. H. and Smith, I. F. C., "Fatigue Life Prediction of Welded Components Based on Fracture Mechanics," *Journal of Testing and Evaluation*, Vol. 8, No. 6, 1980, p. 301.

[152] Reemsnyder, H. S., "Development and Application of Fatigue Data for Structural Steel Weldments," in *Fatigue Testing of Weldments*, ASTM STP 648, ASTM, 1978, p. 3.

[153] Nelson, D. V., "Fatigue Considerations in Welded Structure," SAE Technical Paper 820695, SAE, 1982.

[154] Smith, K. N., El Haddad, M. and Martin, J. F., "Fatigue Life and Crack Propagation Analyses of Welded Components Containing Residual Stresses," *Journal of Testing and Evaluation*, Vol. 5, No. 4, 1977, p. 327.

[155] Dover, W. D., "Fatigue Fracture Mechanics Analysis of Offshore Structures," *International Journal of Fatigue*, April 1981, p. 52.

[156] Fisher, J. W., Yen, B. T. and Frank, K. H., "Minimizing Fatigue and Fracture in Steel Bridges," *Journal of Engineering Materials and Technology*, Transactions, ASME, Vol. 102, ASME, 1980, p. 20.

[157] Leis, B. N., Mayfield, M. E., and Olson, R. J., "Materials Selection Consideration for Marine Risers," in *RISERS, ARCTIC DESIGN CRITERIA, EQUIPMENT RELIABILITY IN HYDROCARBON PROCESSING: A Workbook for Engineers*, American Society of Mechanical Engineers, ASME, pp. 25-32, 1981.

[158] Nelson, D. V., "Analysis of Fatigue Crack Growth in a 1025HR Welded Frame Structure Under Laboratory-Simulated Field Loading," Engineering Report to John Deere, Dubuque Works, July 19, 1979.

[159] Tucker, L. E., "C-11 Field Cultivator (Failure Investigation)," Report No. 4834, Materials Engineering Dept., Deere & Co., February, 1970.

[160] Rooke, D. P. and Cartwright, D. J., *Compendium of Stress Intensity Factors*, Her Majesty's Stationery Office, 1976.

[161] Nelson, D. V. and Fuchs, H. O., "Prediction of Fatigue Crack Growth Under Irregular Loading," ASTM STP 595, ASTM, 1976.

[162] See Ref. 1, pp. 219-236, Dabell, B., "Integrated Design/Fatigue Analysis Study," SAE Paper No. 820697, 1982.

Addresses of Publishers

AIAA, American Institute of Aeronautics & Astronautics, 370 L'Enfant Promenade SW, Washington, DC 20024

Air Force Flight Dynamics Lab, Wright Patterson AFB, OH 45433

ASTM, American Society for Testing & Materials, 1916 Race St., Philadelphia, PA 19103

John Deere Dubuque Works (Deere & Co.), P.O. Box 538, Dubuque, IA 52001

Lehigh University, Bethlehem, PA 18015

NASA, National Aeronautics & Space Administration, 600 Independence Ave., SW, Washington, DC 20546

Naval Publications & Forms Center, 5801 Tabor Ave., Philadelphia, PA 19120

Paris Productions, Inc., 226 Woodbourne Dr., St. Louis, MO 63105

Prentice-Hall, Route 9W, Englewood Cliffs, NJ 07632

Virginia Polytechnic Institute & State University, Blacksburg, VA 24061

John Wiley & Sons, 605 Third Ave., New York, NY 10158

Appendix 10A SAE Cumulative Fatigue Damage Test Program

From 1970 to 1976, the Society of Automotive Engineers (SAE) Fatigue Design and Evaluation Committee conducted a test program to study fatigue under complex loading conditions. The program included the establishment of load histories, specimens (components), materials and testing programs [1,2].

10A.1 Load Histories

Initial load histories were collected that were considered typical of ground vehicle histories, in general. Of the 30 different analog histories considered, only three were finally selected in order to keep the test program to a manageable size. The final three were:

1. Suspension: A history for a suspension part in bending, attached to a vehicle driven over a durability course. It served as an example of random excitation with superimposed maneuvering forces. The mean value of this history was compressive.

2. Transmission: A history for a transmission part attached to a tractor engaged in front-end work. It served as an example of large deviations from a mean value in tension.

3. Bracket: A vibration history for a mounting bracket attached to a vehicle operating on a rough road. It served as an example of random vibration about a mean value.

Each history was originally recorded as an analog signal from an actual component in the field. This signal was then reduced to a series of digital integral values representing the analog peaks between -999 and +999. The absolute maximum corresponded to the value of 999. Each digital history was then further reduced to eliminate any changes less than 20% of the absolute maximum stress.

The order of the peaks was always preserved; however, the absolute maximum peak was designated as the starting point. One complete history was defined as a block. Although the value and order of the peaks were preserved, there was no attempt to model the history frequency . . . that was nominally 1 to 30 Hertz, as measured in the field. Fig. 10A-1 shows the three SAE histories. Rainflow counted histograms of these histories are shown in Tables 10A-1 through 10A-3.

10A.2 Specimens and Materials

The test specimen (component) used for the program was a slotted plate (Fig. 10A-2). Loads were applied to the specimen through a monoball fixture that allowed both tensile and compressive loads to be applied.

This design provided both axial and bending stresses and

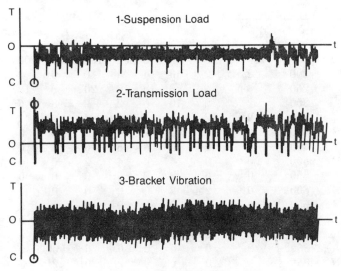

Fig. 10A-1 SAE load histories.

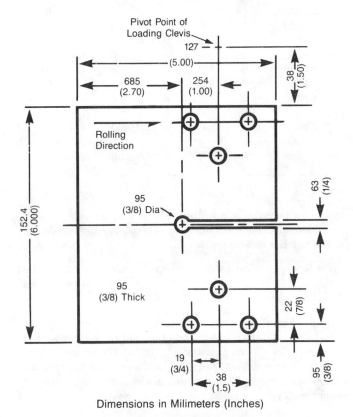

Dimensions in Milimeters (Inches)

Fig. 10A-2 SAE specimen.

components. Also, it permitted the study of both crack initiation and crack propagation.

In addition, the specimen was typical of a simple component design. Specimens were taken from a hot-rolled plate by production machining techniques. The hole was drilled and

Table 10A.1 Rainflow Counted Histogram of Transmission Data

Mean Stress, MPa

STRESS RANGE, MPa		-150	-100	-50	0	50	100	150	200	250	300	350	400	450	500	550	600	650	700	750
	200	8	2	10	4	4	10	0	0	6	12	30	54	58	64	44	24	12	4	0
	250	4	8	6	18	16	20	8	12	4	14	34	74	72	86	66	26	14	2	4
	300	2	2	10	6	2	2	8	6	0	8	26	40	40	46	40	16	12	2	0
	350	2	2	8	4	6	4	0	2	6	4	16	34	32	22	22	14	4	0	0
	400	0	2	2	2	4	2	2	0	0	8	14	30	32	18	16	4	0	0	0
	450	0	2	0	8	6	0	0	0	0	4	2	18	14	4	6	2	0	0	0
	500	0	0	4	4	4	2	0	0	0	4	4	6	6	2	2	2	2	0	0
	550	0	0	2	2	0	0	0	0	0	4	4	8	8	4	0	2	0	2	0
	600	0	2	2	0	0	0	0	0	0	2	2	6	4	2	0	0	0	0	0
	650	0	0	0	0	0	0	0	0	0	0	4	2	0	0	0	0	0	0	0
	700	0	0	0	0	0	0	0	0	0	0	4	0	2	0	0	0	0	0	0
	750	0	0	0	0	0	0	2	0	0	0	2	4	0	0	0	0	0	0	0
	800	0	0	0	0	0	0	0	0	0	0	0	0	0	0	0	0	0	0	0
	850	0	0	0	0	2	0	2	6	6	0	0	0	0	0	0	0	0	0	0
	900	0	0	0	0	0	0	0	0	8	0	0	0	0	0	0	0	0	0	0
	950	0	0	0	0	0	2	0	2	8	2	0	0	0	0	0	0	0	0	0
	1000	0	0	0	0	0	0	10	6	2	0	0	0	0	0	0	0	0	0	0
	1050	0	0	0	0	0	0	6	6	6	0	0	0	0	0	0	0	0	0	0
	1100	0	0	0	0	0	0	0	4	6	0	0	0	0	0	0	0	0	0	0
	1150	0	0	0	0	0	0	0	6	0	0	0	0	0	0	0	0	0	0	0
	1200	0	0	0	0	0	0	2	0	2	2	0	0	0	0	0	0	0	0	0
	1250	0	0	0	0	0	0	0	0	4	0	0	0	0	0	0	0	0	0	0
	1300	0	0	0	0	0	0	0	0	0	0	0	0	0	0	0	0	0	0	0
	1350	0	0	0	0	0	0	2	0	0	0	0	0	0	0	0	0	0	0	0
	1400	0	0	0	0	0	0	0	0	0	0	0	0	0	0	0	0	0	0	0
	1450	0	0	0	0	0	0	0	0	0	0	0	0	0	0	0	0	0	0	0
	1500	0	0	0	0	0	0	0	0	2	0	0	0	0	0	0	0	0	0	0

reamed with no edge preparation, and was then saw cut on one side to provide the notch. The stress concentration factor at the notch was also typical ($K_t = 3$).

Two commonly used structural steels were employed in the program, Man-Ten steel and RQC-100. Material properties for both materials are shown in Table 10A-4.

A number of constant amplitude, fully reversed fatigue tests were conducted to establish the load life curves for the specimen. Results are shown in Fig. 10A-3. Stress and strain-life curves were determined for each material using standard smooth specimens. These results are shown in Figs. 10A-4 through 10A-6.

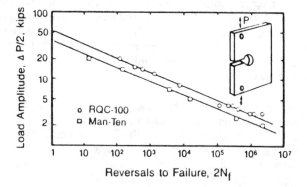

Fig. 10A-3 Load-life data.

Table 10A.2 Rainflow Counted Histogram of Bracket Data

		Mean Stress, MPa														
		-450	-400	-350	-300	-250	-200	-150	-100	-50	0	50	100	150	200	250	300
	200	0	2	2	8	12	22	36	76	52	74	46	32	8	4	2	0
	250	2	6	2	12	20	22	82	122	134	134	90	56	32	8	0	0
	300	0	0	4	12	10	26	74	86	114	112	60	54	26	2	6	2
	350	0	0	0	8	10	30	46	72	106	102	60	38	32	10	2	2
	400	0	0	2	4	6	18	42	94	98	86	42	42	24	10	2	0
	450	0	0	4	2	6	22	34	88	102	88	48	34	12	4	4	0
	500	0	0	0	2	2	18	52	74	104	84	38	18	4	0	0	2
S	550	0	0	0	0	4	12	52	64	116	64	26	10	4	2	0	0
T	600	0	0	0	2	4	18	44	68	84	46	32	12	8	0	0	0
R	650	0	0	0	2	6	8	38	76	70	50	28	6	2	0	0	0
E	700	0	0	0	4	0	20	36	98	76	52	18	8	2	0	0	0
S	750	0	0	0	0	2	6	36	70	76	40	22	6	0	0	0	0
S	800	0	0	0	0	0	12	30	54	38	28	16	4	2	0	0	0
	850	0	0	0	0	2	2	24	46	38	32	4	2	0	0	0	0
R	900	0	0	0	0	0	10	28	34	26	28	2	0	0	0	0	0
A	950	0	0	0	0	0	4	20	46	36	18	0	0	0	0	0	0
N	1000	0	0	0	0	0	2	24	16	22	14	2	0	0	0	0	0
G	1050	0	0	0	0	0	4	12	22	18	8	4	0	0	0	0	0
E,	1100	0	0	0	0	0	4	14	16	6	4	0	0	0	0	0	0
	1150	0	0	0	2	0	4	8	8	6	8	0	0	0	0	0	0
M	1200	0	0	0	0	0	0	6	2	10	2	0	0	0	0	0	0
P	1250	0	0	0	0	0	4	8	10	4	0	0	0	0	0	0	0
a	1300	0	0	0	0	0	0	2	4	0	2	0	0	0	0	0	0
	1350	0	0	0	0	0	0	0	0	4	0	0	0	0	0	0	0
	1400	0	0	0	0	0	0	4	0	0	0	0	0	0	0	0	0
	1450	0	0	0	0	0	0	0	4	0	0	0	0	0	0	0	0
	1500	0	0	0	0	0	0	0	2	0	0	0	0	0	0	0	0
	1550	0	0	0	0	0	0	0	0	2	0	0	0	0	0	0	0
	1600	0	0	0	0	0	0	0	0	0	0	0	0	0	0	0	0
	1650	0	0	0	0	0	0	0	0	0	0	0	0	0	0	0	0
	1700	0	0	0	0	0	0	0	0	0	0	0	0	0	0	0	0
	1750	0	0	0	0	0	0	2	0	0	0	0	0	0	0	0	0

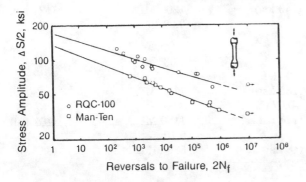

Fig. 10A-4 Stress-life data.

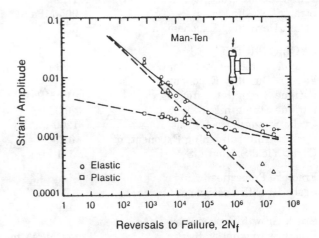

Fig. 10A-5 Strain-life curve for Man-Ten.

Table 10A.3 Rainflow Counted Histogram of Suspension Data

								Mean Stress, MPa							
		−500	−450	−400	−350	−300	−250	−200	−150	−100	−50	0	50	100	150
	200	8	16	24	8	6	66	358	104	6	10	6	6	4	2
	250	2	16	46	10	4	72	564	118	20	10	10	2	2	6
S	300	2	8	22	6	0	36	312	90	14	8	4	0	0	0
T	350	0	0	4	6	0	14	148	36	20	6	2	0	0	0
R	400	0	0	2	10	6	4	64	12	10	6	0	0	0	0
E	450	0	0	0	4	0	2	20	10	4	0	0	2	0	0
S	500	0	0	0	2	2	0	8	12	0	0	0	4	0	0
S	550	0	0	0	0	4	2	8	2	0	2	0	0	0	0
	600	0	0	0	2	0	0	2	0	0	0	0	0	0	0
R	650	0	0	0	2	4	4	2	0	2	0	0	0	0	0
A	700	0	0	0	4	2	0	0	0	0	0	0	0	0	0
N	750	0	0	2	0	0	0	0	0	0	0	0	0	0	0
G	800	0	0	0	2	0	0	0	0	0	0	0	0	0	0
E,	850	0	0	0	4	2	0	0	0	0	0	0	0	0	0
	900	0	0	0	4	4	0	0	0	0	0	0	0	0	0
M	950	0	0	0	2	4	2	0	0	0	0	0	0	0	0
P	1000	0	0	2	0	2	0	0	0	0	0	0	0	0	0
a	1050	0	0	0	0	2	0	0	0	0	0	0	0	0	0
	1100	0	0	0	0	0	0	0	0	0	0	0	0	0	0
	1150	0	0	2	0	0	0	0	0	0	0	0	0	0	0
	1200	0	0	0	0	0	0	0	0	0	0	0	0	0	0
	1250	0	0	0	0	0	0	0	0	0	0	0	0	0	0
	1300	0	0	0	0	0	0	0	0	0	0	0	0	0	0
	1350	0	0	0	2	0	0	0	0	0	0	0	0	0	0

Table 10A.4 Mechanical Properties of Man-Ten and RQC-100

	MAN-TEN		RQC-100	
MONOTONIC PROPERTIES				
Elastic Modulus, E	206 GPa	$(30 \times 10^3$ ksi)	206 GPa	$(30 \times 10^3$ ksi)
Yield Strength, S_Y	324 MPa	(47 ksi)	827 MPa	(120 ksi)
Tensile Strength, UTS	565 MPa	(82 ksi)	863 MPa	(125 ksi)
Reduction in Area, %RA	65%		55%	
True Fracture Strength, σ_f	1000 MPa	(145 ksi)	1190 MPa	(173 ksi)
True Fracture Ductility, ϵ_f	1.19		0.78	
Strength Coefficient, K	965 MPa	(140 ksi)	1200 MPa	(174 ksi)
Strain Hardening Exponent, n	0.21		0.08	
CYCLIC PROPERTIES				
Fatigue Ductility Coefficient, ϵ_f	0.26		1.06	
Fatigue Ductility Exponent, c	-0.47	-0.75		
Fatigue Strength Coefficient, σ_f	917 MPa	(133 ksi)	1160 MPa	(168 ksi)
Fatigue Strength Exponent, b	-0.095	-0.075		
Cyclic Strength Coefficient, K'	1200 MPa	(174 ksi)	1150 MPa	(167 ksi)
Cyclic Strain Hardening Exponent, n'	0.20		0.10	
Cyclic Yield Strength, S_Y'	331 MPa	(48 ksi)	586 MPa	(85 ksi)
Fracture Properties				
Crack Growth Coefficient, C	3.0×10^{-9} mm/cycle MPa^{-m}	(8.6×10^{-11}) in/cycle ksi^{-m}	5.2×10^{-9} mm/cycle MPa^{-m}	(1.5×10^{-10}) in/cycle ksi^{-m}
Crack Growth Exponent, m		3.43		3.25
Fracture Toughness, K_c (10 mm thickness)	121 MPa$\sqrt{m}$	(110 ksi $\sqrt{in}$)	154 MPa$\sqrt{m}$	(140 ksi $\sqrt{in}$)
Load Life				
Intercept	379 kN	85.2 Kips	472 kN	106 Kips
Slope		-0.223		-0.210

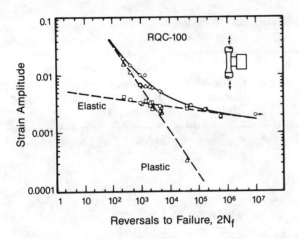

Fig. 10A-6 Strain-life curve for RQC-100.

10A.3 References

[1] Tucker, L. and S. Bussa, "The SAE Cumulative Fatigue Damage Test Program," SAE Technical Paper 750038, SAE, 1975. (See also "Fatigue Under Complex Loading: Analysis and Experiments," SAE, 1977.)

[2] "Fatigue Under Complex Loading: Analysis and Experiments," SAE, 1977.

Addresses of Publishers

SAE, Society of Automotive Engineers, 400 Commonwealth Dr., Warrendale, PA 15096

Fatigue test results for the variable amplitude histories are summarized in Tables 10A-5 and 10A-6 and plotted in Fig. 10A-7.

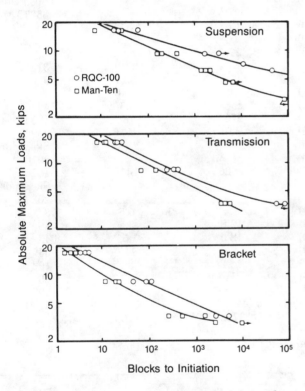

Fig. 10A-7 Summary of experimental results.

Table 10A.5 Variable Amplitude Test Results (RQC-100)

Identification Number	(−) Minimum Load (+) Maximum Load kN	(lbs.)	Load Range kN	(lbs.)	Crack Initiation Blocks	Fatigue Life Crack Propagation Blocks	Total Life** Blocks
SM1-1-FM	−71.2	(-16000)	95.7	(21500)	7.7	2.8	10.5
SM1-2-JD	−71.2	(-16000)	95.7	(21500)	-	-	(Specimen Buckled)
SM1-3-AOS	−71.2	(-16000)	95.7	(21500)	28	20	48
SM2-1-B	−40.0	(-9000)	53.8	(12100)	430*	1790	2200
SM2-2-GM	−40.0	(−9000)	53.8	(12100)	208*	357	565
SM2-3-GM	−40.0	(−9000)	53.8	(12100)	162	605	767
SM3-1-FS	−26.7	(−6000)	35.9	(8060)	1750*	22858	24608
SM3-2-FS	−26.7	(−6000)	35.9	(8060)	2240*	29644	31884
SM3-3-W	−26.7	(−6000)	35.9	(8060)	1410	-	Suspended 25353
SM4-2-MTS	−20.0	(−4500)	26.9	(6050)	4700	19966	24666
SM4-3-MTS	−20.0	(−4500)	26.9	(6050)	-	-	Suspended 6742
SM5-1-FS	−13.3	(−3000)	17.9	(4030)	-	-	Suspended 85370
BM1-1-MTS	−71.2	(-16000)	123.6	(27800)	1.5*	0.7	2.2
BM1-2-B	−71.2	(-16000)	123.6	(27800)	2.6*	0.3	2.9
BM1-3-AOS	−71.2	(-16000)	123.6	(27800)	21.0	3	
BM2-1-MTS	−35.6	(−8000)	61.8	(13900)	20.8	11.0	31.8
BM2-2-FM	−35.6	(−8000)	61.8	(13900)	11.5	8.01	9.5
BM2-3-JD	−35.6	(−8000)	61.8	(13900)	23	21.0	44
BM3-1-MTS	−15.6	(−3500)	27.0	(6080)	1588	3356	4944
BM3-2-MTS	−15.6	(−3500)	27.0	(6080)	270*	784	1054
BM3-3-FS	−15.6	(−3500)	27.0	(6080)	510	2116	2626
BM4-2-FS	−13.3	(−3000)	23.2	(5210)	-	-	Suspended 9910
BM4-3-MTS	−13.3	(−3000)	23.2	(5210)	2666	1410	4076
BM5-1-FS	−11.1	(−2500)	19.3	(4340)	-	-	Suspended 20630
TM1-1-FM	+71.2	(+16000)	106.3	(23900)	8.4	0.5	8.9
TM1-2-JD	+71.2	(+16000)	106.3	(23900)	12.8	3.2	16
TM1-3-W	+71.2	(+16000)	106.3	(23900)	12.5	1.51	4.0
TM2-1-B	+35.6	(+8000)	53.2	(11950)	420	117	537
TM2-2-GM	+35.6	(+8000)	53.2	(11950)	154	39	193
TM2-3-AOS	+35.6	(+8000)	53.2	(11950)	74	12	86
TM3-1-FS	+15.6	(+3500)	23.2	(5230)	5800	1157	6957
TM3-2-FS	+15.6	(+3500)	23.2	(5230)	4270*	1510	5780
TM3-3-MTS	+15.6	(+3500)	23.2	(5230)	3755	2165	5920

*Estimated from crack propagation data of other repetitions of the same test.
**Total Life = Crack Initiation + Crack Propagation

Table 10A.6 Variable Amplitude Test Results (Man–Ten)

Identification Number	(−) Minimum Load (+) Maximum Load		Load Range		Crack Initiation Blocks	Fatigue Crack Propagation Blocks	Life Total Life** Blocks
	kN	(lbs.)	kN	(lbs.)			
SR–1–FM	−71.2	(−16000)	95.7	(21500)	19.9	7.6	27.5
SR1–2–JD	−71.2	(−16000)	95.7	(21500)	24.4	75.6	100
SR1–3–AOS	−71.2	(−16000)	95.7	(21500)	64	154	218
SR–1–B	−40.0	(−9000)	53.8	(12100)	–	–	Suspended 3300**
SR2–2–GM	−40.0	(−9000)	53.8	(12100)	1710	–	5535
SR3–2–FS	−31.1	(−7000)	41.8	(9400)	11200	39924	51124
SR4–1–FS	−26.7	(−6000)	35.9	(8060)	48000	–	Suspended 106732
BR1–1–FS	−71.2	(−16000)	123.6	(27800)	3.3*	2.0	5.3
BR1–2–B	−71.2	(−16000)	123.6	(27800)	5.1	2.3	7.4
BR1–3–MTS	−71.2	(−16000)	123.6	(27800)	4.2	2.4	6.6
BR2–1–FS	−35.6	(−8000)	61.8	(13900)	87.5*	98.5	186
BR2–2–FM	−35.6	(−8000)	61.8	(13900)	47.0	61	108
BR2–3–JD	−35.6	(−8000)	61.8	(13900)	113	99	212
BR3–1–MTS	−15.6	(−3500)	27.0	(6080)	2673	5000	7673
BR3–3–FS	−15.6	(−3500)	27.0	(6080)	5020*	7499	12519
TR1–1–B	+71.2	(+16000)	106.3	(23900)	29.9	5.7	35.6
TR1–2–AOS	+71.2	(+16000)	106.3	(23900)	23.5	2.5	26
TR–3–W	+71.2	(+16000)	106.3	(23900)	22.2	1.8	24
TR2–1–FM	+35.6	(+8000)	53.2	(11950)	269	28	297
TR2–2–JD	+35.6	(+8000)	53.2	(11950)	460*	60	520
TR2–3–GM	+35.6	(+8000)	53.2	(11950)	374	62	436
TR3–1–FS	+15.6	(+3500)	23.2	(5230)	–	–	57090
TR3–2–MTS	+15.6	(+3500)	23.2	(5230)	–	–	Suspended 88020
TR3–3–MTS	+15.6	(+3500)	23.2	(5230)	–	–	Suspended 88020

*Estimated from crack propagation data of other repetitions of the same test.
**Small non-propagating crack present.
***Total Life = Crack Initiation + Crack Propagation

Appendix 10B Stress Strain Simulation at a Notch

10B.1 Introduction

The stress-strain response of a material must be simulated to determine the parameters necessary for a sequential, cumulative damage fatigue analysis. This brief appendix describes some of the key features in modeling the local stress-strain response of a material [1].

Information such as stress amplitude, mean stress, elastic and plastic strain must be determined for each reversal in the load history. The most important feature of a material response model is its ability to correctly describe the history dependence of cyclic deformation. This so-called memory effect can best be illustrated in the stress-strain response shown in Fig. 10B.1. As the material deforms from point a to b, it follows a path described by the cyclic stress-strain curve, magnified by a factor of two, because it represents a hysteresis loop rather than a monotonic loading curve. At point b the load is reversed and the material elastically unloads to point c. When the load is reapplied from c to d, the material elastically deforms to point b, where the material remembers its prior history (i.e., from a to b) and deformation continues along path a to d as if event b-c never occurred.

To perform the analysis, the cyclic stress-strain curve is approximated by a series of straight line segments or elements. The number and size of the elements is arbitrary, depending on the manner in which elements are used. A large number of elements (50-100) can be used. When a smaller number of elements is used, values are usually interpolated within elements to obtain midrange values. The following rules govern the manner in which elements are used:

1. Start with the first element.
2. Use them in order.
3. Skip those elements unavailable for deformation.
4. Continue until the control condition is reached.

For the purpose of illustration, a model consisting of ten elements will be used to simulate the stress-strain response of the spectrum shown in Fig. 10B.1.

First, the largest strain (maximum or minimum) in the history must be determined. This point will lie on the cyclic stress-strain curve and serve as a reference for determining the stress associated with each reversal. The spectrum is arranged in such a manner that the largest strain is the first and last value in the history.

The stress-strain elements are then formed using the cyclic stress-strain curve as illustrated in Fig. 10B.2. Starting at point a, corresponding to a strain of -0.0015 and a stress of -720 MPa, as determined from the cyclic stress-strain curve, the stress and strain elements are doubled, so that the cyclic loading curve is similar to the initial loading curve, but magnified by a factor of two. The availability coefficient of all the elements is set to -1 because the maximum strain was compressive. An availability of -1 means that all elements are available for tensile deformation, but not compressive. Simi-

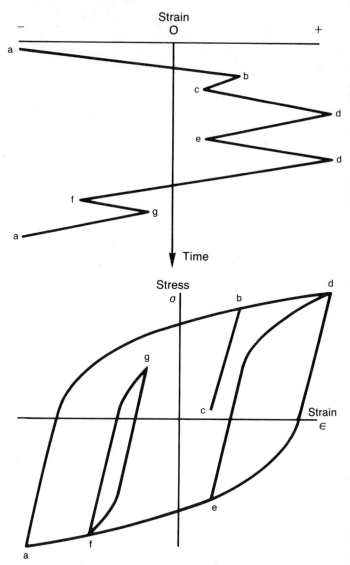

Fig. 10B-1 Stress-strain response during an irregular loading history.

larly, all elements with an availability of $+1$ are available for compressive deformation, but not tensile. Once an element has been used for deformation, its availability is changed to the opposite sign. Simply stated, this means that after an element is used in tension, it must be used in compression before it can be used again in tension.

In traversing from a to b, six and one-half elements would be required, but, since only complete elements can be included, seven full elements are used. The stress and strain are incremented by seven elements with the resulting stress of 640 MPa and strain of 0.006 at b. The error between the calculated and observed strain is, of course, reduced when a greater number of line segments are used to describe the cyclic stress-strain curve. Since the first seven elements were used in tension, their availability sign is changed to $+1$, as shown in Fig. 10B.3.

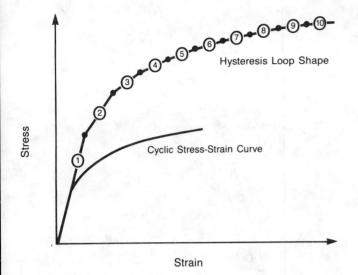

Element	Stress, Δσ	Strain, Δε
1	300	0.0015
2	120	0.0015
3	90	0.0015
4	70	0.0015
5	50	0.0015
6	30	0.0015
7	20	0.0015
8	20	0.0015
9	10	0.0015
10	10	0.0015

Fig. 10B-2 Elemental cyclic stress-strain curve.

	\multicolumn{8}{c}{Reversal}
Element Number	a	b	c	d	e	d	f	g	a
1	−	+	−	+	−	+	−	+	−
2	−	+	+	⊕	−	+	−	+	−
3	−	+	+	⊕	−	+	−	−	⊖
4	−	+	+	⊕	−	+	−	−	⊖
5	−	+	+	⊕	+	⊕	−	−	⊖
6	−	+	+	⊕	+	⊕	−	−	⊖
7	−	+	+	⊕	+	⊕	−	−	⊖
8	−	−	−	+	+	⊕	−	−	⊖
9	−	−	−	+	+	⊕	+	+	−
10	−	−	−	+	+	⊕	+	+	−

◯ Indicates Skipped Element

Fig. 10B-3 Availability matrix used for material response model.

One element is required to go from b to c. The availability is changed to −1 because the elements were used in compression. The stress and strain are also incremented by one element with the result of 40 MPa and 0.003 respectively.

Four elements are required to go from c to d. The availability matrix shows that elements 2 through 7 are unavailable for tensile deformation. Therefore, elements 1, 8, 9 and 10 must be used to make up the four elements required to reach a strain of 0.015. The strain is incremented by 0.012, the stress by 680 MPa, which is the sum of elements 1, 8, 9, and 10. At this point all elements are available for compressive deformation. Reversal d-e requires four elements in compression with the resulting strain and stress of 0.003 and −460 MPa. The availability sign on the first four elements is changed to −1, indicating that they are available for tensile deformation for reversal e-d that uses four elements in tension. All of the elements are once again available for compressive deformation. The last three reversals are treated in the same manner as the first three except that the loading direction is changed, i.e., a-b is tensile, while d-f is compressive. The completed element matrix, Fig. 10B-3 shows all of the elements that were used for each reversal. Fig. 10B-4 shows the simulated stress-strain response for the spectrum that is in excellent agreement with the measured response of Fig. 10B-1.

Besides simulating the stress-strain response of the material, the model provides a convenient and efficient method of cycle counting. There are four closed hysteresis loops in this spectrum: a-d-a, b-c-b, d-e-d, and f-g-f. A close examination of the element matrix will show that when a hysteresis loop is closed, one or more elements have been skipped.

For example, in reversal c-d, elements 2 through 7 were skipped, indicating that a closed hysteresis loop consisting of elements 1 and 2 was formed. Similarly, in reversal e-d, elements 5 through 10 would be skipped if they were required, indicating a closed hysteresis loop of elements 1 through 4 was formed. The key to cycle counting is that a closed hysteresis loop is formed whenever the next element in the availability matrix is unavailable for deformation. Also, there are no skipped elements in a hysteresis loop, which means that every time element 5 was skipped, a loop consisting of elements 1 through 4 was formed.

Once the cycles have been defined and the stress-strain response determined, the appropriate fatigue parameters can be determined so that a damage analysis can be performed.

10B.2 Notch Analysis

In many practical problems engineers and designers are required to evaluate the fatigue resistance of new components while still at the drawing board or prototype stage of development. One method for performing this type of analysis is the so-called "component calibration" technique that requires a relationship between applied load and local strains, such as the one shown in Fig. 10B-5. This type of

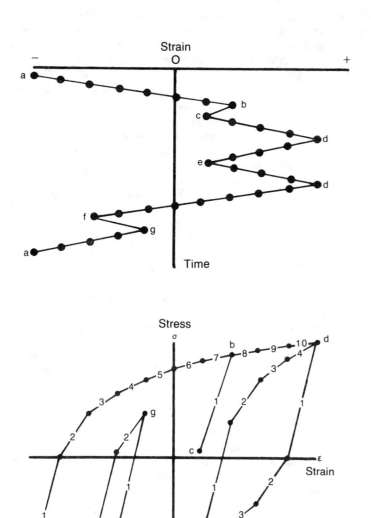

Fig. 10B-4 Stress-strain simulation example.

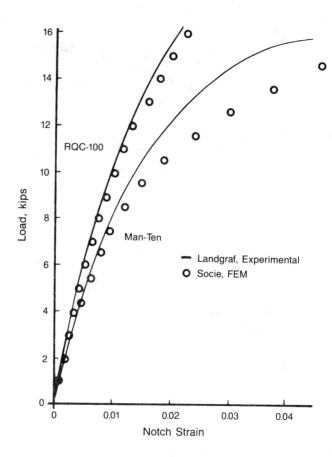

Fig. 10B-5 Load-strain curve for SAE specimen.

information can be obtained analytically with a finite element model or experimentally by testing the component. The component normally would be tested by attaching strain gages to critical locations and applying a load and unload cycle while measuring the load-strain response. However, this type of test will produce inaccurate data because of the cyclic hardening or softening characteristics of the material. For this reason, an incremental step type of test should be used in obtaining load-strain curves from a single component. Similarly, cyclically stable material properties should be used in any analytical calculations.

The conversion of applied load into strain is accomplished in exactly the same manner as strain was converted into stress. The load-strain response has all of the features normally associated with stress-strain response, i.e., hysteresis effects, memory, and cyclic hardening and softening. From a computational viewpoint, this technique is exactly the same as those described in the last section for stress-strain response. In fact, the load-strain and stress-strain response models can be combined, so that the applied load can be converted to both the local stress and strain with one simple computer algorithm.

Neuber's rule may also be used to determine the notch root stresses and strains on a cycle-by-cycle basis to determine the strain ranges and mean stresses at the notch root. Neuber's rule can be directly incorporated into the material response model. There is a one-to-one correspondence between nominal and notched closed hysteresis loops, i.e., for each hysteresis loop of amplitude, ΔS and Δe, there is a hysteresis loop of amplitude, $\Delta\sigma$ and $\Delta\varepsilon$ at the notch root. The elements, ΔS and Δe, have already been determined for the stress-strain response model. Notch stress-strain elements $\Delta\sigma$ and $\Delta\varepsilon$, are then determined for each nominal stress-strain element. The new notch elements are then used in exactly the same manner as the original stress-strain elements. Whenever an element is used in the nominal stress-strain behavior, it is also used in the notch stress and strain response; conversely, whenever an element is skipped in the nominal behavior, it is also skipped in the notch response. The significance of this is that once the nominal stress-strain response is obtained, the notch response is automatically determined. For example in Fig. 10B-4 reversal c-d used elements 1, 8, 9 and 10 to obtain the nominal stress-strain response. Notch elements 1, 8, 9 and 10 would also be used to obtain the notch response.

Material response, cycle counting, component calibration and notch response can be combined into a single model for efficient computation of fatigue damage.

10B.3 References

[1] Richards, F. D., Lapointe, N. R., and Wetzel, R. M., "A Cycle Counting Algorithm for Fatigue Damage Analysis," SAE Technical Paper 740278, SAE, 1974.

Addresses of Publishers

SAE, Society of Automotive Engineers, 400 Commonwealth Dr., Warrendale, PA 15096

Chapter 11 Failure Analysis

11.1 Introduction
    11.1.1 Scope
    11.1.2 Relationship to Other Chapters
    11.1.3 Chapter Plan

11.2 Why Things Break
    11.2.1 Cracking and Fracture
    11.2.2 Strength in the Case of Cracks
    11.2.3 "Brittle" vs. "Ductile" Fractures
    11.2.4 The Reasons for Failures
    11.2.5 The Elements of Failure Analysis

11.3 Fractography for the Engineer
    11.3.1 Scope
    11.3.2 Fractographic Tools
    11.3.3 Cracking Mechanisms and Microscopic Features
    11.3.4 Fracture Mechanisms and Microscopic Features
    11.3.5 Brittle and Ductile Fracture Revisited
    11.3.6 Cleavage vs. Rupture
    11.3.7 Other Fracture Surface Analysis Tools
    11.3.8 Quantitative Fractography

11.4 The Search for the Origin
    11.4.1 Scope
    11.4.2 Distinguishing the Crack and the Fracture
    11.4.3 Crack Features: Beach Marks, Mussel Shells and Sunrise Markings
    11.4.4 Fracture Features: Chevrons, Branches, Shattering
    11.4.5 Crack Origins

11.5 Failure Analysis Procedure
    11.5.1 Preliminaries and General Considerations
    11.5.2 Preliminary-Failure-Analysis Plan and Failure Hypothesis
    11.5.3 Secondary-Failure-Analysis Plan: Verification

11.6 Possible Actions Based on Failure Analysis
    11.6.1 Scope
    11.6.2 Failures During Design Development or Design Verification Testing
    11.6.3 Service Failures

11.7 Summary

11.8 References

11.9 Topical Bibliography
    11.9.1 General Failure Analysis
    11.9.2 Fracture Surface Analysis

11.1 Introduction

11.1.1 *Scope*

ily, failure analysis is thought of as a task to be performed by metallurgists. Thus, the question arises as to why this handbook contains a chapter on failure analysis. The answer to this question is that a failure analysis is usually not complete without the input of, and interpretation by, a structural engineer.

This chapter brings out those aspects of the failure analysis process that typically belong in the domain of structural and design engineers. It also provides a brief survey of the most important terms used in fractographic analysis, for the benefit of design engineers and structural engineers.

As fractography has developed more or less independently, some terms used by metallurgists are used in structural engineering to describe somewhat different phenomena. This may cause confusion and therefore this chapter adheres to a rigorous use of terms.

Defining what this chapter is not may be just as useful as defining what it is. This chapter is *not* a manual on fractography. Such a work would require much more space but, more important, it would be superfluous as several excellent texts are already available to the fractographer [1-4]. Hence, this chapter does not present a survey of all fractographic features. It explains only those features that should be understood by the structural engineer in his communications with the fractographer. This information should facilitate his participation in a failure analysis.

The purpose of the chapter is to show the structural and design aspects of failure analysis and to demonstrate how the analysis results should be used in making decisions regarding design modifications and possible changes in operation.

11.1.2 *Relationship to Other Chapters*

This handbook deals with design against fatigue. Despite even the most conscientious applications of fatigue design analysis and general design rules, failures will continue to occur. More positively stated, because of the great emphasis placed on fatigue resistant design, fatigue failures are a relatively rare occurrence; the few failures that still do occur provide insurance against complacency in design with regard to fatigue.

In order to recognize the truth of the above statement one must go back only half a century and compare the failure reports of that era with the present, taking proper account of the cost increases since then, and the number of structures in service (in particular, automobiles, airplanes and heavy equipment).

The best efforts in fatigue design will not prevent all failures. Nearly every fatigue failure offers a lesson worth learning about fatigue design. When this lesson is properly understood, the problem can usually be remedied and further failures of a similar type prevented. The material discussed

In this context "failure" is not limited to service failures but also includes laboratory failures. Where design development tests are performed to evaluate or compare fatigue performance of design details or components, the end result is almost always a failure. Full scale design life verification tests similarly result more often than not in failure of the structure or at least of one or more components. Such laboratory failures are of equal values as service failures in terms of the information they provide.

11.1.3 *Chapter Plan*

Every failure, in principle, contains all the information about what led to its existence, although this information is sometimes hard to extract. The broader the scope of the failure analysis, the greater the likelihood that the failure scenario and the cause can be reconstructed. This means that design or structural engineers can play a significant role in a failure analysis. This chapter shows what such involvement might entail.

Unfortunately, too often the scope of a failure analysis is limited to a pure fractographic analysis by the metallurgist. From the point of view of the structural engineer, the metallurgist retreats to his lab, with his microscopes, to return with a great number of photographs that (fascinating though they may be) seldom provide much of a clue for the prevention of further failures. On the other hand, fractography in this limited sense is an indispensable part of a failure analysis, since it is usually the only means by which the failure mechanism can be established. Therefore, the design engineer involved in failure analysis should understand the most prominent fractographic features and their significance. Thus, part of this chapter is devoted to an explanation of these features, without all of the details that are of interest to metallurgists only.

Before considering the fractographic features, a foundation is laid by way of a discussion of the reasons why failures occur in the first place. The discussion of fractographic features is followed by a review of broad-scope failure analysis that includes the activities of design engineers and/or structural engineers.

The following subjects are discussed in order:

a. Why things break; types of failures; the difference between cracks and fractures; fractographic features.

b. Fracture surface analysis and quantitative fractography.

c. The significance of the origin of fracture; what anybody can discover without the use of microscopes.

d. Elements of a failure analysis with emphasis on the structural and design aspects.

Through the above, all aspects of failure analysis are reviewed. Subsequent discussion concentrates on the failure analysis procedure and the use of the results.

It should be emphasized at this point that this chapter is not written to be a reference text in the sense that its various sections can be read independently. Rather, it attempts to compose the "total picture" by means of a number of building blocks. Clearly, within this handbook, the subject could not be dealt with exhaustively. In areas where details had to be omitted, references are cited. A topical bibliograhy is also included.

11.2 Why Things Break

11.2.1 *Cracking and Fracture*

Every essential load-bearing structure is designed to carry the maximum anticipated service loads with a certain safety factor. Depending upon the type of structure, the safety factor may vary from 1.5 to 3 (sometimes even 5 or 10). Although there is usually some uncertainty about the maximum anticipated service loads, the safety factor is generally defined in an attempt to include these uncertainties.

As a matter of fact, in most cases the safety factor is really an ignorance factor; it covers uncertainties in loads, inaccuracies in stress analysis, possible below-average material strength or ductility, unknown built-in stresses or strains, lower limit dimensional tolerances, and to some extent small defects that may escape quality control. The design, that is the size of the structural dimensions, is always based on the maximum anticipated service loads times the safety factor. Given the often large safety factor a structure should rarely fail by overload, even at loads higher than the maximum service load. Indeed, true overload failures of uncracked sections are uncommon.

Thus, under normal maximum service loading or even well above that due to the safety factor, a new structure should not fail. Nevertheless, the structure does have a finite strength. If a structure has been designed according to an ultimate strength criteria and an effort is made to intentionally fail the structure, it would only be necessary if physically possible to apply a load equal to the maximum anticipated load times the safety factor. However remote the possibility that such an extreme load would occur in service, the probability is not altogether zero. Hence, any structure even when new will represent a low-probability failure risk when subjected to design loads (a concept rarely understood by lawyers and many lay juries). As already mentioned, this failure risk of the undamaged new structure is so low that "overload failures" are really quite uncommon.

That being the case, why do failures occur? Most of the time failures are the consequence of defects or cracks developing during service. Before considering the rule let us consider some possible exceptions:

a. The small possibility of a true overload failure as discussed above.

b. A "large" pre-existing material defect or mechanical defect not detected during quality control.

c. A poor design with very sharp notches that renders conventional design or stress analysis useless for predicting the safe operating loads in the component.

d. Extreme circumstances, such as temperatures or residual stresses environments, not accounted for during design.

e. "Failure" due to loss of alignment, distortion, or wear that leads to unsatisfactory performance or true failure of other related components.

Of greatest concern in this discussion are failures that are precipitated by cracks. In most cases, within this context, a failure can be taken to mean the point at which the part breaks into two pieces, or fractures. In some cases the development of a crack may precede total failure significantly and still represent failure in terms of part function or established retirement criteria.

Cracks and crack growth can occur by a variety of mechanisms. The word mechanism means the microstructural response to the imposed service. Prominent mechanisms of interest to the ground vehicle industry include:

a. Fatigue

b. Stress corrosion

c. Creep

d. Hydrogen assisted cracking

e. Wear, fretting

f. Combinations of the above or variations on the theme.

By themselves these mechanisms rarely cause the fracture of a component. Rather the fracture occurs as a consequence of a crack that developed by one of the above mechanisms. The cracking mechanism is either ductile, it ruptures, or is brittle, as in cleavage. Ductile behavior implies plastic deformation at the macroscale. Consequently, although fatigue occurs by microscopic plastic strain, the failure shows little evidence of plastic flow. In contrast, true overload failures occur by ductile rupture. Generally, cracking mechanisms as well as fracture can be either transgranular or intergranular. While many mechanisms can contribute to cracking, fracture occurs by only one of two mechanisms: rupture or cleavage.

Cracking and fracture mechanisms will be discussed later in somewhat more detail as this knowledge is necessary for an overall understanding of failure analysis. At this point it is sufficient to distinguish between cracking and fracture.

11.2.2 Strength in the Case of Cracks

Cracks impair the strength of a component, the more so when they are larger (Fig. 11-1). As discussed already, a new structure designed according to an ultimate strength criteria should have a strength equal to the design strength. If P_{max} is the maximum service load and j the safety factor, then the strength of the new structure should be

$$s = j * P_{max} \qquad (11\text{-}1)$$

For an intact structure without cracks, conventional stress and design analysis will closely determine the above strength. Fracture will occur when the load becomes equal to $j * P_{max}$.

If one of the cracking mechanisms becomes operative and a crack develops in time (Fig. 11-1a), the strength of the structure declines [5], the remaining strength, or residual strength will be less than the design strength and fracture will occur at loads less than $j * P_{max}$. At the time when the crack has grown to size a_i (Fig. 11-1b), a fracture will occur at load P_i. Since such a high load is an unlikely event, the structure would normally not fracture with a fatigue "precrack" of this size.

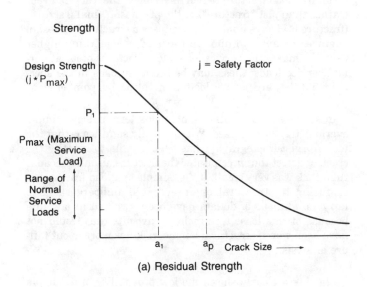

(a) Residual Strength

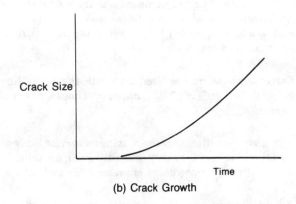

(b) Crack Growth

Fig. 11-1 **Fracture and crack growth.**

When the crack reaches size a_p the strength is further reduced to the point that fracture would occur if the applied load would reach P_{max}, the maximum anticipated service load. P_{max} may or may not occur, but if it does, fracture will result. Therefore, a_p can be considered the maximum permissible crack size because it may precipitate a fracture at loads that are anticipated in service. If, however, P_{max} does not occur, the crack will continue its growth, the strength will decay further and eventually will become so low that almost any service load will cause fracture.

Although this is trivial to a metallurgist, the difference between stable crack growth and fracture (unstable growth) should be restated. Fracture is the final, total separation occurring at a sufficiently high load. The fracture is generally precipitated by a crack, the growth of which takes place by a different mechanism during normal service loading. Likewise, there may be a difference in the mechanism of crack growth and fracture.

Unfortunately, metallurgists often rather loosely use the term "overload fracture" when referring to the fracture event (rather than just "fracture") or the pleonasm "final fracture" (fracture is always final). This confuses a structural or design engineer because, to him, an overload is something higher than the maximum service load. Most fractures occur at normal service loads; these may be the high loads in the spectrum but they are not overloads from the design point of view.

In some cases the rather ambiguous terminology "simple overload fracture" is used. What it really signifies is that the metallurgist-fractographer has been unable to identify the crack or defect that precipitated the fracture, usually because this crack was very small. If the "simple overload fracture" is confusing to structural engineers, it is flabbergasting to lawyers; millions of dollars have been awarded in lawsuits because of this loose terminology, lawsuits that would not have been won if structural engineers knew more about failure analysis.

Fig. 11-1a clearly shows the low probability of a "simple overload" fracture. The only true possibility for a "simple overload" fracture is for a (new) uncracked structure to be subjected to a load equal to j * P_{max}. When it does happen it usually occurs as a result of:

a. Extreme abuse by the user, but with the prevailing safety factor loads of j * P_{max} are many times a *physical impossibility*.

b. A gross underestimate of the maximum service loads so that the structure is grossly underdesigned; an unlikely situation with proper usage of present-day technology.

c. An extremely poor design with extremely sharp notches, misfits, etc.; such that the conventional design analysis was grossly inadequate; an unlikely situation with proper usage of present-day technology.

What is usually meant by "a simple overload" fracture is that the fractographer was unable to identify the crack that precipitated the fracture because the crack was too small. This may occur when the material's toughness is extremely low.

The residual strength curve of Fig. 11-1a is determined by the toughness, i.e., by the equation

$$\sigma_{Fracture} = \frac{K_{Ic} \text{ (or } K_c)}{B\sqrt{\pi a}} \qquad (11\text{-}1)$$

If the toughness K_{Ic} (or K_c) is low, the residual strength curve may look like the one in Fig. 11-2. In this case a_p may be so small, the fractographic features of the crack so ill-defined, that no indication of cracking can be found. All readily discernible microscopic features are those of the final fracture. This does not necessarily signify "a simple overload."

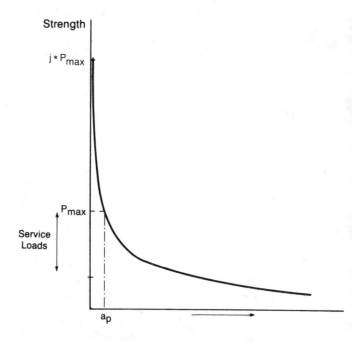

Fig. 11-2 Residual strength in the case of low toughness.

The major point to be made is that simple overload failures are really rather unusual occurrences and anyone involved in a failure investigation should not accept this conclusion without substantive evidence.

11.2.3 *"Brittle" vs. "Ductile" Fractures*

It can safely be said, with the possible exception of fractures in 300-series stainless steels, that all fractures are brittle from the engineering point of view. Yet, probably more than 90 percent of the fractures are ductile as far as the major metallurgist is concerned. Without doubt, this issue is the major cause of confusion between metallurgists and their engineers. Fortunately, the problem is only one of a definition of terms. Does this mean that once other engineers understand the metallurgist's language, the problem is solved? It will be when other engineers also recognize when the loose terms are used.

The metallurgist's definition of a brittle fracture pertains to whether or not plastic deformation is required for fracture (not to whether it occurs or not). Cleavage fracture does not require plastic deformation; the metallurgist calls it brittle fracture although considerable plastic deformation may occur. Rupture is the result of plastic deformation; the metallurgist calls it ductile, although virtually no plastic deformation may occur. Finally, he may use any qualification in the case of limited plasticity. Confining the terms brittle and ductile to their engineering definitions and confining the terms cleavage and rupture to their metallurgical definitions of the actual fracture process would help a lot in avoiding confusion.

What do the terms brittle and ductile mean in an engineering sense? Let us consider Fig. 11-3. If an unnotched bar is pulled to fracture, ample plastic deformation will occur throughout (a). After fracture, large permanent deformation is obvious. Clearly, the fracture is ductile.

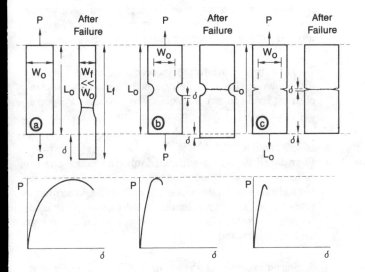

Fig. 11-3 Effects of notch and crack on apparent ductility from the engineering point of view.

A specimen of the same material with mild notches (b) will exhibit a different behavior. The load at fracture will be virtually the same as in case a. However, plastic deformation will be confined to the section with the greatest stress. The overall plastic deformation will be small. After fracture such plastic deformation is discernible only in the vicinity of the fracture path (ductile or brittle?).

Finally, consider case c with sharp notches. The notch will cause a somewhat lower fracture load depending upon toughness. But since plastic deformation is now completely confined to the fracture path, there will be no signs or only weak signs of it after fracture. The engineer calls that a brittle fracture: NO SIGNS OF PLASTICITY. The metallurgist disagrees: the mechanism of fracture was the same in all cases, namely dimple rupture. Hence, nearly all fractures are ductile from a metallurgical perspective.

The case is further illustrated in Fig. 11-4. It shows a piece of brass with holes at incipient fracture. An unnotched piece would show 30 percent or more plastic deformation at fracture. Clearly in Fig. 11-4, plastic deformation is confined mostly to the future fracture path. The fracture is almost brittle from an engineering point of view.

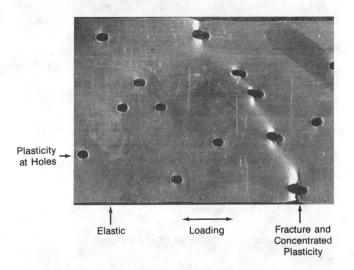

Fig. 11-4 Apparent ductility of highly ductile notched brass.

All real structures contain stress concentrations or notches; therefore, plastic deformation will tend to be confined to the fracture path and, as a result, most fractures will appear brittle or semi-brittle depending upon the acuity of the notch. Even in the absence of a significant stress concentration it has been argued that most fractures are precipitated by cracks. Since any crack represents an extremely sharp notch, all fractures developing from cracks will appear brittle from the engineering point of view.

Plastic deformation of a cracked component is largely if not wholly confined to the fracture path (Fig. 11-4). After failure, only the material immediately adjacent to the fracture will show any signs of plastic deformation. The more ductile the material, the more pronounced these signs will be. However, most structural materials are less ductile than the brass in Fig. 11-4. Hence, almost any fracture will appear brittle to the engineer.

Without going into some detail with regard to fracture mechanisms, it is not possible at this point to explain the metallurgist's definition of brittle and ductile fracture. His definitions are based on the *mechanism* of fracture and these will be discussed later. At this juncture it will suffice if one understands the engineer's definition.

11.2.4 *The Reasons for Failures*

Ultimately, fracture is the direct cause of most failures. However, as explained in the previous sections, the actual culprit is almost always a crack without which the failure would not have occurred. Hence, in analyzing failures one should generally find the cause for the cracking rather than the cause of the fracture.

The first question should be "By which mechanism(s) did the cracking take place?" The most prominent mechanisms were listed in Section 11.2.1. Each mechanism has certain characteristic features by which it can be recognized (Section 11.3). However, the mechanism by itself hardly explains the failure. The really important question is why this mechanism became operative and what caused the crack to start in the first place.

There are at least seven fundamental causes for cracking, namely

a. Material defects

b. Manufacturing defects

c. Poor choice of material or heat treatment

d. Poor choice of production technique

e. Poor (detail) design.

f. Unanticipated service environment or loading

g. Poor materials property data

One might add "poor quality control" to this list, but "poor quality control" is not a fundamental reason; it is merely the secondary reason by which a. or b. can become operative.

Item a, material defects, is of little interest here, because it cannot be remedied by design or structural engineering. Items b through e belong in the domain of "design."

Clearly, if a failure analysis can identify one of the above causes as having been operative in the case of a particular failure, the remedy or solution is at hand. This can best be demonstrated with a few examples:

b. *Manufacturing Defects*
These can be introduced by blunt tools, overheating (grinding cracks), poor welding procedures, etc.

c. *Material and Heat Treatment*
Material selection is so obvious a culprit that it needs little amplification. The material selection may be perfect from the point of view of fatigue and be made for that reason. At the same time the possibility of stress corrosion cracking might have been overlooked; the material (heat treatment combination) may be sensitive to stress corrosion cracking. Local heat treatments such as carburizing, nitriding and surface hardening almost always cause a volume change of the surface layer. These volume changes can introduce residual stresses that may not have been accounted for in a (fatigue) analysis and thereby may cause "unexpected" cracking.

d. *Production Technique*
Electrochemical treatments may generate hydrogen that penetrates into the material or in the case of titanium alloys, the hydrogen level may be outside the specification. The hydrogen may not be able to escape (e.g. in the case of cadmium plating) or not be forced out by subsequent baking. Hydrogen assisted cracking can occur even without the action of external loads.

Fig. 11-5 shows cracks in a forging following the grain flow.

The weak exposed grain boundaries form an excellent path for cracking (Fig. 11-6), while the part was used in bending in such a way that the highest stresses were at A and B (Fig. 11-6). The designer should have prescribed a different manufacturing procedure (rotate dies).

Two alternative production procedures for the same part are shown in Figs. 11-7a and b. In both cases the machining leads to exposed grain boundaries. Proper design can avoid such a problem (as in Fig. 11-7c).

e. *Detail Design*
Poor detail design is without doubt the major cause of failures. That sharp notches should be avoided, that ample radii should be used at section changes, is commonly understood. Yet such mistakes are still commonly made. More important, the hidden stress concentrations are seldom recognized.

A classic example is shown in Fig. 11-8. From a static design point of view the arrangement is perfect. When loaded to failure, plastic deformation of bolt holes will ensure even distribution of the load over the 4 bolts, each bolt carrying 1/4 P; the joint has the necessary design strength of $j * P_{max}$.

However, during service loading at P_{max} or below, the stresses are elastic. Since I and II are joined, A-B must stretch the same amount as C-D (same elongation) but if A-B and 2 * (C-D) have the same cross section, they must carry the same stress to have the same elongation. In that case they each carry half the load. Hence, the two end bolts each carry half the load; the two inner bolts have no function at all. This "unanticipated" stress concentration (factor of 2) causes cracking at the outer bolts.

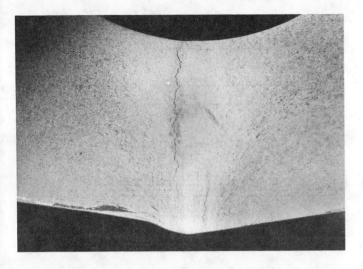

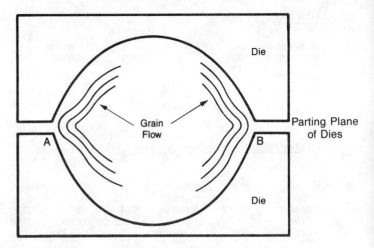

Fig. 11-5 Crack and fracture path following grainflow in aluminum forging. Etched cross section. (3X)

Fig. 11-6 Exposed grainflow after machining of forging.

Proper bolt load distribution requires that the cross sections of I and II are in accordance with the load to be carried, as shown in Fig. 11-8b. The equivalents in full section design are shown in Figs. 11-8c and d. Most people would not let the design of Fig. 11-8c pass, but the design of Fig. 11-8a is common.

Other common detail design errors cause "unanticipated" secondary displacements and/or stresses not accounted for in design analysis or even fatigue analysis. Consider the shrunk fitting in Fig. 11-9. Eventually the shrunk fit will transfer load from the shaft to the shrunk-on part. However, at A, the shrunk-on part is still stress-free (no stress and no strain). The shaft is strained so that there will be relative movement between A and B, causing fretting and fatigue. The same can happen at bolt shafts, under bolt heads, etc.

Secondary stresses are equally troublesome. The loading of parts I and II in Fig. 11-10 causes small displacements of A and therefore of B. B cannot follow these displacements, setting up unanticipated secondary stresses, which, added to the "live" stresses, can cause cracking. Elimination of two bolts would solve the problem in this case.

f. *Unanticipated Service Environment or Loading*
Unanticipated environment or loading is seldom a factor when designing second or third generation components. Design in this instance is more a matter of adapting than designing. When dealing with new situations with unfamiliar materials there always exists the potential for unanticipated environment or loading. Care should be taken in failure analysis to examine these factors as possible causes for cracking. Changes in

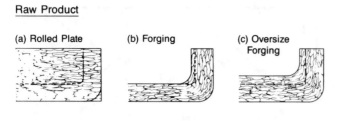

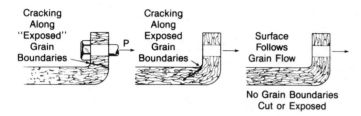

Fig. 11-7 Effect of poor manufacturing design.

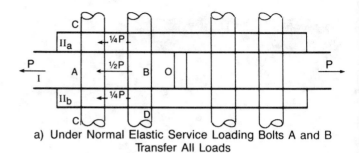

a) Under Normal Elastic Service Loading Bolts A and B Transfer All Loads

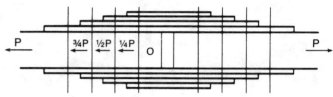

B) All Bolts Transfer Equal Loads in Elastic Case

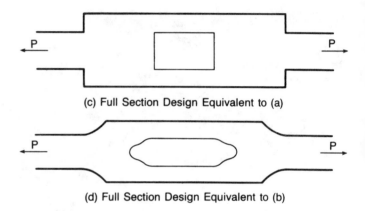

Fig. 11-8 All designs equivalent for static strength with full plasticity, but (a) and (c) are poor at service loads

cracking mechanism from the origin into the macrofracture often signal the role of environment, particularly if the initiation is intergranular. The possibility of a true overload can be estimated in some materials from the size of the shear lip. However, care must be taken with ductile materials (where LEFM principles may be violated) when making such estimates.

g. *Inadequate Materials Property Data*

Many times designers are called upon to design components from new materials or to design parts which serve in situations where fatigue may be coupled with wear mechanisms or corrosive agents. In these situations the designer often lacks the detailed data necessary to ensure the safe operation of the component. The natural reaction is to increase the factor of safety. For new materials this may be a viable option. However, in cases where coupled mechanisms are operative, fatigue strength reduction factors often far exceed anticipated values. As an example, fretting fatigue often causes strength reductions in excess of a factor of 5. Consequently, care should be taken to understand the role of "secondary mechanisms" coupled with fatigue during a failure analysis since the role of these mechanisms may also be very important.

11.2.5 *The Elements of Failure Analysis*

Many more examples of "why things break" can of course be given [3]. It is not the objective of this chapter to review all possible causes of failures but rather to discuss how to pinpoint the cause. The examples in the previous section should adequately illustrate what is needed to pinpoint the cause.

It can now be seen that the determination of the cracking mechanism is only a small though significant part of the failure analysis. The cracking mechanism shows which "forces" were operative. However, in order to determine why these forces could become operative, the starting point of the crack must be found. If the crack had never started, there would be no failure.

After the initiation point has been pinpointed the question can be asked, "Why did a crack initiate there?" All fundamental causes discussed in Section 11.2.4 should be reviewed. Clearly, at this time there is an essential task for the structural and design engineer. Questions such as, "Where did the stresses come from (loads, load path)?," "What are primary and secondary displacements and stresses, etc.?," have to be answered.

One can now more clearly see how a failure analysis should proceed:

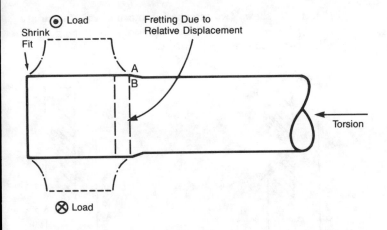

Fig. 11-9 Cracking due to relative displacement.

First Phase - Hypothesis

a. General assessment of situation, of entire component, failure area. It is not just the fracture surface that is of interest. General area deformations (that may have occurred during the failure process) and of adjacent parts will reveal information on acting loads and stresses.

b. Determination of cracking mechanism.

c. Determination of the origin of cracking.

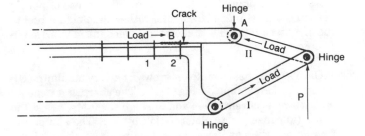

Fig. 11-10 Summary stresses due to displacements. Elimination of bolts 1 and 2 solves the problem.

d. Research at the origin to reveal possible material or production defects.

e. Analysis of general stress field, "unanticipated" stress concentrations, secondary stresses.

f. Establishment of hypothesis for failure scenario.

Second Phase - Proof or rejection of hypothesis

g. Are the loads, stresses, displacements and secondary stresses and displacements compatible with the cracking mechanism? This question should be answered as quantitatively as possible. The use of fracture mechanics may be appropriate.

h. Is the size of crack that finally precipitated the fracture compatible with the scenarios? What is the material's toughness? Given crack size and toughness, is the calculated fracture stress compatible with service stresses? If not, loads and stress analysis should be checked and the assumptions made in the fracture mechanics analysis should be reviewed.

i. Are results of quantitative fractography compatible with analytical results of g and h and with the hypothesis?

j. Confirm or modify hypothesis.

Third Phase - Identify a remedy, assuming cause is not defective material

k. How can secondary stresses, displacements be relieved?

l. How can stress concentration be reduced?

m. How can general loads/stresses be reduced?

n. Should manufacturing procedure be changed? (See, for examples, Figs. 11-8 and 11-9.)

11.3 Fractography for the Engineer

11.3.1 Scope

As argued in the previous section, the role of the engineer in the failure analysis is very significant. However, the engineer's task can be fulfilled only if his expertise is paired with at least an understanding of elementary fractography and fracture mechanics. This section is intended to provide that background. It is written for engineers, not for fractographers.

The most important cracking mechanisms and fracture mechanisms will be discussed. The salient fractographic features associated with each mechanism will be explained.

Finally, quantitative fractography will be briefly reviewed. Emphasis will be on those techniques that are helpful if the fundamental cause of failure is in design in the broad sense. Those techniques that are useful primarily for addressing the material's aspects of a failure are not emphasized since they fall in the metallurgist's domain.

11.3.2 Fractographic Tools

The tools commonly used by the fractographer are

a. Loupe and stereo microscope (recommended to engineers as well)

b. Macrophotographic equipment

c. Optical microscopes (with often forgotten dark field, interference contrast, phase contrast)

d. Transmission electron microscopes

e. Scanning electron microscopes with X-ray analyzer

f. Surface analysis equipment (mostly exotic)

g. Quantimet or comparative image analyzers

It is beyond the scope of this section to discuss the techniques involved in the usage of these devices. The engineer should be aware that an electron-fractograph (high magnification photograph taken usually in microscopes) reveals no more than "one patch of crabgrass in an entire football field." This may be perfectly adequate; however, it can be deceiving.

11.3.3 Cracking Mechanisms and Microscopic Features

The main stable cracking mechanisms are

a. Fatigue

b. Stress corrosion

c. Creep

d. Hydrogen assisted cracking

e. Combinations or variations of above

Most of the following discussion pertains to fatigue. Other sources identified in Section 11.10 should be examined for detailed information on the other cracking mechanisms.

Fatigue damage and fatigue cracking in service almost always takes place under essentially elastic stressing and straining. Nevertheless, a fatigue crack cannot initiate without plastic deformation, however local and minute. Generally, such plastic deformation will occur at the root of a notch or at a stress raiser such as a foreign particle in the material. On a microstructural level, plastic deformation is permanent offset caused by slip of adjacent atomic planes in the material.

Fig. 11-11 shows two alternative mechanisms (more are possible) by which small slip displacements during cyclic loading can result in crack initiation.

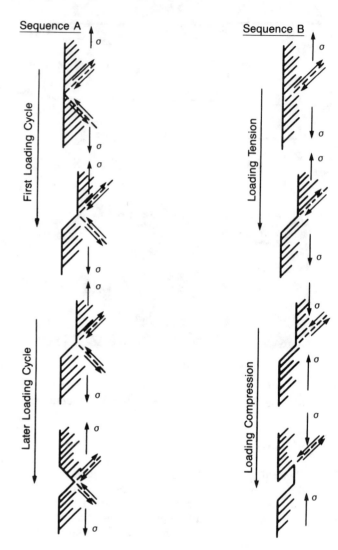

Fig. 11-11 Alternative mechanisms of fatigue crack initiation.

Note that the crack is a geometrical consequence of plastic deformations (slip displacements) rather than actual breaking of material. Fig. 11-12 shows this mechanism in action in micrographs of a metal surface.

Once a crack has initiated it grows by a very similar mechanism [6], for example in Fig. 11-13. The sharp crack causes a high stress concentration and the high shear stresses at the crack tip cause plastic deformation by slip. The slip occurs first on one then on more planes leading to blunting (stages A-D). Note that the crack "grows" by an amount a, and that this growth is again a geometrical consequence of slip. Upon unloading and resharpening the process repeats itself (stages

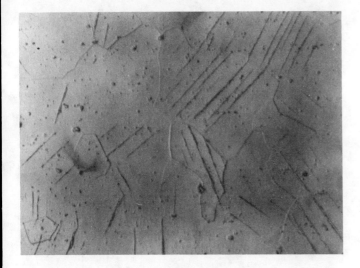

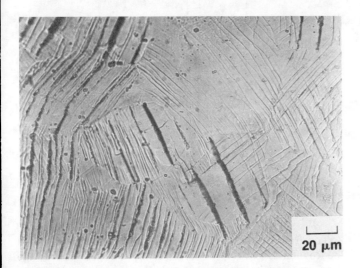

Fig. 11-12 Cyclic stressing causing slip steps (a) which develop into cracks (b). Micrographs of originally smooth surface after increasing number of cycles. (aluminum, 500X)

E-K), each time causing a small a (on the order of 10^{-6} to 10^{-4} inches); small, but enough to add up to substantial crack growth during many cycles.

The regular pattern of blunting and sharpening sometimes causes the formation of distinct lines on the fracture surface (Fig. 11-13b), or fatigue striations. One striation is formed during each cycle, as is confirmed by the electron fractograph in Fig. 11-14; successive batches of 5 load cycles interspersed with one cycle of different magnitude are clearly reflected by the striations. Note that this opens the possibility to estimate the rate of crack growth per cycle from the fractograph by simply measuring the striation spacing.

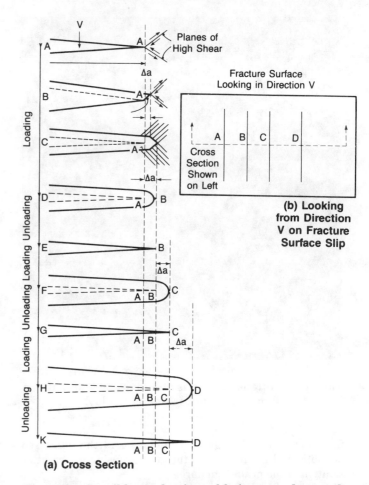

Fig. 11-13 Possible mechanism of fatigue crack growth and striation formation.

Striations as regular as those in Fig. 11-14 can be formed only when the material can accommodate the mechanism of Fig. 11-13 over a sufficiently long distance along the crack front, as in Fig. 11-15.

If the material's deformation possibilities are not sufficient to accommodate the mechanism, the striations become ill-defined, as demonstrated in Fig. 11-16.

This is the case in most steels and the striations might look as in Fig. 11-17, or worse. Note, however, that wherever striations however chopped up are visible, they still provide an indication of the rate of crack growth by their spacing.

Other mechanisms of growth (stress corrosion, creep, hydrogen assisted growth) are more complicated. It is easy to see that in a general sense in stress corrosion, the corrosive environment will assist in creating a crack path through the material. Very often this path follows the grain boundaries. The reason is than the composition at the grain boundaries is generally different than inside the grains (microsegregation). As a consequence an electric cell (battery) is

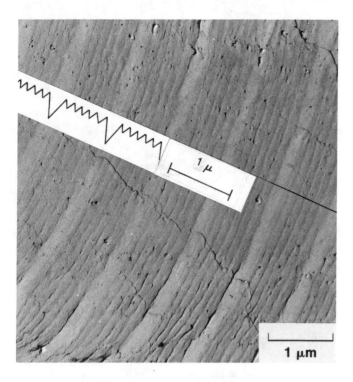

Fig. 11-14 Fatigue striations due to cyclic load pattern shown in insert. (aluminum alloy)

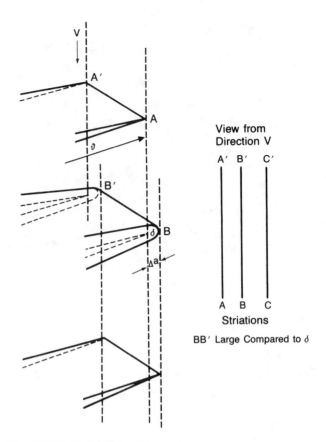

Fig. 11-15 Well defined striations with ample slip possibilities to accommodate equal movement along crack front from A to B and from A' to B', etc.

formed between the grain boundary and the corrosive medium, and the interior of the grain, causing preferential solution of the grain boundaries.

The result is an intergranular fracture that is easily recognizable in the electron microscope, as shown in Fig. 11-18. It should be noted that many variations on the scheme can occur, and that stress corrosion cracks are not always intergranular. However, the fractographer will recognize a number of other features by which to identify a stress corrosion crack.

Hydrogen assisted growth also is often intergranular and in such a case the microscopic surface features resemble those in Fig. 11-18.

11.3.4 Fracture Mechanisms and Microscopic Features

Only two fracture mechanisms exist, namely, cleavage and rupture. The cleavage mechanism is explained in Fig. 11-19. A crack may develop by one of the mechanisms discussed in the previous section and then fracture by cleavage or rupture. Generally, the longer the crack, the more severe the crack tip stress field. Eventually the stresses may become so high that the material will fracture (point D). This fracture may occur by a simple split (cleavage) of a grain as in E. The split generally takes place between planes of widest (atomic) spacing. In an adjacent grain this plane may have a slightly different orientation, so that the cleavage fracture proceeds along E, F and G in different grains (Fig. 11-19).

The fracture facets through the grain are very flat and hence reflect light very well. As a consequence, most cleavage fractures can be recognized with the naked eye or at low magnification, depending upon grain size, because of multifacet glitter, while the other type of fracture is generally dull and unreflective.

The most typical microscopic feature by which cleavage is identified is the river pattern. Due to different orientation of the cleavage planes in adjacent grains, the fracture (in order to maintain continuity) must proceed at different levels when crossing a grain boundary. During further propagation the levels merge, the merging of the steps resembling the tributaries of a river flowing into the main stream. This mechanism is shown schematically in Fig. 11-20.

Actual examples of river patterns are shown in the fractographs of Fig. 11-21.

The alternative fracture mechanism, rupture, is depicted in Fig. 11-22. It is characterized by the cracking or decohesion of foreign particles [7] (inclusions or intermetallic compounds) as in E, F and G.

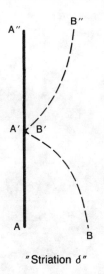

"Striation δ"

B-B' Small Compared to δ

Fig. 11-16 Ill defined striations due to limited slip possibilities preventing equal movement from A to B and from A' to B', etc. (Part 1)

If these particles crack or let loose, new stress concentrations are formed ahead of the crack tip and the plastic deformation (slip) tends to concentrate in planes connecting the broken particles (and the crack tip). This phenomenon can be clearly seen in the surface micrographs (not fracture surface) in Fig. 11-23.

In these regions of high slip and local plastic deformation, the small particles generally break loose [7] just before the fracture takes place (final stage in Fig. 11-22). As a result, an examination of such a fracture surface (insert in Fig. 11-22) will commonly reveal some large holes formed by failure of the large particles surrounded by a large number of smaller holes created by the final separation of the small particles (dimples). Fig. 11-24 illustrates this phenomenon.

Fig. 11-24 is also instructive in that it shows exactly the same area, photographed in a transmission electron microscope and a scanning microscope. The latter image is more "realistic" but the former provides more detailed information, showing that the two microscopes are complementary and should both be used to advantage.

To a certain degree the shape of the dimples depends upon the local stress pattern and gradient. Thus, the dimple shape can be used in some cases to qualitatively assess the nature of the stress field. However, dimple shape can be very deceiving as it depends upon the angle of view. This is clearly demonstrated by two fractographs of the same area taken from different angles in Fig. 11-25.

Fracture by dimple rupture is by far the most common type of fracture. The fracture surface is highly irregular and, in

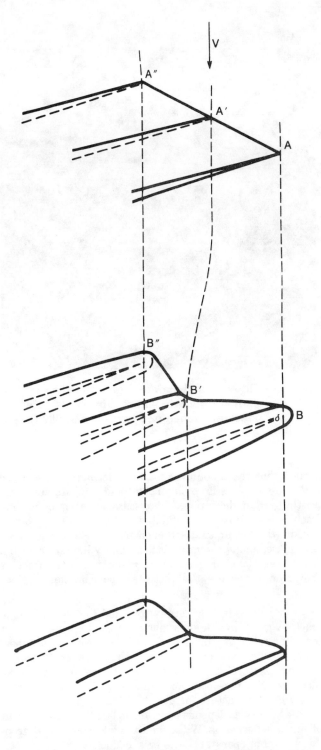

Fig. 11-16 Ill defined striations due to limited slip possibilities preventing equal movement from A to B and from A' to B', etc. (Part 2)

dimple rupture fracture looks dull and gray as shown previously in Fig. 11-20.

11.3.5 Brittle and Ductile Fracture Revisited

The engineering terms for brittle and ductile fracture were

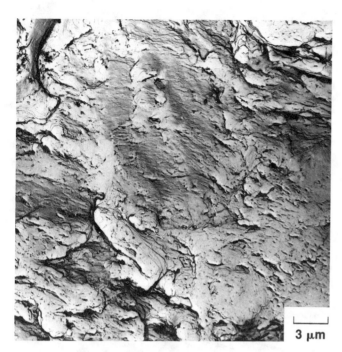

Fig. 11-17 Ill defined striations due to ill-accommodation to crack tip slip requirements. (4340 steel, quenched and tempered at 600 C, 4000X)

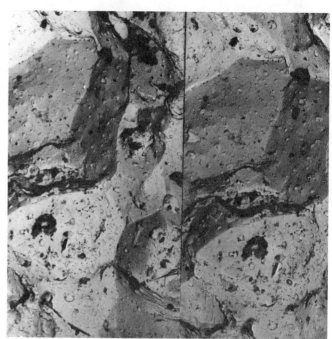

a) Left and right: same area viewed at different tilt angles.

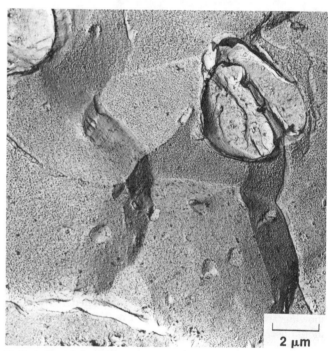

b) Grain boundaries and corrosion products (black dots).

Fig. 11-18 Intergranular stress corrosion. Crack surface. (8000X)

argued that the amount of plastic deformation associated with most failures is so small that nearly all fractures are brittle by that definition. The discussions of cracking and fracture mechanisms presented in the previous two sections support this argument. Although plastic deformation does take place, and is essential, it is normally limited to such a small area along the path of the crack and of the fracture that, macroscopically, the fractures appear brittle.

Nevertheless, it is generally conceded that for a fatigue crack to grow, plastic deformation must occur because, at least in most cases, the growth is a direct geometrical consequence of slip (plastic) deformation. Similarly, for the dimple rupture fracture to occur, plastic deformation must take place. For this reason the metallurgist calls the dimple rupture a ductile fracture (or ductile rupture). The fracture is ductile because it depends upon and is largely a direct result of plastic deformation. Without plastic deformation the mechanism cannot operate. Hence it is a "ductile fracture" even though the actual amount of plastic deformation may be negligible.

Cleavage on the other hand is the splitting of atomic planes. It can occur without any plastic deformation whatsoever. The occurrence of slip is not essential. Therefore, the metallurgist calls this a brittle fracture or brittle cleavage.

Although plastic deformation need not occur, as a rule some does. Some cleavage fractures exhibit more plastic deformation than some dimple ruptures, yet the former is termed brittle and the latter is called ductile.

This mixing of terms need not cause confusion as long as each side defines their terms. But in loose discussions one

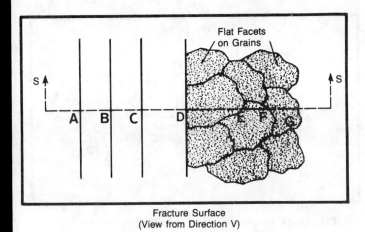

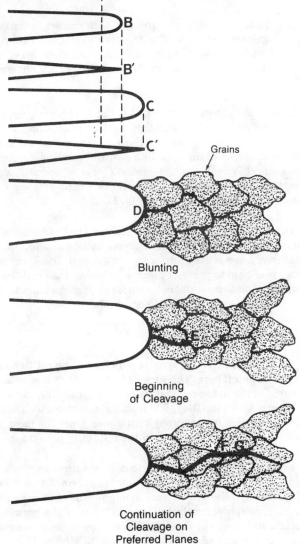

Fig. 11-19 Sequence of events in cleavage (also called brittle fracture).

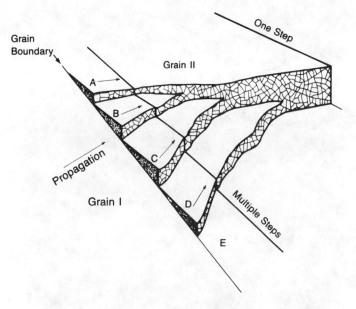

Fig. 11-20 Formation of river pattern in cleavage.

often loses track. Confining the terms brittle and ductile to their engineering definitions and using cleavage and rupture for the actual fracture processes would help a lot in avoiding confusion.

Other confusing terms should be avoided as well, namely, overload fracture and simple overload fracture. It was discussed earlier why these terms are often not appropriate. The fracture surface generally consists of a region of stable crack growth and a region of rapid fracture. The fracture may have occurred at a load higher than the typical service loads. However, what normally precipitates the fracture is the fact that the crack was too long for the structure to sustain whatever the load was. Thus one might call the fracture area, whether cleavage or rupture, the final fracture as is sometimes done, but fracture is always final. Overload is a relative term and can be very misleading.

The term "simple overload" is generally used for dimple ruptures without a discernable zone of stable crack growth. Although the term could be used equally well for cleavage fractures with the same qualification it never is. As was explained previously, a true overload fracture is really quite rare.

11.3.6 *Cleavage vs. Rupture*

Whether fracture occurs by cleavage or rupture depends primarily on the rate of loading, temperature, and state of stress. Roughly speaking, if sufficient plastic deformation can occur to relieve the peak stresses, cleavage will not occur.

At low temperature the yield stress of most materials is higher. At high loading rates the yield stress is higher, at least in rate sensitive materials. If the state of stress is one with high hydrostatic tension, the occurrence of yield is postponed to stresses much higher than the uniaxial yield stress.

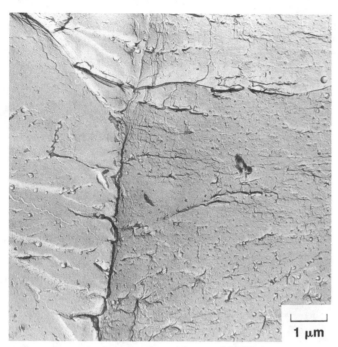

a) River patterns in cleavage of low carbon steel (12,500x).

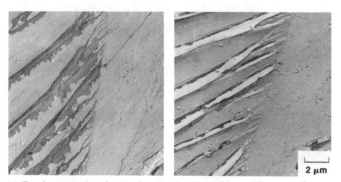

b) Two fractographs of same location but viewed from different angles. Low carbon steel (6000x).

Fig. 11-21 River patterns, the most prominent microscopic fracture of cleavage fracture.

Hence, all three conditions tend to postpone and confine plastic deformation. If in this process the stress peaks at a value above the cleavage stress, then cleavage will occur. In all other cases, local plasticity will be enough to ameliorate conditions and to set the rupture process in motion. In many materials, including most alloy steels, it is virtually impossible to induce a cleavage fracture.

11.3.7 Other Fracture Surface Analysis Tools

The fractographer can use a variety of tools to examine a fracture surface. Often overlooked is the regular (non-inverted) optical microscope in dark field. This instrument may reveal some coarse features better than the scanning microscope, as shown in Figs. 11-26 and 11-27. The optical

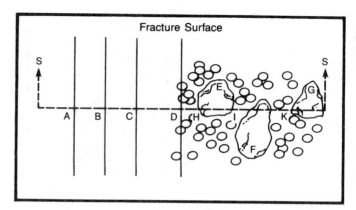

Fig. 11-22 Sequence of events in dimple rupture (also called ductile fracture). (Part 1)

microscope may also be used to check metallographic structures and crack paths.

Fig. 11-27 is of particular interest with respect to the engineering interpretation of a fracture surface. Transverse cracks, such as shown in Fig. 11-27, are an indication of high hydrostatic tension, or rather a high T_z as shown in Fig. 11-27.

The most important other tool is the X-ray analyzer, usually combined with the scanning electron microscope. By analyzing the X-ray emissions of the fracture surface, the chemical composition of the area (or a particle) can be determined. As each chemical element emits X-rays of specific wavelengths, the emission pattern shows which elements are present. This is illustrated in Fig. 11-28.

Use of X-ray analysis, particularly at the crack initiation point, can reveal the presence of unwanted foreign particles that may have caused initiation. In general, the X-ray analyzer is of help to prove or disprove whether the crack initiation was due to material deficiencies or perhaps manufacturing deficiencies as opposed to design deficiencies.

11.3.8 Quantitative Fractography

Quantitative fractography is too large a subject to be discussed in detail here. The various techniques available are of specific value in the evaluation of the engineering aspects of a failure. Unfortunately, few fractographers are well versed in quantitative fractography techniques. Only the most useful techniques are reviewed in the following paragraphs.

Fracture surface topography can be obtained in much the same way as terrestrial topography is obtained from aerial photography. The projectional displacement of identical features in two photographs taken at different angles permit the determination of the height of the feature through measurement of the size of the viewing angle difference [8,9]. The larger the viewing angle difference, the larger the relative displacement, and the more accurate the procedure. Normal

Section S-S Events

During Cracking

Final Event

Blunting

Large Particles Cracking

Concentrated Slip Between Cracked Large Particles

Final Give Void Formation and Coalescence at Small Particles Fracture

Fig. 11-22 Sequence of events in dimple rupture (also called ductile fracture). (Part 2)

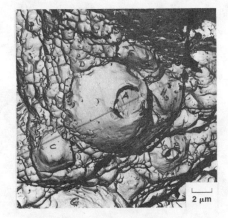

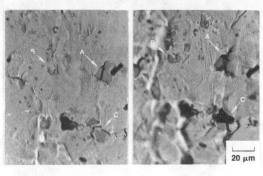

a: 0% strain
b: 3% strain
c: 15% strain
d: 24% strain

Fig. 11-23 Cracking of large particles and concentration of slip. (surface micrographs of aluminum alloy)

stereo pairs will not do. They provide a qualitative impression of topography, but the angular difference is too small for accurate measurement of fracture surface topography.

An example of the results is shown in Fig. 11-29. It should be noted that the topography is based on three photographs at three different angles, only one of which is shown. (It is also of interest to mention that Fig. 11-24 presents a higher magnification of part of this same area.)

The most fruitful application of this technique is probably in the area of transition between stable cracking and fracture [10]. To illustrate this, reference is made first to Figs. 11-19 and 11-22, and subsequently to the examples in Figs. 11-30 and 11-31. A topographic measurement will provide an estimate of the size of the Crack Tip Opening Displacement (CTOD). Note that single angle photographs will be deceiving because they will show the blunting to be of a different size

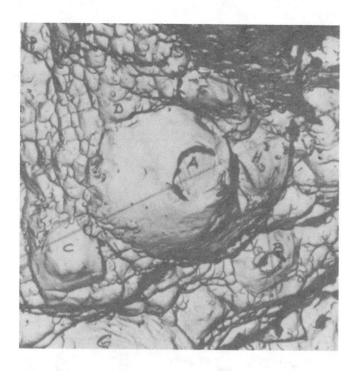

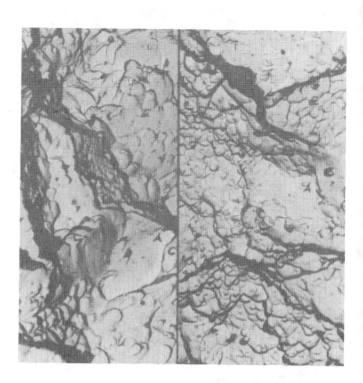

Fig. 11-25 Example of dimple rupture (partly intergranular) showing different appearance of dimples due to different viewing angles. (aluminum alloy; 5,000X) Micrographics show how deceiving a fractograph can be in either microscope (TEM or SEM).

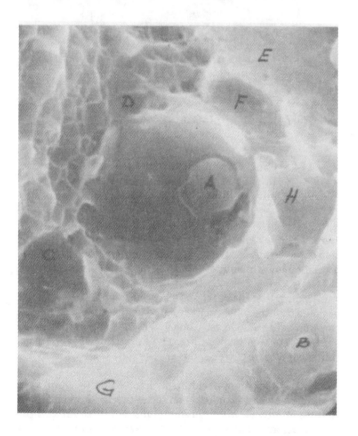

Fig. 11-24 Large holes (dimples) around fractured particles (A & B). Small dimples in final stage of fracture. (aluminum alloy)

than it actually is (Fig. 11-31). These different sizes are exactly the reason why the actual height differences can be determined.

If the conditions for the usage of fracture mechanics are satisfied as described in Chapter 10 on Life Prediction, the CTOD can be related to the applied stress. In other words, measurements of the CTOD in the stable cracking to unstable fracture transition zone may allow inference of the applied stress required to cause fracture. Clearly, multiple CTOD measurements on both mating fracture surfaces are necessary to apply this technique.

It may seem logical that the fracture topography could be measured in a more direct way from cross sections of the fracture viewed in an optical microscope. This method has several drawbacks. First, there is the unavoidable rounding of the edges during polishing, but more important are the limited *depth of field* and the limited resolution of optical microscopes. Therefore, a scanning or transmission electron microscope is a better instrument to use for CTOD measurements [11].

To utilize the TEM a plastic replica must be made of the fracture surface. Carbon should be deposited as is common for

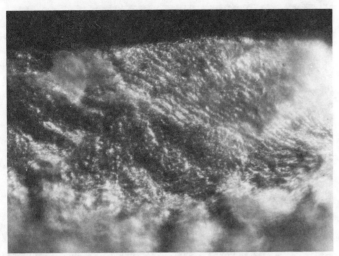

a) at outer surface; 200x

b) batches of striation; 200x

Fig. 11-26 View of a fracture surface through a regular (non-inverted) optical microscope in a dark field.

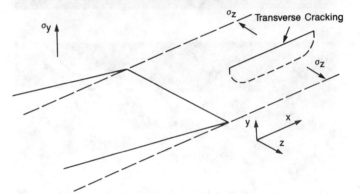

Fig. 11-27 Transverse crack (perpendicular to fracture surface). (optical microscope; 200X)

the production of carbon replicas. However, instead of dissolving the plastic to free the carbon replica, a new plastic layer should be placed on top. The sandwich should then be sectioned in an ultra-microtome and the sections viewed in the transmission microscope. Fine detail, certainly to the size of the CTOD, but even to the height of striations can be resolved and measured. An example is shown in Fig. 11-32.

Finally, striation counts should be mentioned as a source of invaluable information. This is probably the only quantitative fractographic technique practiced by most fractographers. Needless to say then, it is also the most controversial technique. However, the technique can provide valuable information in many materials if properly applied. The technique is time consuming and costly. It requires not ten, but dozens, preferably hundreds of fractographs. In many cases the photographing can be done semi-automatically; a modern microscope (1970 vintage) can easily produce 50-60 fractographs per hour. Special specimen holders that accommodate 1/4 inch replicas (with a TEM) or samples (with a SEM) will further economize the procedure.

It should be clear from Figs. 11-13 through 11-17 that there is basically one striation per load cycle. Whether striations occur only here or there, or chopped up or regular, the measurement of their spacing does provide a good indication of the local rate of fatigue crack propagation. The crack does not grow uniformly, all along its front; this is why many measurements and many fractographs are necessary.

A typical result of the exercise is shown in Fig. 11-33. The figure shows an enlarged "plan" of the failure surface. The arrows indicate the direction of crack growth (perpendicular to the local striations). The numbers indicate the number of striations per micron, the inverse of which is the spacing, and hence, the propagation rate.

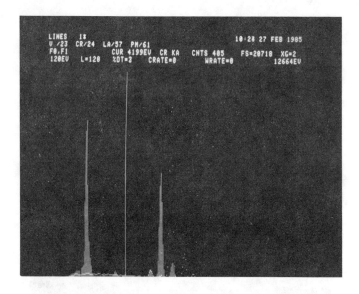

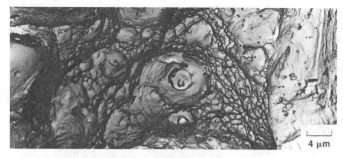

(a) Transmission Electron Micrograph (3000x)   Rupture ↑ Blunting ↑ Fatigue

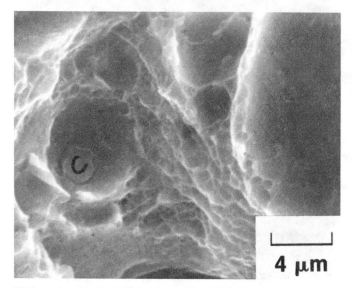

(b) Scanning Electron Micrograph of Same Area (3000x).

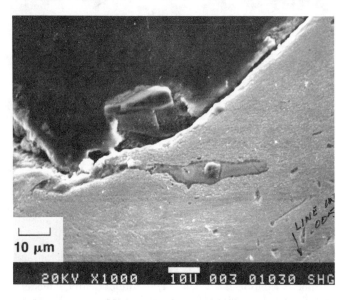

Fig. 11-28 X-ray spectrum of particle, identifying particle as manganese-sulfide inclusion.

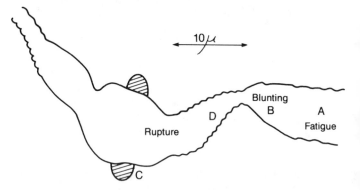

Fig. 11-29 Fracture surfaces topography; quantitative fractography.

The arrows, showing the direction of propagation all point back to the crack origin. This is of help in pinpointing the origin. Integration of the rates provides an estimate of the crack propagation curve as a function of the number of cycles. This curve can be compared with the one obtained by crack growth analysis using the design service load spectrum.

If discrepancies are found between "predicted" and actual growth rates, are they due to the fractographic procedure, or are the predictions wrong?

The answer to this question should be based on the clarity and the consistency of the crack growth striations. If the striations are very well defined, they probably offer the most reliable estimate of the duration of stable crack growth. However, in order to translate these striations into an actual service life estimate, it is necessary to accurately reconstruct the timing of significant service loadings that actually produced increments of crack extension.

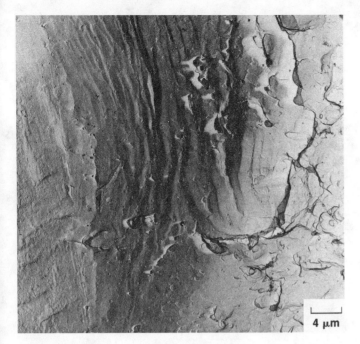

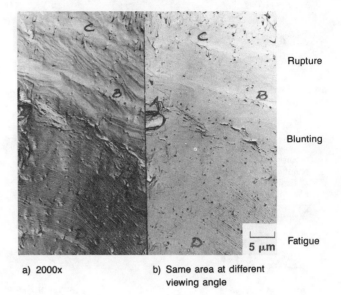

a) 2000x   b) Same area at different viewing angle

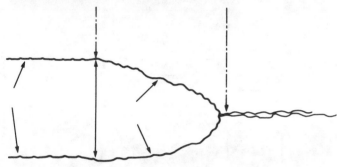

Fig. 11-30 Stretched some at beginning of final fracture.

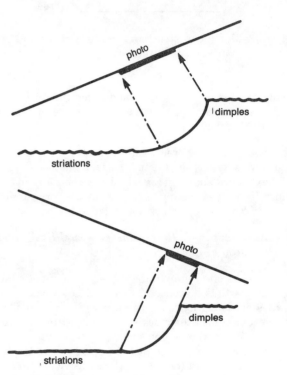

c) Effect of viewing angle (tilt) on projected size.

Fig. 11-31 Size of stretched done depending upon viewing angle.

11.4 The Search for the Origin

11.4.1 Scope

Despite the valuable information obtainable from the crack surface and fracture surface, the true failure cause can only be found at the crack origin. It is, therefore, of considerable importance to pinpoint the crack origin. There are many features that indicate where the origin is located. Many of these features are macroscopic and can be observed with the naked eye or with only slight magnification using stereo binoculars. Thus, the engineer involved in a failure analysis can recognize these features and draw his own conclusions. This section provides a brief survey of typical macroscopic features.

11.4.2 Distinguishing the Crack and the Fracture

As discussed already, the fracture is either glittering and shiny (cleavage), or, in most cases, dull gray (rupture). It should be noted here that in fine grained material a cleavage fracture may be rather dull and optical magnification is needed for its distinction. Furthermore, on a severely corroded fracture surface, the distinction is often impossible until the surfaces are descaled.

Whether cleavage or rupture, the fracture surface is generally rather rough. The crack surface on the other hand is relatively smooth, at least in the case of fatigue. As a consequence, it has a greyish luster, making it often easily distinguishable from the fracture.

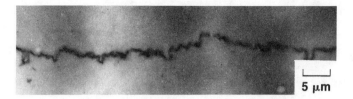

Fig. 11-32 Cross section of dimples. (2000X)

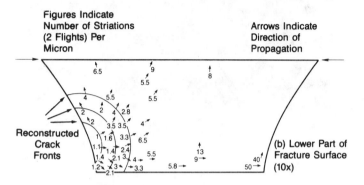

Fig. 11-33 Measured striations spacing and crack propagation direction.

In many cases involving fatigue, and in most materials, a fatigue fracture surface can be rather easily interpreted with the aid of the chart in Fig. 11-34 [12].

At low stresses, longer cracks can be sustained so that the crack surface is larger, the fracture surface smaller. Representative examples, shown in Fig. 11-35, were obtained in the laboratory. The keyed specimens show a surface very similar to the one in Fig. 11-36, that is from a failed turbine rotor.

11.4.3 Crack Features: Beach Marks, Mussel Shells and Sunrise Markings

Fatigue cracks often grow rather slowly over a period of time. Often growth is suspended for long periods because the structure is not being loaded dynamically (not in use). The already present part of the crack will be subject to environmental influences and may undergo slight discoloration not present on the cracked surface that is formed subsequently. Thus a ringshaped mark (beach mark) may be visible on the crack surface, delineating the shape of the crack (front) at a given point in time.

Changes in the severity of the loading will change the rate of crack growth. Higher growth rates (higher loads) result in a somewhat rougher crack surface than low rates (low loads). A smoother crack surface has more luster because it is more reflective. Again this will result in a delineation (difference in luster) on the crack surface: a beach mark.

Beach marks are common on all crack surfaces, whether fatigue, stress corrosion, creep or hydrogen assisted growth was the primary cracking mechanism. They always signify a change in circumstances, i.e., a change in load (higher, lower or complete rest) or in environment. A total absence of beach marks is an indication of uniform circumstances throughout the cracking phase. Only a slight change in surface roughness causes a beach mark, because of different reflectivity, but the difference may be slight. Letting the lights "play" from different directions may help bring out these markings when using stereo binoculars.

Pronounced beach marks are shown in Figs. 11-37 and 11-38. Since a beach mark signifies the position of the crack front at a certain time, it must necessarily "surround" the origin. Hence, multiple beach marks such as in Figs. 11-38 and 11-39 unfailingly betray the area of the origin; they are concentric around the origin. For this reason multiple beach marks are also referred to as mussel-shell marks.

Initial crack growth is usually slow as the crack is "trying to find its way" by growing on slightly different levels or

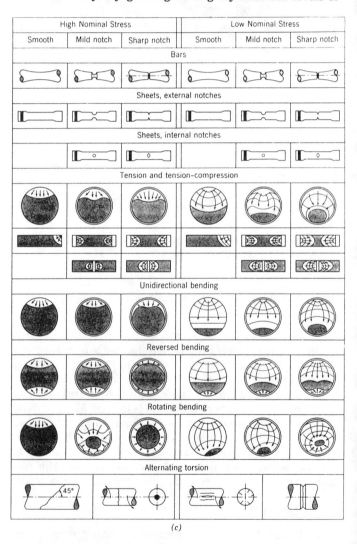

Fig. 11-34 Appearance of fatigue failures on macroscopic scale.

planes simultaneously. This sometimes gives rise to a sunrise pattern as shown in Fig. 11-39. Eventually the various levels link-up or the general roughness of the crack surface becomes of the same order as the step height. In both cases, the rays fade out as shown. Naturally, the center of the rays is the origin.

A less distinctive sunrise appears in Fig. 11-36. The crack initiated at the keyway and initially grew at two different levels. Subsequent initiation at the protruding corner leads to parallel rays.

11.4.4 Fracture Features: Chevrons, Branching, Shattering

Fracture is a fast process. The macroscopic fracture features are reflections of fracture speed. Many fracture surfaces (both cleavage and rupture) show chevron marks as depicted in Fig. 11-40. Note in Fig. 11-40 the small smooth crack area to the left. The points of the chevron point in the direction of the start of the fracture; hence, in a general sense, in the

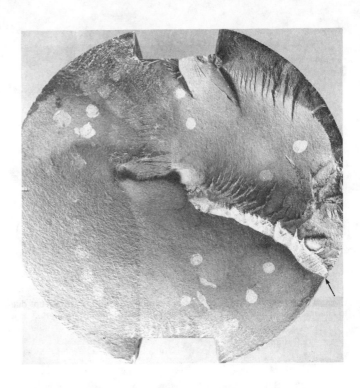

Fig. 11-36 Fatigue crack and fracture of turbine rotor.

direction of the origin of the crack. The sharper the point of the chevron, the faster the fracture.

The chevrons show increasing height towards the outer surface. This feature is what makes them stand out in the first place. In the interior the state of stress is usually triaxial (high T_z as in Fig. 11-27), but at the surface T_z is always zero. Hence, there is a gradual decrease of the hydrostatic tension towards the surface. The higher the hydrostatic tension, the lower the fracture resistance. Thus, the crack tends to run easier and faster in the center giving a smoother fracture surface. At the outer surface the fracture must keep up with the fracture front in the center. This tends to slow the fracture down, but the fracture at the free surface is accelerated from what would be the case in the complete absence of T_z. Hence, the chevrons are not indicative of the shape of the fracture front. Rather they are almost a mirror image of the fracture front.

If the stresses are high or if the material's fracture resistance is low or both, the energy release during fracture is high. When sufficient energy is available, two or more fractures can be maintained simultaneously and a branch is formed. The higher the energy surplus, the more pronounced is the branching (Fig. 11-41). High energy surplus occurs in the case of low fracture resistance, e.g., glass.

High energy surplus also means that the fracture speed is high. Thus, branching and high fracture speeds are usually paired as both are caused by the same circumstances. Note that branching is not a result of high speed.

Fig. 11-35 Fatigue specimens. Large arrows indicate crack nucleation sites. Small arrows delineate final fracture.

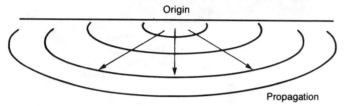

Shell rings (growth rings; beach marks) showing positions of crack front due to change in circumstances due to which surface roughness changed.

Fig. 11-37 Mussel shell marking of crack growth.

Branching provides an indication of the fracture origin and of the location of the initial crack. This is particularly helpful when shattering has occurred, as happens in some cases with sheet structures subjected to rapidly increasing loads.

11.4.5 *Crack Origins*

Often an engineer picks up a failed component, examines the fracture features and draws conclusions about the cause of the failure. Many times the same engineer fits the mating fracture faces together perhaps to better appreciate the failure and its relationship to the component's geometry and function. Such an engineer demonstrates his naivete on both counts. First, a failure cannot occur without a point of origin which means that the macroscopic fracture features will normally tell a person little about the true cause of failure. Second, the joining of mating fracture faces will destroy valuable evidence that would otherwise be observable through fractography. Unfortunately, such a process may result in conclusions that mislead the investigation. If features are damaged near the origin, the cause of initiation may not be discernable.

Previous sections have described macroscopic features useful in identifying the origin of the failure. These included both fracture features involved in the final unstable growth of the crack and crack features involved with the stable extension of the crack. For example, the focus of beach marks defines the origin as does the chevron nature of the herringbone pattern from brittle fracture.

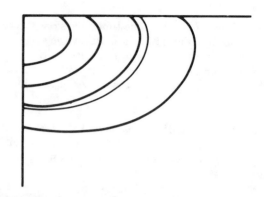

Fig. 11-38 Mussel shell pattern of crack growth.

Origins are significant. If the origin is blocked out, the cause of the crack is probably lost. Origins can be "built in" as a result of material processing or component fabrication or origins can be due to poor detail design, inadequate design data, or inadequate repairs that lead to service induced crack initiation. Origins also can be ascribed to the failure to anticipate the loading on a component (direction, magnitude, repetition) or to an unanticipated aggressive service environment. Of course, cracks can also nucleate as a result of combinations of the above.

Examples of built-in origins in material processing or fabrication include:

a. Forging segregation

b. Exposure of endgrains

c. Laps or folds in material

d. Tears in forming or stamping

e. Weld problems such as volumetric defects (porosity)

f. Planar defects such as lack of fusion (LOF), lack of penetration, (LOP) or poor fitup

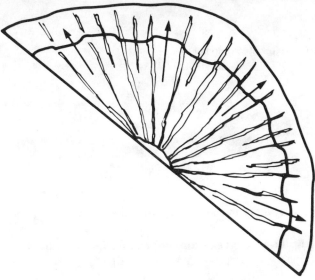

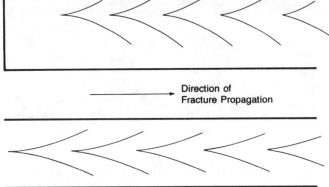

Fig. 11-39 Sunrise pattern of crack growth.

Fig. 11-40 Chevron pattern in fracture part of fracture surface.

g. Casting porosity and cold shuts

h. Inadequate sintering or densification in powder metals
Fig. 11-42 is an example of segregation at a critical location in a forging resulting from poor billet design and chemistry control. In this case, the problem was compounded by subsequent machining that exposed the segregation at a steep strain gradient in a notch. Exposed end grain has a similar deleterious effect on life.

Fig. 11-43 shows an example of poor fitup that was not accounted for as required by the applicable design code.

Here, the stress was elevated by a lack of cross section and aggravated by local bending. Multiple thumbnail origins are common in such cases. Several examples of casting porosity are provided in Fig. 11-44.

Volumetric weld defects are often similar in appearance to casting porosity, except of course, that they are located in welds and their shape may be strongly effected by localized shrinkage. Finally, cracking from an area of marginal densification and sintering is illustrated in Fig. 11-45.

An example of cracking from an inadequate repair is illustrated in Fig. 11-46. The initial defect was missed during an inspection for reworked prior to repair. The repair involved a weld overlay to build up a journal, that then closed off the service induced crack. (Note the gas bubbles dotting the edge of the crack and the large bubble just below the surface.) Service induced cracks tend to focus at sites of high stress. If caught before component failure, the crack initiation features are often well defined.

Fig. 11-47 shows several views of cracks at blade roots in a rotor with integrally machined blades. Although found before failure the fracture features were nondescript because of the nature of the microstructure, so that the cause of crack initiation was uncertain. However, by benchmarking fracture features with laboratory experiments, comparison features can

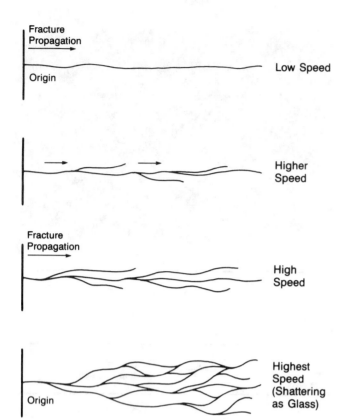

Fig. 11-41 Effect of fracture speed on branching or shattering. High speed: high loads, low toughness; Low speed: low loads, high toughness.

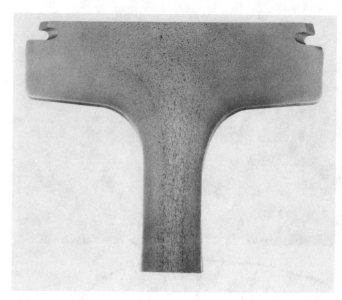

a) Overview of forging failure

b) Cross section through a) showing failure at exposed segregation

Fig. 11-42 Segregation and exposed end-grains in a forging.

be developed for candidate loadings and initiation mechanisms. By matching service cracking features to loading and mechanism, one can better understand the cause of cracking and so determine how to avoid it.

Difficultings during heat treatment produced a series of hard spots along a shaft, failure through which is represented by the fracture features shown in Figure 11-48a. Further study, presented in Figures 11-49b and c, indicates intergranular cracking that for the present situation were most consistent with a quench-cracking scenario.

An example of a fatigue crack origin due to an aggressive environment is shown in Fig. 11-49 for a multi-element leaf spring. Pitting from an aggressive environment may be extreme, as shown in the figure. Pitting acts to focus initiation which, in this case, resulted from corrosion fatigue. Environmentally assisted cracking may also develop in the absence of pits due to a stress-assisted dissolution mechanism or a stress-assisted hydrogen embrittlement. This may be intergranular or transgranular.

Often, the origin of a fatigue crack (as illustrated above) is easy to find macroscopically, but the site is badly damaged, mechanically or environmentally. Crack origins are often hard to interpret. A proper interpretation must bring together a treatment of surface stresses, environment, and material. Sometimes a material has a strong microstructural dependence on processing or fabrication. This dependence serves to complicate and confound a failure analysis even further. In any case, analyses of crack origins requires patience and often calls for some benchmark testing to determine conclusively the mode, mechanism and root cause. This work is crucial, however, because future problems cannot be avoided by redesign nor the significance of the failure to an operating fleet be assessed without a clearcut definition of the root cause.

11.5 Failure Analysis Procedure

11.5.1 *Preliminaries and General Considerations*

In the following section, the word fracture surface will be used in the generic sense. When confusion is possible the words crack area and fracture area will be used.

Fig. 11-43 Poor flange fitup which led to greatly elevated stresses and joint failure.

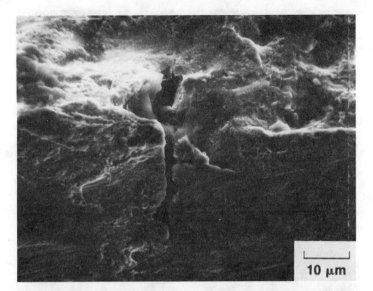

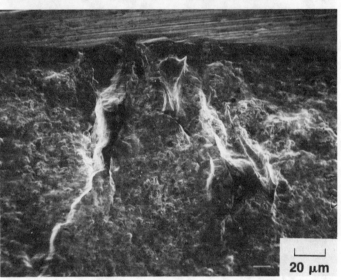

Fig. 11-44 Views of casting porosity. (Part 1)

More often than not, the failure analyst is presented with a piece of material, cut close to the fracture, that contains the fracture surface or only part of the fracture surface. To make matters worse, the fracture surface is often damaged and/or corroded.

Much information can be gained from the general condition of the broken component and surrounding structure. Deformations that occurred after fracture and secondary cracks and fractures can be used to reconstruct the fracture event and provide information on the direction and sometimes the magnitude of the loads. In general, as much of the surrounding structure should be preserved for the analysis as practical.

The fracture features observable in electron microscopes as discussed in Section 11.3 can be distinguished because of their topography. Considering the magnifications used, it is clear that the topographic (height) differences that make the features distinguishable are generally an order of magnitude smaller than the feature itself. Thus, the slightest damage to the fracture surface will obliterate them. For this reason it is very important not to try to fit the fracture pieces back together. The slight acid deposit of a fingerprint on the fracture will also cause damage.

One of the most important preliminaries to a failure analysis is the protection of the fracture surface. Clear acrylic lacquers or plastic spray coatings can be used for this purpose or, if these are not available, a clean, acid-free oil will do. These protective coatings can be removed with organic solvents when this becomes necessary for microscopic examination. The metallurgist/fractographer will know how to deal with this problem. Corrosion products and scratches are more difficult to remove.

The coating will prevent environmental damage. Mechanical damage should be prevented by wrapping the surfaces in cloth, held in place with paper and tape. As already mentioned, *none should try to fit the fractured pieces together; any contact may severely damage the fracture surface.* If a reconstruction is to be made on the spot for some reason, all pieces should be laid out (without the fracture surfaces touching) in the position from whence they came. Photographs from different angles will provide the necessary documentation. If it is not strictly necessary to perform the reconstruction in the field, it should be done in the laboratory; the pieces should be protected as discussed above and shipped.

Fig. 11-44 Views of casting porosity. (Part 2)

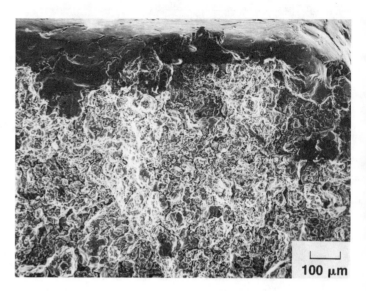

Fig. 11-45b Close-up view of porosity in part a)

11.5.2 *Preliminary-Failure-Analysis Plan and Failure Hypothesis*

A review of the circumstances of the fracture will generally permit the establishment of a more or less complete preliminary-analysis plan. Most possible ingredients of the analysis are shown in Table 11-1. Not all of these will have to be included in every analysis. The analysis plan should contain those that are likely to be useful for the case at hand. As the analysis proceeds, the plan and ingredients may have to be changed or expanded.

Once the preliminary analysis is completed, it is usually possible to establish a failure hypothesis according to Table 11-2. The hypothesis should include a reconstruction of the failure scenerio and provide explanations (hypotheses) for all observations, including the extent of fracture deformation, sequence of separation of parts, position after failure, etc. Subsequent failure analyses can then be directed toward the substantiation or disproof of these hypothesis.

11.5.3 *Secondary-Failure-Analysis Plan: Verification*

Depending upon the failure hypothesis, the second part of the failure analysis may follow three or more alternative routes. In Tables 11-3 to 11-5 these alternative analysis plans are summarized.

Fig. 11-45a View of macro porosity.

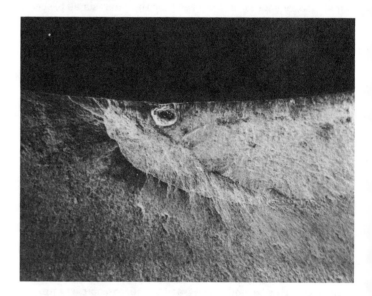

Fig. 11-46 Fatigue cracking from an inadequate repair.

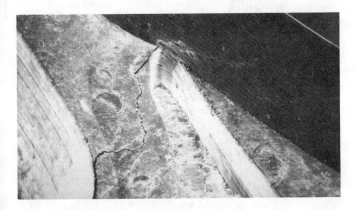

Fig. 11-47 Multiple views of fatigue cracking at a blade root in a rotor. (Part 1)

(a) Macroscopic View of Failure Site

100 μm

(b) Mechanical Damage Destroys Microscopic Features

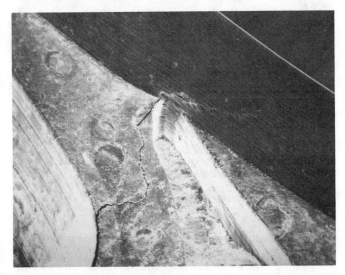

Fig. 11-47 Multiple views of fatigue cracking at a blade root in a rotor. (Part 2)

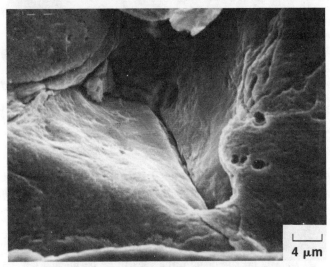

4 μm

(c) Evidence of Intergranular Cracking at Surface Intergranular Origin

Fig. 11-48 Fatigue fracture caused by a local hard spot.

View of Typical Corrosion Pit Leading to Corrosion-Fatigue Crack Growth

Fig. 11-49 Fracture of a multielement leaf spring due to an aggressive environment (road salt-water)

If the results of the subsequent analyses fail to support the hypothesis, it must be concluded that the hypothesis cannot be maintained. A new hypothesis must then be developed, and the appropriate plan pursued (Tables 11-3 to 11-5).

Details on what is involved in each step of the fractographic analysis have been provided in previous sections of this chapter. Details on the design analyses to be performed can be found in other parts of this handbook.

Table 11-1 Preliminary Analysis

1. *Documentation*

    Reconstruction photographs.
    Detail photographs.
    General documentation of fracture surfaces.

2. *Information Gathering*

    Engineering drawings.
    Design analysis, load and stress analysis.
    Fatigue analysis.
    Material properties as per specification.
    Other material properties, K_{Ic}, K_c, J_c, da/dN, C_v from handbook.
    Reports on previous failures if any.

3. *Establish Crack and Crack Origin Area*

    Macroscopic examination (including stereoscopic); fracture chevrons, crack, branching; beachmarks, mussel-shells, sunrise patterns.
    Photographic documentation.
    Characterize fracture, amount of plasticity, shear lips, secondary cracks, transverse cracks.

4. *Identify Cracking and Fracture Mechanisms*

    Microscopic examination (SEM, TEM, optical).
    Establish fracture mechanism (cleavage or rupture).
    Establish cracking mechanism.

5. *Pinpoint Crack Origin*

    Microscopic examination (SEM, TEM, optical).
    Work backward from crack-fracture front, using any possible indication of crack direction, microscopic beachmarks, striations, etc. Extensive rotation/tilting is required, because many features show up only at specific viewing angles[13]. (See visibility of striations in Figure 11-32.)
    Photographic mapping after best viewing angle has been determined.
    Preparation of montages.
    High magnification documentation of origin.

6. *Metallographic Substantiation*

    General chemical analysis of piece remote from fracture to verify chemical composition is as specified.
    Examination of polished and etched cross sections to check micro-structure (heat treatment) and inclusion content.
    Hardness measurements to approximately verify specified mechanical properties.

7. *Preliminary Design Analysis Check*

    Given crack size at fracture (now known), calculate fracture stress and load using fracture mechanics.
    Check whether these are within design spectrum.

8. *Reproduce Fracture*

    Depending upon the cost of failures (including lawsuits) as compared to the cost of a test, reproduce the failure in the laboratory and perform items 3 through 6 on laboratory failure as well.

Table 11-2 Failure Hypothesis and Failure Scenario

Hypothesis of Cause

a. Material defect or material selection problem
b. Manufacturing problem or production problem
c. Design deficiency

Reconstruction

Scenario of sequence of events that would lead to the failure as observed.

a. Are deformations and fracture path compatible with load distribution and load path?
b. Are crack location and path compatible with stress distribution?
c. If a fatigue failure is suspected, are preliminary fractographic fatigue features compatible with load spectrum?

Questions to be Answered by Subsequent Analysis

a. Retrofit?
b. Replacement? (different material, production procedure, design?)
c. Continue operation (of similar structures)?
d. Inspection?
e. Retirement?

Table 11-3 Material Defect or Material Selection Problem

1. *Verify Crack Initiation Point*

   Confirm microscopically (SEM, TEM) that initiation point contains defect.

2. *Characterize General "Defect Content"*

   Document inclusion and particle content. Use Quantimet or equivalent instrument and polished (and etched) cross sections.
   Compare with "normal" material.

3. *Perform X-ray Analysis*

   X-ray analysis of initiation area and/or particles (inclusions) in that area. For comparison establish general X-ray footprint.

4. *Local Surface Treatments*

   Carburizing, nitriding, plating. Measure hydrogen content (if applicable) before and after baking a piece of the material.
   Perform micro-hardness traverses from surface inward to check with standard.
   Measure residual stresses with X-ray diffraction (short range).

5. *Measure Properties*

   Cut specimens from failed part or identical new part; measure mechanical properties, including crack growth properties relevant to mechanism of service cracking.

6. *Toughness*

   If part too small for toughness testing, measure toughness indirectly from stretched zone. Perform residual strength analysis.

Table 11-4 Production or Manufacturing Problem

1. *Defects*

   Trace production procedures, cutting speed, casting or forging procedure, inclusion content or porosity. Perform cutting surface topography. Establish size of possible defects. Establish size of permissible defects using crack growth analysis and residual strength analysis.

2. *Residual Stresses*

   Establish cooling rates at thin/thick section interfaces during casting/forging/rolling, surface treatments, machining (grinding). Micro-hardness traces may help.

3. *Grainflow*

   Map grainflow, map calculated stresses (2-axes) on grainflow diagram.

4. *Wear/Tear*

   Due to insufficient hardness or due to extreme relative (secondary) displacements. In the latter case it is a design problem (Table 11-5).

Table 11-5 Design Deficiency

1. *Primary Stresses*

   Check stress analysis and design loads. Was design based on assumption that there would be "even" stress distribution at the maximum load ($j \cdot P_{max}$)? Would elastic stress distribution at normal operating loads have high gradients (be non-uniform) due to notches, local loads, eccentricities?

Table 11-5 Design Deficiency (Cont)

2. *Secondary Stresses*

   Check sources of secondary stresses; include deformation after fracture (also of adjacent structure) in consideration. Consider misfits, tolerances. Were the normal deformations due to normal loads possible? If not, there probably were secondary stresses.

3. *Relative/Secondary Displacements*

   Were there relative displacements between adjacent parts due to difference in loading/stresses? Did they cause wear/tear or fretting?

4. *Design Assumptions vs. Service*

   Service environment/temperature vs. design.
   Is crack size at fracture compatible with design loads (residual strength analysis)?
   If fatigue,

   a. Check fatigue analysis and load spectrum
   b. Perform crack growth analysis
   c. Perform striation counts and integrate
   d. Compare crack growth analysis with striation count results for EQUAL amounts of crack growth

   Evaluate effects of possible secondary stresses.
   Update design analysis; if necessary measure service loads/stresses.

Details on what is involved in each step of the fractographic analysis have been provided in previous sections of this chapter. Details on the design analyses to be performed can be found in other parts of this handbook.

Clearly, the design or structural engineer should play a prominent role in a typical failure analysis, especially in the second part. Fractographic analysis will provide accurate information on the crack size at fracture. A residual strength analysis using fracture mechanics is an excellent means to check service loads (stresses) against design assumptions. Of course, this analysis should take proper account of the approach of net section yield or limit load conditions. Note that, contrary to general belief, linear elastic fracture mechanics will tend to overestimate the actual fracture stress (K_c is infinite for $a = 0$).

If striation counts are possible or if the cracking times can be obtained otherwise, a crack growth analysis will also provide an excellent means to verify design load spectra against service spectra. The question as to what the service load spectrum would have to be to produce the growth as it actually occurred is of particular relevance for the decisions on subsequent actions. A combination of residual strength analysis and crack growth analysis will provide the necessary information on the feasibility of fracture control by inspection. Fracture mechanics analysis can be of help also in verifying crack growth direction, fracture planes, crack branching and secondary cracking.

11.6 Possible Actions Based on Failure Analysis

11.6.1 *Scope*

The ultimate purpose of a failure analysis from the engineer's perspective is not to cast blame but to determine how further failures can be prevented. The choices are often obvious and perhaps even trivial. However, the choice will depend upon whether the failure occurred in the laboratory during design and development testing or design life verification testing or during actual service. These cases will be briefly reviewed in this section. In either case, the alternatives are as shown in Table 11.6.

Table 11-6 Action

1. *Incidental Case?*

   Probably no action.

2. *Symptomatic?*

   Replace
   Select other material
   Redesign
   Change manufacturing procedure

3. *Risk Involved?*

   Replace all immediately
   Prescribe inspection on the basis of detectable crack size and maximum permissible crack size (based on crack growth analysis; replace or retrofit when crack is detected)

11.6.2 *Failures During Design Development or Design Verification Testing*

The reason for performing usually expensive tests for design and development or verification is always that the part or component is a critical one in terms of the consequences of failure. These consequences can always be measured on an economic scale. The cost of fracture control measures should not exceed the cost of potential failures.

If the test shows adequate life to failure no action may be necessary at all. In all other cases (if the incidental failure due to a rare material defect can be ruled out) where the life to failure is inadequate, there are essentially four alternatives to proceed:

a. Change material

b. Redesign

c. Limit service life and prescribe replacement or retirement

d. Prescribe periodic inspections and repair or replacement when cracks are found.

Any one of these measures (or combinations) may bring the anticipated cost of failure down to an acceptable level. Clearly, the choice between a. and b. will depend upon the results of the failure analysis.

If a redesign is indicated because of generally too high a stress level, an increased size would be required. In case size is dictated by function or surrounding structure, the choice may refer back to a. (stronger material). The latter is a potentially undesirable alternative because a stronger material often has lower ductility and fracture resistance and, hence, this choice will require a thorough evaluation.

In many cases redesign will only be concerned with detail design:

a. Elimination of secondary stresses (increase flexibility)

b. Reduce relative displacements

c. Reduce stress concentrations

d. Bypass loads (stresses)

Almost any change in detail design will chase the problem to a different location (sometimes nearby, sometimes remote). As long as life at these other locations is adequate the problem is solved, but a thorough analysis is certainly indicated.

Limiting service life or prescribing inspections puts a burden on the operator. The economical disadvantages of these choices need not be elaborated upon.

11.6.3 *Service Failures*

Essentially the possible actions following service failures are identical to those discussed in Section 11.6.2 if the structure is one of a kind. If many are in service, redesign or changing material will require retirement and replacement which is always costly. Also in this case, the cost of fracture control must be in line with the anticipated cost of new failures.

Barring redesign or a change of materials, which mean replacement, remaining alternative are:

a. Retrofit (load bypass or other kind of "repair")

b. Life limitation

c. Inspection followed by repair upon crack detection

All these alternatives are equivalent in the sense that they will put part of the burden on the operator, even if this burden is limited to the losses from down-time. A straight life limitation may be preferable if the cost of the part or component and the cost of replacement are low. For large costly parts or components, a retrofit may be the only alternative to redesign and replacement. This is particularly true if the operator is the "general public," since inspection is only an alternative for professional equipment.

The possibility of taking no action at all should not be overlooked; the cost of fracture may be less than the cost of fracture control.

11.7 Summary

Failure analysis is too often limited to metallographic and fractographic examinations. Although these are essential, the knowledge that fracture was caused by fatigue or stress corrosion does not solve any problems. In concert with the intention of this handbook as a design handbook, the eminent role of the design/structural engineer in failure analysis was highlighted. Fractographic features and techniques were discussed to the degree they should be understood by the designer; in particular, the confusing terms and issues were explained.

The purpose of the failure analysis is to arrive at "solutions" that will prevent subsequent failures. The great majority of failures are due to design and production deficiencies and very few indeed are caused by material defects. Hence, the remedies are usually to be affected by design and structural engineers. The most important criterion in the selection of the remedy is an economical one; the cost of fracture control should be less than the cost of fracture. The fact that the cost of fracture is often hard to quantify does not invalidate this statement.

Since this is a handbook on fatigue, emphasis was on fatigue failures. Quantitative fractographic analysis and fracture mechanics analysis were also emphasized. Sources of additional information are cited in Sections 11.8 and 11.9.

11.8 References

[1] Ryder, D. A., "Elements of Fractography," AGARDograph 155-71, 1971.

[2] Phillips, A. et. al., "Electron Fractography Handbook," AFML-TDR-64-416, Air Force Materials Lab, 1965.

[3] "Failure Analysis and Prevention Metals Handbook," Vol. 10, 8th Ed., ASM, 1975.

[4] Broek, D., "Some Contributions of Electron Fractography to the Theory of Fracture." *International Metal Reviews,* Review 185, Vol. 9, 1974, pp. 135-181.

[5] Broek, D., "Elementary Engineering Fracture Mechanics," Nyhoff, 1982.

[6] Bowles, C. Q., Broek, D., "On the Formation of Fatigue Striations." *International Journal of Fracture Mechanics,* Vol. 8, 1972, pp. 75-85.

[7] Broek, D., "The Role of Inclusions in Ductile Fracture and Fracture Toughness," *Engineering Fracture Mechanics*, Vol. 5, 1973, pp. 55-66.

[8] Nankisell, J. F., "Minimum Differences in Height Detectable in Electron Stereo Microscopy," *British Journal of Applied Physics*, Vol. 13, 1962, pp. 126-128.

[9] Wells, O. C., "Correction of Errors in Electron Stereo Micrography," *British Journal of Applied Physics*, Vol. 11, 1960, pp. 199-201.

[10] Broek, D., "Correlation Between Stretched Zone Size and Fracture Toughness," *Engineering Fracture Mechanics*, Vol. 6, 1974, pp. 173-182.

[11] Broek, D., Bowles, C. Q., "The Study of Fracture Surface Profiles in the Electron Microscope," *International Journal of Fracture*, Vol. 6, 1970, pp. 321-322.

[12] Jacoby, *Experimental Mechanics*, 1965, pp. 5.

[13] Broek, D., "A Critical Note on Electron Fractography," *Engineering Fracture Mechanics*, Vol. 2, 1970, pp. 691-695.

11.9 Topical Bibliography

The following bibliography is limited to publications considered particularly useful in general failure analysis. References dealing with specialized subjects are limited because they are of little relevance to the engineer. The reader will notice that most entries are "old" by present day standards (5-10 years or more). It should be noted that little has been added to the knowledge of fractography since then; the most important developments occurred immediately after the introduction of electron microscopes. Any new major leaps are not anticipated until the introduction of the X-ray microscope.

11.9.1 *General Failure Analysis*

[14] Polushkin, E. P., "Defects and Failures of Metals," Elsevier Science Publishing Co., 1956.

[15] Pohl, E. J., "The Face of Metallic Fractures," 2 Volumes, Munich Reinsurance Company, 1964.

[16] "Failure Analysis and Prevention Metals Handbook," Vol. 10, 8th Ed., ASM, 1975.

[17] Weinstein, et. al., "Products Liability and the Reasonably Safe Product," John Wiley & Sons, 1978.

[18] Wulpi, D. J., "How Components Fail," ASM, 1966.

11.9.2 *Fracture Surface Analysis*

[19] Ryder, D. A., "Elements of Fractography," AGARDograph 155-71, 1971.

[20] Broek, D., "Some Contributions of Electron Fractography to the Theory of Fracture," *International Metal Reviews*, Review 185, Vol. 9, 1974, pp. 135-181.

[21] Nankivell, J. F., "Minimum Differences in Height Detectable in Electron Stereo Microscopy," *British Journal of Applied Physics*, Vol. 13, 1962, pp. 126-128.

[22] Broek, D., Bowles, C. Q., "The Study of Fracture Surface Profiles in the Electron Microscope," *International Journal of Fracture*, Vol. 6, 1970, pp. 321-322.

[23] Broek, D., "A Critical Note on Electron Fractography," *Engineering Fracture Mechanics*, Vol. 1, 1970, pp. 691-695.

[24] "Electron Fractography," ASTM STP 436, ASTM, 1968.

[25] Phillips, A., et. al., "Electron Fractography Handbook," AFML-TDR-64- 416, Air Force Materials Lab, 1965.

[26] Beachem, C. D., Pelloux, R. M. N., "Electron Fractography," ASTM STP 381, ASTM, 1965, pp. 210-244.

Addresses of Publishers

Air Force Materials Lab, Wright Patterson AFB, OH 45433

ASM International, 9639 Kinsman Road, Metals Park, OH 44073

ASTM, American Society for Testing & Materials, 1916 Race Street, Philadelphia, PA 19103

Elsevier Science Publishing Co., 52 Vanderbilt Avenue, New York, NY 10017

John Wiley & Sons, 605 Third Avenue, New York, NY 10158

Chapter 12 Case Histories

12.1   Introduction
    12.1.1   Scope
    12.1.2   Relationship to Other Chapters
    12.1.3   Chapter Plan

12.2   Case No. 1 - Automotive Wheel Assembly Design Using High Strength Sheet Steel
    12.2.1   Introduction
    12.2.2   Case History

12.3   Case No. 2 - Design and Development of Components for a Suspension System
    12.3.1   Introduction
    12.3.2   Case History

12.4   Case No. 3 - Heavy Duty Cast Axle Housing
    12.4.1   Introduction
    12.4.2   Case History

12.5   Case No. 4 - Forged Connecting Rod
    12.5.1   Introduction
    12.5.2   Case History

12.6   Case No. 5 - Cast Steel Axle Box for a Railway Vehicle
    12.6.1   Introduction
    12.6.2   Case History

12.7   Case No. 6 - Axle Shaft Problem on a Scraper Vehicle
    12.7.1   Introduction
    12.7.2   Case History

12.8   References

12.1   *Introduction*

12.1.1   *Scope*

This chapter contains case histories that illustrate how various methods, approaches and information contained in the other chapters have been utilized in fatigue design and evaluation. The cases are condensed versions of engineering reports or published papers. Details of computations, of how tests were conducted, etc., are not included so that this chapter is of reasonable length while still retaining the important aspects of each case. Applicable references are cited for each case.

12.1.2   *Relationship to Other Chapters*

Each case history has a brief introduction noting the other chapters that have topics considered in that case.

12.1.3   *Chapter Plan*

The order in which the cases appear is arbitrary and does not indicate any priority. A list of the other chapters is given below along with the numbers of those cases that contain material related to a given chapter. Chapter 1 (Overview) and Chapter 2 (General Fatigue Design Considerations) are omitted from the list since all of the cases relate to these chapters in one way or another. The number of cases appearing in the list for a particular chapter is not meant to imply a lack of importance or relevance of the topics in that chapter, since it was not possible to obtain a selection of cases from those available that could provide comparable coverage of the topics in each chapter. Some other cases of potential interest, but not included in this chapter, are cited in the list of references (Section 12.8).

Chapter 3   (Fatigue Properties) - Case Nos. 1, 3, 4
Chapter 4   (Effects of Processing on Fatigue Performance) - Case Nos. 1, 3, 4, 6
Chapter 5   (Service History Determination) - Case Nos. 1 through 6
Chapter 6   (Vehicle Simulation) - Case Nos. 1, 3, 4
Chapter 7   (Strain Measurement and Flaw Detection) - Case Nos. 3 through 6
Chapter 8   (Numerical Analysis Methods) - Case Nos. 1, 3, 4
Chapter 9   (Structural Life Evaluation) - Case Nos. 2, 5, 6
Chapter 10 (Fatigue Life Prediction) - Case Nos. 1, 3, 5
Chapter 11 (Failure Analysis) - Case Nos. 5, 6

12.2   **Case No. 1 - Automotive Wheel Assembly Design Using High Strength Sheet Steel [1]**

12.2.1   *Introduction*

This case history is concerned primarily with strain-based crack initiation life predictions (Chapter 10), of fatigue properties (Chapter 3), effects of processing (i.e., cold working) on those properties (Chapter 4), specification of the service environment (Chapter 5), and structural analysis by finite elements (Chapters 6 and 8).

12.2.2   *Case History*

An opportunity was available to down-gauge an automotive wheel assembly (Fig. 12-1) through the use of high strength sheet steel. High strength steel sheets figure prominently in current efforts to reduce weight in ground vehicle structures. The substitution of these steels for conventional hot-rolled low-carbon steel is not without problems. High strength steels are more difficult to fabricate, may necessitate changes in welding practice, and are possibly more vulnerable to in-service corrosive attack due to the use of a thinner gauge and typically higher strength (hardness).

A crucial aspect of this re-design exercise was development of a comprehensive understanding of the performance of these steels in terms of fatigue resistance and the effects of manufacturing processes and the environment. Steels considered included two high strength low alloy steels (HSLA), a dual phase (DP) steel and a baseline hot-rolled low-carbon steel (HRLC) SAE 1010. Material properties were required as

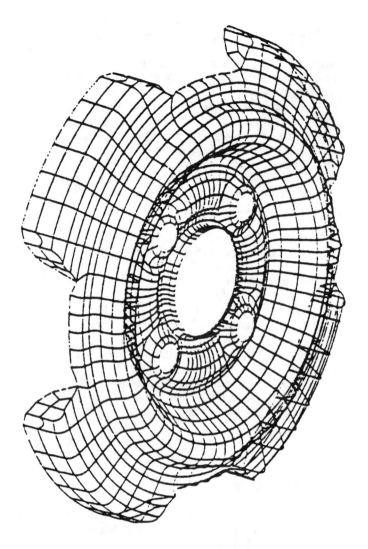

Fig. 12-1 Finite element representation of wheel.

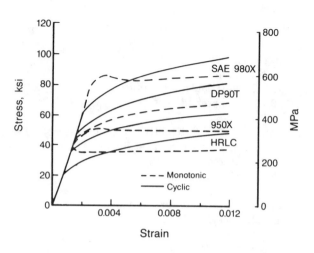

Fig. 12-2 Monotonic and cyclic stress-strain curves for representative sheet steels.

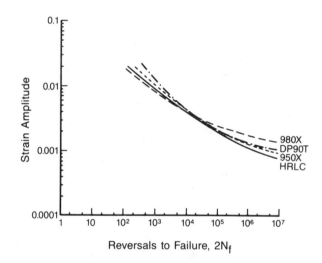

Fig. 12-3 Strain-life curves for representative sheet steels.

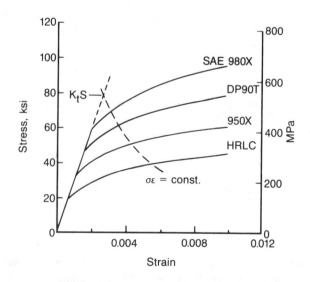

Fig. 12-4 Neuber analysis construction for estimating notch root stress and strain.

basic design input data. Data were generated from strain-controlled tests on smooth specimens to establish relationships between cyclic stress and strain, and a strain-life relationship from which fatigue damage assessments would be made. In the as-received condition the HRLC and HSLA steels exhibited cyclic softening at low strains and cyclic hardening at high strains, while the dual phase steel cyclically hardened at all strains (Fig. 12-2).

The fatigue test results indicated that increasing cyclic strength was accompanied by improved long life fatigue resistance but a decreasing transition fatigue life (Fig. 12-3) (that was indicative of increasing notch sensitivity). Since design decisions can rarely be based on average material properties, the degree of variability in the material performance is important. Standard statistical procedures were used to establish lower bound values for fatigue analysis (95% confidence levels).

Critical locations, such as notches, are clearly evident in the wheel spider design. To provide estimates of local stress-strain response at these locations, the measured data were modified using an approach based on Neuber's Rule (Fig. 12-4).

For the same loading and geometry (K_t), higher strength steels experience lower local strain than lower strength steels. The cyclic deformation response of a steel, as well as fatigue resistance, dictate notched fatigue behavior. Neuber parameter $(\Delta\sigma\Delta\epsilon E)^{1/2}$ - life curves were developed from the strain-controlled smooth specimen data (Fig. 12-5). Since the Neuber parameter is equivalent to (K_f S), and the value of this parameter can be calculated, the notched fatigue life can be estimated. These life curves account for differences in both strain-life and stress-strain behavior and provide more realistic material ratings than simple strain-life curves.

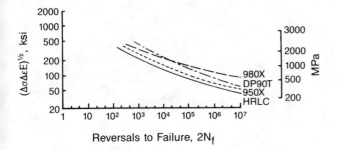

Fig. 12-5 Neuber parameter - life curves for high strength steels.

Effects of cold working also had to be considered since the large deformations during metal forming can result in significant hardness increases and residual stresses. The energy of distortion theory was used to reduce the effects of different deformation processes (e.g., rolling, stretching, biaxial stretching) to an effective prestrain. While large increases in monotonic yield strength result from prestraining, cyclic strength is much less affected because heavily cold worked structures cyclically soften. Prestrain increases effective fatigue resistance at long lives (Figs. 12-6 and 12-7) but may degrade low cycle resistance.

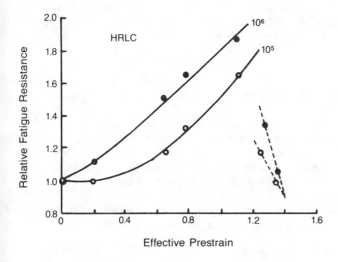

Fig. 12-6 Influence of prestrain on properties of HRLC steel.

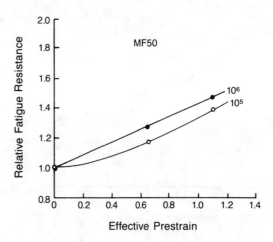

Fig. 12-7 Influence of prestrain on fatigue resistance of SAE 950X steel.

All three steels considered exhibited these tendencies. A useful correlation has been developed between long life fatigue strength and as-formed hardness of sheet steels (Fig. 12-8). Hardness measurements at critical regions can thus be used to estimate long life fatigue improvements. Residual stresses due to forming were not considered as a major influence in these types of sheet steels because of their tendency to relax even at low cyclic strains.

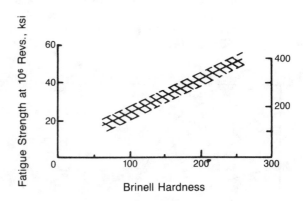

Fig. 12-8 Correlation between hardness and long life fatigue strength.

Consideration was also given to the effects of corrosion on fatigue performance. Two factors are known to be influential: section thinning (that increases operating stresses) and pitting (that can act as crack initiation sites). The latter appeared more significant, based on available test data, and reductions in fatigue strength were noted at long lives (Fig. 12-9). Coating the steel (galvanizing) minimized corrosion effects. With increasing strength, steels tend to become more sensitive to notches and/or pits. The associated fatigue strength reduction factor or fatigue notch factor (K_f) is used in conjunction with the Neuber parameter-life curves (Fig. 12-5) to estimate fatigue resistance when pits are present.

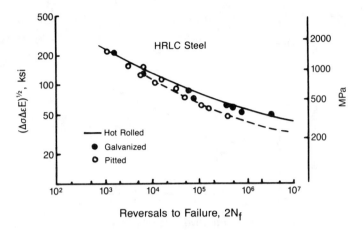

Fig. 12-9 Influence of corrosion pits and galvanizing on the fatigue behavior of HRLC steel.

Another critical aspect of the re-design process is knowledge of the mechanical (loading) environment. Acceptance of a wheel design in the U.S. depends heavily on the satisfactory completion of accelerated laboratory fatigue tests. These include a rotary (or cornering) test and a rim roll test. The former is *the* critical test for the wheel spider because it effectively applies a bending moment through the wheel hub. In contrast to the U.S. procedure, European design philosophy promotes the use of synthesized service usage spectra. These spectra are derived from many years of experience in measuring vehicle road loads under a variety of driving conditions. Thus, it was possible to define a series of usage spectra for different types of drivers (Fig. 12-10).

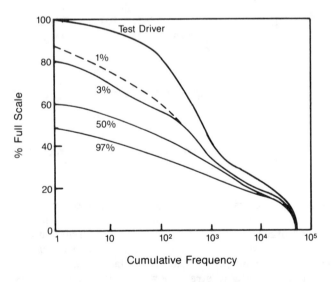

Fig. 12-10 Wheel loading spectra for various types of drivers (50% = average severity driver).

With an understanding of the materials and loads, attention was next focused on establishing stress-strain behavior at critical regions on the wheel spider. Using finite element analysis (FEA), a model of the wheel spider was developed to relate local stresses and strains to the applied loading. Once the load history was specified, a reversal-by-reversal, local stress-strain analysis was performed (using the appropriate cyclic stress-strain relationship), damage assessed (using the Neuber parameter-life relationship), and fatigue life estimated using well-proven computer programs. Component design curves were developed from life predictions for different critical regions using the appropriate notch geometry (K_t) subjected to the specified nominal load history (Fig. 12-11).

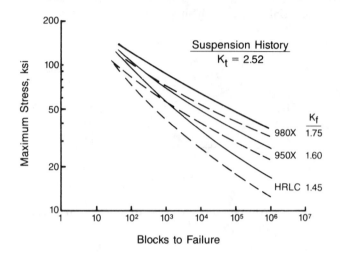

Fig. 12-11 Design curves for evaluating performance of notched member undergoing suspension history.

Predictions were made with the history scaled to different maximum values and plotted as maximum load versus blocks of history to failure. The variable notch sensitivity of the different steels was accounted for through the use of K_f factors. These component design curves allow a designer to make a rapid assessment of relative material performance for a component; in this instance, the wheel spider. As a result, the permissible degree of down-gauging was quantified for each of the high strength steels. In addition, a convenient way to incorporate both material and service usage variability in design analysis is to use cumulative damage analysis, as above, to obtain life estimates for different types of drivers. The results, when plotted on probability paper, provided an indication of expected performance in service (Fig. 12-12).

Based on the above analysis, high strength prototype wheels were manufactured and subjected to the standard laboratory fatigue tests. Results were found to correlate closely with the initial predictions and a satisfactory wheel assembly was obtained with the first prototype built.

12.3 Case No. 2 - Design and Development of Components for a Suspension System [2]

12.3.1 *Introduction*

This case history is concerned primarily with specification of the service environment (Chapter 5), laboratory and proto-

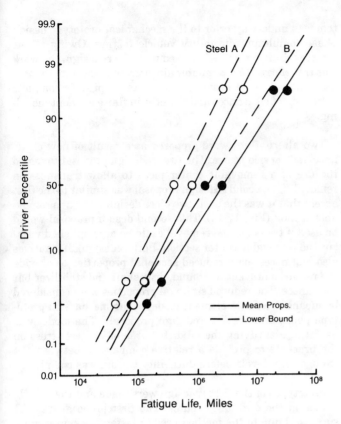

Fig. 12-12 Estimated service performance of two steels.

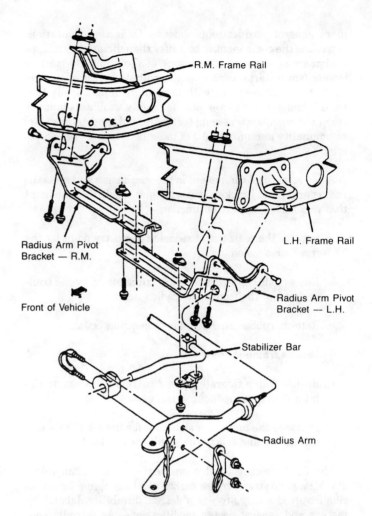

Fig. 12-13 Exploded view of a radius arm pivot bracket, stabilizer bar and radius arm assembly.

type testing (Chapter 9), and structural analysis by finite elements (Chapters 6 and 8). It also discusses the interaction between fatigue design considerations and other design factors (e.g., stiffness, ultimate load capacity, functional constraints).

12.3.2 Case History

This case history reviews the evolution of a radius arm pivot bracket (Fig. 12-13), illustrating the complexities involved in designing for fatigue. Packaging and clearance requirements, ultimate strength, assembly, complexity, vehicle program timing, test facilities, vehicle load data and resources all play a role in confounding what might otherwise be a straightforward fatigue analysis. Fatigue plays an important role, but it was only one of many factors confronting the design engineer.

The environment that can damage or fatigue a component and in which a vehicle is expected to operate and survive is the first thing to be defined. Test objectives can then be established to estimate adequate component life. To establish vehicle usage factors a survey was conducted of 1400 potential customers. The steps in using this survey to establish customer correlation to analytical, laboratory and vehicle test loading are listed below:

a. Defined expected market for vehicle.

b. Conducted potential customer survey in this market to define all significant usage factors and selected an appropriate test vehicle.

c. Conducted vehicle data acquisition on customer roads and a proving ground. In this case, data were recorded from all major body and chassis components, on over 40 independent channels using multiplexing equipment.

d. Established correlation between customer usage and proving ground events using fatigue analysis methods. Loads and frequency of occurrence of potentially damaging events were established from the customer sample and used to define vehicle test parameters on the proving ground and the objectives for fatigue performance in the laboratory.

Having established the performance objectives of the vehicle, the system and component design characteristics could now be developed. The suspension component designs not only had to meet the fatigue durability requirements but also more general product requirements. Deflection limitations can cause the component to be stiffer than dictated by fatigue design considerations for reasons of handling. Specialized events (snap-starts, curb impacts, pothole braking, etc.) require the component to withstand high impact loads that would seldom occur in vehicle life. Safety standards must be complied with, system complexity limited, and inter-model commonality maximized. All of these factors must be considered.

A key structural component in the suspension system is the radius arm pivot bracket. It is a fatigue sensitive component that performs a variety of functions:

a. Locates the rearmost suspension geometry point for the front suspension.

b. Reacts axial and vertical loads transmitted by the trailing arm of the front axle (radius arm).

c. Retains rubber bushings for suspension isolation.

d. Resists frame siderail torsion.

e. Resists frame siderail inboard/outboard flex due to the lateral force component of radius arm.

f. Provides the mounting structure for the aft of axle stabiliser bar and reacts the stabiliser bar loads.

The final, release level component was a 3.2mm thick, 220 MPa yield strength, low carbon steel stamping. Its evolution involved a long process of design, durability, laboratory testing and several design modifications. As initially conceived, the bracket/crossmember was to be an integral part of the frame. However, due to interference with the frame during stacking and shipment, suspension geometry and fastening requirements, the intention was to attach the crossmember to the frame during vehicle assembly. In addition, due to frame siderail-to-siderail tolerance, it was necessary to have a left and a right hand bracket that would overlap in the center and provide a slip joint rather than a single piece crossmember.

Initial design work to support the mechanical prototype vehicle testing phase was conducted using conventional, mechanical drafting means. Results were transformed into functional prototypes using "soft" tooling, and installed on durability vehicles for exposure to defined events at the proving grounds. The mechanical prototype design phase successfully passed all of the structural vehicle durability requirements. During the three month period required to complete the initial prototype testing at the proving grounds with mechanical prototype vehicles, the vehicle design was continuously being refined. These refinements included providing package space for a larger diameter front driveshaft for a 4 × 4 version of the vehicle, improved ground clearance, and provision for the removal of a newly added diesel engine/transmission for servicing. Each of these issues affected the bracket design, and redesign of the bracket/crossmember system was underway prior to the mechanical prototype design phase completing durability vehicle testing. Owing to the complex drafting issues involved with the redesign, the work was transferred to a computer aided graphics system to facilitate the review of design alternatives, to provide complete definition of clearances and to assist in finite element modeling.

Two alternatives were proposed as a result of new clearance and service issues. The first redesign proposal involved the use of a removable center piece to allow transmission removal. The second redesign proposal was similar to the first except that it was the basic two piece design without a removable section (Fig. 12-14). This second design proposal would be used if transmission removal could be accomplished without the removable center section. Both redesign alternatives also had necessarily reduced sectional properties to provide the required bracket to ground, driveshaft and stabilizer bar clearances. The reduced sectional properties were considered by means of a finite element model of the mechanical prototype phase and the two redesign proposals. The model was useful in identifying the predicted area of highest stress on the brackets to provide a relative comparison between the three design levels before durability testing was completed.

Prototypes of the first redesign were made and the vehicle tested on the durability routes. The first proposal was not successful due to the method used to fasten the center piece to the left and right hand brackets and the bracket/crossmember system failed in a mode not related to the mode analytically predicted. However, the diesel engine/transmission removal trials were successful using the original two piece design and the three piece proposal was dropped in favor of the second redesign proposal (two piece design). By this time in the program, major design changes such as the ones occurring with the subject brackets could affect tooling lead time and produce some risk of not meeting production timing requirements. In view of this, a laboratory systems test was begun that included the front half of a frame, one crossmember, cab mount loads, radius arm loads and stabilizer bar loads (Fig. 12-15).

The test was intended to allow a relatively rapid iteration of design alternatives if additional redesign was required. Prototypes of the second redesign did fail in fatigue (prior to completion of the vehicle durability testing) in the area where the finite element model predicted the greatest stress (Fig. 12-16).

At this point in the program, the computer aided graphics system, the finite element model, and the laboratory test could be utilized to optimize the redesign with sufficient sectional properties and clearances to meet the durability requirements. With the baseline performance established, conservative life predictions could be made based upon a Weibull statistical analysis of the multiple sample tests of the various bracket design improvements (Fig. 12-17). The proposed redesign was optimized, and prototypes made that surpassed all vehicle durability requirements prior to release for production.

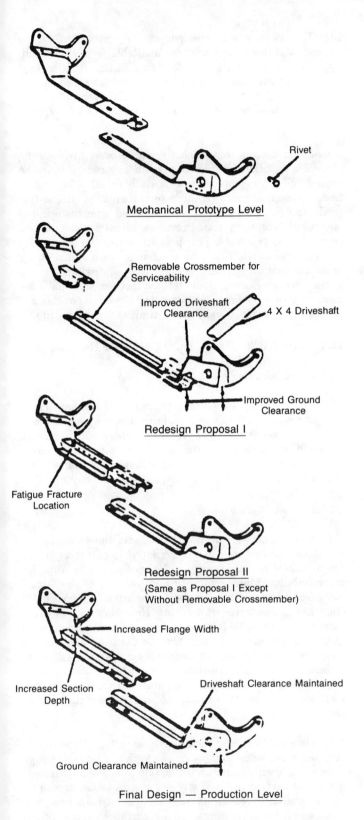

Fig. 12-14 Evolution of the radius arm pivot brackets.

Fig. 12-15 Radius arm pivot bracket system laboratory test fixture.

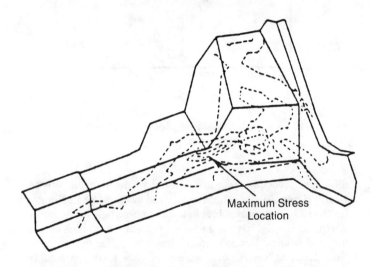

Fig. 12-16 Radius arm pivot bracket stress contour plots from finite element model.

While the fatigue life of the metal stamping was being defined and improved, the fasteners that attached the brackets to the vehicle were also being evaluated. Fastener fatigue failures are every bit as important as the fatigue failures of the stampings they attach, yet fastener design (in the automotive industry) is generally dictated by ultimate load and assembly constraints more often than by fatigue. The cold upset rivet is cost effective, reliable (cannot be undertorqued), good in shear applications (due to fill of the holes), and the easiest to inspect. In suspension systems it is also much more likely to fail from ultimate loading due to impact (e.g., potholes, curb impact, etc.) than from fatigue. The ra-

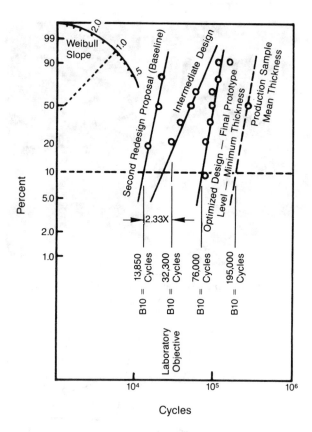

Fig. 12-17 Weibull statistical plots of various bracket design improvements.

dius arm pivot brackets are secured to the siderail web by 10mm rivets. Rivets could not be used at the siderail flanges for the radius arm pivot brackets due to the angle of the frame during assembly. Riveting was not considered reliable on the inclined siderail flange surface (due to a higher frequency of "nail heading") and bolts were used due to this assembly related constraint. Rivets could not be used to attach the left and right hand brackets at the vehicle center due to the assembly requirement that the joint have slotted holes to accommodate frame tolerance. Once again, bolts were employed because of assembly-related constraints. In each case where bolts were used, they were designed with a great enough margin in clamping load to perform as required under ultimate loading conditions within the lowest range of specified torque. From a fatigue standpoint, all the fasteners were "overdesigned". This is typically the case in automotive suspension systems because ultimate load and assembly constraints predominate.

Following the success of the proposed radius arm pivot bracket redesign and associated fasteners, a product decision was made to provide for a 28.6mm diameter stabilizer bar in addition to the tested 25.4mm diameter bar. Additional design and laboratory testing was now easily conducted and it was determined that the bracket fatigue life was very sensitive to the stabilizer bar loading. However, a 4.0mm increase in the flange thickness in the area where a fracture developed during laboratory tests nearly doubled the predicted fatigue life. This change was also successfully made before the 28.6mm stabilizer bar was made available for sale. In addition, the superior fatigue performance of this final bracket design permitted its use on all 4 × 4 vehicles based upon a comparison of measured vehicle loads. This successful commonization was verified by durability testing of a 4 × 4 vehicle.

One final step in this bracket design process involved the stamping supplier. In order to prevent any reduction in the fatigue life of the components as initially released for production, the supplier was required to periodically fatigue test production samples using a test fashioned after the original successful laboratory test. Prototype samples were used to correlate the two tests. This practice alerts the supplier to any changes due to tool wear or changes in manufacturing processes that may adversely effect the required fatigue life of the bracket. Such periodic fatigue test practices have become increasingly important as automotive components are reduced in cost and weight and optimized for fatigue life.

12.4 Case No. 3 - Heavy Duty Cast Axle Housing [3,4]

12.4.1 *Introduction*

This case history is concerned primarily with generation of fatigue properties (Chapter 3), effects of processing (i.e., machined vs. as-cast surfaces) on those properties (Chapter 4), load and strain data acquisition (Chapters 5 and 7), structural analysis by finite elements (Chapters 6 and 8), and strain-based crack initiation life predictions (Chapter 10).

12.4.2 *Case History*

Investigations were initiated to assess the feasibility of using nodular cast iron as an alternative to cast steel for the manufacture of heavy duty axle housings. The superior machinability, high castability and low cost of the former compared with cast steel makes it an attractive possibility from a manufacturing standpoint. However, serious doubt existed concerning its fatigue performance and consistency of quality. Three cast materials, two nodular irons and one cast steel, were compared on the basis of their ability to withstand typical service loading histories experienced by an axle housing on off-highway vehicles. Computer-based fatigue analysis methods were used to evaluate the suitability of the nodular cast iron.

In some applications, such as crankshafts, all the highly stressed areas where fatigue cracks typically originate (e.g., fillet radii) are machined and often roller burnished to impart beneficial surface residual stresses and residual stress profiles as well as to reduce surface roughness. In other applications, such as axle housings, the cast surface skin may be left intact at sites of stress concentration. The cast and machined surfaces will produce different component fatigue strengths, even if the matrix material and the stress state in the critical areas are identical. It therefore becomes essential to thoroughly understand the factors influencing the fatigue properties of both machined and as-cast surfaces of cast materials.

To provide representative load data, three types of vehicles were examined: cable and grapple skidders used in the logging industry and a shovel loader. Loads experienced in axle housings fitted to these vehicles were measured during service operation in forests and quarries respectively. Special axles were assembled, having strain gages strategically attached to each wheel spindle to measure and differentiate between vertical load, longitudinal load and torsional load inputs at each of the four wheels (Fig. 12-18).

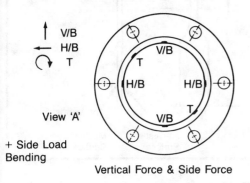

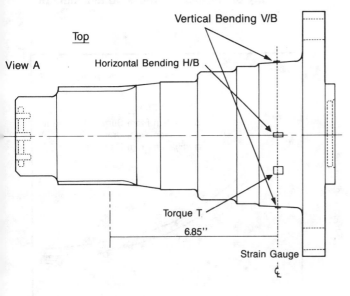

Fig. 12-18 Location of strain gages to measure different loadings.

These axles were fitted to each vehicle in turn and loads measured during all expected types of operation. A typical load history is shown in Fig. 12-19. The instrumentation consisted of an analog magnetic tape high-g recorder and associated signal conditioning equipment, battery power source, etc. During all tests, a strip-chart recorder was used to monitor each data channel and to ensure that the signal integrity and quality was maintained. After digitization of the analog recordings into a computer, the generated data files were combined to represent daily duty cycles prior to subsequent fatigue analysis. A preliminary "rainflow cycle counting" analysis established simple torque and loading spectra for each axle application and provided the designer with an indication of relative severities in terms of axle loading.

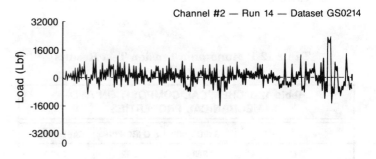

Fig. 12-19 Sample of load history.

To identify critical areas and establish relationships between applied loadings and behavior at these areas, finite element models of the axle housing were developed in the computer (Fig. 12-20). It was possible to identify the most severe and critical loading inputs and establish appropriate stress concentration factors to use in subsequent more detailed fatigue analyses.

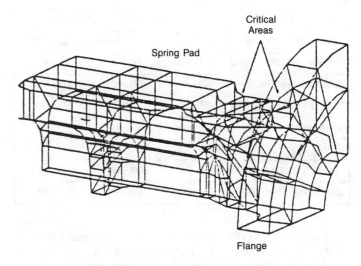

Fig. 12-20 Finite element mesh for critical section of axle housing.

Basic material properties were evaluated for machined and as-cast test specimens. A total of eighteen axle housings were cast; twelve in nodular iron from two sources and six in cast steel. A fully ferritic structure was required in both cases. To achieve this, one source used an inoculation process while the other used heat treatment. Four locations were selected in each casting and test specimens were machined from each location. Plate specimens with gage length cross-sections of 6×6mm were machined, in some cases retaining the cast surface on one face of the specimen (Fig. 12-21). Fatigue tests were conducted in strain-control on a computer-controlled, servo-hydraulic test system. Constant amplitude tests provided strain-life performance data for both machined and as-cast surface specimens over a wide range of lives. Cyclic stress-strain behavior of the material was also obtained from these test results. Monotonic tension tests results are given in Table 12-1.

Table 12-1 Monotonic Tension Test Results

Table 12-1 CHEMICAL COMPOSITION AND MECHANICAL PROPERTIES			
	S G IRON "S"	S G IRON "H"	CAST STEEL
C	3.63	3.92	0.23
Si	2.30	2.60	0.32
Mn	0.06	0.30	0.84
S	0.012	0.01	—
P	0.03	0.052	—
Mg	0.045	0.049	—
Ni	—	—	0.15
Cr	—	—	0.06
Mo	—	—	0.06
Fe	Balance	Balance	Balance
Modulus of Elasticity, MPa	159,464 - 158,410	156,612 - 163,400	199,523 - 208,958
Yield Strength, 0.2% S_y, MPa	280 - 338	334 - 340	285 - 415
Ultimate Strength S_u, MPa	422 - 441	422 - 460	513 - 524
True Fracture Strength σ_f, MPa	487 - 525	524	675 - 750
True Fracture Ductility ε_f	0.157 - 0.322	0.19 - 0.27	0.46 - 0.81
Cyclic Yield Strength, 0.2% S_y', MPa	385 - 395	410 - 420	345 - 360
Cyclic Strain H'g Exp. n'	0.08 - 0.10	0.066 - 0.075	0.116 - 0.121
Cyclic Strength Coeff, K', MPa	540 - 726	638 - 542	707 - 755

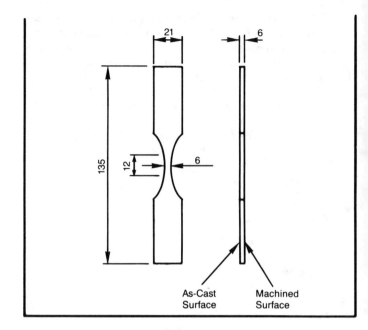

Fig. 12-21 Geometry of fatigue specimens used for the study of the cast surfaces (dimensions in mm).

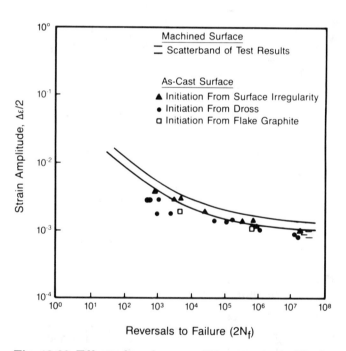

Fig. 12-22 Effect of surface condition on strain-life fatigue behavior.

Fatigue results (Fig. 12-22) showed that the surface irregularities and pits left by the molding sand did not appreciably reduce the fatigue life below that of a machined surface because the machined surfaces also contained notches in the form of pores, nodules, etc.

Larger dross defects (generally caused by impurities in the mold) lowered fatigue life substantially, reducing the fatigue strength to 20% below the lowest measured for a machined surface. This reduction in fatigue strength had to be accounted for in the design stresses. More traditional methods of comparing materials based on *static* test results would have resulted in the conclusion that cast steel is superior to the nodular irons because it has a greater ultimate tensile strength and true fracture stress. Yield point comparisons would have been inconclusive. When cyclic stress-strain relationships were used as the basis for comparison, the following two trends emerged; first, the remarkable lack of scatter in the cyclic yield values and second, the slight superiority of the cast nodular irons (in terms of cyclic yield strength). Other bases for comparison were the stress-life and strain-life relationships. Stress (load) controlled fatigue tests (Fig. 12-23) showed that, in terms of endurance and variability, there were slight differences between the three materials.

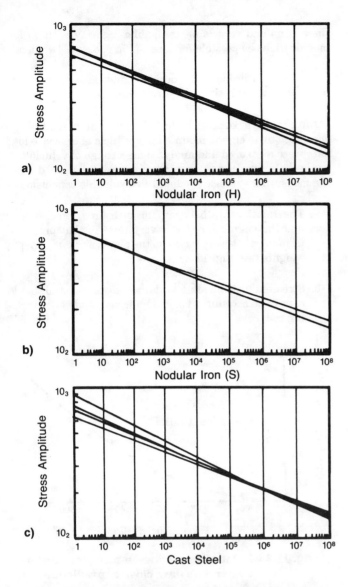

Fig. 12-23 Stress-life curves for each material (a) nodular iron (H), (b) nodular iron (S), (c) cast steel.

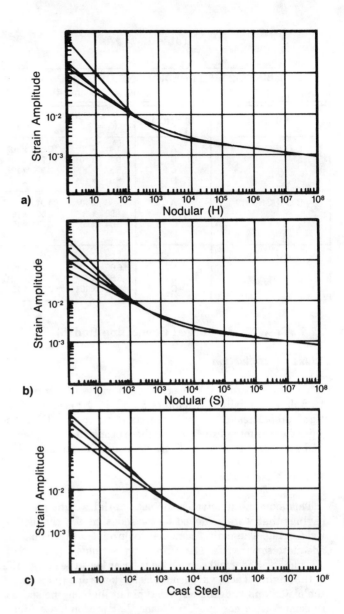

Fig. 12-24 Strain-life curves for each material (a) nodular iron (H), (b) nodular iron (S), (c) cast steel.

In the absence of other data, nodular cast iron would be chosen. However, this contradicted the trends in the strain-life data (Fig. 12-24). The cast steel was obviously superior in the presence of very high strains, but the irons had longer lives at strain levels near the fatigue limit. This feature of the materials had to be considered in the design.

The initial picture that emerged was one of confusion, reinforcing the view that realistic comparison had to be based upon all three variables that influence the fatigue life of a component: the level and nature of the service loads, component geometry and the properties of the material. The complete axle housing was assessed using the local stress-strain approach with all the aforementioned as input. Effects of gross section changes, local geometry adjustments and material substitutions were evaluated to determine the most appropriate design to satisfy operational requirements. Fatigue life estimates were made with the computer for a range of alterations, effectively made at the drawing board prior to making expensive prototype commitments. These assessments indicated that simply changing from cast steel to cast iron, while retaining the same detail geometry, reduced the durability. However, attention to improving local detail (i.e., reducing K_t values) indicated a significant improvement (Table 12-2). Overall, by considering both shape and material, it was possible to improve the design of these axle housings and use nodular cast iron for manufacture. The validity of using the same axle housings for each application was also confirmed. Nodular cast iron axle housings are now in successful production with no reported field failure problems.

Table 12-2

Table 12-2 LIFE PREDICTIONS OBTAINED AT TWO CRITICAL AREAS (K_t = 2.5 and 4)			
MATERIAL	SECTION	LIFE IN BLOCKS K_t = 2.5	K_t = 4
S G IRON "S"	A	1.41×10^5	2.08×10^3
	B	8.15×10^4	1.77×10^3
	C	6.37×10^4	8.92×10^2
	D	5.81×10^4	1.15×10^3
S G IRON "H"	A	5.66×10^4	1.01×10^3
	B	1.88×10^4	3.60×10^2
	C	3.64×10^4	8.97×10^2
	D	2.01×10^4	8.06×10^2
CAST STEEL	A	1.03×10^5	3.39×10^3
	B	6.05×10^4	2.95×10^3
	C	5.84×10^4	2.28×10^3
	D	5.05×10^4	2.24×10^3

12.5 Case No. 4—Forged Connecting Rod [5]

12.5.1 *Introduction*

This case history is concerned primarily with strain and load data acquisition (Chapters 5 and 7), generation of fatigue properties (Chapter 3), effects of processing (i.e., machined vs. as-forged surfaces) on those properties (Chapter 4), and structural analysis by finite elements (Chapters 6 and 8).

12.5.2 *Case History*

This example illustrates the substantial weight and cost savings that were achieved by the redesign of a connecting rod forging. The main functional requirement of this part is adequate service life. The connecting rod must endure both normal and abnormal service loads for at least the design life of the engine. These vary from tension loads due to inertia of the piston and gudgeon (or wrist) pin at high engine speeds, to compressive gas pressure loads that predominate under high torque conditions.

The measurement of loads in a high-speed engine required the means of transmitting or transferring strain gage signals from the connecting rod within the engine to an external recorder. A preliminary survey of experimental devices suited to this application led to two possible routes, telemetry or mechanical linkage. Telemetry offered the best long-term solution, particularly as development of existing techniques could resolve familiar problems associated with the use of telemetry for connecting rod load measurement, i.e., added inertia effects due to the attachment of heavy and bulky batteries and a transmitter, interference from the ignition, limited numbers of channels, etc. However, in view of the time constraints associated with this project, emphasis was switched to the design of a mechanical linkage system to support the signal cable. The linkage system had to operate continuously within the geometry constraints of a 1.6 litre engine, at rotational speeds of up to 6500 rpm. In addition, any installation was constrained to allow normal operation of the engine and vehicle on the public highway with as few modifications as possible for a period in excess of two hours.

A two-beam linkage was designed and made from a titanium alloy to meet these requirements.

After minor modifications to the engine sump, the linkage was fitted to an engine mounted in a vehicle and service load data were measured and analyzed for a range of vehicle maneuvers. In addition to acquiring representative load data, the measuring exercise produced two valuable conclusions:

a. The relationship between the measured tensile loads and the engine speed was very close to that predicted from inertia theory based on the piston and gudgeon pin weights (see Fig. 12-25).

b. Stresses due to bending loads are substantial and should always be accounted for in any design study of connecting rods.

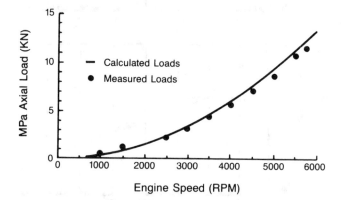

Fig. 12-25 Relationship between measured loads and engine data close to prediction.

Cyclic material properties were determined from specimens, having at least one as-forged surface, taken from connecting rods. As the connecting rod in service experiences a high number of load cycles, interest focused mainly on the fatigue limit of the material. Again the effect of the forged surface was substantial. Fig. 12-26 shows how the data from machined and ground specimens compares with the forged surface data; the fatigue limit is reduced by approximately 30%. This effect must be accounted for in any design optimization exercise. In addition, the large differences in fatigue performances achieved by machined specimens from steels treated to substantially different static strength levels was not evident when comparing the same materials in the as-forged state.

Theoretical analysis also played an important part in this case study, in both the early and later stages of the design process. A connecting rod is basically a pin-jointed strut, and its resistance to buckling must be considered as an important design factor. Buckling characteristics can be predicted early

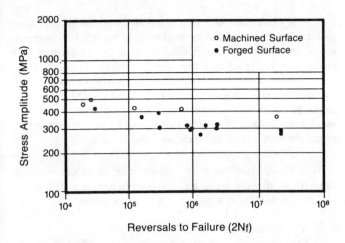

Fig. 12-26 Effect of forged surface on fatigue strength.

in the design process using computer structural software. In long, slender connecting rods, measured buckling loads relate to the simple Euler buckling curve (Fig. 12-27), and high bending stiffness is necessary in the design. In connecting rods that have a low slenderness ratio, buckling occurs when the yield strength of the material is reached, and is therefore resisted by axial stiffness. High bending stiffness, normally achieved by adopting an I-section in the shaft, is not necessary to resist buckling in smaller connecting rods. The new design incorporated a more rounded I-section that retained adequate bending stiffness and which was desired by the forging company to improve metal flow and minimize die wear.

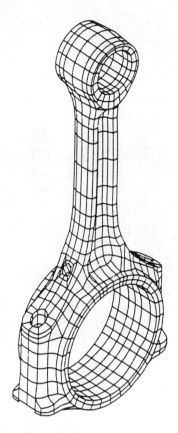

Fig. 12-28 Finite element mesh of a connecting rod.

The applied loads included a bolt tightening load, a gudgeon pin/small end interference load, a compressive gas pressure load and a tensile inertia load. Some of these loads are fairly static during the service life of the component, whereas the others are variable and difficulty arises when attempts are made to relate predicted stresses with material fatigue properties. Stresses for each load case were combined and transformed into an equivalent stress function that was comparable directly with cyclic material property data. As a result, a single plot of safety factor as a function of fatigue strength was produced as shown in Fig. 12-29. From this, regions prone to fatigue damage could be identified, as could areas where factors of safety were exceptionally high, and where metal could be removed without substantial risk of failure. Other factors such as bearing support, fretting resistance, draft angles and the oil squirt hole were accounted for in the redesign. A replacement connecting rod that showed a 23% weight savings on the forged component and 14% on the finished machined connecting rod, was designed to utilize a less expensive forging material in line with the findings during the basic material evaluation phase.

12.6 Case No. 5—Cast Steel Axle Box for a Railway Vehicle [6]

12.6.1 Introduction

This case history is concerned primarily with acquisition of load and strain data (Chapters 5 and 7), stress- and strain-

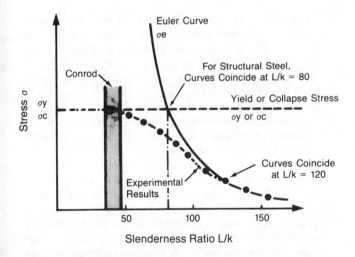

Fig. 12-27 Comparison of experimental buckling results with Euler curve.

A detailed finite element analysis of the new design was carried out to predict the distribution of stress and strain for a range of load cases. The finite element mesh is shown in Fig. 12-28.

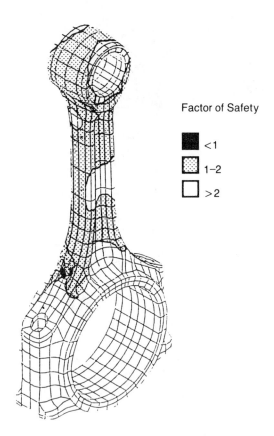

Fig. 12-29 Predicted factors of safety compared to the material's estimated fatigue strength.

based fatigue life predictions as well as crack propagation analysis (Chapter 10), failure analysis (Chapter 11), flaw detection (Chapter 7) and laboratory testing (Chapter 9).

12.6.2 Case History

Many of the procedures commonly used in fatigue design were used in evaluating this cast steel component. The axle box consists of a bearing housing for the wheelset and a horizontal spring tray, stiffened above by part-cylindrical walls and below by vertical ribs. The loads on the box are mainly vertical and include reactions from the coil spring, inertia loads from the wheelset and the box itself, and the forces from the transverse, Panhard rod.

Two axle boxes fractured in service at unknown mileage. Both had collapsed in a safe manner and examination showed that both had initially cracked by fatigue. The fatigue crack had started at a sharp notch where the machined face for the bearing-retaining ring met the curved wall of the spring pocket. The preliminary investigation established that many of the castings contained this feature. Some also had casting defects in the same general area from which small fatigue cracks were growing. One casting exhibited a fatigue crack in the same area even though it had no machined notches or defects.

It was clear that the two fractures were unlikely to be isolated occurrences and several actions were taken immediately. Inspection of the critical area of the in-service axle boxes was initiated and the supplier whose boxes had the least defects was asked to produce more components. As an interim measure, defective material was cut out of the critical areas of boxes with defects, and shaped pieces of wrought steel were butt-welded into this area. At the same time, a complete redesign of the axle box was initiated.

All of these activities were supported by, or depended upon, knowledge of the fatigue and fracture behavior of the material in the design. A number of questions arose:

a. *Will it break?*
   Attempts were made to relate (by means of linear elastic fracture mechanics) the stress immediately prior to fracture to the crack size at that instant. In this case, the complexity of the structure and of the crack path and the toughness of the material made it difficult to calculate the stress at fracture, and no reliable estimate was obtained.

b. *How long would a "good" box last?*
   Only two boxes had failed. It was tempting to think that in-service inspection would eliminate the boxes liable to fail and that improved quality control at manufacture would prevent any more from getting into service. If this were so, replacement of bad boxes would be a steady process avoiding the disruption and cost of a redesign and replacement campaign. Hence the fatigue life of a good box was calculated to give a baseline for judgement.

   Typical load and strain information was obtained by in-service measurements using strain gages. Strain-life fatigue data were obtained from strain-controlled testing of small "defect-free" specimens from axle boxes. The strain cycles were extracted from the service history by the rainflow counting procedure and combined with the fatigue data in a computer-based calculation of life. The calculation showed that the machined notch would give a fatigue life approximately equivalent to a quarter million miles (or roughly two years of service) and a box with no notches would have a life approximately equivalent to six million miles. Although the actual lives of the failed boxes were not known, the first calculation agreed with experience. The second calculation suggested a service life that would be too short to be safe, considering the problems with quality control in castings. The decision was made to redesign the component.

c. *What should the inspection frequency be?*
   In the meantime, before the new boxes became available, close inspection was considered necessary. The inspection frequency depends upon two things; the capability of the inspection procedure to locate and size cracks and the rate of crack growth.

The fatigue life prediction method used in this study relies on crack propagation data based on constant-amplitude testing. Normally, straightforward integration is used to combine the data with stress cycles extracted from the variable-amplitude history by the rainflow counting procedure. In this case, although the crack growth data could be obtained readily from tests on small, notched specimens, the complex shape of the axle box made the stress-intensity factor for the growing crack very difficult to determine. Accordingly, the cracks were grown in axle boxes under a simulation of the measured service stress history. An initial constant-amplitude loading test to produce the crack showed that its visible length increased very rapidly as it spread along the surface. The crack progressed more slowly through the wall thickness, and, with this and the difficulties of inspection in mind, lengths of 25 to 30mm were used as the starting points of the two variable amplitude tests. The finish point, when the crack was growing rapidly, was taken as 95mm in the first instance, corresponding to a life of 7,000 miles, and 125mm for the second, corresponding to 16,000 miles. The apparent differences in growth rates reflected the considerable variation in wall thickness between the two castings.

In practice it was found that cracks were very much smaller (about 3mm) than the initial test lengths that could be detected reliably; the inspection period for the crack-free box was kept at 18,000 miles (one month), or three times that figure, if the notch had been "dressed out". This policy was successful in that no further failures occurred.

d. *How long would the weld-modified boxes last?*
The determination of the likely life of a weld-repaired box was undertaken as a matter of urgency. For this purpose, a nominal stress approach was used with data from a fatigue code of practice for bridges. Here the stresses include the effects of general stress raisers, but ignore local concentrations at weld toes, etc. that are accounted for by the classification of the weld detail; different classes, based on overall geometry, have different fatigue strengths.

The first problem was to classify the repair, since it did not match any of the standard classes, and was in an area of very high stress gradient. Constant amplitude fatigue tests on complete weld-modified boxes gave lives corresponding to a certain class and this was used in the service life predictions, together with the service strain history. The predicted life was found to be only 33,000 miles, for a 50% probability of failure, and a weekly inspection was instituted immediately. In practice, the boxes had a mean life of about 108,000 miles showing the prediction to be correct in forecasting a short life, and rather safe.

At the same time, before the redesigned axle box was ready and while there was a considerable shortage of replacement boxes, some of the axle boxes being manufactured were found to contain surface defects. These defects were gouged out and repaired by local welding, a practice that is very common with castings. Constant amplitude fatigue testing of repaired boxes indicated a different class having a stress-life line with a more shallow slope. The fatigue life prediction was about the same as for the original, poorly machined box (a quarter of a million miles) and the boxes were allowed to stay in service on the same basis as the unrepaired ones.

e. *How safe is the redesigned box?*
For a primary suspension component there are two design requirements. The chance of a fatigue crack initiating in a box of reasonable quality, subject to foreseeable loads, must be extremely small and if a crack initiates due to overloads, damage or a gross defect, the rate of propagation should be so low that normal non-specific inspection will reveal it.

In this case, the "fatigue limit" approach was used. The maximum service stresses were not easily defined, but an effort was made to define them based on a worst-case combination of expected loadings. Both the maximum service stresses and the fatigue limit are probabilistic concepts. Since there is seldom enough information to make a numerical assessment of probability, judgement is required. In this case, on-track stress measurements on the redesigned box showed the maximum ranges to be comfortably below the fatigue limit, determined from fatigue tests on complete boxes.

12.7 Case No. 6—Axle Shaft Problem on a Scraper Vehicle [7]

12.7.1 *Introduction*

This case history is concerned primarily with strain and load data acquisition (Chapter 5), laboratory testing (Chapter 9), effects of processing (i.e., shot-peening) on fatigue strength (Chapter 4), and failure analysis (Chapter 11).

12.7.2 *Case History*

Axle shafts were failing on scraper type tractors after only six months in service, although they had survived specified new product durability tests in the laboratory prior to release. At first sight, to improve the strength of the axles without increasing their size appeared difficult. The consequence of increasing size would be a costly requirement for larger bearings and, hence, larger axle housings.

The first stage towards understanding why the axle shafts were failing in service was to determine the actual loads in service. A shaft was instrumented with strain gages for use as a torque meter in the vehicle during regular field service (Fig. 12-30). The field testing equipment consisted of the instrumented shaft, a 110-volt generator unit, a strip-chart recorder, an amplifier and several mercury slip rings. The instrumented shaft was installed in a tractor that was being used with a bowl scraper. The necessary recording equipment was bolted onto a piece of plywood and this was attached to the fender of the tractor with C-clamps.

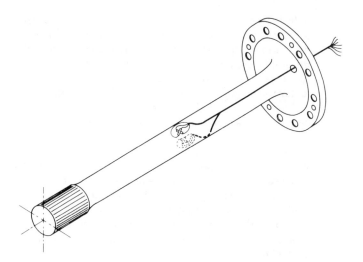

Pair of strain gages mounted 180 degrees apart.

Fig. 12-30 Strain gage configuration used on axle shaft for field testing.

Analysis of the measured data revealed the problem immediately. When the tractor required some assistance from a push tractor, the torque in the axle shafts of the original test tractor was reversed. As the test tractor was being pushed, its wheels accelerated the engine causing the axle shafts to experience a reversed torque. Strain measurements indicated that this maximum reversed torque was equivalent to 20,300N-m. The original shaft was designed to withstand only forward "slip" torque of 47,500N-m in fatigue.

The reason for the failures was now known but a solution to improve shaft strength was needed without a costly replacement of bearings and axle housings (which would have to accompany the use of larger axle shafts). A literature review revealed no data on torsional testing of truck axle shafts but data were found that suggested an improvement in bending fatigue strength through shot-peening. As a consequence, a test stand was designed and built to test any new axle shaft designs for their torsional fatigue strength. The machine was hydraulically operated and powered by a 40 HP electric motor. An oil cooler was used to cool the working hydraulic fluid. Axle shafts were tested by cycling from zero to 67,800N-m torque. (With the available actuators and controllers this was easier to achieve than reversed cycles and the tests were being conducted for comparative purposes. In addition, it was recognized that stress range that governs torsional fatigue; the torsional mean stress should not be important according to Sine's criterion for multiaxial fatigue and available test data.)

Fatigue test results were reviewed and a decision was made to strengthen the axle shafts by changing the material from SAE 8645 steel to SAE 86B45 steel. The major difference between the steels was the addition of boron in the latter which increases hardenability. The spline design was also modified to provide a full radius at the root of each spline. In addition, the shafts were heavily shot-peened to increase their fatigue strength.

Results showed a significant difference in life between unpeened and peened shafts. Ten unpeened shafts had lives ranging from 17,000 cycles to 48,000 cycles of torque loading and 16 peened shafts of the same size and material had lives ranging from 80,000 cycles to 325,000 cycles. The problem was successfully solved in that the modified axle shafts performed satisfactorily in service.

12.8 References

[1] Landgraf, R. W., "Fatigue Considerations in Use of High-Strength Sheet Steel," SAE Technical Paper 820700, SAE, 1982.

[2] Bickerstaff, D. J., Birchmeier, J. E., and Tighe, W. R., "Overview of Design Approaches for Optimizing Fatigue Performance of Suspension Systems," SAE Technical Paper 820676, SAE, 1982.

[3] Starkey, M. S. and Irving, P. E., "A Comparison of the Fatigue Strength of Machined and As-Cast Surfaces of SG Iron," *International Journal of Fatigue*, July 1982, pp. 129-136.

[4] Watson, P., Dabell, B., Hill, S., and Rebbeck, R., "A Realistic Comparison of Nodular Iron and Cast Steel Using Computer-Based Fatigue Analysis Techniques," Closed Loop, MTS Systems Corp., May 1979, pp. 15-26.

[5] McRobert, S. C. and Watson, P., "Design Optimization of Forgings," Techwrite Services Ltd., Wolverhampton, January 1984.

[6] McLaster, R., "Case Study: The BT10 Axle Box," British Rail Technical Centre, Darby, England, 1984.

[7] Todd, R. H., "Sometimes It Gets Pushed," Engineering Case Library, ECL 179R, Leland Stanford Junior University, 1971.

References That Include Other Case Histories of Potential Interest

[8] Kaufman, P. and Kershienik, M., "Case History—A High Speed Turbine Wheel," *Mechanical Engineering*, July 1984, pp. 38-45.

[9] Sherman, A. M., "Application of Fatigue Life Prediction Techniques to Material Substitution," *Journal of Testing and Evaluation*, JTEVA, Vol. 7, No. 2, March 1979, pp. 111-116.

[10] Sailors, R. H., "Presentation of Failure Analysis Data by the Fatigue-Fracture Mechanics Diagram," ASME Publication No. 81-PVP-11, ASME, 1981.

[11] Thakkar, R., "Strain-Life Predictions Using PVC Scale Models," unpublished work by the author, 1986.

[12] Birchmeier, J. E. and Smith, K. V., "Optimization of a Light Truck Rough Road Durability Procedure Using Fatigue Analysis Methodology," SAE Technical Paper 820693, SAE, 1982.

[13] Dabell, B., "Integrated Design/Fatigue Analysis Study," SAE Technical Paper 820697, SAE, 1982.

[14] Rice, R. C. and Smith, C. E., "Fatigue and Fracture Tolerance Evaluation of Tall Loran Tower Eyebolts," *Design of Fatigue and Fracture Resistant Structures*, ASTM STP 761, ASTM, 1982, pp. 424-444.

[15] Mitchell, M. R., Meyer, E. M., Nguyen, N. Q., "Fatigue Considerations in Use of Aluminum Alloys, SAE Technical Paper 820699, SAE, 1982.

Addresses of Publishers

ASME, American Society of Mechanical Engineers, 345 E. 47th Street, New York, NY 10017

ASTM, American Society for Testing & Materials, 1916 Race Street, Philadelphia, PA 19103

MTS Systems Corp., P.O. Box 24012, Minneapolis, MN 55242

SAE, Society of Automotive Engineers, 400 Commonwealth Drive, Warrendale, PA 15096

APPENDIX A Definitions

The terminology used in fatigue design and analysis has been standardized in most cases by the American Society for Testing and Materials (ASTM). The definitions presented here are consistent with current ASTM accepted terminology [1-7]. Pertinent additional terminology was also reproduced from SAE J1099 [8]. The terms are listed in alphabetical order and are numbered for easy reference. In these definitions the term load (P) is taken to represent stress (S), strain (ϵ), force (F), load factor (g), stress intensity factor (K) or any other expression or function of loading. All symbols are listed and defined at the front of this handbook.

1. **alternating load**—see loading amplitude.

2. **bearing load, F**—a compressive load on an interface.

3. **block**—in fatigue loading, a specified number of constant amplitude loading cycles applied consecutively, or, a spectrum loading sequence of finite length which is repeated identically.

4. **breaking load, F**—the load at which fracture occurs.

   NOTE: When used in connection with tension tests of thin materials or materials of small diameter for which it is often difficult to distinguish between the breaking load and the maximum load developed, the latter is considered to be the breaking load.

5. **clipping**—in fatigue spectrum loading, the process of decreasing or increasing the magnitude of all loads that are, respectively, above or below a specified level, referred to as clipping level; the loads are decreased or increased to the clipping level (see Fig. A-1).

6. **compressive strength**—the maximum compressive stress that a material is capable of sustaining. Compressive strength is calculated from the maximum load during a compression test and the original cross-sectional area of the specimen.

   NOTE: In the case of a material that fails in compression by a shattering fracture, the compressive strength has a very definite value. In the case of materials which do not fail in compression by a shattering fracture, the value obtained for compressive strength is an arbitrary value depending upon the degree of distortion which is regarded as indicating complete failure of the material.

7. **confidence interval**—an interval estimate of a population parameter computed so that the statement "the population parameter lies in this interval" will be true, on the average, in a stated proportion of the times such computations are made based on different samples from the population.

8. **confidence level (or coefficient)**—the stated proportion of the times the confidence interval is expected to include the population parameter.

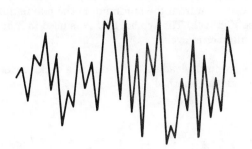

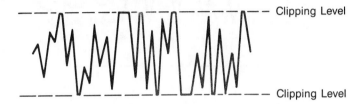

Fig. A-1 Clipping of fatigue spectrum loading.

9. **confidence limits**—the two statistics that define a confidence interval.

10. **constant amplitude loading**—in fatigue loading, a loading in which all of the peak loads are equal and all of the valley loads are equal.

11. **constant life diagram (in fatigue)**—a plot (usually on rectangular coordinates) of a family of curves, each of which is for a single fatigue life, N, relating S, S_{max} and/or S_{min} to the mean stress S_m. The constant life fatigue diagram is usually derived from a family of S-N curves each of which represents a different stress ratio (A or R) for a 50 percent probability of survival.

12. **corrosion fatigue**—synergistic effect of fatigue and aggressive environment acting simultaneously, that leads to a degradation in fatigue behavior.

13. **counting method**—in fatigue spectrum loading, a method of counting the occurrences and defining the magnitude of various loading parameters from a load-time history; (some of the counting methods are: level crossing, peak, mean crossing peak, range, range-pair, rainflow, race-track).

14. **crack size, a, c or l**—a lineal measure of a principal planar dimension of a crack, commonly used in the calculation of quantities descriptive of the stress and displacement fields, and often also termed "crack length."

15. **cumulative frequency spectrum**—see exceedances spectrum.

16. **cumulative occurrences spectrum**—see exceedances spectrum.

17. **cycle (in fatigue)**—under constant amplitude loading, the load variation from the minimum to the maximum load. (See Fig. A-2.) The symbol n or N is used to indicate the number of cycles.

    NOTE: In spectrum loading, definition of a cycle varies with the counting method.

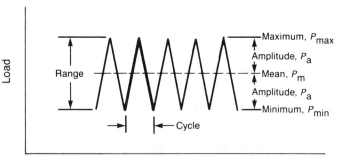

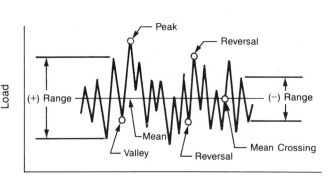

(a) Constant Amplitude Loading
(b) Spectrum Loading

Fig. A-2 Fatigue loading basic terms.

18. **cycles endured,** n—the number of cycles of specified character (that produce fluctuating load) which a specimen or component has endured at any time in its load history.

19. **cycle ratio,** C—the ratio of cycles endured, n, to the estimated fatigue life, N, obtained from the S-N or -N diagram, for cycles of the same character, that is, C = n/N.

20. **cyclic loading**—see fatigue loading.

21. **cyclic strain hardening exponent,** n'—the power to which "true" plastic strain amplitude must be raised to be proportional to "true" stress amplitude. It is taken as the slope of the log $\Delta\epsilon_p/2$ and $\Delta\sigma/2$ versus log $\Delta\sigma/2$ plot, where $\Delta\epsilon_p/2$ and $\Delta\sigma/2$ are measured from cyclically stable hysteresis loops.

$$\Delta\sigma/2 = K' (\Delta\epsilon_p/2)n'$$

where $\Delta\epsilon_p/2$ = "true" plastic strain amplitude. The line defined by this equation is illustrated in Fig. A-3.

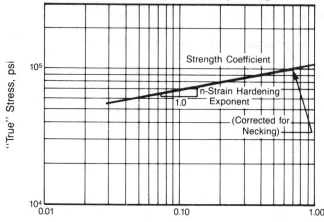

Fig. A-3 Cyclic stress-plastic strain plot (1020 H.R. steel).

22. **cyclic strength coefficient,** K'—the "true" stress at a "true" plastic strain of unity. It may be necessary to extrapolate as indicated in Fig. A-3.

23. **cyclic stress-strain curve**—the locus of tips of stable "true" stress-strain hysteresis loops obtained from companion test specimens. A typical stable hysteresis loop is illustrated in Fig. A-4 and a set of stable loops with a cyclic stress-strain curve drawn through the loop tips is illustrated in Fig. A-5. As illustrated, the height of the loop from tip-to-tip is defined as the stress range ($\Delta\sigma$). For completely reversed testing one-half of the stress range is generally equal to the stress amplitude while one-half of the width from tip-to-tip is defined as the strain amplitude ($\Delta\epsilon/2$). Plastic strain amplitude is found by subtracting the elastic strain amplitude ($\Delta\epsilon_e/2$) from the strain amplitude as indicated below.

$$\Delta\epsilon_p/2 = \Delta\epsilon/2 - \Delta\epsilon_e/2$$

According to Hooke's law,

$$\Delta\epsilon_e/2 = \Delta\sigma/2E$$

where E = modulus of elasticity,

$$\Delta\epsilon_p/2 = \Delta\epsilon/2 - \Delta\sigma/2E$$

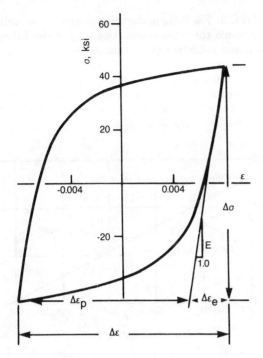

Fig. A-4 Stable stress-strain hysteresis loop.

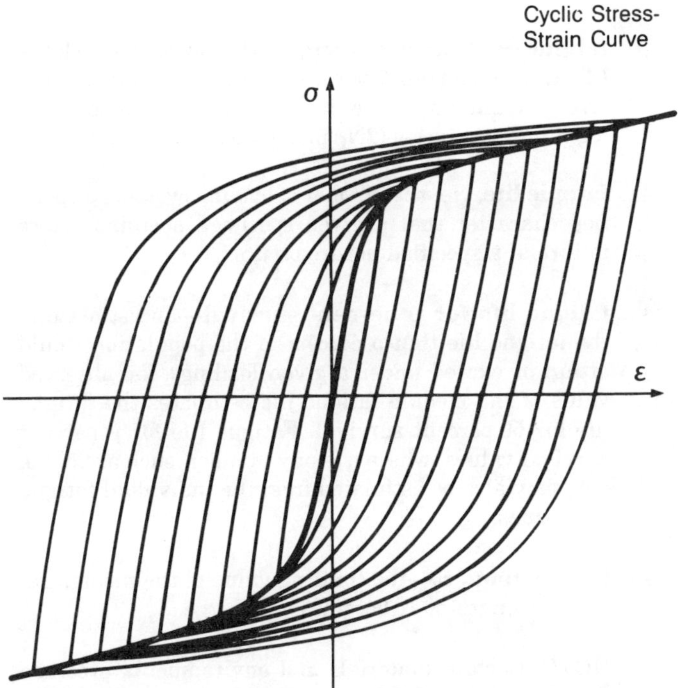

Fig. A-5 Cyclic stress-strain curve drawn through stable loop tips.

A more complete discussion of the cyclic stress-strain curve and other methods of obtaining the curve are given in STP 465 [10] and in Ref. [11].

24. **cyclic yield strength, 0.2% σ_{ys}'**—the stress to cause 0.2% inelastic strain as measured on a cyclic stress-strain curve. It is usually determined by constructing a line parallel to the slope of the cyclic stress-strain curve at zero stress through 0.2% strain and zero stress. The stress where the constructed line intercepts the cyclic stress-strain curve is taken as the 0.2% cyclic yield strength.

25. **ductility**—the ability of a material to deform plastically before fracturing.

    NOTE 1: Ductility is usually evaluated by measuring (1) the elongation or reduction of area from a tension test; (2) the depth of cup from a cupping test; (3) the radius or angle of bend from the bend test; or (4) the fatigue ductility from the fatigue ductility test.

    NOTE 2: Malleability is the ability to deform plastically under repetitive compressive forces.

26. **elastic constants**—see modulus of elasticity and Poisson's ratio.

27. **elastic limit**—the greatest stress which a material is capable of sustaining without any permanent strain remaining upon complete release of the stress.

    NOTE—Due to practical considerations in determining the elastic limit, measurements of strain, using a small load rather than zero load, are usually taken as the initial and final reference.

28. **elastic strain amplitude**—see cyclic stress-strain curve.

29. **elongation, El**—the increase in gage length of a body subjected to a tension force, referenced to a gage length on the body. Usually elongation is expressed as a percentage of the original gage length.

    NOTE 1: The increase in gage length may be determined either at or after fracture, as specified for the material under test.

    NOTE 2: The term elongation, when applied to metals, generally means measurement after fracture; when applied to plastics and elastomers, measurement at fracture. Such interpretation is usually applicable to values of elongation reported in the literature when no further qualification is given.

    NOTE 3: In reporting values of elongation the gage length shall be stated.

    NOTE 4: Elongation is affected by specimen geometry; length, width, thickness of the gage section and adjacent regions; and test procedure, such as alignment and speed of pulling.

30. **environment**—in fatigue testing, the aggregate of chemical species and energy that surrounds a test specimen.

31. **estimate**—in statistical analysis, the particular value or values of a parameter computed by an estimation procedure for a given sample.

32. **estimation**—in statistical analysis, a procedure for making a statistical inference about the numerical values of one or more unknown population parameters from the observed values in a sample.

33. **exceedance spectrum**—in fatigue loading, representation of spectrum loading contents by the number of times specified values of a particular loading parameter (peak, range, etc.) are equaled or exceeded (also known as cumulative occurrences or cumulative frequency spectrum).

34. **extensometer**—a device for measuring linear strain.

    NOTE: Devices for measuring decreases in length are sometimes called "compressometers."

35. **fatigue**—the process of progressive localized permanent structural change occurring in a material subjected to conditions which produce fluctuating stresses and strains at some point or points and which may culminate in cracks or complete fracture after a sufficient number of fluctuations.

36. **fatigue crack growth rate**, da/dN or a/N—in fatigue, the rate of crack extension caused by fatigue loading, expressed in terms of average crack extension per cycle.

37. **fatigue crack growth threshold**, K_{th}—that asymptotic value of K at which da/dN approaches zero. For most materials an "operational," though arbitrary, definition of K_{th} is given as that K which corresponds to a fatigue crack growth rate of 10^{-10} m/cycle.

    NOTE: The intent of this definition is not to define a "true" threshold, but rather to provide a practical means of characterizing a material's fatigue crack growth resistance in the near-threshold regime. Caution is required in extending this concept to design.

38. **fatigue ductility**, D_f—the ability of a material to deform plastically before fracturing, determined from a constant-strain amplitude, low-cycle fatigue test.

    NOTE 1: Fatigue ductility is usually expressed in percent in direct analogy with elongation and reduction of area ductility measures.

    NOTE 2: The fatigue ductility corresponds to the fracture ductility. Elongation and reduction of area represent the engineering tensile strain after fracture.

NOTE 3: The fatigue ductility is used for metallic foil for which the tension test does not give useful elongation and reduction of area measures.

39. **fatigue ductility coefficient**, (ϵ'_f)—the "true" strain required to cause failure in one reversal. It is taken as the intercept of the log ($\Delta\epsilon_p/2$) versus log ($2N_f$) plot at $2N_f = 1$ (see Figs. A-6 and A-7).

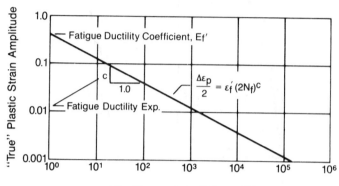

Fig. A-6 Plastic strain amplitude vs. reversals to failure (1020 H.R. steel).

40. **fatigue ductility exponent**, c—the power to which the life in reversals must be raised to be proportional to the "true" strain amplitude. It is taken as the slope of the log ($\Delta\epsilon_{p/2}$) versus log ($2N_f$) plot. (See Figs. A-6 and A-7.).

41. **fatigue life**, N—the number of loading cycles of a specified character that a given specimen sustains before failure of a specified nature occurs.

42. **fatigue life for p percent survival**—an estimate of the fatigue life that p percent of the population would attain or exceed under a given loading. The observed value of the median fatigue life estimates the fatigue life for 50 percent survival. Fatigue life for p percent survival values, where p is any number, such as 95, 90, etc., may also be estimated from the individual fatigue life values.

43. **fatigue limit**, S_f—the limiting value of the median fatigue strength as N becomes very large.

    NOTE: Certain materials and environments preclude the attainment of a fatigue limit. Values tabulated as "fatigue limits" in the literature are frequently (but not always) values of S_N for 50 percent survival at N cycles of stress in which the mean stress, S_m, equals zero.

44. **fatigue limit for p percent survival**—the limiting value of fatigue strength for p percent survival as N becomes very large; p may be any number, such as 95, 90, etc.

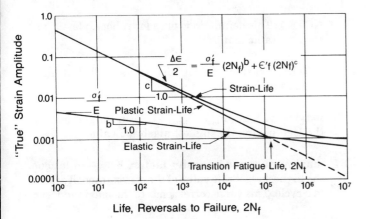

Fig. A-7 Strain amplitude vs. reversals to failure (1020 H.R. steel).

45. **fatigue loading**—periodic or non-periodic fluctuating loading applied to a test specimen or experienced by a structure in service. (Also known as cyclic loading.)

46. **fatigue notch factor, K_f**—the ratio of the fatigue strength of a specimen with no stress concentration to the fatigue strength at the same number of cycles with stress concentration for the same conditions.

   NOTE 1: In specifying K_f it is necessary to specify the geometry and the values of S_a, S_m and N for which it is computed.

   NOTE 2: K_f was originally termed the fatigue limit (endurance limit) reduction factor. Early data pertained almost exclusively to mild steels, namely, to S_a-N curves with knees. Later the term was generalized to fatigue strength reduction factor; but nevertheless the K_f values tabulated in the literature still pertained almost exclusively to very long (infinite) fatigue lives where the notched and unnotched S_a-N curves where almost parallel and almost horizontal. Otherwise, the K_f data were not consistent and were markedly dependent on the type of notch, the fatigue life of interest, and the value of the mean stress.

   NOTE 3: Virtually no K_f data exist for percentiles other than (approximately) fifty percent. Nevertheless, it is obvious that K_f is highly dependent on the percentile of interest.

47. **fatigue notch sensitivity, q**—a measure of the degree of agreement between K_f and K_t.

   NOTE 1: The definition of fatigue notch sensitivity is $q = (K_f - 1)/(K_t - 1)$.

   NOTE 2: q was originally termed the fatigue notch sensitivity index.

   NOTE 3: Virtually all q data and q curves found in the literature pertain to very long (infinite) fatigue lives where the notched and unnotched S_a-N curves are almost parallel and almost horizontal, as well as to tests in which $S_m = 0$. Thus, these values should not normally be extrapolated to $S_m = 0$ or "finite" life situations.

48. **fatigue strength at N cycles, S_N**—a value of stress for failure at exactly N cycles as determined from an S-N diagram. The value of S_N thus determined is subject to the same conditions as those which apply to the S-N diagram.

   NOTE: The value of S_N which is commonly found in the literature is the value of S_{max} or S_a at which 50 percent of the specimens of a given sample could survive N stress cycles in which $S_m = 0$. This is also known as the median fatigue strength for N cycles.

49. **fatigue strength coefficient, σ_f'**—the "true" stress required to cause failure in one reversal. It is taken as the intercept of the log $\Delta\sigma/2$ versus log $(2N_f)$ plot at $2N_f = 1$ (see Figs. A-7 and A-8).

50. **fatigue strength exponent, b**—the power to which life in reversals must be raised to be proportional to "true" stress amplitude. It is taken as the slope of the log $\Delta\sigma/2$ versus log $(2N_f)$ plot (see Figs. A-7 and A-8).

51. **fatigue strength for p percent survival at N cycles**—an estimate of the stress level at which p percent of the population would survive N cycles; p may be any percent, such as 95, 90, etc.

52. **fracture ductility, ϵ_f**—the true plastic strain at fracture.

53. **fracture strength, S_f**—the normal stress at the beginning of fracture. Fracture strength is calculated from the load at the beginning of fracture during a tension test and the original cross-sectional area of the specimen.

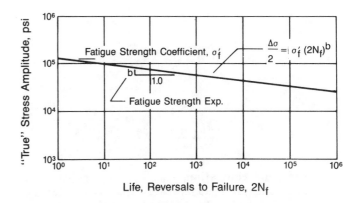

Fig. A-8 Stress amplitude vs. reversals to failure (1020 H.R. steel).

54. **frequency distribution**—the way in which the frequencies of occurrence of members of a population, or a sample, are distributed according to the values of the variable under consideration.

55. **gage length** [L]—the original length of the specimen over which strain or change of length is determined.

56. **group**—in fatigue, the specimens of the same type tested at one time, or consecutively, at one stress level. A group may comprise one or more specimens.

57. **hysteresis diagram**—the stress-strain path during the cycle.

58. **interval estimate**—the estimate of a parameter given by two statistics, defining the end points of an interval.

59. **irregularity factor**—in fatigue loading, the ratio of the number of zero crossings with positive slope (or mean crossings) to the number of peaks or valleys in a given, load-time history.

60. **irregular loading**—see spectrum loading.

61. **level crossings**—in fatigue loading, the number of times that the load-time history crosses a given load level with a positive slope (or a negative slope, or both, as specified) during a given length of the history.

62. **loading amplitude,** P_a, S_a, or ϵ_a—in fatigue loading, one-half of the range of a cycle (see Fig. A-2). Also known as alternating load.

63. **loading (unloading) rate**—the time rate of change in the monotonic increasing (decreasing) portion of the load time function.

64. **load ratio,** R or A—in fatigue loading, the algebraic ratio of the two loading parameters of a cycle; the most widely used ratios are:

$$R = \frac{\text{minimum load}}{\text{maximum load}} = \frac{P_{min}}{P_{max}} \text{ or } \frac{S_{min}}{S_{max}}$$

or

$$R = \frac{\text{valley load}}{\text{peak load}}$$

and

$$A = \frac{\text{loading amplitude}}{\text{mean load}} = \frac{P_a}{P_m} \text{ or } \frac{S_a}{S_m}$$

65. **maximum elongation,** El_{max}—the elongation at the time of fracture, including both elastic and plastic deformation of the tensile specimen.

NOTE 1: This definition is used for rubber, plastic, and some metallic materials.

NOTE 2: Maximum elongation is also called ultimate elongation or break elongation.

66. **maximum load,** P_{max}, S_{max}, ϵ_{max}—in fatigue loading, the load having the highest algebraic value.

67. **maximum stress-intensity factor,** K_{max}—in fatigue, the highest algebraic value of the stress-intensity factor in a cycle. This value corresponds to the maximum load.

68. **mean crossings**—in fatigue loading, the number of times that the load-time history crosses the mean load level with a positive slope (or a negative slope, or both, as specified) during a given length of the history (see Fig. A-2).

69. **mean load,** P_m (or S_m or ϵ_m)—in fatigue loading, the algebraic average of the maximum and minimum loads in constant amplitude loading, or of individual cycles in spectrum loading,

$$P_m = \frac{P_{max} + P_{min}}{2} \text{ or}$$

$$S_m = \frac{S_{max} + S_{min}}{2} \text{ or}$$

$$\epsilon_m = \frac{\epsilon_{max} + \epsilon_{min}}{2} \text{ or}$$

or the integral average of the instantaneous load values of a spectrum loading history.

70. **mechanical hysteresis**—the energy absorbed in a complete cycle of loading and unloading.

NOTE: A complete cycle of loading and unloading includes any stress cycle regardless of the mean stress or range of stress.

71. **mechanical properties**—those properties of a material that are associated with elastic and inelastic reaction when force is applied, or that involve the relationship between stress and strain.

NOTE: These properties have often been referred to as "physical properties," but the term "mechanical properties" is preferred.

72. **mechanical testing**—the determination of mechanical properties.

73. **median fatigue life**—the middle value of the observed fatigue lives, arranged in order of magnitude, of the individual specimens in a group tested under identical

conditions. If the sample size is even, it is the average of the two middlemost values.

NOTE 1: The use of the median instead of the arithmetic mean (that is, the average) is usually preferred.

NOTE 2: In the literature, the abbreviated term "fatigue life" usually has meant the median fatigue life of the group. However, when applied to a collection of data without further qualification the term "fatigue life" is ambiguous.

74. **median fatigue strength at N cycles**—an estimate of the stress level at which 50 percent of the population would survive N cycles.

NOTE 1: The estimate of the median fatigue strength is derived from a particular point of the fatigue life distribution, since there is no test procedure by which a frequency distribution of fatigue strengths at N cycles can be directly observed.

NOTE 2: This is a special case of the more general definition of fatigue strength for p percent survival at N cycles.

75. **minimum load,** P_{min}, Smin, or ϵ_{min}—in fatigue loading, the load having the lowest algebraic value.

76. **minimum stress-intensity factor,** K_{min}—in fatigue, the lowest algebraic value of the stress-intensity factor in a cycle. This value corresponds to the minimum load when the load ratio (R) is greater than zero and is set equal to zero when R is less than or equal to zero.

77. **modulus of elasticity**—the ratio of stress to corresponding strain below the proportional limit.

   a. tension or compression, E—Young's modulus (modulus in tension or modulus in compression).

   b. shear or torsion, G—commonly designated as modulus of rigidity, shear modulus, or torsional modulus.

NOTE 1: The stress-strain relations of many materials do not conform to Hooke's law throughout the elastic range, but deviate therefrom even at stresses well below the elastic limit. For such materials the slope of either the tangent to the stress-strain curve at the origin or at a low stress, the secant drawn from the origin to any specified point on the stress-strain curve, or the chord connecting any two specified points on the stress-strain curve is usually taken to be the "modulus of elasticity." In these cases the modulus should be designated as the "tangent modulus," the "secant modulus," or the "chord modulus," and the point or points on the stress-strain curve described. Thus, for materials where the stress-strain relationship is curvilinear rather than linear, one of the four following terms may be used:

   a. initial tanent modulus—the slope of the stress-strain curve at the origin.

   b. tangent modulus—the slope of the stress-strain curve at any specified stress or strain.

   c. secant modulus—the slope of the secant drawn from the origin to any specified point on the stress-strain curve.

   d. chord modulus—the slope of the chord drawn between any two specified points on the stress-strain curve.

NOTE 2: Modulus of elasticity, like stress, is expressed in force per unit of area (pounds per square inch, etc.).

78. **modulus of rupture in torsion**—the value of maximum shear stress in the extreme fiber of a member of circular cross section loaded to failure in torsion computed from the equation:

$$S_s = Tr/J$$

where: T = maximum twisting moment,
r = original outer radius, and
J = polar moment of inertia of the original cross section.

NOTE 1: When the proportional limit in shear is exceeded, the modulus of rupture in torsion is greater than the actual maximum shear stress in the extreme fiber, exclusive of the effect of stress concentration near points of application of torque.

NOTE 2: If the criterion for failure is other than fracture or attaining the first maximum of twisting moment, it should be so stated.

79. **monotonic strain hardening exponent,** n—the power to which the "true" plastic strain must be raised to be proportional to "true" stress. It is generally taken as the slope of log $\sigma\epsilon_p$ versus log σ_p plot as shown in Fig. A-3.

$$\sigma = K\epsilon_p^n$$

80. **monotonic strength coefficient,** K—the "true" stress at a "true" plastic strain of unity as shown in Fig. A-3. If fracture ductility is less than 1.0, it is necessary to extrapolate (see Fig. A-3).

81. **necking**—the localized reduction of the cross-sectional area of a specimen which may occur during loading.

82. **occurrences spectrum**—in fatigue loading, representation of spectrum loading contents by the number of times a particular loading parameter (peak, range, etc.)

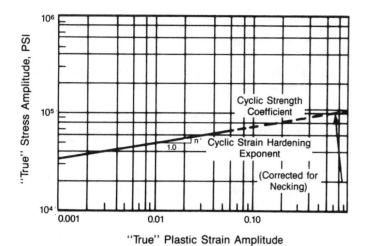

Fig. A-9 "True" stress-plastic stain plot (1020 H.R. steel).

occurs within each specified loading interval between lower and upper bound values.

83. **parameter**—in statistics, a constant (usually to be "estimated") defining some property of the frequency distribution of a population, such as a population median or a population standard deviation.

84. **peak**—in fatigue loading, the point at which the first derivative of the load-time history changes from a positive to a negative sign; the point of maximum load in constant amplitude loading (see Fig. A-2).

85. **plastic strain amplitude**—see cyclic stress-strain curve.

86. **point estimate**—the estimate of a parameter given by a single statistic, e.g., sample average (see sample average).

87. **Poisson's ratio**—the absolute value of the ratio of transverse strain to the corresponding axial strain resulting from uniformly distributed axial stress below the proportional limit of the material.

NOTE 1: Above the proportional limit, the ratio of transverse strain to axial strain will depend on the average stress and on the stress range for which it is measured and, hence should not be regarded as Poisson's ratio. If this ratio is reported, nevertheless, as a value of "Poisson's ratio" for stresses beyond the proportional limit, the range of stress should be stated.

NOTE 2: Poisson's ratio will have more than one value if the material is not isotropic.

88. **population (or universe)**—in fatigue testing, the totality of the set of test specimens, real or conceptual, that could be prepared in the specified way from the material under consideration.

89. **power spectral density**—the limiting mean-square value (for example, of acceleration, velocity, displacement, stress, or other random variable) per unit bandwidth of frequency, that is the limit of the mean-square value of a given rectangular bandwidth divided by the bandwidth, as the bandwidth approaches zero.

90. **precision**—the closeness of agreement between randomly selected individual measurements or test results.

NOTE 1: The standard deviation of the error of measurement may be used as a measure of "imprecision."

NOTE 2: The estimate of precision usually will contain a component of variance due to variability in the material being tested as well as a random measurement component of variance due to application of the test method. Special test setups may permit the separation of these two components of variance.

91. **proportional limit**—the greatest stress which a material is capable of sustaining without any deviation from proportionality of stress to strain (Hooke's law).

NOTE: Many experiments have shown that values observed for the proportional limit vary greatly with the sensitivity and accuracy of the testing equipment, eccentricity of loading, the scale to which the stress-strain diagram is plotted, and other factors. When determination of proportional limit is required, the procedure and the sensitivity of the test equipment should be specified.

92. **random loading**—in fatigue loading, a spectrum loading where the peak and valley loads and their sequence is a result of a random process; the loading is usually described in terms of its statistical properties, such as the probability density function, the mean, the root mean square, the irregularity factor, and others as appropriate.

93. **random-ordered loading**—in fatigue loading, a spectrum loading that is generated from a distinct set of peak and valley loads into a loading sequence by using a specific random sequencing process; a sequence of finite length is usually repeated identically.

94. **range**—in fatigue loading, the algebraic difference between successive valley and peak loads (positive range or increasing load range) or between successive peak and valley loads (negative range or decreasing load range). See Fig. A-2.

NOTE 1: In spectrum loading, range may have a different definition, depending on the counting method used; for example, "overall range" is defined by the algebraic difference between the largest peak and the smallest valley of a given load-time history.

NOTE 2: In cycle counting by various methods, it is common to employ ranges between valley and peak loads, or between peak and valley loads, which are not necessarily successive events. The word "range" is used in this broader sense when dealing with cycle counting.

95. **range of stress-intensity factor, ΔK**—in fatigue, the algebraic difference between the maximum and minimum stress-intensity factors in a cycle, that is:

$$\Delta K = K_{max} - K_{min}$$

NOTE 1: The loading variables, R, ΔK, and K_{max} are related such that specifying any two uniquely defines the third according to the following relationship: $\Delta K = (1 - R) K_{max}$ for $R > 0$ and $\Delta K = K_{max}$ for $R \leq 0$. The relationship follows directly from the operational definition of K_{min}.

NOTE 2: The operational stress-intensity factor definitions (K_{max}, K_{min}, ΔK) do not include local crack-tip effects; for example, they do not account for crack closure, residual stress, and crack tip blunting or branching. While the operational definition of ΔK states that ΔK does not change for a constant value of K_{max} when $R \leq 0$, increases in fatigue crack growth rates can be observed when R becomes more negative. Excluding the compressive loads in the calculation of ΔK does not alter the material's response since this response (da/dN) is independent of the operational definition of ΔK. For predicting crack growth lives generated under various R conditions, the life prediction methodology must be consistent with the data reporting methodology.

96. **reduction of area**—the difference between the original cross-sectional area of a tension test specimen and the area of its smallest cross section (Note 1). The reduction of area is usually expressed as a percentage of the original cross-sectional area of the specimen.

NOTE 1: The smallest cross section may be measured at or after fracture as specified for the material under test.

NOTE 2: The term reduction of area when applied to metals generally means measurement after fracture; when applied to plastics and elastomers, measurement at fracture. Such interpretation is usually applicable to values for reduction of area reported in the literature when no further qualification is given.

97. **response curve for N cycles**—a curve fitted to observed values of percentage survival at N cycles for several stress levels, where N is the preassigned number such as 10^6, 10^7, etc. It is an estimate of the relationship between applied stress and the percentage of the population that would survive N cycles.

NOTE 1: Values of the median fatigue strength at N cycles and the fatigue strength for p percent survival at N cycles may be derived from the response curve for N cycles if p falls within the range of the percent survival values actually observed.

NOTE 2: Caution should be used in drawing conclusions from extrapolated portions of the response curves. In general, the curves should not be extrapolated to other values of p.

98. **reversal**—in fatigue loading, the point at which the first derivative of the load-time history changes sign. See Fig. A-2.

NOTE: The number of reversals in constant amplitude loading, is equal to twice the number of cycles.

99. **sample**—the specimens selected from the population for test purposes.

NOTE: The method of selecting the sample determines which statistical inferences or generalizations can be made about the population.

100. **sample average (arithmetic average)**—the sum of all the observed values in a sample divided by the sample size. It is a point estimate of the population mean.

101. **sample median**—the middle value when all observed values in a sample are arranged in order of magnitude if an odd number of items (units) are tested. If the sample size is even, it is the average of the two middlemost values. It is a point estimate of the population median, or 50 percent point.

102. **sample percentage**—the percentage of observed values between two stated values of the variable under consideration. It is a point estimate of the percentage of the population between the same two stated values. (One stated value may be "minus infinity or plus infinity".)

103. **sample standard deviation, s**—the square root of the sample variance. It is a point estimate of the population standard deviation, a measure of the "spread" of the frequency distribution of a population.

NOTE: This value of s provides a statistic that is used in computing interval estimates and several test statistics. For small sample sizes, s underestimates the population standard deviation. (See a text on statistics that gives an unbiased estimate of the standard deviation of a normal population.)

104. **sample variance, s^2**—the sum of the squares of the differences between each observed value and the sample average divided by the sample size minus one. It is a point estimate of the population variance.

NOTE: This value of s^2 provides both an unbiased point estimate of the population variance and a statistic that is used in computing the interval estimates and several test-statistics. Some texts define s^2 as "the sum of the squares of the differences between each observed value and the sample average divided by the sample size" but this statistic is not as useful.

105. **shear fracture**—a mode of fracture in crystalline materials resulting from translation along slip planes which are preferentially oriented in the direction of the shearing stress.

106. **shear strength**—the maximum shear stress that a material is capable of sustaining. Shear strength is calculated from the maximum load during a shear or torsion test and is based on the original dimensions of the cross section of the specimen.

107. **significant**—statistically significant. An effect or difference between populations is said to be present if the value of a test-statistic is significant, that is, lies outside of predetermined limits.

   NOTE: An effect that is statistically significant may or may not have engineering significance.

108. **significance level**—the stated probability (risk) that a given test of significance will reject the hypothesis (that a specified effect is absent) when the hypothesis is true.

109. **S-N curve for 50 percent survival**—a curve fitted to the median values of fatigue life at each of several stress levels. It is an estimate of the relationship between applied stress and the number of cycles-to-failure that 50 percent of the population would survive.

   NOTE 1: This is a special case of the more general definition of S-N curve for p percent survival.

   NOTE 2: In the literature, the abbreviated term "S-N Curve" usually has meant either the S-N curve drawn through the mean (averages) or the medians (50 percent values) for the fatigue life values. Since the term "S-N Curve" is ambiguous, it should be used in technical papers only when adequately described.

110. **S-N curve for p percent survival**—a curve fitted to the fatigue life for p percent survival values of each of several stress levels. It is an estimate of the relationship between applied stress and the number of cycles-to-failure that p percent of the population would survive, p may be any percent, such as 95, 90, etc.

   NOTE: Caution should be used in drawing conclusions from extrapolated portions of the S-N curves. In general, the S-N curves should not be extrapolated significantly beyond observed life values.

111. **S-N diagram**—a plot of stress against the number of cycles to failure. The stress can be S_{max}, S_{min}, or S_a. The diagram indicates the S-N relationship for a specified value of S_m, A, or R and a specified probability of survival. For N a log scale is almost always used. For S a linear scale is used most often, but a log scale is sometimes used.

112. **specimen temperature**—in fatigue testing, the average temperature of the test specimen in the test section during constant temperature tests and, the temperature of the test specimen in the test section at any instant of time during cyclic temperature tests.

113. **spectrum loading**—in fatigue loading, a loading in which all of the peak loads are not equal or all of the valley loads are not equal, or both (also known as variable amplitude loading or irregular loading).

114. **statistic**—a summary value calculated from the observed values in a sample.

115. **strain**, e—the per unit change, due to force, in the size or shape of a body referred to its original size or shape. Strain is a nondimensional quantity, but it is frequently expressed in inches per inch, metres per metre, or percent.

   NOTE 1: Strain, as defined here is sometimes called "engineering strain," to emphasize the difference from true strain.

   NOTE 2: In this standard, "original" refers to dimensions or shape of cross section of specimens at the beginning of testing.

   NOTE 3: Strain at a point is defined by six components of strain; three linear components and three shear components referred to a set of coordinate axes.

   NOTE 4: In the usual tension, compression, or torsion test it is customary to measure only one component of strain and to refer to this as "the strain." In a tension or a compression test this is usually the axial component.

   NOTE 5: Strain has an elastic and a plastic component. For small strains the plastic component can be imperceptibly small.

   NOTE 6: Linear thermal expansion, sometimes called "thermal strain," and changes due to the effect of moisture are not to be considered strain in mechanical testing.

   a. linear (tensile or compressive) strain—the change per unit length due to force in an original linear dimension.

      NOTE: An increase in length is considered positive.

   b. axial strain—linear strain in a plane parallel to the longitudinal axis of the specimen.

   c. transverse strain—linear strain in a plane perpendicular to the axis of the specimen.

NOTE: Transverse strain may differ with direction in anisotropic materials.

d. angular strain—use shear strain.

e. shear strain—the tangent of the angular change, due to force, between two lines originally perpendicular to each other through a point in a body.

f. true strain, ϵ—in a body subjected to axial force, the natural logarithm of the ratio of the gage length at the moment of observation to the original gage length.

g. elastic true strain, ϵ_e—elastic component of the true strain.

h. plastic true strain, ϵ_p—the inelastic component of true strain.

i. macrostrain—the mean strain over any finite gage length of measurement large in comparison with interatomic distances.

NOTE 1: Macrostrain can be measured by several methods, including electrical-resistance strain gages and mechanical or optical extensometers. Elastic macrostrain can be measured by X-ray diffraction.

NOTE 2: When either of the terms macrostrain or microstrain are first used in a document, it is recommended that the physical dimension or the gage length which indicate the size of the reference strain volume involved, be stated.

j. microstrain—the strain over a gage length comparable to interatomic distances.

NOTE 1: These are the strains being averaged by the macrostrain measurement. Microstrain is not measurable by existing techniques. Variance of the microstrain distribution can, however, be measured by X-ray diffraction.

NOTE 2: When either of the terms macrostrain or microstrain are first used in a document, it is recommended that the physical dimension or the gage length, which indicate the size of the reference strain volume involved, be stated.

k. residual strain—strain associated with residual stress.

NOTE: Residual strains are elastic.

116. **stress**—the intensity at a point in a body of the forces or components of force that act on a given plane through the point. Stress is expressed in force per unit of area (pouands-force per square inch, megapascals, etc.).

NOTE: As used in tension, compression, or shear tests prescribed in product specifications, stress is calculated on the basis of the original dimensions of the cross section of the specimen. This stress is sometimes called "engineering stress," to emphasize the difference from true stress.

a. engineering stress, s—the stress calculated on the basis of the original dimensions of theh specimen.

b. nominal stress, S—the stress at a point calculated on the net cross section by simple elastic theory without taking into account the effect on the stress produced by geometric discontinuities such as holes, grooves, fillets, etc.

c. normal stress—the stress component perpendicular to a plane on which the forces act. Normal stress may be either:

—tensile stress—normal stress due to forces directed away from the plane on which they act, or

—compressive stress—normal stress due to forces directed toward the plane on which they act.

NOTE: Tensile stresses are positive and compressive stresses are negative.

d. shear stress—the stress component tangential to the plane on which the forces act.

e. torsional stress—the shear stress on a transverse cross section resulting from a twisting action.

f. true stress—in a tension or compression test the axial stress calculated on the basis of the cross-sectional area at the moment of observation instead of the original cross-sectional area.

g. principal stress (normal)—the maximum or minimum value of the normal stress at a point in a plane considered with respect to all possible orientations of the considered plane. On such principal planes the shear stress is zero.

NOTE: There are three principal stresses on three mutually perpendicular planes. The states of stress at a point may be:

—uniaxial—a state of stress in which two of the three principal stresses are zero.

—biaxial—a state of stress in which only one of the three principal stresses is zero, or

—triaxial—a state of stress in which none of the principal stresses is zero.

—multiaxial—biaxial or triaxial.

h. fracture stress—the true normal stress on the minimum cross-sectional area at the beginning of fracture.

NOTE: This term usually applies to tension tests of unnotched specimens.

i. residual stress—stress in a body which is at rest and in equilibrium and at uniform temperature in the absence of external and mass forces.

117. **stress-intensity factor,** K, K_1, K_2, or K_3—the magnitude of the ideal-crack-tip stress field (a stress-field singularity) for a particular mode in a homogeneous, linear-elastic body.

NOTE: Value of K for the modes 1, 2, and 3 are given by:

$$K_1 = \lim_{r \to 0} [\sigma_y (2 \pi r)^{1/2}],$$

$$K_2 = \lim_{r \to 0} [\tau_{xy} (2 \pi r)^{1/2}], \text{ and}$$

$$K_3 = \lim_{r \to 0} [\tau_{yz} (2 \pi r)^{1/2}],$$

where r = a distance directly forward from the crack tip to a location where the significant stress is calculated.

118. **stress-intensity factor range**—see range of stress-intensity factor.

119. **stress-strain diagram**—a diagram in which corresponding values of stress and strain are plotted against each other. Values of stress are usually plotted as ordinates (vertically) and values of strain as abscissas (horizontally).

120. **tensile strength,** Su—the maximum tensile stress that a material is capable of sustaining. Tensile strength is calculated from the maximum load during a tension test carried to rupture and the original cross-sectional area of the specimen.

121. **test of significance**—a test that, by use of a test-statistic, purports to provide a test of the hypothesis that an effect is absent, for example, that an imposed treatment in an experiment has no effect.

NOTE: Recognizing the possibility of false rejection, the rejection of the hypothesis being tested usually indicates that an effect is present.

122. **test-statistic**—a function of the observed values in a sample that is used in a test of significance.

123. **theoretical stress concentration factor (or stress concentration factor)** K_t—the ratio of the greatest stress in the region of a notch or other stress concentrator as determined by the theory of elasticity (or by experimental procedures that give equivalent values) to the corresponding nominal stress.

NOTE: The theory of plasticity should not be used to determine K_t.

124. **tolerance interval**—an interval computed so that it will include at least a stated percentage of the population with a stated probability.

125. **tolerance level**—the stated probability that the tolerance interval includes at least the stated percentage of the population. It is not the same as a confidence level but the term confidence level is frequently associated with tolerance intervals.

126. **tolerance limits**—the two statistics that define a tolerance interval. (One value may be "minus infinity or plus infinity".)

127. **total elongation,** El_t—the elongation determined after fracture by realigning and fitting together of the broken ends of the specimen.

NOTE: This definition is usually used for metallic materials.

128. **transition fatigue life,** $2N_t$—the life where elastic and plastic components of the total strain are equal. It is the life at which the plastic and elastic strain-life lines cross (see Fig. A-8).

129. **trough**—see valley.

130. **true fracture ductility,** ϵ_f—the "true" plastic strain after fracture (see Fig. A-10).

$$\epsilon_f = \ln (A_o/A_f) = \ln (100/(100 - \% \text{ RA}))$$

131. **true fracture strength** $[\sigma_f]$—the "true" tensile stress required to cause fracture (see Fig. A-10).

$$\sigma_f = P_f/A_f$$

where: P_f = load at failure
A_f = minimum cross sectional area after failure

The value must be corrected for the effect of triaxial stress present due to necking. One such correction suggested by Bridgman [9] is illustrated in Fig. A-11. In this figure, the ratio of the corrected value to the uncorrected value ($\sigma_f/(P_f/A_f)$) is plotted against true tensile strain.

132. **truncation**—in fatigue spectrum loading, the exclusion of cycles with values above, or the exclusion of cycles with values below a specified level (referred to as

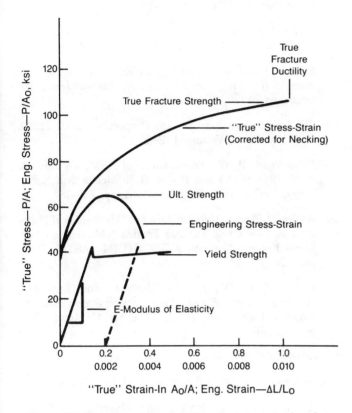

Fig. A-10 Engineering and "true" stress-strain plot (1020 H.R. steel).

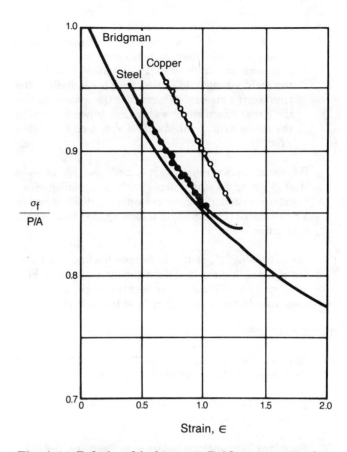

Fig. A-11 Relationship between Bridgman correction factor $\sigma_f/(P/A)$ and true tensile strain.

truncation level) of a loading parameter (peak, valley, range, etc.).

133. **uniform elongation, El_u**—the elongation determined when the maximum load is reached.

    NOTE: This definition applies to materials whose cross section decreases uniformly along the gage length up to maximum load.

134. **valley**—in fatigue loading, the point at which the first derivative of the load-time history changes from negative to positive sign; (also known as trough); the point of minimum load in constant amplitude loading (see Fig. A-2).

135. **variable amplitude loading**—see spectrum loading.

136. **wave form**—the shape of the peak-to-peak variation of a controlled mechanical test variable (for example, load, strain, displacement) as a function of time.

137. **yield point, YP**—the first engineering stress in a test in which stresses and strains are determined for a material that exhibits the phenomenon of discontinuous yielding, at which an increase in strain occurs without an increase in stress.

    NOTE 1—For materials that do not exhibit a yield point, yield strength serves the same purpose as yield point.

    NOTE 2—Testing parameters can influence the phenomenon of discontinuous yielding.

138. **yield strength, S_y or YS**—the engineering stress at which a material exhibits a specified limiting deviation from the proportionality of stress to strain.

    NOTE 1—Yield strength determination is only necessary for materials that do not have a yield point, YP. For these materials the defined quantity yield strength takes the place of the yield point.

    NOTE 2—It is customary to determine yield strength by one of the following methods:

    a. specified offset yield strength (usually an offset strain of 0.2% is specified)—the engineering stress determined at the intersection of the stress-strain curve with a line drawn in the diagram with a slope equal to the modulus of elasticity and offset by the specified strain.

b. specified extension under load yield strength (usually a strain of 0.5% is specified although higher values of strain may be necessary for elastomers, polymers, and high-strength materials in order for the yield strength to exceed the elastic limit)—the engineering stress determined at the intersection of the stress-strain curve with a line drawn parallel to the stress axis at the specified strain on the strain axis.

Whenever yield strength is specified, the type of yield strength must be stated along with the specified offset or extension under load, for example, YS (0.2% offset) or YS (3% EUL). The values obtained by the two methods may differ.

139. **zero crossing,** N_o or n_o—in fatigue loading, the number of times that the load-time history crosses zero load level with a positive slope (or negative slope, or both, as specified) during a given length of the history.

REFERENCES

[1] "Standard Definitions of Terms Relating to Methods of Mechanical Testing," ASTM E6, ASTM.

[2] "Standard Definitions of Terms Relating to Statistical Methods," ASTM E456, ASTM.

[3] "Standard Terminology Relating to Fracture Testing," ASTM E616, ASTM.

[4] "Standard Test Method for Measurement of Fatigue Crack Growth Rates," ASTM E647, ASTM.

[5] "Standard Practice for Statistical Analysis of Linear or Linearized Stress-Life (S-N) and Strain-Life (ϵ-N) Fatigue Data," ASTM E739, ASTM.

[6] "Standard Practices for Cycle Counting in Fatigue Analysis," ASTM E1049, ASTM.

[7] Standard Definitions of Terms Relating to Fatigue, American Society for Testing and Materials, ASTM Standard E1150.

[8] "Technical Report on Fatigue Properties," SAE J1099, SAE Handbook, Volume 1, SAE.

[9] Bridgeman, P. W., Transactions of ASM, Volume 32, ASM, 1944, p. 553; also Dieter, G. E., Mechanical Metallurgy, McGraw-Hill Book Co., 1961, pp. 250-254.

[10] Raske, D. T., and Morrow, JoDean, "Mechanics of Materials in Low Cycle Fatigue Testing," Manual on Low Cycle Fatigue Testing, ASTM STP465, ASTM, 1969, pp. 1-26.

[11] Landgraf, R. W., Morrow, JoDean, and Endo, T., "Determination of the Cyclic Stress-Strain Curve," Journal of Materials, Vol. 4, No. 1, ASTM, March 1969, pp. 176-188.

Addresses of Publishers

ASM International, 9639 Kinsman Road, Metals Park, OH

ASTM, American Society for Testing & Materials, 1916 Race Street, Philadelphia, PA 19103

McGraw-Hill Book Co., 1221 Avenue of the Americas, New York, NY 10020

SAE, Society of Automotive Engineers, 400 Commonwealth Drive, Warrendale, PA 15096

INDEX

Accelerated testing, 100
Accelerometer, 104, 106-107
Acoustic emission testing, 151-152
Acousto elastic stress measurement, 159-160
Aging
  crack propagation, 36
  cyclic strength, 27
  fatigue, 68
Aliasing, 115
Alloy steels
  crack propagation
    stress ratio, 29-31
  fatigue data, 16, 18, 34-35
  wrought, 24
Almen intensity, 160
Aluminum
  fatigue
    prestrain, 95
Aluminum alloys
  corrosion fatigue, 44, 53-54
  crack propagation
    environment, 46-47, 54
    fabrication, 72-73
    stress ratio, 36, 48
  crack retardation, 264
  fatigue, 29, 31-32, 41
    prestrain, 94
    salt water, 261
  stress strain curves, 29, 41
  stress strain properties, 22
  wrought, 29, 41
Amplitude based summarization, 119
Amplitude distributions, 228
Analog to digital converters, 109, 112
Analyzers, 117
Annealing
  fatigue, 68
Anodizing
  fatigue, 68
Austenite, 33, 43
Automobiles; see also Prototypes
  testing, 218-222
  usage
    customer survey, 335
Axial bending, 260
Axial loading, 260
Axial torsion, 260
Axles
  boxes
    cast steel
      fatigue, 343-345
  fatigue, 11, 13, 24, 272-273
  housings
    cast iron
      fatigue, 338-341
    cast steel, 338-341
    fatigue, 272-273
    load history, 339
  shafts
    fatigue analysis, 345-346
Ball joints, 100
  forces, 141-142
Barkhauser noise, 160
Bars
  prismatic
    shear stresses, 201-202
Beach marks
  crack propagation, 318, 320-321
Beam dissection method
  residual stress measurement, 157

Beams
  deflection, 200
Bending
  alloy steels, 16, 18
  axial, 260
Blind hole drilling method
  residual stress measurement, 158
Blocks
  engine
    structural analysis, 186-187
Bodies (Automobile)
  fatigue, 225
Boring and turning method
  residual stress measurement, 157-158
Boundaries, 198-199, 206
  elastic-plastic, 199
Boundary element method, 203-208
Boundary elements, 205
Boundary integral equations, 203, 208
  computer programs, 207
Box-bean
  crack propagation life prediction, 255-256
Brackets
  load history, 285
  radius arm
    fatigue analysis, 335-338
Brag's law, 154
Branching, 319-320, 322
Brass
  ductility, 301
Brittle coatings, 144, 160
Brittle fracture, 299-305, 309-310
Buckling
  connecting rods, 343
Cables
  noise, 103-104
Carbon steels; see also Steels
  fatigue, 34-35, 86
  hot rolled
    fatigue, 241
  strain-life curves, 35
  stress-strain curves, 35
  wrought, 24
    crack propagation, 29-31
Carburizing
  residual stresses, 82
Case depth, 80
Case hardening
  crack propagation, 35
Castigliano's theorum, 172
Casting
  fatigue, 67
  porosity
    cracks, 321, 324
Cast iron, 27-28, 38-40; see also Iron
  axle housing
    fatigue analysis, 338-341
  crack propagation
    stress ratio, 35-36, 46, 48
  strain-life curves, 341
  stress-life curves, 341
Cast steel, 27-28, 36-38, 40; see also Steels
  axle box
    fatigue analysis, 343-345
  axle housing
    fatigue analysis, 338-341
  crack propagation
    fatigue, 345
    stress ratio, 34-35, 43
  strain-life curves, 341
  stress-life curves, 341

Charpy toughness, 24
Chevrin marks, 318, 321
Cleavage, 310
  fracture, 308, 311-312
Closed-loop control, 222-223
Coating, 74
  fatigue, 98
Coatings
  brittle, 144
Coil springs
  fatigue
    shot peening, 76, 78
Cold forming
  fatigue, 93-97
  hardness, 93
  residual stresses, 93
  strain, 93
Cold work
  stresses, 67, 69
Common node noise, 108
Component testing, 11, 100-101
Components
  service load, 100-101
Composite materials
  stress analysis, 202
Computer amplitude loadings, 240-242
Computer models, 219
Computer programs
  boundary integral analysis, 207
  cycle counting, 124-132
  finite difference method, 203
  finite element method, 180-181, 192
  peaks and valleys, 124-132
Computers
  digital, 109-110
Connecting rods
  forged
    buckling, 343
    fatigue analysis, 342
    load measurement, 342
Constant amplitude frequency sweeping, 117
Copper alloys
  fatigue
    prestrain, 94
Corrosion fatigue, 44, 47, 53-54, 333
  galvanizing, 333-334
Crack closure, 45-46, 265-266
Crack growth; See Crack propagation
Crack initiation, 21, 70, 75, 232, 235-249, 253
  creep, 40, 52
  environment, 44, 53, 261, 304-305, 322, 326
  fabrication, 72, 320-321
  fretting, 262-263
  loads, 290-291
  materials, 43
  measurement, 151-152
  notches, 36, 39-40, 236
  plastic deformation, 306-307
  processing, 263-264, 320-321
  residual stresses, 81
  slip displacement, 306-307
  strain, 42, 52, 236
  stress corrosion, 307
  stress relaxation, 40, 49, 52, 236
Crack initiation life prediction, 235-249, 253, 258, 259-272
Crack life defects, 188-189
Crack origin, 317, 320-322

Crack propagation, 16, 21-24, 70, 75-77, 174, 202, 232, 235-236, 265, 306-308, 315-319
  alloy steels
    stress ratio, 29-31
  aluminum alloys
    aging, 36
    fabrication, 72-73
    microstructure, 36, 50-51
    stress ratio, 36, 48-49
  axle boxes
    fatigue, 345
  beach marks, 318, 320
  carbon steels, 29-31, 43
  cast iron
    stress ratio, 35-36, 46-48
  cast steel, 36, 48
    fatigue, 345
    stress ratio, 34-35, 43-45
  circular hole, 268-269
  crack size, 256
  cyclic stress, 254
  electrochemical conditions, 46, 54
  environment, 46-47, 54, 261
  frames, 273-277
  frequency, 46, 54
  from notches, 268-270
  heat treatment, 33, 35, 43
  high strength low alloy steels, 43
    stress ratio, 32-33
  iron
    ductile, 36, 48
    gray, 36, 48
  load sequence, 264
  loads, 254, 256, 290-291
  measurement, 151-154
  microstructure, 33, 65-66
  multiaxial loading, 261
  mussel shells, 318, 320
  notch elastic-plastic strain, 269-270
  overloads, 77, 264
  plastic deformation, 310
  processing, 263-264
  quenching, 33, 43
  residual stresses, 64-65, 70, 81, 266-268, 277
  self stresses, 64-65, 71
  slip, 234
  steels, 42-43, 254-255
    case hardening, 35
    embrittlement, 31, 41
    SAE 1045
      fabrication, 72
  stress, 54
  stress ratio, 29-49
  sunrise markings, 318, 321
  tensile stress, 252
  turbine rotors, 318-319
  welds, 270-272
  wrought steel
    stress ratio, 34
Crack propagation life prediction, 249-272
Crack retardation, 264-265
  overloads, 264
  stress ratio, 264
Crack surface, 317-319
Crack tip elements, 191-192
Crack tip opening displacement, 313-314
Crack tip stresses, 21-24, 29, 174, 188-192, 235, 250-251, 253, 306, 308-309
Crack tip yield zone, 265
Cracks; see also Crack initiation, Crack propagation
  casting porosity, 321, 324
  causes, 302-305
  design, 302-304
  detection, 148-152
  displacement, 305
  ductility, 301
  edge, 251-252
  failure, 298-302
  fatigue, 302-305
  forging, 302-303
  fracture, 299
  heat treatment, 302, 327
  machining, 302-303
  multiaxial loading, 260
  non-propagating, 269
  origins, 317, 321-322
  processing, 302, 327
  repair, 321, 324
  rotor blades, 321, 325
  service induced, 321-322
  size, 253, 258-259
  stresses, 235, 269
  surface, 251-253, 256
  welds, 270-272, 321
Crankshafts
  residual stresses, 156, 338
Creep
  steel
    crack initiation, 40, 52
Crosstalk noise, 108
Crystallography, 153
Cumulative damage analysis, 9, 11, 13, 285-291
Curve fitting, 22
Customer environmental sampling, 100
Customer survey
  vehicle usage, 335
Cycle counting, 118, 120-122, 133-136, 257
  computer program, 124-132
  level crossing, 121
  peak, 121-122
  rainflow, 122, 133-136, 257, 286-288
  range-pair, 133
  simple range, 121, 133-136
Cyclic loads, 306
  yield, 181
Cyclic plastic deformation, 15-16
Cyclic softening, 70, 76
Cyclic strain
  fatigue, 175
  slip, 234
Cyclic strength
  aging, 27
  aluminum alloys, 29, 41
  hardness, 26
Cyclic stress
  crack propagation, 54, 254
    aluminum alloys, 54
  fatigue
    steel, 332
Cyclic strength coefficient, 27
Cyclic stress-plastic strain plot
  SAE 1010 steel, 350
Cyclic stress relaxation curves, 18
Cyclic stress-strain, 9, 18-20, 24, 26, 182, 246, 292-293, 351
  alloy steels, 34
  aluminum alloys, 29, 41
  carbon steels, 35
  dual phase steels, 332
  high strength low alloy steels, 27, 35
  hot rolled low carbon steels, 332
  low carbon steels, 24, 26, 33
  quenched steels, 34
  SAE 9262 steel, 27, 36-37
  tempered steels, 34
Cyclic yield strength, 27
Cyclic yield stress, 75
Data acquisition, 103-110, 226, 272-273
Damage analysis, 238-240, 250
Data evaluation, 117-122
Data recording, 110-112, 225-226
Data reduction, 110-112
Data storage, 110-112
Deep drawing
  fatigue, 67
Defects, 232
  detection, 148-152
  failure, 298-302
Definitions
  fatigue design and analysis, 349-362
Deflections
  analysis, 173
  beams, 200
  measurement, 146-148
Design analysis; see also Engineering design
  life prediction in, 272-277
Diffractometers, 155
Digital computers, 109-110
Digital techniques, 109-110
Digitized signals
  compression, 119
Dislocation cell structures, 26-27
Displacement
  crack initiation, 305-306
  failure, 303, 305
  measurement, 146-148
  stresses, 305
Displacement field, 175-176
Driveshafts
  loads, 101
Ductile fracture, 299-305, 309-310, see also Fracture
Ductility
  brass, 301
  cracks, 301
  notches, 301
Durability testing, 11
  vehicles, 220, 224-228
Dye penetrant inspection, 149
Dynamic
  structures, 228-229
Dynamic modeling
  suspension systems, 138-139
Eddy current inspection, 150
Elastic bending, 16
Elastic constant matrix, 177
Elastic modulus, 20
Elastic plastic analysis, 181-192
Elastic plastic boundary, 199
Elastic plastic matrix, 182, 184-185
Elasticity theory, 189
Elastostatics
  integral equation, 204
Electrical equipment, 104
Electrochemical conditions
  crack propagation, 46, 54
Electron microscope
  crack tip opening displacement, 314
Electronic circuits, 104
Electrostatic shielding, 107-108
Element stiffness matrix, 175-178
Embrittlement
  steel
    crack propagation, 31, 41
Endurance limit, 21
Engine blocks
  structural analysis, 186-187

Engine mounts
  loads, 100-101
Engineering design, 222-223; see also Design analysis
  failure, 302-304, 327-328
Environment
  crack initiation, 44, 53, 261, 303-304, 322, 326
  crack propagation, 46-47, 55
  durability testing, 219, 224
  fatigue, 44, 53, 261
  prototype testing, 220
  slip, 234
Equilibrium equation, 177
Euler buckling curve
  connecting rods, 343
Extrusion, 74
  fatigue, 67
Fabrication
  crack initiation, 320-321
    SAE 1045 steel, 72
  crack propagation
    aluminum alloys, 72-73
  fatigue, 67, 261
Failure
  causes, 302-304
  cracks, 302-305
  design, 302-304, 327-328
  displacement, 303-305
  service, 329
  stresses, 329
Failure analysis, 297-330
Failure hypothesis, 324, 327
Failure modes, 6
Fast Fourier Transform analyzers, 117
Fasteners
  fatigue, 338
Fastress, 155
Fatigue, 221, 228, 231-295, 318
  alloy steels, 35
  aluminum alloys, 29, 41
    prestrain, 94
    salt water, 261
  annealing, 71
  annodizing, 68
  axles, 11, 13, 24
    housings, 212-213
  bodies (automobiles), 228
  carbon steels, 34-35, 86
    hot rolled, 241
  casting, 67
  cast iron, 27, 29, 338-341
  cast steels, 34-35, 43
    crack propagation, 345
  coating, 98
  coil springs
    shot peening, 76, 78
  cold forming, 93-97
  copper alloys
    prestrain, 94
  corrosion, 333-334
  cyclic strain, 175
  deep drawing, 67
  environment, 44, 53, 333-334
  extruding, 67
  fabrication, 67, 72, 261-262
  fasteners, 338
  forging, 67, 69-70
  forming, 67, 93-97
  frequencies, 117
  gears
    shot peening, 76, 78
  hardness, 27, 76-78, 94, 96-97, 333
  heat treatment, 80-84

high strength low alloy steel, 86
loads, 222-223, 285-291
low alloy steels, 34
machining, 67, 89-92
measurement, 161-162
microstructure, 65-69
monotonic tension, 15-16
multiaxial stress-strain, 260-261
nickel alloys, 89, 91-92
orientation, 71
overloading, 76-77
overstrain, 42
peening, 69, 87
piston rods
  forging hammer, 273
plating, 68, 98
prestrain, 93-95, 333
processing, 63-98
quenching, 68
rolling, 69, 76
salt water, 261, 263
simulation, 235
slip, 234
spot welds, 82
springs, 22
  shot peening, 76, 78
steels, 42, 52-53
  salt water, 261
strain, 8-9, 245-249
  steel, 42, 52-53
stress and strain, 8-9
stresses, 8-9, 15, 75-76, 80, 91-92, 243-245
surface finish, 64-66, 73, 89
  cast iron, 340
  cast steel, 340
surface roughness, 64-66
suspension systems, 225
temperature, 261
testing, 217-230
thermal processing, 68
threads
  rolling, 76
titanium alloys, 91-92
vehicles, 24
welding, 67-68
wheels
  hot rolled low carbon, 331-334
Fatigue analysis
  axle box
    cast steel, 343-345
  axle housing, 338-341
  axle shaft
    scrapers, 345-346
  brackets
    radius arm, 335-338
  connecting rods, 342-343
  wheels, 331-334
Fatigue crack growth; See Crack propagation
Fatigue design, 5-13, 29
Fatigue ductility coefficient, 27
Fatigue ductility exponent, 27
Fatigue failure, 22, 29
Fatigue life, 231-295
Fatigue notch factor, 237, 241, 333-334
Fatigue strength coefficient, 27
Fatigue strength exponent, 27
Fatigue striations, 307-310, 315-316
Fatigue testing, 11, 15-16, 18-24, 64-66, 217-234, 285-291
  as-cast surfaces, 339-340
  machine surfaces, 339-340
  wheels, 334
Ferrite, 35, 254

Field testing, 224
  prototypes, 221-222
Finite difference method, 192-194, 203, 207
Finite element method, 175-180
  axial housing loading, 339
  computer programs, 180-181, 192
  connecting rod stresses, 343
  radius arm pivot bracket stresses, 336-337
  stiffness derivative, 190-191
  wheel stresses and strains, 334
Flame hardening
  stresses, 81-82
Flaws, 258-259
  detection, 152, 163-164
  fasteners, 191
  stress intensity factor, 271
Fluid dynamics analysis
  finite element method, 199
Fluid mechanics analysis
  finite element method, 199
Fluoroscopy, 150
Force
  slip, 234
Forging, 74
  cracking, 302-303
  fatigue, 67, 342-343
  orientation, 69-70
Forging hammer
  piston rods
    fatigue, 273
Forming, 74
  fatigue, 67, 93-97
Fractography, 297, 305-306, 312-316
Fracture
  brittle, 299-305, 309-310
  cleavage, 308, 311-312
  cracks, 299-302
  ductile, 299-305, 309-310
  notches, 301
  plastic deformation, 301
  rupture, 308-309, 311-312, 314
  turbine rotors, 318-319
Fracture ductility, 20-21
  aluminum alloys, 22
  steel, 22, 27
Fracture mechanics, 21, 174, 188-192, 202, 249, 254
Fracture strength, 20-21
  aluminum alloys, 22
  steel, 22
Fracture surface diagram, 16-17
Fracture toughness, 23-24, 32, 252-253
  aluminum alloys, 31-32
  multiaxial loads, 261
  steels, 32, 253
  tests, 23-24
  titanium, 32
Frames
  crack propagation, 274-277
Frequencies
  fatigue, 118
  measurement, 147
  vibration, 118
Frequency response structure, 228-229
Fretting
  crack initiation, 262-263
  fatigue, 262-263
Fuel tanks
  brackets
    loads, 101
Galvanizing
  hot rolled low carbon steel
    fatigue, 333-334

Gears
  fatigue
    shot peening, 76, 78
Gel electrode method, 152-153
Graphite gray iron, 27-28
Grinding
  residual stresses, 90-91
Ground loop, 104-108
Ground noise, 104-107
Grounding
  shields, 104-108
Guassian, 227
Haigh diagram, 75
Hammer peening
  fatigue, 87
Hardening, 16, 183
  furnace
    stresses, 80-82
  low carbon steels, 26, 33
Hardness, 80
  cold forming, 93
  cyclic strength, 26
  fatigue, 27, 76-78, 94, 96-97, 333
  stresses, 81
  testing, 150
Harshness
  prototypes, 220
Heat affected zone, 86-87
Heat transfer analysis
  finite difference method, 199
Heat treatment, 74
  crack propagation, 33, 35, 43, 302-303, 322, 325
  fatigue, 80-84
Hencky's theory, 184
High strength low alloy steel, 26
  crack propagation
    stress ratio, 33, 43
  fatigue, 86
  hardness, 26
Histograms, 286
Holographic interferometry, 146-148, 150-151
Hook's law, 181
Hot working, 67, 69
Housings
  axle
    fatigue, 272-273
      cast iron, 338-341
      cast steel, 338-341
      load history, 339
Hysteresis, 102
Hysteresis loops, 23, 351
Hysteresis rejection, 120
Inconel
  fatigue, 89
Induction hardening
  stresses, 81-82
Inspection, 148-149
Instrumentation
  shielding, 103-104
Interference fringes, 146-148
Interferometry, 146-148
Iron
  ductile
    crack propagation, 35-36, 46-47
    fatigue, 27, 40
    graphite gray, 27-28, 40
    gray, 28, 40
      crack propagation, 36, 48
    nodular, 28, 40
Isochromatics, 145
Isoclinics, 145

Isoentatic, 144
Isoparametric finite elements, 178-180
Isostatics, 144
J-integral, 24, 190
Joints
  failure, 321, 323
Kinematic modeling
  suspension systems, 138
Layer removal method
  residual stress measurement, 157
Leaf springs
  crack initiation
    environment, 322, 326
Level crossing counting, 121
Life prediction, 234-295
  crack initiation, 240-249
  crack propagation, 249-250
  design analysis application, 272-277
Linear elastic fracture mechanics 21, 174, 188-192, 202, 250, 258-259
Linearity, 102
Load cell, 101-102
Load histories, 101, 285
  axle housings, 339
Load-life method, 240-247, 250
Load sequence
  crack propagation, 260
  crack retardation, 260
Load-strain curves, 294
Loaders
  axle housings
    fatigue, 272-273, 339-341
Loads
  connecting rods
    measurement, 342
  crack initiation, 256-258, 290-291
  cyclic
    yield, 181
  fatigue, 285-292
  life, 240-243
  multiaxial
    crack propagation, 260-261
    out-of-phase, 260
  out-of-phase variable amplitude
    fatigue, 9
  residual stresses, 174-175
  service, 8-9, 24, 32-33, 99-136, 247-348
Longitudinal waves, 159
Low alloy steels
  fatigue, 44
Low carbon steels
  cyclic stress-strain responses, 24, 33-34
  fatigue, 34
Machining, 74
  cracking, 302-303
  fatigue, 67, 89-92
  residual stresses, 89-90
Macrostress, 80
Magnetic particle inspection, 149
Mapping, 178
Martensite, 27, 33, 36-39, 43, 80-81
  hardness, 26
Material defects
  in failure analysis, 327
Matrix of elastic constants, 177
Mean stresses, 238, 249
  fatigue, 15, 19, 22, 75, 80, 246-248
Mechanical prestressing, 74-78
Microcrack nucleation, 16
Microscopes
  fracture surface analysis, 312, 315
Microstress, 80
Microstructure

crack propagation
  aluminum alloys, 36, 50-51
  fatigue, 9, 65-66
  fabrication, 67, 262-263
  mechanical processing, 69
  quenching, 69
  thermal processing, 68
Modal analysis
  dynamics, 219
  vibration, 219, 228-229
Models, 219, 229
  product design, 221-222
  semi-empirical
    crack retardation, 264
  structural, 180-181
  suspension systems, 138-139
  vehicles, 139
  Wheeler's, 264-265
Modulus of elasticity, 21
  aluminum alloys, 22, 32
  steel, 22, 32
  titanium, 32
Monotonic stress-strain curves
  steels, 27, 35
  SAE 9262, 27, 36-37
Monotonic stress-strain properties, 21
Monotonic tension properties, 20
  aluminum alloys, 22
  cast iron, 340
  fatigue, 15-16, 18-19
  steels, 22, 340
Monotonic yield strength, 181, 277
Mounts
  engine
    loads, 80-81
Multiaxial loads
  crack propagation, 260-261
  fracture toughness, 261
Multiaxial stresses
  fatigue, 9
Multiaxial stress-strain
  fatigue, 260-261
Multiplexing
  frequency division, 108
Mussel shells
  crack propagation, 318, 320
Neuber's rule, 237, 245, 294, 332-333
Nickel alloys
  fatigue, 89, 91
Nil ductility temperature, 253
Nitriding
  residual stresses, 82
Nodal shape function, 178-179
Noise, 8, 9
  cables, 103-104
  common mode, 108
  crosstalk, 111
  ground loop, 108
  magnetic, 107
  prototypes
    testing, 220
  static, 108
Nominal strain life
  analysis, 172-175, 247, 250
Nominal stress-strain, 294
Non-destructive testing
  flaws, 148-156, 159
Notch field, 259
Notch root stress-strain, 247-250, 260, 269, 294, 306, 332
Notch sensitivity, 18
Notch strength, 18

Notched specimens
  fatigue, 333
  residual stresses, 81
Notches
  crack initiation, 36, 39-40, 236-237
  crack propagation, 268-270
  ductility, 301
  elastic-plastic strain
    crack propagation, 269-270
  fracture, 301
  stress intensity factor, 268
  stresses, 236-248, 269
Numerical analysis, 171-215
Off-highway vehicles
  axle housings
    fatigue, 232-273, 338-341
On-board data collection, 112
Orientation
  forging, 69, 71
Oscilloscopes, 111
Overloading
  crack propagation, 77
  crack retardation, 264
  fatigue, 76-77
  fracture, 300
Overstrain
  fatigue, 42
Peak counting, 121
Peak-valley computer programs, 124-132
Peak-valley sequencing, 120
Pearlite, 26-28, 35, 36-39
Peening
  fatigue, 67, 69, 346
Phrase transformation, 67
Photoelastic coatings, 160
Photoelasticity, 143-146
Piston rods
  forging hammer
    fatigue, 273
Pitting (Corrosion)
  hot rolled low carbon steel
    fatigue, 333-334
Plane strain fracture toughness, 24
Plane stress toughness, 24
Plastic deformation, 67, 182-184
  crack initiation, 306-307
  crack propagation, 309-310
  cyclic
    fatigue, 15-16
  fracture, 301
Plastic flow, 182-183
Plastic strain amplitude
  SAE 1010 steel, 352
Plasticity, 174, 181-182, 184, 301
Plating, 74
  fatigue, 68, 98
Polariscope, 145-146
Portable X-Ray Analyzer for Residual Stress (PARS), 155
Porta/Rapid/Stress
  stress measurement, 155
Positive Sensitive Scintillation Detector (PSSD), 155
Potential drop method, 152
Power spectra density, 118-119
Power supply transients, 104-105
Precipitation hardening, 27
Prestrain
  fatigue resistance, 93-95
  SAE 950X steel, 333
  hot rolled low carbon steel, 333
Processing
  crack initiation, 263-264, 320-321
  crack propagation, 263-264

fatigue, 63-98
  internal stresses, 263
  microstructure, 67-69, 262-264
  residual stresses, 67-69
  surface finish, 67-69, 262-264
  surface roughness, 67-69
Product design, 272-277
Product development, 5-6, 217-221, 231-232
Product evaluation, 7
Product planning, 5-6
Product usage, 6
Production
  cracking, 302, 327
  vehicles
    testing, 220
Programs; see Computer programs
Prototypes
  field testing, 221-222
  testing, 219-220, 222
Proving ground tests, 220
Pseudo-force method, 185
Pulse code modulation, 109
Quality control
  vehicles, 221
Quenching
  crack propagation, 33, 43
  fatigue, 68-69
  microstructure, 69
  stresses, 80
Radio frequency transmission, 108-109
Radiography, 149-150
Radius arm pivot bracket
  fatigue analysis, 335-338
Railroad vehicles
  axle box
    fatigue analysis, 343-345
Rainflow cycle counting, 122, 133-136, 257, 286-288
  stress-strain response, 12
Raleigh waves, 159
Random processes, 118
Range-pair counting, 133
Recorders, 111-112
R-curve method, 24
Remote parameter control, 229
Repair
  cracking, 321
Residual stresses, 75-77, 80, 270, 277, 286-288
  AISI 4340, 90-91
  carburizing, 82
  cold forming, 93
  crack initiation, 75, 81
  crack propagation, 64-65, 70, 75, 81, 266-268, 277
  crankshafts, 156, 338
  fatigue, 75-76, 80-81, 91-92
  furnace hardening, 80-94
  loads, 174-175
  machining, 89-91
  measurement, 153-160, 165, 166
  nickel alloys, 89, 91-92
  nitriding, 82
  processing, 67-69
  shot peening, 76-77, 153, 160
  stress intensity factor, 267
  thermal processing, 68
  titanium alloys, 92
  welding, 87
Resonant frequencies, 146
Retardation, 45-46
  aluminum alloys, 53
Road load
  simulation, 140

Rods
  axial forces, 100
Rolling
  fatigue, 69, 76
Rotors
  blades
    crack initiation, 321, 325
  turbine
    fatigue failure, 318-319
Rupture
  fracture, 308, 311-312, 314
SAE Keyhole specimen, 187-188
Salt water
  fatigue, 261, 263
Sample rate definition, 112-114
Scrapers
  axle shafts
    fatigue analysis, 345-346
Sectioning
  residual stress measurement, 156
Self-stresses; see Residual stresses
Service history, 223-224
  simulation, 217
Service life, 221, 224
Service loads, 8-9, 99-136
  axle shafts
    scrapers, 345
Service strain history, 101
Service tests, 24, 32-33, 220-221
Shear stress
  prismatic bar, 200, 202
Shear waves, 159
Sheet steel
  spot welded
    fatigue, 86
Shell analysis
  finite difference method, 199
Shielding
  electrostatic, 107-108
  instrumentation, 103-104
  thermocouple cables, 103-104
Shot peening
  fatigue, 67, 69, 76, 78, 87, 346
  residual stresses, 76-77, 153, 160
Signal conditioning, 103
Signal transmission, 107-109
Simple overload fracture, 300
Simple range counting, 121, 133
Simulation
  fatigue, 235
  road load, 140
  service history, 217
  service tests, 32-33
  suspension abuse, 140
  vehicles, 137-142
Simulator
  tire/road, 226
Sine waves, 113
Skidders
  axle housing
    fatigue, 272-273, 339-341
Slip
  crack initiation, 306-307
  crack propagation, 234
  environment, 234
  fatigue, 234
  force, 234-235
Smooth specimen cyclic properties, 27
Softening, 27
  high strength low alloy steels, 27, 35
Specimen testing, 224
Speckle interferometry, 146-148
Spectral density, 221, 227-228
Spectrum analysis, 116-118

Spindles
  motion, 226
Spot welds
  fatigue, 82
Springs
  fatigue, 20
    shot peening, 76, 78
  leaf
    crack initiation
      environment, 322, 326
Stability
  tempering, 26-27
Static noise, 108
Stationary potential energy, 177
Statistical analysis
  loading histories, 118
Steels
  AISI 4340
    residual stresses, 90-91
  alloy
    fatigue, 27, 35
  carbon
    fatigue, 27, 35, 86
    hot rolled
      fatigue, 241
  cast
    strain life curves, 27, 38
  corrosion fatigue, 44, 53
  crack propagation, 29, 41-43, 254-255
  crack retardation, 264
  dual phase, 331
    stress-strain curves, 332
  fatigue, 16, 18, 85-86
    hardness, 95-96
    salt water, 261
  fracture toughness, 253
  high strength
    wheels, 331
  high strength low alloy, 26
    crack propagation, 32-33
    fatigue, 86
    stress-strain curve, 332
  hot rolled low carbon
    corrosion fatigue, 333-334
    SAE 1010, 331
      stress-strain curve, 332
  low carbon
    cyclic stress-strain, 24, 26, 33
  Man-ten
    properties, 288
  Maxiform, 52
    fatigue
      prestrain, 94
  monotonic stress-strain properties, 22
  R-curve behavior, 31
  RQC-100
    cyclic stress-strain curve, 26
    properties, 288
    strain-life curves, 289
  sheet
    fatigue
      spot welding, 86
  SAE 1008
    fatigue
      prestrain, 95
  SAE 1010, 331
    cyclic stress-plane strain plot, 350
    fatigue
      prestrain, 94
    plastic strain amplitude, 352
    strain amplitude, 353
  SAE 1020
    stress-plastic strain plot, 356
    stress-strain curve, 361

SAE 1045
  crack initiation
    fabrication, 72
  cyclic creep, 40, 52
  fatigue, 22
  stress relaxation, 40, 49
SAE 9262
  strain amplitude, 27, 37
  stress-strain curves, 27, 36-37
  strain hardening rates, 27
  strain life curve, 26
  stress life fatigue, 22
  stress relaxation, 40, 52, 247
  stress-strain curves, 11
  tensile strength, 85-86
  wrought
    crack propagation, 35
    strain cycling, 29, 39
Steering axle
  fatigue, 11
Stiffness derivative finite element technique, 190-191
Stiffness matrix, 199
Strain
  cold forming, 93
  components, 101
  damage, 238-240
  elastic, 174
  fatigue, 9, 20, 26, 245-249
  measurement, 101-102, 143-169
  plastic, 174
  pre-; see Prestrain
Strain amplitude
  SAE 1010 steel, 353
Strain cycling fatigue tests, 19-21, 23
Strain cycling properties, 19-21
Strain energy release rate, 190
Strain gages, 101-103, 144-145, 339
Strain hardening, 181, 183
  exponents, 20-21
    aluminum alloys, 22
    high strength low allow steel, 27
    steel, 22
  rates
    aluminum alloys, 29, 41
    high strength low alloy steel
      hardness, 26-27
Strain histories, 101
Strain life analysis, 245-250, 333
Strain life curves, 19, 333
  aluminum alloys, 29, 41
  cast iron, 27, 38-39, 341
  cast steel, 27, 38, 341
  nodular iron, 341
  steel, 26-27, 35, 333
  RQC-100, 289
Strain life techniques, 21, 243-245
Strain stiffness matrix, 199
Strength coefficient, 20
Strength of materials, 172-173
Stress analysis, 8-9, 172-173
  notches, 236-238
  steering axle, 10
Stress corrosion cracks, 307-308, 310
Stress cracks, 29-49, 52, 54, 64, 75, 235-236
Stress cycling fatigue, 15-16, 18-19
Stress intensity factor, 23-24, 174, 188-189, 202, 250-252, 259, 266, 268, 270-271, 274-275
  crack closure, 265-266
  residual stresses, 267
Stress-life analysis, 243, 245, 250
Stress-life curve
  cast iron, 341

  cast steel, 341
Stress raisers, 173-174, 235
  plastic deformation, 306
Stress ratio
  crack propagation, 29-44, 255
Stress relaxation, 40
  crack propagation
    steels, 40, 49, 52
  steels, 82
Stress relaxation curves, 18
Stress strain, 9, 18-20, 24, 26, 181
  alloy steels, 34
  aluminum alloys, 29, 41
  carbon steels, 35
  fatigue, 8-9, 245-249
  high strength low alloy steels, 27, 35
  low carbon steel, 26, 33
  microstructure, 70
  multiaxial
    fatigue, 260-261
  notch root, 247-248, 294
  notches
    simulation 292-295
  quenched steels, 34
  rainflow cycle counting, 12
  simulation, 292-295
  SAE 9262 steel, 27, 36-37
  tempered steels, 34
  wheels, 334
Stress strain curves, 286-287, 292, 351
  steel, 11, 19
    dual phase, 332
    high strength low alloy, 332
    hot rolled low carbon, 332
    RQC-100, 20, 26
    SAE 950, 19
    SAE 1010, 350
    SAE 1020, 356, 361
Stress strain properties
  aluminum alloys, 22
Stresses; See also Mean stresses, Residual stresses
  brackets
    radius arm, 336-337
  carburizing, 82
  composite materials, 202
  crack propagation, 29-36
  crack tip, 21-23, 29, 174, 188-192, 235, 250-251, 253
  cracks, 235, 302-304
  damage, 238-240
  displacement, 305
  failure, 302-304
  fatigue, 8-9, 15-16, 18-19, 22, 26, 64-65, 72, 75-77, 243-245
  flame hardening, 81-82
  furnace hardening, 80-84
  induction hardening, 81-82
  internal, 206
    processing, 262-263
  measurement, 159-160
  nitriding, 82
  residual; see Residual stresses
  shot peening, 76-78
  steel, 82
  welds, 270
Strip chart recorders, 111
Structural analysis, 19, 175-180
  engine blocks, 186-187
Structural elements
  stress analysis, 172-173
Structural life testing, 217-230
Structural stiffness matrix, 178
Subparametric element, 179

Sunrise pattern
  crack propagation, 319, 321
Superparametric element, 179
Surface finish
  inspection, 149
  fatigue, 64-66, 73, 90
    cast iron, 340
    cast steel, 340
  processing, 67-68
Surface rolling
  fatigue, 76
Surface roughness
  fatigue, 64-65
  processing, 67-68
  thermal processing, 69
Surface yielding, 69
Suspension geometry, 138
Suspension systems
  abuse tests, 141
  computer simulation, 137, 142
  fatigue, 225
  load histories, 285
  radius arm pivot bracket
    fatigue analysis, 335-338
Tangent modulus method, 185
Tape recorders, 111-112
Tardy method, 145-146
Telemetry, 108-109
Temperature
  fatigue, 261
  nil ductility, 253
Tempering
  crack propagation, 33, 43
  cyclic stability, 26-27
Tensile mean stress
  creep, 40, 52
  fatigue, 19
Tensile strength, 20-21
  aluminum alloys, 22
  steel, 22
Tensile tests, 18
Tensile yield strength, 21
  aluminum alloys, 22
  steel, 22, 80
Tensile yield stress, 75
Tension tests, 18
Terminology
  fatigue design analysis, 349-362
Test site sampling, 100
Test specimens, 70
Testing
  component load, 100
  durability, 11, 227-228
    axles, 13
    vehicles, 220, 224-228
  fatigue, 15-16, 101
  field
    prototypes, 221
  prototypes, 219-220, 222
  tension, 18
Theory of elasticity, 173
Thermal processing
  fatigue, 68
Thermocouple cables
  shielding, 103-104
Thiessen's polygon, 195, 197
Threads
  fatigue
    rolling, 76
Time compression analyzers, 117
Time histories, 227-228
Tire/road simulator, 226
Titanium alloys
  fatigue, 92

Toe cracks
  welds, 271
Tomography, 149
Torsion
  axial, 260
Total life, 235
Transducers, 101-103
Transmissions
  load histories, 285
Tresca yield criterion, 183, 260
Turbine rotors
  fatigue fracture, 318-319
Ultimate tensile strength, 20-21
  aluminum alloys, 22
  steel, 22
Ultrasonic inspection, 148-149
Ultrasonic methods
  residual stress measurement, 159-160
Variable amplitude
  cyclic, 243-245
  loading, 242-248
Vector of strain, 176
Vector of stress, 177
Vehicles; see also Automobiles, Loaders, Off-highway vehicles, Prototypes, Scrapers, Skidders
  quality control, 221
  testing, 218-222
  usage
    customer survey, 335
Vibration, 8, 10
  frequencies, 117
  measurement, 104, 146-148, 228-229
  modal analysis, 217, 228-229
  prototypes
    testing, 220
Virtual work principle, 177
Von Nises criteria, 183-185, 260
Waveforms, 114-115
Wave-front solution, 180
Waves, 159
Weibull statistical analysis
  radius arm pivot bracket, 336-338
Welding, 74
  fatigue, 67
  residual stresses, 87
Welds
  crack propagation, 270-272
  fatigue, 85-88, 191, 261-262
Wheels
  high strength low alloy steel
    fatigue analysis, 331-334
X-ray analyzer
  fracture surface analysis, 312, 316
X-ray diffraction, 153-156
X-Y recorders, 111
Yield, 181-183
  aluminum alloys, 181
  cyclic loads, 181
Yield strength, 20, 27, 184, 277, 311-312
  aluminum alloys, 32
  prestraining, 333
  steels, 32
  titanium alloys, 32
Yield stress; see Yield strength